DEFINITIONS

newton—force that will give 1-kg mass an acceleration of 1 m/sec^2
joule—work done by a force of 1 N over a displacement of 1 m
1 newton per sq m (N/m^2) = 1 pascal
1 kilogram force (kgf) = 9.807 N
1 gravity acceleration (g) = 9.807 m/sec^2
1 are (a) = 100 m^2
1 hectare (ha) = 10,000 m^2
1 kip (kip) = 1000 lb

Probability Concepts in Engineering Planning and Design

Probability Concepts in Engineering Planning and Design

VOLUME II   DECISION, RISK, AND RELIABILITY

ALFREDO H-S. ANG
Professor of Civil Engineering
University of Illinois at Urbana-Champaign

WILSON H. TANG
Professor of Civil Engineering
University of Illinois at Urbana-Champaign

JOHN WILEY & SONS
New York   •   Chichester   •   Brisbane   •   Toronto   •   Singapore

Dedicated to Myrtle and Bernadette

Library of Congress Cataloging in Publication Data:

(Revised for volume 2)

Ang, Alfredo Hua-Sing, 1930–
    Probability concepts in engineering planning and
design.

    Includes bibliographical references and indexes.
    Contents: v. 1. Basic principles—v. 2. Decision,
risk, and reliability.
    1. Engineering—Statistical methods.   2. Probabilities.
I. Tang, Wilson H.
TA340.A5   620'.00422'015192   75-5892
ISBN 0-471-03200-X (v. 1)
ISBN 0-471-03201-8 (v. 2)

Printed in the United States of America

10 9 8 7 6 5 4 3 2 1

preface

Like Volume I, this volume emphasizes the applications of probability and statistics in engineering. It is built on the basic principles contained in Volume I but additional more advanced tools are developed, including statistical decision analysis, Markov and queueing models, the statistics of extreme values, Monte Carlo simulation, and system reliability. Again, the necessary concepts are introduced and illustrated within the context of engineering problems. Indeed, the development of the logical concepts and the illustration of established principles in engineering applications are the main objectives of the volume. The tools provided should form the quantitative bases for risk evaluation, control, and management, and intelligent decision making under uncertainty.

In some instances, problems in several diverse areas are used to illustrate the same or similar principles. This was purposely done to demonstrate the universal application of the pertinent concepts and also to provide readers with wider choices of illustrative applications.

Much of the mathematical material is, of course, available in the literature and some of it is in text form; for example, Gumbel (1958) on extreme value statistics; Rubinstein (1981) on Monte Carlo simulation; Raiffa (1970) and Schlaifer (1969) on statistical decision analysis; and Parzen (1962) and Saaty (1961) on Markov and queueing processes. However, the concepts that are especially useful for engineering planning and design are emphasized; moreover, these are presented purposely in the context of engineering significance and in terms that are more easily comprehensible to engineers (supplemented with numerous illustrative examples). Much of the material in Chapters 6 and 7 on basic and systems reliability is, for the most part, only currently available in scientific and technical journals—in this regard, only typical references to the available literature are cited, as there is no intention to be exhaustive.

The material is suitable for two advanced undergraduate or graduate level courses. One would cover engineering decision and risk analysis, based on Chapters 2, 3, and 5 of the present volume supplemented by Chapter 8 of Volume I; the other would focus on system reliability and design and may be developed using the material contained in Chapters 4, 6, and 7. Both courses would require a prerequisite background in introductory applied probability and statistics, such as the material in Volume I.

During the preparation and development of the material for this volume, the authors are indebted in many ways to colleagues and students—to former students who had to endure imperfect and incomplete versions of the material; to our colleague Professor Y. K. Wen for numerous discussions and suggestions over the years; and to Professors A. der Kiureghian, M. Shinozuka, E. Vanmarcke, and J. T. P. Yao for constructive suggestions and incisive reviews of the manuscript. A number of former graduate students also contributed to the development of some of the material including numerical calculations of the examples; in particular, the contributions and assistance of R. M. Bennett, C. T. Chu, H. Pearce,

J. Pires, I. Sidi, M. Yamamoto, and A. Zerva are greatly appreciated. Finally, our thanks to Claudia Cook for her patience and expert typing of several versions of the manuscript material, and to R. Winburn for his professional artwork of the figures.

A. H-S. Ang AND W. H. Tang

contents

1. Introduction

1.1 AIMS AND SCOPE OF VOLUME II

The principal aims of the present volume parallel those of Volume I; namely, the modeling of engineering problems containing uncertainties and the analysis of their effects on system performance, and the development of bases for design and decision making under conditions of uncertainty.

The basic principles and elementary tools introduced in Volume I are supplemented here with additional advanced tools and concepts, including statistical decision theory, Markov and queueing processes, and the statistical theory of extreme values. The discussion of numerical tools for probability calculations would not be complete without mention of the Monte Carlo simulation methods; one chapter summarizes the elements of this calculational procedure. One of the most important applications of probability concepts in engineering is in the evaluation of the safety and reliability of engineering systems, and the formulation of associated design criteria. Recent developments for these purposes are the topics of the final two chapters, with the last chapter devoted specifically to the analysis of the reliability of technological systems.

The role of judgment is emphasized when dealing with practical problems; engineering judgments are often necessary irrespective of theoretical sophistication. However, the proper place and role for judgments are delineated within the overall analysis of uncertainty and of its effects on decision, risk, and reliability. Moreover, consistent with a probabilistic approach, judgments have to be expressed in probability or statistical terms; wherever expert judgments are expressed in conventional or deterministic terms (which is often the case), they have to be translated into appropriate probabilistic terms. Methods for these purposes are suggested.

1.2 ESSENCE AND EMPHASIS

The main thrust and emphasis in each of the ensuing chapters may be summarized as follows.

1.2.1. Decision Analysis (Chapter 2)

The goal of most engineering analysis is to provide information or the basis for decision making. Engineering decision making could range from simply selecting the size of a column in a structure, to selecting the site for a major dam, to deciding whether nuclear power is a viable energy source. Unfortunately, uncertainties are invariably present in practically all facets of engineering decision making. In this light, some measure of risk is unavoidable in any decision made (i.e., alternative selected) during the planning and design of an engineering system. A systematic framework for decision analysis under uncertainty, in which the feasible alternatives are identified and the respective consequences evaluated, could be very useful. The tools for such decision analysis are the subject of Chapter 2.

The decision tree model is introduced to identify the necessary components of a decision problem, consisting of the feasible alternatives, the possible outcomes associated with each alternative and respective probabilities, and the potential consequences associated with each alternative. In short, the decision tree provides an organized outline of all the information relevant to a systematic decision analysis.

Various engineering examples are used to illustrate the concepts involved in formal decision analysis. The concept of *value of information* is discussed with respect to whether or not additional information should be gathered before making a final decision. The problem associated with sampling, namely, determining the parameter estimates for design, and the development of optimal sampling plans may also be formulated as a decision problem.

Elementary concepts of *utility theory* are introduced as a generalized measure of value, on the basis of which the relative significance of various potential consequences of a decision may be evaluated. Specific case studies of complex engineering decision problems are discussed, including those involving multiple objectives.

1.2.2. Markov, Queueing, and Availability Models (Chapter 3)

The performance characteristics of an engineering system may frequently be classified into different discrete states. For example, the potential effects of an earthquake on a structure may be classified under several states corresponding to distinct damage levels; the financial status (e.g., in terms of cash flow) of an engineering contractor may represent the various states, including bankruptcy; and the number of vehicles waiting (queue length) at a toll booth could represent a state of a toll system. The conditions of these systems may change or move from one state to another in accordance with some probability law. The probability that the system will be in a particular state after a given number of moves may then be of interest.

The *Markov chain* is a probability model specifically designed to analyze a system with multiple states. The basic concepts of the discrete parameter homogeneous Markov chain model are first introduced. Queueing models are then shown to be examples of a continuous parameter Markov chain. Steady state queue length probabilities are derived for specific queueing systems with Poisson arrivals.

The performance of an engineering system may often be divided into two states; for example, safe and unsafe states, or operating and nonoperating states. The study of the transient behavior of this two-state system is known as the *availability problem*. Problems of system availability (system in a safe state) may include a maintenance program, that is, inspection and repair at regular intervals, in which case the *renewal theory* is a useful model. Availability problems of interest in engineering would include the availability at a given time for a system with or without maintenance.

A number of engineering applications of the Markov process, including the queueing and availability models, are illustrated in Chapter 3.

1.2.3   Statistical Theory of Extremes (Chapter 4)

Extreme values or extremal conditions of physical phenomena are of special interest in many engineering problems, especially in those concerned with the safety and reliability of engineering systems, and/or involving natural hazards. The statistical theory of extreme values, therefore, is of special significance in engineering problems concerned with risk and reliability.

Much of the theory of statistical extremes is available in the literature; however, the available material is primarily of a mathematical nature and is hardly accessible to many engineers. One of the purposes of Chapter 4 is to summarize and highlight the essential concepts of statistical extremes, and to emphasize the practical significance of these concepts in engineering problems. In particular, the asymptotic property of the statistics of extremes enhances the significance of the extremal theory; understanding this concept is essential for many applications. The asymptotic property and its associated concepts are, therefore, stressed throughout the chapter, and illustrated with numerous examples of engineering problems.

1.2.4   Monte Carlo Simulation (Chapter 5)

As the complexity of an engineering system increases, the required analytical model may become extremely difficult to formulate mathematically unless gross idealization and simplifications are invoked; moreover, in some cases even if a formulation is possible, the required solution may be analytically intractable. In these instances, a probabilistic solution may be obtained through Monte Carlo simulations. Monte Carlo simulation is simply a repeated process of generating deterministic solutions to a given problem; each solution corresponds to a set of deterministic values of the underlying random variables. The main element of a Monte Carlo simulation procedure is the generation of random numbers from a specified distribution; systematic and efficient methods for generating such random numbers from several common probability distributions are summarized and illustrated.

Because a Monte Carlo solution generally requires a large number of repetitions, particularly for problems involving very rare events, its application to complex problems could be costly. There is, therefore, good reason to use the Monte Carlo approach with some caution; generally, it should be used only as a last resort, that is, when analytical or approximate methods are unavailable or inadequate. Often, Monte Carlo solutions may be the only means for checking or validating an approximate method of probability calculations.

A number of case study problems are illustrated, to demonstrate the simulation procedure as well as the type of information that may be derived from a Monte Carlo calculation.

1.2.5   Reliability and Reliability-Based Design (Chapter 6)

The safety and/or performance of an engineering system is invariably the principal technical objective of an engineering design. In order to achieve some desired level of reliability, proper methods for its evaluation are, of course, required. As this must invariably be done in the presence of uncertainty, the proper measure of reliability or safety may only be stated in the context of probability. Indeed, consistent levels of safety and reliability may be achieved only if the criteria for design are based on such probabilistic measures of reliability.

Engineering reliability and its significance in engineering design is a rapidly growing field; the most recent and practically useful developments are presented in this chapter. Applications in the various fields of engineering are illustrated with emphasis on civil engineering. It goes without saying that problems of safety and reliability arise because of uncertainty in design. The quantification and analysis of uncertainty are, therefore, central issues in the evaluation of reliability and the development of associated reliability-based design. The statistical bases and methods for these purposes are widely illustrated in this chapter.

1.2.6 Systems Reliability (Chapter 7)

Although the reliability of the total system is of principal concern, system reliability is nonetheless a function of the reliabilities of its constituent components (that are covered in Chapter 6); that is, the determination of system reliability must invariably be based on reliability information for the components. Moreover, engineering designs are invariably performed at the component level, from which the reliability of the system is evaluated through analysis; techniques for this purpose are presented in Chapter 7.

From a reliability standpoint, a system is characterized by multiple modes of failure, in which each of the potential failure modes may be composed of component failure events that are in series or in parallel, or combinations thereof (especially for a redundant system). In the case of a general redundant system, the redundancy may be of the standby or active type. Depending on whether the redundancies are active or standby, the reliability of the system as well as its analysis will be different.

In the case of complex systems, the identification of the potential modes of failure may be quite involved, requiring a systematic procedure for identification, such as the *fault tree* model. Moreover, the potential consequences of a failure (or initiating event) of a system may vary, depending on the subsequent events that follow the particular initiating event. The *event tree* model may be used to facilitate the systematic identification of all potential consequences. The applications of the fault tree and event tree models are illustrated with a variety of engineering examples in this final chapter.

2. Decision Analysis

2.1 INTRODUCTION

Making technical decisions is a necessary part of engineering planning and design; in fact, the primary responsibility of an engineer is to make decisions. Often, such decisions have to be based on predictions and information that invariably contain uncertainty. Under such conditions, risk is virtually unavoidable. Through probabilistic modeling and analysis, uncertainties may be modeled and assessed properly, and their effects on a given decision accounted for systematically. In this manner, the risk associated with each decision alternative may be delineated and, if desired or necessary, measures taken to control or minimize the corresponding possible consequences.

Decision problems in engineering planning and design often also require the consideration of nontechnical factors, such as social preference or acceptance, environmental impact, and sometimes even political implications. In these latter cases, the selection of the "best" decision alternative cannot be governed solely by technical considerations. A systematic framework that will permit the consideration of all facets of a decision problem is the *decision model*.

The elements of the decision model and the analyses involved in a decision problem are developed in this chapter. Both single- and multiple-objective engineering decision problems are discussed and illustrated.

2.1.1 Simple Risk-Decision Problems

In any decision analysis, the set of decision (or design) variables should first be identified and defined. For example, in the design of storm sewers, the engineer may select a pipe diameter to provide a desired flow capacity. Because of variability in the future storm runoff, there is a probability that the flow capacity of this pipe may not be sufficient. Since this probability is related to the pipe size, the probability level may also be the decision variable instead of the pipe diameter.

In order to rank various designs, an objective function is usually defined in terms of the decision variables. This function is frequently expressed in monetary units representing total benefit or total cost. It is obvious that the optimal design will be determined by the values of the decision variables that will maximize the benefit or minimize the loss function. In some cases, in which the objective function is a continuous function of the decision variables, the calculus of maximization and minimization provides a convenient tool for optimization. Formally, if $X_1, X_2, \ldots, X_n$ denote the set of decision variables, the optimal design will be

5

given by values of the decision variables satisfying the following set of equations:

$$\frac{\partial F\,(X_1, \ldots, X_n)}{\partial X_i} = 0; \qquad i = 1, 2, \ldots, n \qquad (2.1)$$

where $F(X_1, \ldots, X_n)$ is the objective function. The second partial derivatives may be examined to determine whether Eq. 2.1 yields the maximum or minimum objective function. The procedure is illustrated in the following examples.

EXAMPLE 2.1   (excerpted from Shuler, 1967)

A contractor is preparing a bid for a construction project. Based on experience, it is his judgment that the probability of winning a bid depends on his bid ratio, R, as follows:

$$p = 1.6 - R; \qquad 0.6 \le R \le 1.6$$

in which R is the ratio of his bid price to the total estimated cost (Fig. E2.1).

Obviously, the decision variable here is the bid price B that the contractor will submit. The corresponding objective function may be the expected profit from the project;

$$X = (B - C)p + 0 \cdot (1 - p)$$

$$= \frac{B - C}{C}\,pC$$

$$= (R - 1)pC$$

$$= (-R^2 + 2.6R - 1.6)C; \qquad \text{for } 0.6 \le R \le 1.6$$

where C is the estimated cost of construction. The optimal decision requires the maximization of X with respect to R; thus, setting

$$\frac{dX}{dR} = C(-2R + 2.6) = 0$$

we obtain $R = 1.3$. Since $d^2X/dR^2 < 0$ at $R = 1.3$, the bid ratio of 1.3 will maximize X. Therefore, the optimal bid for the contractor should be 1.3 times the estimated cost.

The above solution assumes that there is no monetary gain or loss if the contractor does not win the job. Suppose that the construction crew would be idle and the contractor will be operating at a loss L if he fails to win this job. In such a case, the objective function should include an expected loss; suppose $L = 0.1C$. Then, the expected overall gain would be

$$X = (R - 1)pC - (1 - p)L$$
$$= [(R - 1)(1.6 - R) - (R - 0.6) \times 0.1]C$$
$$= [-R^2 + 2.5R - 1.54]C$$

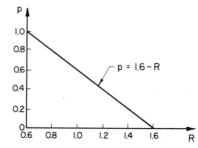

Figure E2.1   Probability of winning versus bid ratio.

Then,

$$\frac{dX}{dR} = -2R + 2.5 \equiv 0$$

yields $R = 1.25$. Hence, the optimal bid becomes 1.25 times the estimated cost. A lower bid is required here so that the contractor will increase his chance of winning the bid, thus reducing the chance of idling his construction crew.

EXAMPLE 2.2

The construction of a bridge pier requires the installation of a cofferdam in a river. Suppose the occurrence of floods follows a Poisson distribution with a mean occurrence rate of 1.5 times per year, and the elevation of each flood is exponentially distributed with a mean of 5 feet above normal water level. Each time the cofferdam is overtopped, the expected loss resulting from a possible delay in construction and pumping cost is estimated to be $25,000. Since the flood elevation can be predicted sufficiently ahead of time to effect the evacuation of personnel working inside the cofferdam, the possibility of loss due to workmen being trapped may be neglected. Suppose that the construction cost of the cofferdam is given by

$$C_c = C_o + 3000h$$

where C_o includes the cost of a cofferdam foundation and necessary construction to dam the river to normal water level, and h is the height of the cofferdam (in feet) above normal water level. Determine the optimal height of the cofferdam if it is expected to be used over a two-year period.

A convenient choice of the objective function here is the expected monetary loss C_T, which would consist of the construction cost and the expected total loss from the flooding of the cofferdam during the two years of service. The expected loss in each flood is

$$C = E(\text{loss}|\text{overtopping})P(\text{overtopping})$$

$$= 25,000 \int_h^\infty \frac{1}{5} e^{-x/5}\, dx$$

$$= 25,000 e^{-h/5}$$

Hence, the expected total loss from all floods during the two-year period (the time value of money neglected) is

$$C_f = \sum_{i=0}^\infty E(\text{loss}|i \text{ floods})P(i \text{ floods in 2 years})$$

$$= \sum_{i=0}^\infty i \cdot (25,000 e^{-h/5}) \frac{3^i e^{-3}}{i!}$$

$$= 25,000 e^{-h/5} \sum_{i=0}^\infty i \frac{3^i e^{-3}}{i!}$$

$$= 75,000 e^{-h/5}$$

The total expected monetary loss becomes

$$C_T = C_c + C_f$$
$$= C_o + 3000h + 75,000 e^{-h/5}$$

Clearly, h is the only decision variable, whereas C_o is a constant independent of h. Differentiating C_T with respect to h, we obtain the optimal cofferdam height as follows:

$$\frac{dC_T}{dh} = 3000 + 75,000 e^{-h/5}(-\tfrac{1}{5}) \equiv 0$$

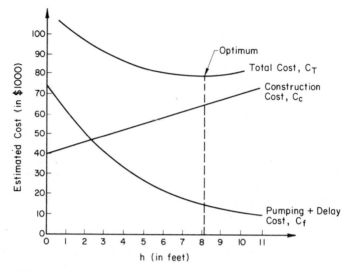

Figure E2.2 Costs as functions of cofferdam elevation above normal water level.

or

$$e^{-h/5} = \tfrac{1}{5}$$

Thus,

$$h_{\text{opt}} = 5 \ln (5) = 8.05 \text{ ft}$$

Therefore, the optimal height of the cofferdam should be approximately 8 feet above the normal water level of the river. Figure E2.2 shows the various cost components as functions of the decision variable h. It may be observed that as h increases, the construction cost increases; whereas the expected flood loss decreases. The total expected loss takes on a minimum when the slopes of these two functions are equal but of opposite signs. At this point, the marginal increase in C_c is balanced by the marginal decrease in C_f. The value of C_o, which was assumed to be \$40,000 in Fig. E2.2, simply adds a constant value to the ordinate at each value of h, and does not affect the determination of the optimal value of h. The probability that a cofferdam with $h = 8.05$ ft will be overtopped in a flood is given by

$$P(\text{flood level} > 8.05') = \int_{8.05}^{\infty} \frac{1}{5} e^{-x/5} \, dx = 0.2$$

Alternatively, the objective function may be formulated using the probability of overtopping, p, as the decision variable. Since the construction cost is a function of h, we need to express h in terms of the overtopping probability p. Based on the exponential distribution assumed for the flood level, it can be shown that

$$p = \int_{h}^{\infty} \frac{1}{5} e^{-x/5} \, dx$$

or

$$h = -5 \ln p$$

Hence, the total expected cost becomes

$$C_T = C_o + 3000(-5 \ln p) + 3(25{,}000)p$$
$$= C_o - 15{,}000 \ln p + 75{,}000p$$

Differentiating C_T with respect to p, we obtain the optimal probability as follows:

$$\frac{dC_T}{dp} = -15{,}000 \frac{1}{p} + 75{,}000 \equiv 0$$

from which

$$p_{opt} = 0.2$$

which is the same as obtained earlier. Based on this value of p_{opt}, the corresponding optimal elevation of the cofferdam (above normal water level) is determined as

$$
\begin{aligned}
h_{opt} &= -5 \ln p_{opt} \\
&= -5 \ln (0.2) \\
&= 8.05 \text{ ft}
\end{aligned}
$$

The first approach described above directly gives the optimal value of the design variable, whereas the second formulation directly yields the optimal probability level. In cases involving several cofferdams, each subject to a different distribution of floods, expressing the overall objective function in terms of a common risk level could simplify the optimization process. In any case, the two methods should lead to identical optimal designs, provided there is a correspondence between the design variables and the probability of overtopping.

2.1.2 Characteristics of a General Decision Problem

The applicability of the optimization procedure presented above, based on calculus, is limited; first of all, the objective function must be expressed as a continuous function of the decision variables. Unfortunately, this may not be the case for many engineering decision problems. Consider a dam that has been proposed as a possible solution for flood control in a certain drainage basin. Its elevation may be a decision variable; the dam site may also need to be determined. Moreover, other forms of flood control strategies such as diversion channels, levees, multiple reservoirs may be possible alternatives to be considered in the decision analysis. Thus, it will be difficult, if not impossible, to obtain a continuous objective function in terms of all the decision variables.

Sometimes, a decision may not be based solely on the available information. If time permits, additional information may be collected prior to the final selection among the feasible design alternatives. In engineering problems, additional data could take the form of laboratory or field tests, or further research. Since there could be a variety of such schemes for collecting additional information, this would further expand the spectrum of alternatives.

In short, in more general decision problems, a framework for systematic analysis is required. Specifically, the decision analysis should at least include the following components:

(1) A list of all feasible alternatives, including the acquisition of additional information, if appropriate.
(2) A list of all possible outcomes associated with each alternative.
(3) An estimation of the probability associated with each possible outcome.
(4) An evaluation of the consequences associated with each combination of alternative and outcome.
(5) The criterion for decision.
(6) A systematic evaluation of all alternatives.

A decision model that considers all these basic components is presented in Section 2.2.

2.2 THE DECISION MODEL

2.2.1 Decision Tree

The various components of a decision problem may be integrated into a formal layout in the form of a *decision tree*, consisting of the sequence of decisions—namely, a list of feasible alternatives; the possible outcomes associated with each alternative; the corresponding probability assignments; monetary consequences and utility evaluations (see Section 2.4). In other words, the decision tree integrates the relevant components of the decision analysis in a systematic manner suitable for an analytical evaluation of the optimal alternative. Probability models of engineering analysis and design may be used to estimate the relative likelihoods of the possible outcomes, and appropriate value or utility models evaluate the relative desirability of each consequence.

Figure 2.1 shows a generic example of a decision tree with three alternatives, in which the third alternative involves performing an experiment to gather additional information prior to any final decision. The word "experiment" should be interpreted in a broad sense, covering any method of gathering additional data. The following notations are used in Fig. 2.1:

a_i = Alternative i.

θ_j = Outcome j.

e_k = Experiment k designed to gather additional information.

z_l = Experimental outcome l.

$u(a_i, \theta_j)$ = Utility value corresponding to alternative a_i and outcome θ_j; if the utility depends on experiment e_k and the corresponding experimental outcome z_l, it will be denoted as $u(e_k, z_l, a_i, \theta_j)$.

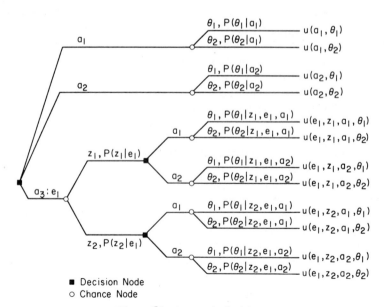

Figure 2.1 A generic decision tree.

The decision tree in Fig. 2.1 begins with a square node, called a *decision node*, at which point there are alternatives a_1, a_2, and a_3. With each alternative, there may be several possible outcomes, shown as branches originating from a circular node, called a *chance node*. At each chance node, nature controls what will occur subsequently. In this example, θ_1, θ_2 are the two possible outcomes associated with alternatives a_1 and a_2. These outcomes may or may not depend on the preceding alternative; in general, their probabilities are conditional on the preceding alternatives; for example, $P(\theta_j | a_i)$. Alternative a_3 is followed by experimental outcomes, whose probabilities will generally depend on the experiment. Hence, the probabilities of (z_l) are also conditional on e_k; that is, $P(z_l | e_k)$. Another decision node follows each experimental outcome, denoting that a decision between alternatives a_1 and a_2 will be required after observing the specific additional information. The probability of θ_j in the subsequent branches would be updated based on the particular experimental outcome. Hence, in general, this probability is expressed as $P(\theta_j | z_l, e_k, a_i)$.

The outcomes from a chance node are mutually exclusive and collectively exhaustive; thus, the sum of the conditional probabilities at each chance node should add up to unity. Sometimes there may be a continuous spectrum of outcomes originating from a chance node, such as those given by the values of a continuous random variable. In such cases, the PDF of the random variable will be used to denote the relative likelihoods of these branches. The desirability of the consequence of each sequence or path in the tree is measured by its "utility value" recorded at the end of the sequence, such as $u(a_i, \theta_j)$ or $u(e_k, z_l, a_i, \theta_j)$.

The following examples illustrate the application of the decision tree model in engineering. In Section 2.2.8, several in-depth examples of complete decision analysis further illustrate the decision tree model.

EXAMPLE 2.3

Two alternative designs are considered for the structural scheme of a building. Design A is based on a conventional procedure, whose probability of satisfactory performance is 99%, and costs $1.5 million. Design B is based on modern concepts and will reduce the cost of the building to $1 million. The reliability of design B is not known; however, the structural engineer estimates that if his assumptions are correct, the reliability of satisfactory performance will be 0.99; whereas, if his assumptions are not valid, the reliability is only 0.9. According to his judgment, he is only 50% sure of the validity of his assumptions. Suppose the cost of unsatisfactory performance is $10 million. The decision tree would be as shown in Fig. E2.3; the utilities, in terms of losses in million $, are indicated for each sequence of the tree.

EXAMPLE 2.4

In order to prevent or minimize thermal pollution from the operation of a nuclear power plant, cooling lakes are commonly used to allow the heat from the plant to dissipate through a natural process. At one location, the dikes for the lake were constructed on a relatively pervious ground where water seepage is possible. Such seepage can cause a loss of water as well as become a nuisance to the surrounding area. As a remedy, pumps may be installed on the downstream slope to pump back the water to the cooling lake, as shown schematically in Fig. E2.4a. Based on a theoretical seepage model and soil data from the site, the flow of water is estimated, for simplicity, to be either $Q_1 = 100$ gal/min or $Q_2 = 120$ gal/min, with relative likelihoods of 9 to 1.

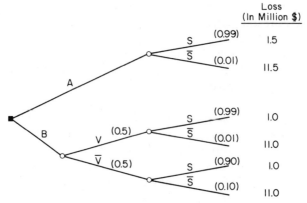

Figure E2.3   Decision tree for structural design problem.

There are two pump system capacities that the engineer may order, namely pump system A with 100 gal/min or pump system B with 120 gal/min, which cost 17 and 20 units (1 unit = $100), respectively. If pump system A is insufficient, an additional small pump system C with 20 gal/min will be ordered and the overall additional expense, including inconvenience and temporary disruption of performance, is estimated to be 8 units. Over the expected service period, the total operational (OMR) cost, including maintenance and repair costs, discounted to present values are 22, 21, and 5 units for pump systems B, A, and C, respectively. Two units of the OMR cost of pump B could be discounted if the flow Q is actually 100 gal/min.

Another feasible alternative is to install a barrier of bentonite seal beneath the upstream end of the embankment. The total material and installation cost of the seal will be 41 units, and there are negligible maintenance costs over the years.

The decision tree for this example is shown in Fig. E2.4b. The costs associated with each path of the decision tree are clearly dependent on the particular alternative and the associated outcome.

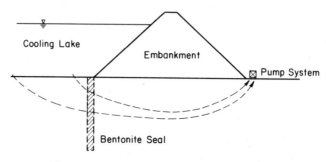

Figure E2.4a   Cross-section of embankment.

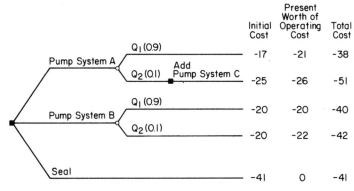

Figure E2.4b   Decision tree for seepage problem.

EXAMPLE 2.5 (*excerpted from Haefner and Morlok, 1971*)

Improving the safety of a short section of a rural road with a complex alignment through geometric design changes is proposed. For simplicity, only one improvement scheme is considered; namely, to improve the alignment and profile and install a median barrier. The decision alternatives are simply whether to adopt this improvement scheme or to make no improvement. The decision tree is shown in Fig. E2.5 (following the branch indicating "No Prediction Model"). The outcome is described by the various states of accident and fatality rate encountered. To simplify the problem, accident and fatality rates are assumed to be either high or low, thus leading to four possible combinations of outcomes—namely, $H_A H_F$ (high accident rate, high fatality rate), $H_A L_F$, $L_A H_F$, and $L_A L_F$. The probabilities for each of these outcomes (as shown in Fig. E2.5) have been assessed with existing information. The total cost associated with each path consists of the road improvement cost, if any, and the accident and fatality losses.

It is common in highway engineering to employ a prediction model, such as the Kihlberg and Tharp model or the Mann and Dart (Louisiana) model, to predict accident rates for a given design alternative. The models provide information on whether or not the proposed road improvement will result in a significant reduction in total accidents and fatalities.

There are two stages in the decision process. The first decision is to select the particular prediction model. Based on the model prediction result, namely $H'_A H'_F$ or $H'_A L'_F$, or $L'_A H'_F$ or $L'_A L'_F$, a decision is then made whether or not to improve the road condition. A representative path of the decision tree for the Tharp model is shown in Fig. E2.5.

Observe that the probabilities associated with various fatality and accident combinations do not change if no improvement is adopted. The Tharp model applies only to the prediction associated with the improved scheme. Depending on the model prediction, the updated probability for each fatality and accident combination will be different (see Problem 2.11). The cost for performing the prediction model is also included in the total cost for appropriate paths.

2.2.2   Decision Criteria

The objective of decision analysis is to make the "best" decision. Only in rare occasions would a decision maker be comparing alternatives whose monetary values are known deterministically. In this case, he would obviously choose the alternative with the highest value. However, most decisions are made under conditions of uncertainty, in which the possible utility values of $u_1, u_2, \ldots, u_n$ of a given alternative may be stated only in terms of respective probabilities $p_1, p_2, \ldots, p_n$. Also, depending on the temperament, experience, and degree of

Total Cost
(In 10^4 Dollars)

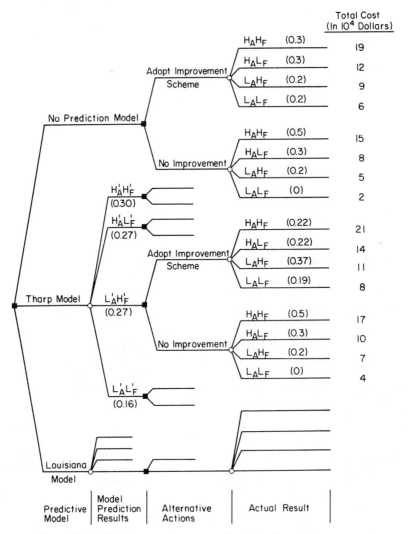

Figure E2.5   Decision tree for geometric design in road improvement.

risk-aversiveness of the decision maker, the "best" decision may mean different things to different people and at different times. Nevertheless, some general criteria for a decision can be identified.

Consider a simple decision problem, in which an owner has an option to install an emergency electrical power system to guard against potential utility power failure. The additional cost for the system will be $2000 per year; whereas if power failure occurs and no emergency power supply is available, he could incur a loss of $10,000. Based on his experience and judgment, he believes that the annual probability of power failure is only 10%. Assume that the chance of more than one power failure in a year is negligible. The decision tree for this example is shown in Fig. 2.2. If the owner installs the emergency system, he would have to spend $2000;

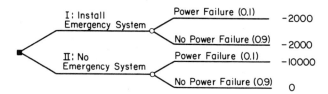

Figure 2.2   Decision tree for installation of emergency power system.

if he does not, he has a 90% chance of not spending any money but also a 10% chance of losing $10,000.

If the owner is extremely pessimistic (believing that nature always works against him), he would try to minimize his loss by installing the emergency system, because the maximum loss in this alternative is only $2000, relative to $10,000 for the other alternative. On the other hand, if he is an extremely optimistic person (believing that luck is always on his side), he would not install the emergency power system, since this alternative will yield the maximum gain; namely, it may cost the owner nothing compared to $2000 for the other alternative. The first basis of selecting the optimal alternative is called the *mini-max* criterion, in which the decision maker selects the alternative that minimizes the maximum loss. In the latter case, he tries to maximize his maximum possible gain among the alternatives, referred to as the *maxi-max* criterion.

Neither of these two criteria are practical in the long run. If the mini-max decision maker consistently follows this criterion, he will never venture into a decision that may result in substantial gain as long as there is a finite (though small) chance that he may be incurring a loss, whereas the maxi-max decision maker will never choose the realistic alternative provided that there is a finite chance for a more attractive outcome. Any rational decision maker should consider the relative likelihood associated with the gain or loss in each alternative, rather than strictly following these criteria.

Maximum Expected Monetary Value Criterion   When the consequences associated with each alternative in a decision analysis can be expressed in terms of monetary values, a widely used criterion for decision is the *maximum expected monetary gain*. Suppose d_{ij} denotes the monetary value of the jth consequence associated with alternative i, and p_{ij} is the corresponding probability; then the expected monetary value of the ith alternative is

$$E(a_i) = \sum_j p_{ij} d_{ij} \tag{2.2}$$

The optimal alternative according to this max EMV criterion is the one whose expected monetary value is

$$d(a_{\text{opt}}) = \max_i \left\{ \sum_j p_{ij} d_{ij} \right\} \tag{2.3}$$

EXAMPLE 2.6

Suppose the owner referred to in Section 2.2.2 follows the max EMV decision criterion. He would compute the EMV of the two alternatives as (see the decision tree in Fig. 2.2):

$$E(I) = 0.1 \times (-2000) + 0.9 \times (-2000)$$
$$= -2000 \text{ dollars}$$

and

$$E(II) = 0.1 \times (-10{,}000) + 0.9 \times (0)$$
$$= -1000 \text{ dollars}$$

Comparing the two expected monetary values, the owner should not install the emergency power supply system.

Both the maxi-max and mini-max criteria ignore the probabilities of the possible outcomes of an alternative. The available information is, therefore, not fully utilized in the decision analysis. With the maximum EMV criterion, a decision maker is systematically weighing the value of each outcome by the corresponding probability.

It is conceivable that an event with an occurrence probability of 99.99 % may still fail to occur, in spite of the extreme odd. Nevertheless, if a decision maker consistently bases his decisions on the max EMV criterion, the total monetary value obtained from all his decisions (in the long run) will be maximum.

Another common decision criterion known as the *maximum expected utility* criterion is presented in Section 2.4.6.

2.2.3   Decision Based on Existing Information—Prior Analysis

Once the decision tree has been established, with the probabilities of the possible outcomes at the chance nodes and the consequences of all respective paths evaluated based on existing information, the expected monetary value of each alternative may be computed according to Eq. 2.2. A decision analysis based entirely on available prior information is called *prior analysis*; if this analysis is updated subsequently with additional new information, the latter is called *terminal analysis* (described in Section 2.2.4).

Sometimes a given alternative may involve a series of subsequent chance nodes such as that shown in Fig. 2.3. Suppose at chance node A, the outcomes are the possible values of a discrete random variable X; whereas at chance node B, the outcomes involve a continuous random variable Y.

The expected value calculation for alternative a_i starts at the last chance node; namely, the chance node B in Fig. 2.3. At node B, the expected value of alternative a_i, given $X = x_j$, is

$$E(a_i|x_j) = \int_{-\infty}^{\infty} E(a_i|x_j, y) f_Y(y)\, dy$$

$$= \int_{-\infty}^{\infty} d(x_j, y, a_i) f_Y(y)\, dy$$

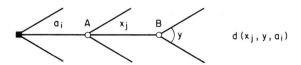

Figure 2.3 Decision tree with series of chance nodes.

where $f_Y(y)$ is the PDF of Y. Subsequently, at node A,

$$E(a_i) = \sum_j E(a_i|x_j)p_j$$

where p_j is the PMF of X. Sometimes the complete probability distribution of the random variable may not be necessary to calculate the expected value. For instance, if the cost function is quadratic in y, namely

$$d(x_j, y, a_i) = ay^2 + by + c$$

then, at node B,

$$E(a_i|x_j) = \int_{-\infty}^{\infty} (ay^2 + by + c)f_Y(y)\,dy$$

$$= aE(Y^2) + bE(Y) + c$$

$$= a[\text{Var}(Y) + E^2(Y)] + bE(Y) + c$$

indicating that only the mean and variance of Y are required to compute the expected cost at node B.

EXAMPLE 2.7

A new waste treatment process is designed for installation in a local community. Because the proposed waste treatment process is based on a completely new concept, its efficiency is yet unknown. The project engineer has a choice of designing a small unit A or a larger unit B. Assume that the efficiency could be classified into two levels; namely, EH and EL for high and low efficiencies, respectively. If the efficiency is actually high, unit A will be the appropriate design, whereas unit B is basically designed for low efficiency. The engineer estimates a probability of 70% that the efficiency of the proposed treatment process will be high. The decision tree is shown in Fig. E2.7, in which the relative monetary loss (in $1000) for each path in the tree has been evaluated.

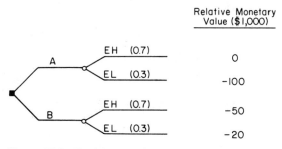

Figure E2.7 Decision tree for waste treatment system.

The expected monetary value for the two designs are

$$E(A) = 0.7 \times 0 + 0.3 \times (-100) = -30$$

and

$$E(B) = 0.7 \times (-50) + 0.3 \times (-20) = -41$$

With the given information, design A should be selected as the proposed design; the corresponding expected monetary loss is $30,000.

EXAMPLE 2.8   (Sensitivity to Loss Functions)

After feasibility studies on the possible sites for a proposed earth dam, two locations (sites A and B) remain for final consideration. From a cost-benefit analysis of the project, it is shown that locating the dam at site A will yield an overall return of $22 million, whereas the dam at site B will give an overall return of $20 million. However, the earthdams at these sites are also susceptible to earthquake damages. From a seismic study of the area (that takes into account the existing fault lines, geological conditions, and available earthquake records), the annual mean occurrence rate of earthquakes with damaging intensity at site A is estimated to be 0.01 or 0.02 with equal likelihood, and the corresponding occurrence rate for site B is 0.005 or 0.01 per year with equal likelihood. The occurrence of an earthquake may be modeled by a Poisson process. Assume that earthquake damages to the dam will be repaired after each earthquake and the dam restored to its original condition. Let X be the number of earthquakes occurring in the life of the dam, assumed to be 50 years.

Determine the expected earthquake loss for each of the following cases. All losses are measured in terms of million dollars.

Case 1

The earthquake damage function is

$$L = 5X$$

The expected loss given that site A is selected with a mean rate of earthquake occurrence v_j is

$$E[L|A, v_j] = \sum_{x=0}^{\infty} 5x \cdot P_X[x|v_j]$$

$$= 5E[X|v_j]$$

$$= 5(50v_j)$$

$$= 250v_j$$

Hence,

$$E[L|A] = \sum_{j=1}^{2} E[L|A, v_j]P[v_j|A]$$

$$= (250 \times 0.01)(0.5) + (250 \times 0.02)(0.5)$$

$$= \$3.75 \text{ million}$$

Similarly,

$$E[L|B] = \sum_{j=1}^{2} E[L|B, v_j]P[v_j|B]$$

$$= (250 \times 0.005)(0.5) + (250 \times 0.01)(0.5)$$

$$= \$1.875 \text{ million}$$

Case 2

Suppose the earthquake damage function is

$$L = 5X^2$$

Then,

$$E[L|A, v_j] = \sum_{x=0}^{\infty} 5x^2 P_X[x|v_j]$$

$$= 5E[X^2|v_j]$$

$$= 5[\text{Var}(X|v_j) + E^2(X|v_j)]$$

$$= 5[(50v_j) + (50v_j)^2]$$

$$= 250v_j + 12{,}500v_j^2$$

Hence, in this case,

$$E[L|A] = \sum_{j=1}^{2} (250v_j + 12{,}500v_j^2)P[v_j|A]$$

$$= (250 \times 0.01 + 12{,}500 \times 0.01^2)(0.5) + (250 \times 0.02 + 12{,}500 \times 0.02^2)(0.5)$$

$$= \$6.875 \text{ million}$$

$$E[L|B] = (250 \times 0.005 + 12{,}500 \times 0.005^2)(0.5) + (250 \times 0.01 + 12{,}500 \times 0.01^2)(0.5)$$

$$= \$2.656 \text{ million}$$

Case 3

Finally, if the earthquake damage function follows an exponential loss function,

$$L = c^X$$

in which $c = 3$. Observe that $L = 1$ when $X = 0$; the small loss here could denote the maintenance cost through the years. Then,

$$E[L|A, v_j] = \sum_{x=0}^{\infty} c^x \frac{(50v_j)^x e^{-50v_j}}{x!}$$

$$= \sum_{x=0}^{\infty} \frac{(50cv_j)^x (e^{-50v_jc})}{x!} \cdot \frac{e^{-50v_j}}{e^{-50v_jc}}$$

$$= e^{50v_j(c-1)}$$

$$E(L|A) = 0.5e^{50(0.01)(3-1)} + 0.5e^{50(0.02)(3-1)}$$

$$= 0.5e + 0.5e^2$$

$$= \$5.06 \text{ million}$$

Similarly,

$$E(L|B) = 0.5e^{50(0.005)(2)} + 0.5e^{50(0.01)(2)}$$

$$= 0.5e^{0.5} + 0.5e$$

$$= \$2.186 \text{ million}$$

The decision tree is shown in Fig. E2.8.

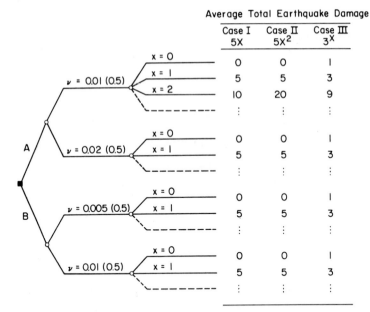

Average Total Earthquake Damage

	Case I 5X	Case II 5X^2	Case III 3^X
x = 0	0	0	1
x = 1	5	5	3
x = 2	10	20	9
	⋮	⋮	⋮
x = 0	0	0	1
x = 1	5	5	3
	⋮	⋮	⋮
x = 0	0	0	1
x = 1	5	5	3
	⋮	⋮	⋮
x = 0	0	0	1
x = 1	5	5	3
	⋮	⋮	⋮

Figure E2.8   Decision tree for dam site selection.

Table E2.8 summarizes the expected benefit and damage for the three cases considered. Depending on the specific loss function, *A* or *B* may be the optimal site for the dam.

Table E2.8   Expected Benefits and Damages for the Dam Sites

Site Location	Benefit	Damage Cost			Net Benefit		
		Case 1	Case 2	Case 3	Case 1	Case 2	Case 3
A	22	3.75	6.875	5.06	18.25[a]	15.125	16.94
B	20	1.875	2.656	2.186	18.125	17.344[a]	17.814[a]

[a] Indicates optimal alternative.

EXAMPLE 2.9

Suppose at a certain stage of construction, there are only 30 days left before a project's scheduled completion date. Because of bad weather in the first phase of the project, construction has not been proceeding at the anticipated rate. Based on the contractor's judgment, the project could still meet the deadline if exceptionally good weather (condition *G*) prevails in the next 30 days. He estimates that this may occur with 20% probability. However, if normal weather (*N*) prevails, completion will be delayed with a mean delay of 5 days and a standard deviation of 3 days beyond the scheduled completion date; if bad weather (*B*) continues, the corresponding mean and standard deviation of delay will be 15 and 5 days, respectively. Assume that normal and bad weather conditions are equally likely to occur in the remaining time required and the cost of delay is $(0.1x^2 + x)$ in thousands of dollars, where x is the length of delay in days. This cost function implies that the loss per day of delay increases as the delay becomes longer.

The contractor has an option to launch a crash program, at an additional cost of $10,000. With this crash program, he estimates that he can improve the chances of completing the project on schedule in the events of normal (N) and bad (B) weather to 90% and 70%, respectively. Also, if a delay does occur, its mean and standard deviation will both be reduced to 1 day under normal weather, and 5 and 2 days, respectively, under bad weather.

Should the contractor launch the crash program based on the maximum EMV criterion?

Solution

From the decision tree shown in Fig. E2.9, the expected loss if no crash program is launched is

$$E[L|\bar{C}] = E[L|\bar{C}G]P(G) + E[L|\bar{C}N]P(N) + E[L|\bar{C}B]P(B)$$
$$= 0 \times 0.2 + E[L|\bar{C}N] \times 0.4 + E[L|\bar{C}B] \times 0.4$$

where

$$E[L|\bar{C}N] = E[(0.1X^2 + X)|\bar{C}N]$$
$$= 0.1\{Var[X|\bar{C}N] + E^2[X|\bar{C}N]\} + E[X|\bar{C}N]$$
$$= 0.1\{9 + 25\} + 5$$
$$= \$8.4 \text{ thousand}$$

$$E[L|\bar{C}B] = 0.1\{25 + 225\} + 15 = \$40 \text{ thousand}$$

Hence,

$$E[L|\bar{C}] = 0 + 0.4 \times 8.4 + 0.4 + 40 = \$19.36 \text{ thousand}$$

Similarly, for the crash alternative

$$E[L|CG] = \$10 \text{ thousand}$$
$$E[L|CN] = 0.9 \times 10 + 0.1\{10 + 0.1[Var(X|CN) + E^2(X|CN)] + E(X|CN)\}$$
$$= 0.9 \times 10 + 0.1\{10 + 0.1(1 + 1) + 1\}$$
$$= \$10.12 \text{ thousand}$$

$$E[L|CB] = 0.7 \times 10 + 0.3\{10 + 0.1(4 + 25) + 5\}$$
$$= \$12.37 \text{ thousand}$$

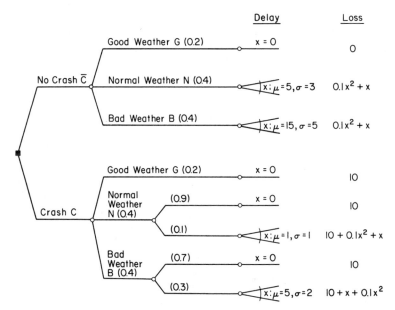

Figure E2.9 Decision tree for completing construction project.

and

$$E[L|C] = 0.2 \times 10 + 0.4 \times 10.12 + 0.4 \times 12.37$$
$$= \$11.00 \text{ thousand}$$

The criterion of maximizing the expected monetary value is equivalent to that of minimizing the expected monetary loss. According to this analysis, the contractor should launch a crash program to complete the construction project on schedule.

EXAMPLE 2.10

A distance AB is to be measured. With instrument (I), the surveyor must measure 10 separate segments. The error in each segment is random with mean zero and standard deviation σ_1; σ_1 is estimated to have a mean of 1 inch and c.o.v. of 50%. Another instrument (II) is available, which only requires 5 separate segments to measure the distance AB; the error in each segment has mean zero and standard deviation σ_2 with a mean of 3 inches and c.o.v. of 60%. The cost of performing the required measurements are $150 and $100 with instruments (I) and (II), respectively. The penalty of inaccurate measurement is proportional to the square of the total error, or

$$L = k(X_1 + X_2 + \cdots + X_n)^2$$

where $k = 2$, and X_i is the measurement error (inch) in segment i. Which of the instruments should the surveyor use based on the maximum EMV criterion?

Solution

The expected loss from inaccurate measurement based on the above quadratic penalty function can be expressed in terms of the standard deviation of the error in each segment as

$$E(L|\sigma) = E[k(X_1 + X_2 + \cdots + X_n)^2 | \sigma]$$
$$= k\left[\sum_{i=1}^{n} E(X_i^2 | \sigma) + \sum_{i \neq j}\sum E(X_i X_j | \sigma) \right]$$

If each measurement is assumed to be statistically independent and identically distributed,

$$E(L|\sigma) = knE(X^2 | \sigma)$$
$$= kn\sigma^2$$

and

$$E(L) = \int_{-\infty}^{\infty} kn\sigma^2 f(\sigma) \, d\sigma$$
$$= knE(\sigma^2)$$
$$= kn[\text{Var}(\sigma) + E^2(\sigma)]$$

For instrument (I), the expected total cost, including operating and inaccuracy costs, is

$$E(C_1) = 150 + 2 \times 10[0.5^2 + 1^2]$$
$$= \$175$$

The corresponding total cost for instrument (II) is

$$E(C_{II}) = 100 + 2 \times 5[1.8^2 + 3^2]$$
$$= \$222.5$$

Therefore, on the basis of the maximum EMV criterion, the surveyor should use instrument (I).

EXAMPLE 2.11   (excerpted from Revell, 1973)

(a) The existing Grapefruit Dam on the Grapefruit River is a multipurpose project including water supply, power supply, recreation, and flood control.

After a detailed economic analysis, the cost of the damage as a function of the discharge level of the river is tabulated in Table E2.11a. Determine the expected annual cost of damage.

Table E2.11a Damage Cost at Given Discharge Level

Discharge (1000 cfs)	Cost of Damage in $1000
60	0
80	1
100	101
124	5200
132	7635
170	17,000
185	24,500
222	40,180
240	59,500
255	92,125
> 255	92,125

Solution

The annual cost of damage depends on the level of maximum discharge in that year. Since the level of maximum discharge is a random variable (denoted as X below), the expected annual cost of damage is

$$E(C) = \int_{-\infty}^{\infty} c(x) f_X(x) \, dx$$

Since the damage cost (Table E2.11a) is given only for discrete levels of discharge, the expected annual cost of damage may be computed approximately as follows:

$$E(C) = \sum_i \frac{c(x_i) + c(x_{i+1})}{2} P(x_i < X \le x_{i+1})$$

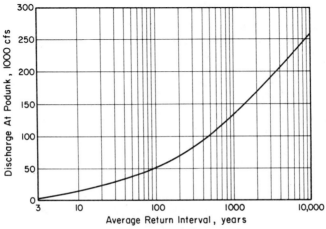

Figure E2.11   Flow-frequency, Grapefruit River at Podunk.

where x_i and x_{i+1} represent two successive discretized discharge levels. From the flow frequency curve of Fig. E2.11, the probability that each tabulated discharge value will be exceeded in a year can be obtained. Table E2.11b summarizes the computations for the expected cost; these results yield the expected annual damage cost of $32,810 for the existing Grapefruit Dam. In the next part of this example (see below), other alternatives for improving the spillway capacity are considered. The expected annual damage cost for each alternative is also computed using the same technique.

Table E2.11b   Summary of Computations of Expected Annual Risk Cost

Discharge (1000 cfs)	Damage Cost (in $1000)	Return Period (in years)	Probability of Exceedence	$P(x_i < x < x_{i+1})$	Average Damage Cost (in $1000)	Annual Risk Cost (in $1000)
60	0	150	0.00667		0.5	0
80	1	250	0.00400	0.00267	51.0	0.10
100	101	500	0.00200	0.00200	2650.5	1.99
124	5200	800	0.00125	0.00075	6417.5	1.60
132	7635	1000	0.00100	0.00025	12,317.5	6.16
170	17,000	2000	0.00050	0.00050	20,750	3.53
185	24,500	3000	0.00033	0.00017	32,340	4.20
222	40,180	5000	0.00020	0.00013	49,840	2.99
240	59,500	7000	0.00014	0.00006	75,812	3.03
255	92,125	10,000	0.00010	0.00004	92,125	9.21
> 255	92,125	∞	0.0	0.00010		

$$\Sigma = 32.81.$$

EXAMPLE 2.11   (Continued)

(b)   It is feared that the existing spillway capacity of the Grapefruit Dam may be inadequate to handle the flood discharge. To improve the capacity, three alternative designs are considered:

a_1:   Lengthening the spillway.

a_2:   Lengthening the spillway, lowering the crest, and installing flashboard.

a_3:   Lengthening the spillway, considerable crest lowering, and installing radial gates.

The capital costs required in the three alternatives are $1.04, 1.3, and 3.9 million, respectively. Alternatives a_2 and a_3 would require annual operation and maintenance costs of $2000 and $10,000, respectively. Each alternative design is expected to have a 50-year service life, and the annual interest rate is assumed to be 6%.

The total annual expected damage cost, including losses in recreation, structural and property damages, loss of power revenue, human injuries and deaths, etc., for alternatives a_1, a_2, and a_3 are $5.56, 1.95, 2.97 thousand, respectively; whereas, the damage cost for the existing design (a_0) is $32.81 thousand.

Determine the best alternative based on the minimum expected cost criterion.

Solution

The equivalent annual cost for the initial investment of each alternative design can be obtained by multiplying the capital recovery factor corresponding to a service life of 50 years at an annual interest rate of 6%, which is 0.0634. Table E2.11c compares the annual costs for the various alternatives. The existing design is associated with the minimum total annual cost; hence, no improvement should be implemented on the existing structure.

Table E2.11c Summary of Annual Costs (in Dollars)

Alternatives	Annual Average Equivalent Capital Cost	Annual Operation and Maintenance Cost	Total Annual Expected Risk Cost	Total Annual Cost
a_0	0	0	32,810	32,810
a_1	66,000	0	5560	71,560
a_2	82,000	2000	1950	85,950
a_3	247,000	10,000	2970	259,970

2.2.4 Decision with Additional Information—Terminal Analysis

In order to improve the existing state of information or to reduce the levels of uncertainty, it is common in engineering to conduct research studies or perform laboratory or field tests before a final decision is made. This occurs especially when a decision involves unusual and innovative design concepts, or when time is available for collecting additional information. The acquisition of additional information, of course, will require the expense of time, energy, and financial resources; in spite of these added costs, the additional information may be valuable and will generally improve the chance of making a better decision. The added cost for this new information should be included or reflected in the decision analysis.

In general, the additional information would not eliminate all the uncertainties in a decision problem. At best, based on the experimental outcome, the probabilities in the decision tree would be updated by applying Bayes' Theorem (see Chapters 2 and 8, Vol. 1). A decision analysis with additional information (commonly referred to as *terminal analysis*) is similar to the prior analysis, except that the updated probabilities (probabilities conditional on the experimental outcome) are used in the computation of the expected monetary value or utility. The following examples illustrate the terminal decision analysis.

EXAMPLE 2.12

Suppose the engineer in Example 2.7 feels that it is too risky to adopt the smaller design, that is, unit A; whereas choosing the conversative design, unit B, would mean a large initial investment that may be wasted if the efficiency actually turns out to be high. After conferring with his client, the engineer decided to find out more about the proposed waste treatment process before adopting a specific design. The engineer ordered the construction of a laboratory model to simulate the proposed waste treatment process. The cost of constructing the experiment is $10,000, and the reliability of the experimental results is as follows: if the efficiency of the proposed treatment process is actually high (EH), the probability that the experimental results will indicate a high efficiency rating is 0.8, and the corresponding probabilities for medium and low efficiency ratings are 0.15 and 0.05, respectively. On the other hand, if the efficiency of the process is actually low, the probabilities that the experimental results will show high, medium, and low efficiency ratings are 0.1, 0.2 and 0.7, respectively. In short, the conditional probabilities are as follows:

$$P(HR|EH) = 0.8, P(MR|EH) = 0.15, P(LR|EH) = 0.05$$

$$P(HR|EL) = 0.1, P(MR|EL) = 0.2, P(LR|EL) = 0.7$$

where HR, MR, and LR denote events of high, medium, and low efficiency ratings, respectively.

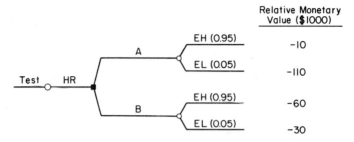

Figure E2.12a   Terminal analysis for HR test outcome.

Suppose the experiment was performed and the results indicated a high rating. In such a case, the updated probability of *EH* is given by Bayes' Theorem as (see Example 2.7 for prior probabilities):

$$P''(EH) = P(EH|HR)$$

$$= \frac{P(HR|EH)P'(EH)}{P(HR|EH)P'(EH) + P(HR|EL)P'(EL)}$$

$$= \frac{0.8 \times 0.7}{0.8 \times 0.7 + 0.1 \times 0.3}$$

$$= 0.95$$

In other words, the probability that the process will actually have high efficiency is increased from 0.7 to 0.95. The corresponding updated probability of *EL* is 0.05.

The decision tree for this terminal analysis is shown in Fig. E2.12a, where the probabilities for all the paths are revised in accordance with the *HR* test outcome.

The expected monetary value of the two alternative designs are, respectively,

$$E(A|HR) = 0.95 \times (-10) + 0.05 \times (-110) = -15$$

$$E(B|HR) = 0.95 \times (-60) + 0.05 \times (-30) = -58.5$$

Hence, if the test results indicate a high efficiency rating, design *A* should be chosen. It is interesting to observe that the expected monetary loss is only $15,000, including the cost of the experiment (compare this with a $30,000 loss if no experiment is performed).

Similarly, if the test outcome indicates a medium rating (*MR*), the updated probabilities of *EH* and *EL* are, respectively, 0.637 and 0.363 as indicated in Fig. E2.12b. The corresponding

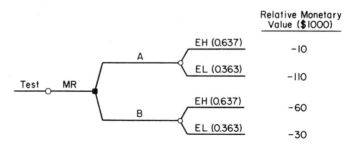

Figure E2.12b   Terminal analysis for MR test outcome.

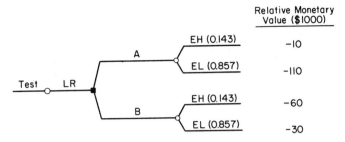

Figure E2.12c   Terminal analysis for LR test outcome.

revised expected monetary values for the two designs will be:

$$E(A|MR) = 0.637 \times (-10) + 0.363 \times (-110) = -46.3$$

$$E(B|MR) = 0.637 \times (-60) + 0.363 \times (-30) = -49.11$$

Although design A is still slightly preferable to design B, the expected loss of \$46,300 exceeds the expected loss with no experiment. In this case, the experimental evidence is not very useful; the test results fall in a fuzzy region. Of course, the chance of this occurring may be reduced if a more discriminating experiment is performed.

Finally, in Fig. E2.12c, if the test indicates a low rating (LR), the updated probabilities of EH and EL will be 0.143 and 0.857, respectively; in such a case, design B should be chosen, with an expected loss of \$34,300.

EXAMPLE 2.13

In Example 2.8, suppose that additional data were obtained revealing that over a 20-year period, one earthquake of damaging intensity had occurred in site A, whereas no earthquakes were recorded at site B. Would the decision on the optimal dam site be different given this new piece of information?

Assume that the average earthquake damage costs are $5X$ (million dollars), where X is the number of earthquake occurrences (i.e., *Case 1* in Example 2.8).

First, the prior probabilities on the mean rate of quake occurrences at sites A and B, namely v_A, v_B, should be updated with respect to the additional earthquake data. For site A,

$$P''(v_A = 0.01) = P(v_A = 0.01 | X = 1 \text{ in } 20 \text{ yr})$$

$$= \frac{P(X = 1 \text{ in } 20 \text{ yr} | v_A = 0.01)P'(v_A = 0.01)}{P(X = 1 \text{ in } 20 \text{ yr} | v_A = 0.01)P'(v_A = 0.01) + P(X = 1 \text{ in } 20 \text{ yr} | v_A = 0.02)P'(v_A = 0.02)}$$

$$= \frac{\dfrac{(0.01 \times 20)^1 e^{-0.2}}{1!} \times 0.5}{\dfrac{(0.01 \times 20)e^{-0.2}}{1!} \times 0.5 + \dfrac{(0.02 \times 20)^1 e^{-0.4}}{1!} \times 0.5}$$

$$= \frac{0.08187}{0.08187 + 0.13406}$$

$$= 0.378$$

$$P''(v_A = 0.02) = 1 - 0.378 = 0.622$$

Similarly for site B,

$$P''(v_B = 0.005)$$

$$= \frac{P(X = 0 \text{ in } 20 \text{ yr}|v_B = 0.005)P'(v_B = 0.005)}{P(X = 0 \text{ in } 20 \text{ yr}|v_B = 0.005)P'(v_B = 0.005) + P(X = 0 \text{ in } 20 \text{ yr}|v_B = 0.01)P'(v_B = 0.01)}$$

$$= \frac{\dfrac{(0.005 \times 20)^0 e^{-0.1}}{0!} \times 0.5}{\dfrac{(0.005 \times 20)^0 e^{-0.1}}{0!} \times 0.5 + \dfrac{(0.01 \times 20)^0 e^{-0.2}}{0!} \times 0.5}$$

$$= 0.524$$

$$P''(v_B = 0.01) = 1 - 0.524 = 0.476$$

The expected seismic damage loss for each alternative is now computed based on these updated probabilities. For site A,

$$E''(L|A) = \sum_{j=1}^{2} E(L|A, v_j)P''(v_j|A)$$

$$= (250 \times 0.01)(0.378) + (250 \times 0.02)(0.622)$$

$$= \$4.055 \text{ million}$$

Similarly, for site B,

$$E''(L|B) = (250 \times 0.005)(0.524) + (250 \times 0.01)(0.476)$$

$$= \$1.845 \text{ million}$$

Combining the benefit data in Example 2.8, the net expected benefits for site A and site B are, respectively, $17.945 and $18.155 million. Therefore, with the additional information, site B becomes the preferred location for the dam.

2.2.5   Preposterior Analysis

In a terminal analysis, the analysis assumes that the additional information (or experimental outcome) is available. However, of equal or greater interest may be the question: "Should additional information be obtained?" Of course, additional information, such as performing an experiment, involves the added cost of time and money; therefore, the answer to this question depends on the benefit of the new information relative to the added cost. The added cost may be justified if it eliminates a significant part of the uncertainty, thus leading to a lower expected loss. Decisions involving whether additional information should be acquired calls for *preposterior analysis*.

Figure 2.4 shows a typical decision tree for preposterior analysis. A two-stage decision problem is involved. At stage A, a decision on whether or not to proceed with an experiment is required; that is, a choice between an experiment (a_3) and no experiment (a_1, a_2). If alternative a_3 (experiment) is chosen, another decision at stage B is needed, depending on the particular experimental outcome z_j. A subtree may be constructed for each experimental outcome z_j at stage B, as shown in Fig. 2.4.

In a preposterior analysis, terminal analyses are first performed on each of the subtrees as indicated in Fig. 2.4. The optimal alternative of each subtree is first

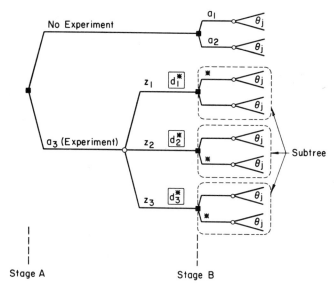

Figure 2.4 Decision tree for preposterior analysis.

identified as marked in the figure. The maximum expected monetary value associated with the optimal alternative of each subtree, namely d_i^*, is then used to compute the EMV of the experiment alternative; thus,

$$E(a_3, \text{experiment}) = \sum_l d_i^* P(z_l) \tag{2.4}$$

in which $P(z_l)$ may be obtained through

$$P(z_l) = \sum_j P(z_l|\theta_j)P'(\theta_j) \qquad l = 1, 2, \ldots \tag{2.5}$$

where $P'(\theta_j)$ is the prior probability of outcome θ_j (i.e., before the experimental outcome is observed). Comparing the expected utilities of a_1, a_2 with that of a_3 (experiment), the optimal alternative at stage A can be selected. If the alternative a_3 is selected, the decision maker must wait until a specific experimental outcome is obtained; at this point the corresponding optimal alternative (as indicated by an asterisk) is selected.

EXAMPLE 2.14

For the decision problem of Example 2.12, should the experiment be conducted in the first place?

The decision tree for a preposterior analysis is shown in Fig. E2.14. The probabilities that the test will indicate *HR*, *MR*, and *LR* are computed by applying Eq. 2.5. For example,

$$P(MR) = P(MR|EH)P'(EH) + P(MR|EL)P'(EL)$$
$$= 0.15 \times 0.7 + 0.2 \times 0.3$$
$$= 0.165$$

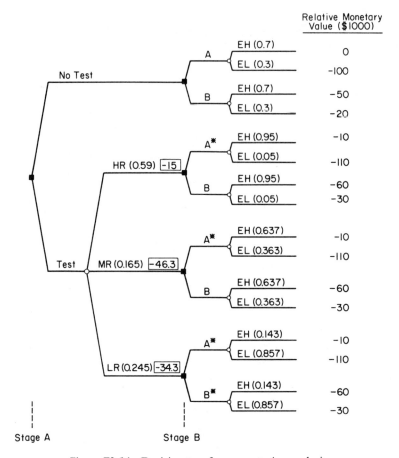

Relative Monetary
Value ($1000)

Figure E2.14   Decision tree for preposterior analysis.

The subtrees following the respective test outcomes are also shown in Fig. E2.14. In each of the subtrees, the optimal alternative is denoted by the asterisk and the corresponding optimal monetary value is shown; for example, the EMV for the subtree with outcome *HR* is −15. The expected monetary value of the test alternative is thus

$$E(\text{Test}) = 0.59 \times (-15) + 0.165 \times (-46.3) + 0.245 \times (-34.3)$$
$$= -24.86$$

Comparing this with the expected monetary values of alternatives *A* and *B* assuming no test is performed, namely −30 and −41, respectively (from Example 2.7), the engineer should opt for the experiment. Then, depending on the test outcome, design *A* or *B* may be selected accordingly, as indicated in Fig. E2.14.

2.2.6   Value of Information

In purely expected monetary terms, the value of an experiment may be measured by

$$VI = E(T) - E(a^*) \tag{2.6}$$

where $E(T)$ is the EMV of a test alternative excluding the cost of the experiment and a^* is the optimal alternative if the test was not performed. Hence, if VI exceeds the experimental cost, the test alternative should be selected. The value of the additional information, however, is bounded by a limit referred to as the "value of perfect information" (VPI). VPI is simply the difference between $E(a^*)$ and the expected monetary value of a "perfect test"; that is, a test with 100% reliability. Hence,

$$VPI = E(PT) - E(a^*) \qquad (2.7)$$

Therefore, VPI represents the maximum cost that the decision maker should allow for acquiring any additional information.

EXAMPLE 2.15

Consider the waste treatment process design in Example 2.14. Suppose a perfect test is available that will definitely tell whether the efficiency of the treatment process is high or low. The corresponding subtree is shown in Fig. E2.15; the cost of conducting the perfect test is not included.

The events EH_0, EL_0 represent the possible perfect test results; accordingly, the subsequent conditional probabilities of EH and EL are either one or zero. The expected monetary value for the perfect test is thus

$$E(PT) = 0.7 \times 0 + 0.3 \times (-20)$$
$$= -6$$

and Eq. 2.7 yields

$$VPI = -6 - (-30)$$
$$= 24$$

Therefore, the cost of acquiring any additional information should not exceed $24,000. For the specific test program studied in Example 2.14, the value of information from that test can be shown as

$$VI = E(T) - E(a^*)$$
$$= (-24.86 + 10) - (-30)$$
$$= 15.14$$

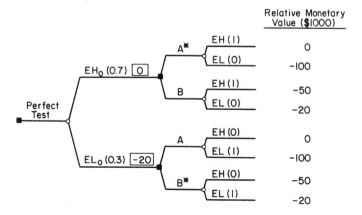

Figure E2.15   Subtree of a perfect test.

Since that test program costs only $10,000, which is less than its value of information, namely $15,140, the test should be performed. Observe that the value of this test is still considerably less than the value of perfect information. Hence, more sophisticated test programs could be considered for reducing the uncertainties of the efficiency of the treatment process, as long as such test programs cost less than $24,000. Otherwise, the amount of saving resulting from reductions in expected loss through such additional information will not be sufficient to pay for the cost of acquiring such information in the first place.

2.2.7 Sensitivity of Decision to Error in Probability Estimation

The prior probabilities in a decision analysis are often based on or supplemented by the subjective judgments of the decision maker. Errors in the estimated probabilities, therefore, are unavoidable. The effects of these errors on a decision are naturally of interest. In particular, the sensitivity of the optimal alternative to the values of the calculated probabilities is of interest. For instance, if the probability estimates are off by 10%, will this alter the optimal alternative? The following example illustrates the nature of this problem and shows how a sensitivity analysis may be performed.

EXAMPLE 2.16

Consider the decision problem of Example 2.7 again. Instead of assigning 0.7 as the prior probability of high efficiency, let this probability be p. The corresponding probability of low efficiency is $(1 - p)$. Based on the decision tree of Fig. E2.7 (expressed in terms of p), the EMV for the two alternative designs are

$$E(A) = p(0) + (1 - p)(-100) = -100(1 - p)$$
$$E(B) = p(-50) + (1 - p)(-20) = -(20 + 30p)$$

Figure E2.16 shows a plot of these two EMV as functions of p. It can be observed that for $p < 0.62$, $E(B) > E(A)$, hence, design B is the preferred alternative. On the other hand, for $p > 0.62$, design A would be preferred. Therefore, according to this analysis, design A should be selected if $p > 0.62$; it is not necessary that p be determined very precisely.

Moreover, based on the decision tree in Fig. E2.15, the expected monetary value of a perfect test is

$$E(PT) = p(0) + (1 - p)(-20) = -20(1 - p)$$

which is also plotted in Fig. E2.16. Observe that for $p = 0$ or 1, a perfect test will indicate the same conclusion; hence, a perfect test would be of negligible value.

The value of perfect information (VPI) is given by the difference in ordinates between $E(PT)$ and the max EMV at a given value of p. It attains the highest value at $p = 0.62$ in this example. Around $p = 0.62$, the optimal alternative is quite sensitive to p, and any additional information can significantly increase the EMV of the decision. For the experiment proposed in Example 2.13, the EMV of the test alternative, excluding the experiment cost, is denoted by the curve $E(T)$ in Fig. E2.16. In general, the EMV of any imperfect test will not be linear with p and it should lie between $E(PT)$ and max EMV, that is, within the triangular shaded region shown in Fig. E2.16. The difference in the ordinates of $E(T)$ and the max EMV is VI.

In short, if the number of alternatives is few, small errors in the estimated probability may not affect the selection of the optimal alternative, especially if the probability is not close to the critical value (such as 0.62 in Example 2.16). The sensitivity analysis illustrated above can be extended to decision problems with more alternatives; however, in such cases, the analysis would naturally be more complex.

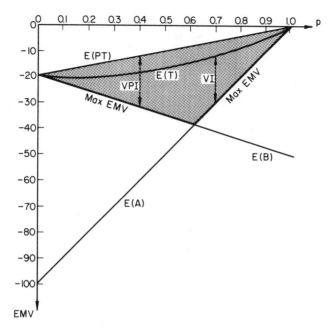

Figure E2.16 EMV of alternatives as functions of *p*.

2.2.8 More Examples

EXAMPLE 2.17 (*excerpted from Howard et al, 1972*)

The hurricane has long been a devastating force along the coastal regions of the United States around the Gulf of Mexico; the average annual property damage during the 1960s was approximately $440 million. A decision analysis was performed to determine the possibility of mitigating the destructive force of hurricanes by seeding them with silver iodide before a hurricane hits the coast. The decision tree for the study is shown in Fig. E2.17a.

With the present seeding technology, the effect of seeding remains uncertain. Based on subjective judgment and limited data from the seeding of Hurricane Debbie, the probabilities for given (discretized) changes in wind speed are estimated. It can be observed from these probabilities that wind speed tends to decrease with seeding; however, there is some probability that the hurricane wind speed may actually increase after seeding. For this reason, the consequence of a seeding decision must include a "government responsibility cost." If the government decides not to seed, the government will not be held responsible for damages, since people will accept hurricanes as natural disasters. On the other hand, if the government should order seeding and the hurricane wind speed actually grows stronger because of it, the government may be subject to lawsuits for hurricane-induced damages. These costs have been estimated and included in Fig. E2.17a.

Based on the information given in the decision tree of Fig. E2.17a, the expected monetary value for the *seeding* and *no seeding* alternatives are, respectively,

$$E(\text{Seed}) = 0.038 \times (-503.95) + 0.143 \times (-248.65)$$
$$+ 0.392 \times (-105.25) + 0.255 \times (-46.95) + 0.172 \times (-16.55)$$
$$= -19.15 - 35.56 - 41.26 - 11.97 - 2.85$$
$$= -\$110.8 \text{ million}$$

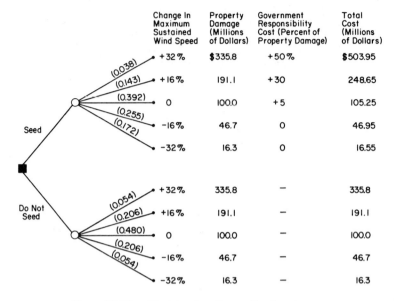

Change In Maximum Sustained Wind Speed	Property Damage (Millions of Dollars)	Government Responsibility Cost (Percent of Property Damage)	Total Cost (Millions of Dollars)
+32%	$335.8	+50%	$503.95
+16%	191.1	+30	248.65
0	100.0	+5	105.25
-16%	46.7	0	46.95
-32%	16.3	0	16.55
+32%	335.8	—	335.8
+16%	191.1	—	191.1
0	100.0	—	100.0
-16%	46.7	—	46.7
-32%	16.3	—	16.3

Figure E2.17a Decision tree for seeding hurricane

$$E(\text{No Seed}) = 0.054 \times (-335.8) + 0.206 \times (-191.1)$$
$$+ 0.48 \times (-100) + 0.206 \times (-46.7) + 0.054 \times (-16.3)$$
$$= -18.1 - 39.3 - 48 - 9.62 - 0.88$$
$$= -\$115.9 \text{ million}$$

According to this analysis, seeding should be implemented, in spite of the potential liability cost to the government, because of the strong likelihood that property damages will be greatly reduced.

A third alternative is to perform seeding experiments before deciding on whether or not to seed. Figure E2.17b shows the corresponding decision tree that includes the seeding experiments as an alternative. The possible changes in the wind speed resulting from the experiment are discretized into five outcomes—namely, +32%, +16%, 0, -16%, -32%. Corresponding to each experimental outcome, the probability of a particular change in wind speed in future hurricanes will be updated from the previously estimated values. Representative paths of the decision tree are shown in Fig. E2.17b; here the results of the preposterior analysis are also summarized.

EXAMPLE 2.18

(a) For the structural design of Example 2.3, which scheme should be adopted?

(b) Suppose the performance of design B can be investigated with a model test at a cost of $50,000. If the assumptions of design B are not valid, the model structure will fail in the test with 90% probability; if the assumptions are valid, the failure probability of the model structure is only 30%. Should the test be performed?

(c) At most how much should be spent to check if the assumptions are valid?

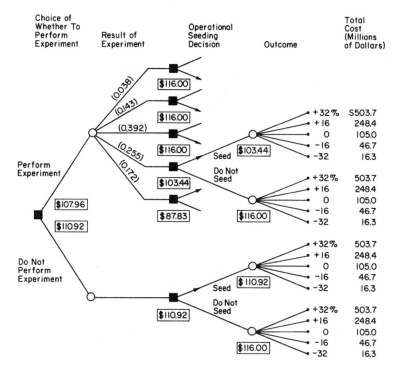

Choice of
Whether To
Perform
Experiment

Result of
Experiment

Operational
Seeding
Decision

Outcome

Total
Cost
(Millions
of Dollars)

(0.038)
(0.143)
(0.392)
(0.255)
(0.172)

$116.00
$116.00
$116.00
$103.44
$87.83

Perform
Experiment

$107.96
$110.92

Seed $103.44

Do Not
Seed
$116.00

+32% $503.7
+16   248.4
 0   105.0
-16   46.7
-32   16.3

+32%   503.7
+16   248.4
 0   105.0
-16   46.7
-32   16.3

Do Not
Perform
Experiment

$110.92

Seed $110.92

Do Not
Seed
$116.00

+32%   503.7
+16   248.4
 0   105.0
-16   46.7
-32   16.3

+32%   503.7
+16   248.4
 0   105.0
-16   46.7
-32   16.3

Figure E2.17b   Decision analysis including value of seeding experiment.

Solutions

(a) The expected monetary loss associated with designs A and B are, respectively,

$$E(A) = 0.99 \times 1.5 + 0.01 \times 11.5 = \$1.6 \text{ million}$$

$$E(B) = 0.5 \times 0.99 \times 1.0 + 0.5 \times 0.01 \times 11.0 + 0.5$$
$$\times 0.9 \times 1.0 + 0.5 \times 0.1 \times 11.0 = \$1.55 \text{ million}$$

Based on the max EMV criterion, design B should be adopted.

(b)   For the test alternative, the corresponding decision tree is given in Fig. E2.18a. Observe that the test result will only affect the probabilities of whether the assumptions are valid or not, whereas the probability of satisfactory performance of the actual structure depending on the validity of the design assumptions are the same as those in part (a). With the given reliability level of the model test, that is,

$$P(F \mid \overline{V}) = 0.9, \ P(\overline{F} \mid \overline{V}) = 0.1, \ P(F \mid V) = 0.3, \ P(\overline{F} \mid V) = 0.7$$

and $P(V) = P(\overline{V}) = 0.5$, we have

$$P(F) = P(F \mid V)P(V) + P(F \mid \overline{V})P(\overline{V})$$
$$= 0.3 \times 0.5 + 0.9 \times 0.5$$
$$= 0.6$$

$$P(V \mid F) = \frac{P(F \mid V)P(V)}{P(F)} = \frac{0.3 \times 0.5}{0.6} = 0.25$$

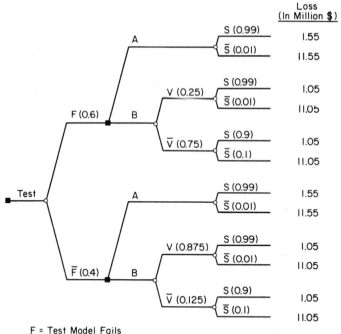

Figure E2.18a   Decision tree for test alternative.

and $P(\overline{V}|F) = 0.75$. Similarly, $P(\overline{F}) = 0.4$.

$$P(V|\overline{F}) = \frac{P(\overline{F}|V)P(V)}{P(\overline{F})} = \frac{0.7 \times 0.5}{0.4} = 0.875$$

and $P(\overline{V}|\overline{F}) = 0.125$.

If it is assumed that the test model failed, the expected losses for the designs are, respectively,

$E(A|F) = 0.99 \times 1.55 + 0.01 \times 11.55 = \$1.65$ million

$E(B|F) = 0.25 \times (0.99 \times 1.05 + 0.01 \times 11.05) + 0.75 \times (0.9 \times 1.05 + 0.1 \times 11.05)$
$\qquad = 0.25 \times 1.15 + 0.75 \times 2.05$
$\qquad = \$1.825$ million

Hence, A is the better design if the test model failed. On the other hand,

$$E(A|\overline{F}) = \$1.65 \text{ million}$$
$$E(B|\overline{F}) = 0.875 \times 1.15 + 0.125 \times 2.05$$
$$= \$1.2625 \text{ million}$$

indicating that B is the better design if the test model survives. The expected value of the test is

$$E(T) = 0.6 \times 1.65 + 0.4 \times 1.2625$$
$$= 1.455$$

Since the expected loss of the test alternative is less than that of the optimal design in (a), namely design B, the test should be performed.

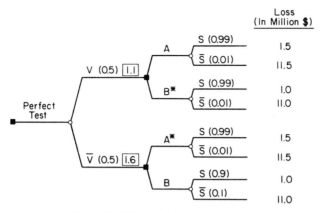

Figure E2.18b Decision tree for perfect test.

(c) The maximum cost of the test that can be justifiably spent may be determined by considering the perfect test. The decision tree for this alternative is shown in Fig. E2.18b.

Performing the expected loss calculation, we have

$$E(A|V) = 0.99 \times 1.5 + 0.01 \times 11.5 = \$1.6 \text{ million}$$

$$E(B|V) = 0.99 \times 1.0 + 0.01 \times 11.0 = \$1.1 \text{ million}$$

$$E(A|\overline{V}) = 0.99 \times 1.5 + 0.01 \times 11.5 = \$1.6 \text{ million}$$

$$E(B|\overline{V}) = 0.9 \times 1.0 + 0.1 \times 11.0 = \$2.0 \text{ million}$$

Therefore,

$$E(PT) = E(B|V)P(V) + E(A|\overline{V})P(\overline{V})$$
$$= 1.1 \times 0.5 + 1.6 \times 0.5$$
$$= \$1.35 \text{ million}$$

The value of perfect information is

$$VPI = -[E(PT) - E(B)]$$
$$= -(1.35 - 1.55)$$
$$= \$0.20 \text{ million}$$

Hence, $200,000 is the maximum amount that may be spent for checking the validity of the assumptions.

EXAMPLE 2.19

The proposed site for a structure overlies a layer of saturated sand (see Fig. E2.19a) that is vulnerable to liquefaction under earthquake excitation. To estimate the probability of liquefaction, the following simplified model is suggested. Let S be the maximum earthquake load over the proposed lifetime of the structure, $S = N(100, 40)$ at a critical point in the stratum. Let R be the resistance capacity of the sand against liquefaction, such that

$$R = \left(1 + \frac{d}{10}\right)X$$

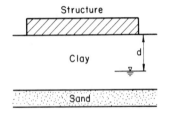

Figure E2.19a   Profile of proposed site.

where X is $N(150, 50)$ and d is the depth (in feet) of the water table below the ground surface. Liquefaction occurs if S exceeds R. The present water table is at the ground surface. However, the water table may be lowered by as much as 10 feet to improve the soil resistance against liquefaction. Suppose the cost of lowering the water table is

$$C_1 = 0.01d^2 \qquad \text{(in \$ million)}$$

and the loss in the event of liquefaction is \$5 million.

(a) If the water table is to be changed, how much should it be lowered?

As the depth of the water table d increases, the resistance capacity of the sand against liquefaction increases, thereby reducing its risk of liquefaction during earthquakes; however, the associated cost of lowering the water table also increases rapidly. The optimal value of d can be obtained by minimizing the sum of the cost of lowering the water table and the expected loss from liquefaction. For a given value of d, the mean and standard deviation of R can be determined as

$$\mu_R = \left(1 + \frac{d}{10}\right)\mu_X = 150\left(1 + \frac{d}{10}\right)$$

$$\sigma_R = \left(1 + \frac{d}{10}\right)\sigma_X = 50\left(1 + \frac{d}{10}\right)$$

Since R and S may be assumed to be statistically independent, the probability of liquefaction is

$$P(S > R) = P(R - S < 0) = \Phi\left[\frac{0 - (\mu_R - \mu_S)}{\sqrt{\sigma_S^2 + \sigma_R^2}}\right] = \Phi\left[\frac{\mu_S - 150(1 + d/10)}{\sqrt{\sigma_S^2 + 50^2(1 + d/10)^2}}\right]$$

Substituting the values of μ_S and σ_S yields the expected total cost

$$C_T = 0.01d^2 + 5\Phi\left[\frac{100 - 150(1 + d/10)}{\sqrt{1600 + 2500(1 + d/10)^2}}\right]$$

A trial and error solution, as shown below, reveals that C_T is minimum with $d \simeq 4$ ft. Therefore, if the water table is to be lowered, 4 ft will be the optimal depth.

d	C_1	Expected Liquefaction Loss	C_T
0	0	1.09	1.09
2	0.04	0.67	0.71
3	0.09	0.54	0.63
4	0.16	0.44	0.60
5	0.25	0.36	0.61
6	0.36	0.29	0.65

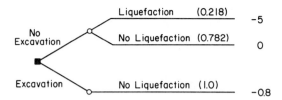

Figure E2.19b Decision tree on excavation.

(b)   Upon further investigation, any lowering of the water table will not be permitted by a regulatory agency. However, the saturated sand can be excavated and replaced by material with a negligible chance of liquefaction. The excavation will cost \$0.8 million. Should the soil be excavated or left untouched?

The decision tree for this case is shown in Fig. E2.19b. If no excavation is performed, $d = 0$, and the probability of liquefaction is

$$P(S > R) = \Phi\left[\frac{100 - 150}{\sqrt{1600 + 2500}}\right] = 0.218$$

The corresponding expected monetary value is

$$\text{EMV(no excavation)} = 0.218(-5) = -\$1.09 \text{ million}$$

which is a greater loss than the cost of excavation. Therefore, excavation is the preferred alternative, at a cost of \$0.8 million.

(c)   Suppose the soil resistance parameter X consists of two components; namely,

$$X = Y + W$$

where $Y = N(100, 40)$, $W = N(50, 30)$, and Y and W are statistically independent. Moreover, the exact value of W may be obtained by conducting a test, with which the decision to excavate or not is then made. Determine the value of this test alternative.

Since the test will yield a given value of W, the random variable X becomes

$$X = Y + w$$

which is $N(100 + w, 40)$. After the test, if no excavation is performed, the probability of liquefaction becomes

$$P(S > R) = \Phi\left[\frac{100 - (100 + w)}{\sqrt{1600 + 1600}}\right]$$

$$= \Phi\left[\frac{-w}{40\sqrt{2}}\right]$$

The decision tree for the test alternative is shown in Fig. E2.19c.
   The distribution of W at the chance node A is $N(50, 30)$. The decision analysis starts with node B, where the expected monetary values of the two alternatives are

$$\text{EMV(no excavation)} = -5\Phi\left(\frac{-w}{40\sqrt{2}}\right)$$

and

$$\text{EMV(excavation)} = -0.8$$

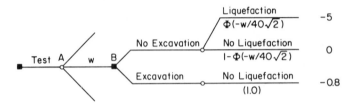

Figure E2.19c Decision tree for test alternative.

Therefore, whether or not to excavate depends on the value of w obtained from the test. The critical value of w occurs at

$$-5\Phi\left(\frac{-w}{40\sqrt{2}}\right) = -0.8$$

or

$$w = -40\sqrt{2}\,\Phi^{-1}\left(\frac{0.8}{5}\right)$$

$$= 56.29$$

If w is less than 56.29, EMV(excavation) > EMV(no excavation), and the optimal alternative is to excavate, with a corresponding expected monetary value of -0.8. On the other hand, if w exceeds 56.29, EMV(excavation) < EMV(no excavation), the preferred alternative is not to excavate, with a corresponding expected monetary value of $-5\Phi(-w/40\sqrt{2})$. At node A, the expected monetary value of the test alternative is thus

$$\text{EMV(test)} = \int_{-\infty}^{56.29} -0.8\, f_W(w)\, dw + \int_{56.29}^{\infty} -5\Phi\left(\frac{-w}{40\sqrt{2}}\right) f_W(w)\, dw$$

where $f_W(w)$ is the PDF of $N(50, 30)$. After simplification,

$$\text{EMV(test)} = -0.8\Phi\left(\frac{56.29 - 50}{30}\right) - 5\int_{56.29}^{\infty} \Phi\left(\frac{-w}{40\sqrt{2}}\right) \frac{1}{\sqrt{2\pi}\,30} \exp\left[-\frac{1}{2}\left(\frac{w - 50}{30}\right)^2\right] dw$$

$$= -0.47 - 0.16$$

$$= -0.63$$

where the integral has been evaluated numerically. When we compare this with the expected monetary value of the optimal alternative without the test, from part (b), the value of the test in monetary terms is

$$VI = -0.63 - (-0.8)$$
$$= \$0.17 \text{ million}$$

EXAMPLE 2.20 (excerpted from Chamberlain, 1970)

The water quality in the Great Lakes is gradually being degraded because of industrial pollution. Three alternatives have been proposed by the Great Lakes Management to improve the situation; namely:

a_1: Build a new efficient treatment plant for $10 million.
a_2: Conduct research to develop a secondary processor; the outcome of the research may be one of the following:

(i)   developed processor in 2nd year at a cost of $0.5 million with 0.3 probability;
(ii) developed processor in 3rd year at a cost of $0.75 million with 0.4 probability;
(iii) developed processor in 4th year at a cost of $1.0 million with 0.2 probability;
(iv) failure to develop processor in 4 years at $1.0 million with 0.1 probability.

a_2: Stock the lakes with robust species of bacteria that are more resistant to pollution at a cost of $0.1 million.

A fourth alternative, a_4, is to do nothing. Because of uncertainties in the environment, particularly with respect to the potential sources of pollutants from industry, the level of future water quality prior to treatment may be idealized as either q_1 or q_2, where q_1 is better than q_2. The relative likelihoods of these two water qualities are subjectively estimated to be 3 to 2. For all possible combinations of alternatives and states of water quality prior to treatment, five levels of improvement in water quality are possible, namely I, II, III, IV, V, in decreasing degree of improvement.

Suppose the monetary value associated with each of these five levels of water quality are assessed by a group of planners as follows:

$$d(\text{I}) = x$$

$$d(\text{II}) = 0.9x \text{ if the processor is developed in 2nd year}$$
$$= 0.8x \text{ if the processor is developed in 3rd year}$$
$$= 0.6x \text{ if the processor is developed in 4th year}$$

$$d(\text{III}) = 0.8x \text{ if the processor is developed in 2nd year}$$
$$= 0.7x \text{ if the processor is developed in 3rd year}$$
$$= 0.5x \text{ if the processor is developed in 4th year}$$

$$d(\text{IV}) = 0.2x$$

$$d(\text{V}) = 0.01x$$

where the value of x in million dollars remains to be assigned.

Table E2.20 summarizes the conditional probabilities that the alternatives will achieve the

Table E2.20 Conditional Probabilities

Alternative	Pretreatment Water Quality	Improved Water Quality Level				
		I	II	III	IV	V
a_1	q_1	1	0	0	0	0
	q_2	0.8	0.2	0	0	0
a_2 (with processor developed)	q_1	0	0.7	0.3	0	0
	q_2	0	0.5	0.5	0	0
a_2 (without processor)	q_1	0	0	0	0	1
	q_2	0	0	0	0	1
a_3	q_1	0	0	0	1	0
	q_2	0	0	0	1	0
a_4	q_1	0	0	0	0	1
	q_2	0	0	0	0	1

corresponding water quality. For example, $P(\text{II}|a_1q_2)$ denotes the probability of achieving level II in water quality if alternative a_1 is chosen and the water quality prior to treatment is q_2; $P(\text{II}|a_1q_2) = 0.2$.

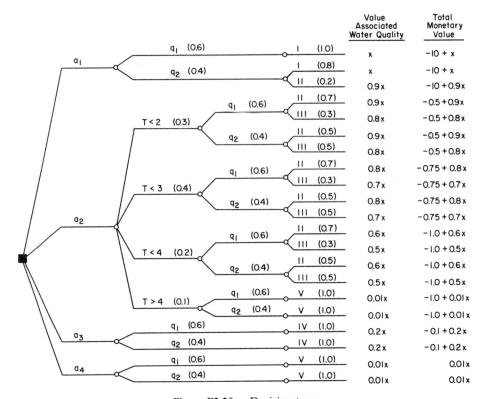

	Value Associated Water Quality	Total Monetary Value

Figure E2.20a   Decision tree.

The complete decision tree is given in Fig. E2.20a, where T is the development time under alternative a_2. The expected monetary values are $-10 + 0.992x$, $-0.75 + 0.677x$, $-0.1 + 0.2x$, and $0.01x$ for alternatives a_1, a_2, a_3 and a_4, respectively. It can be observed from Fig. E2.20b that each of the four alternatives could be the optimal one, depending on the value of x. For x less than 0.5, a_4 is the optimal alternative; a_3 becomes optimal for $0.5 < x < 1.4$; beyond which a_2 becomes optimal. For very large values of $x (> 29.4)$, a_1 will become the optimal alternative, implying that if water quality is extremely valuable, it would pay to build a new and efficient treatment plant even at a great cost.

EXAMPLE 2.21

A stream is subject to floods each spring. Because of frequent flood damages to an adjacent town, a plan to increase the present levee elevation of 10 feet to either 14 or 16 feet is being investigated. The corresponding costs of construction are $2 and $2.5 million, respectively, and the reconstructed levee will last 20 years. The average annual damage caused by the inadequate level of protection is $2 million. Based on the data collected from previous years, the annual flood level in the stream can be estimated quite accurately as having a median of 10 feet and a c.o.v. of 20%; however, both the normal and lognormal distributions appear to fit the data equally well. Suppose the true distribution could equally likely be normal or lognormal. Assume a discount rate of 7% per year.

(a) Determine the optimal decision based on EMV criterion.
(b) How much is it worth to verify the true distribution of the annual flood level?

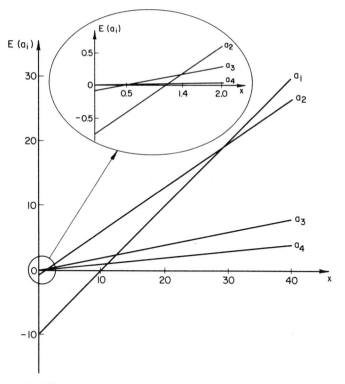

Figure E2.20b   Expected monetary value of alternatives as function of *x*.

Solution

The decision tree for the levee improvement is shown in Fig. E2.21a. The alternatives are

a_0: Levee elevation remains at 10 feet.

a_1: Increase levee to 14 feet.

a_2: Increase levee to 16 feet.

For each alternative the probability distribution of the annual maximum flood level, X, could either be lognormal (LN) or normal (N); on these bases, the probability of inadequate flood protection (event F) may be computed for each alternative as follows:

$$P(F|a_0, LN) = 0.5$$

$$P(F|a_0, N) = 0.5$$

$$P(F|a_1, LN) = 1 - \Phi\left(\frac{\ln 14 - \ln 10}{0.2}\right) = 0.0462$$

$$P(F|a_1, N) = 1 - \Phi\left(\frac{14 - 10}{2}\right) = 0.02275$$

$$P(F|a_2, LN) = 1 - \Phi\left(\frac{\ln 16 - \ln 10}{0.2}\right) = 0.00939$$

$$P(F|a_2, N) = 1 - \Phi\left(\frac{16 - 10}{2}\right) = 0.00135$$

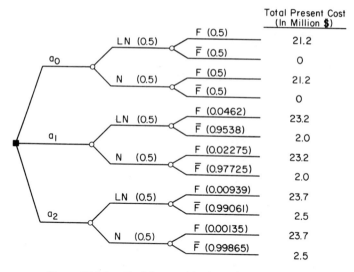

Figure E2.21a Decision tree for levee improvement.

The total present cost of each alternative is the sum of the initial investment and the equivalent present value of the annual damage cost. The present worth factor for 20 years at 7% interest is 10.594. As an example, for path $a_0 - LN - F$ of Fig. E2.21a, the total present cost is

$$C_{a_0-LN-F} = 0 + 2 \times 10.594 = \$21.188 \text{ million}$$

whereas

$$C_{a_1-LN-F} = 2 + 2 \times 10.594 = \$23.188 \text{ million}$$

$$C_{a_2-LN-\bar{F}} = 2.5 + 0 \times 10.594 = \$2.5 \text{ million}$$

The expected total present costs for each of the alternatives are:

$$E(C_{a_0}) = 0.25 \times 21.2 + 0.25 \times 21.2 = 10.594$$

$$E(C_{a_1}) = 0.5 \times (0.0462 \times 23.2 + 0.9538 \times 2.0) + 0.5(0.02275 \times 23.2 + 0.97725 \times 2.0)$$
$$= 2.731$$

$$E(C_{a_2}) = 0.5 \times (0.00939 \times 23.7 + 0.99061 \times 2.5) + 0.5(0.00135 \times 23.7 + 0.99865 \times 2.5)$$
$$= 2.614$$

Therefore, the levee should be increased to 16 feet if the expected total present cost is to be minimized; the corresponding expected total present cost is \$2.614 million.

(b) For perfect information on the probability distribution of the annual flood level, the corresponding decision tree is shown in Fig. E2.21b. Presumably the true distribution will be lognormal or normal with equal likelihood. Given that the distribution is lognormal, the probabilities of flooding are 0.5, 0.0462, and 0.00939 for alternatives a_0, a_1, and a_2, respectively. The expected total costs for the respective alternatives are

$$E(C_{a_0}|LN) = 0.5 \times 21.2 = \$10.6 \text{ million}$$

$$E(C_{a_1}|LN) = 0.0462 \times 23.2 + 0.9538 \times 2.0 = \$2.979 \text{ million}$$

$$E(C_{a_2}|LN) = 0.00939 \times 23.7 + 0.99061 \times 2.5 = \$2.699 \text{ million}$$

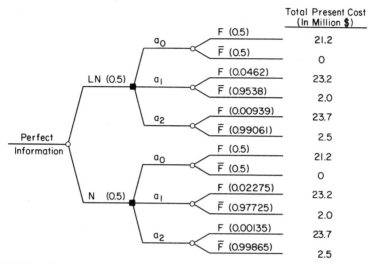

Figure E2.21b Decision tree for perfect information on distribution of annual flood level.

Hence, a_2 is optimal if the true distribution is lognormal. Similarly, if we assume that the true distribution is normal, the probabilities of flooding are 0.5, 0.02275, and 0.00135 for alternatives a_0, a_1, and a_2, respectively. The expected total costs are then

$$E(C_{a_0}|N) = 0.5 \times 21.2 = \$10.6 \text{ million}$$

$$E(C_{a_1}|N) = 0.02275 \times 23.2 + 0.97725 \times 2.0 = \$2.482 \text{ million}$$

$$E(C_{a_2}|N) = 0.00135 \times 23.7 + 0.99865 \times 2.5 = \$2.529 \text{ million}$$

This implies that a_1 is optimal if the true distribution is normal. Hence, the expected present cost of the perfect information alternative is

$$E(C_{PI}) = 0.5 \times 2.699 + 0.5 \times 2.482 = 2.590$$

Therefore, the maximum fund that may be spent to verify the true distribution of the annual flood level is

$$\text{VPI} = 2.614 - 2.590 = \$0.024 \text{ million} = \$24,000$$

EXAMPLE 2.22 (excerpted from Davis and Dvoranchik, 1971)

A 500-foot bridge is proposed over the flood dikes at Rillito Creek, near Tucson, Arizona. The bridge will rest on four piers, each supported by 25 piles. Part of the bridge may be lost in a flood as a result of scour undercutting the piles. If this occurs, the cost is estimated to be $150,000. Suppose the cost of driving each pile is $4/foot. What is the optimal depth of the pier foundation?

Suppose p_F is the probability that the bridge will need to be replaced in a year. The expected annual replacement cost is thus $150,000p_F$. Assume an expected life span of t years for the bridge and an interest rate of $i\%$ per annum. The present value of the expected total replacement cost over t years is $150,000p_F$ multiplied by the present worth factor (pwf) for interest rate $i\%$ per annum. Therefore, the objective function is the sum of the cost for the pile foundation and the expected replacement cost, given by

$$C = 4h(4 \times 25) + (150,000p_F) \times pwf(i\%, t)$$

where h is the depth of the pier foundation.

The annual probability of failure, p_F, is the probability that the annual maximum scour depth D exceeds the pier foundation depth. To determine p_F, the annual maximum stream flow is first studied. Based on past records of stream flows, the annual maximum flow Y may be assumed to be lognormally distributed with parameters λ, ζ. Based on n years of record and a uniform (diffuse) prior distribution, the posterior joint distribution of these parameters is

$$L(\lambda, \zeta) = \frac{1}{\sqrt{2\pi}\zeta/\sqrt{n}} \exp\left[-\frac{1}{2}\left(\frac{\lambda - \bar{x}}{\zeta/\sqrt{n}}\right)^2\right] \left[\frac{\left(\frac{n-1}{2}\right)^{n+1/2}}{\Gamma\left(\frac{n+1}{2}\right)} \left(\frac{s^2}{\zeta^2}\right)^{n-1/2} \exp\left(-\frac{n-1}{2}\frac{s^2}{\zeta^2}\right)\right]$$

(see Table 8.1 of Vol. 1), where $\bar{x}$ and s^2 are the sample mean and variance of the logarithm of the n stream flow values. Hence, incorporating the effect of uncertainties in the distribution parameters, we compute the probability of failure p_F corresponding to a design pier depth h as

$$p_F(h) = \int_{y(h)}^{\infty} \left[\int_{\lambda} \int_{\zeta} f_Y(y|\lambda, \zeta) L(\lambda, \zeta)\, d\lambda\, d\zeta\right] dy$$

where

$$f_Y(y|\lambda, \zeta) = \frac{1}{\sqrt{2\pi}\, y\zeta} \exp\left[-\frac{1}{2}\left(\frac{\ln y - \lambda}{\zeta}\right)^2\right]$$

$y(h)$ is the stream flow that has a scour depth of h.

The annual expected total costs based on 10 years of data (1920–1929, with $\bar{x} = 3.667$ and $s^2 = 0.1314$) and 40 years of data (1920–1959, with $\bar{x} = 3.684$, $s^2 = 0.977$) are computed accordingly. The results are plotted in Fig. E2.22 as a function of the design depth h. The optimal pier depth based on 10 years of data occurs at 16.5 feet with an annual expected cost of $7845, whereas the optimal depth based on 40 years of data is 11 feet with an expected cost of $4680. The difference in the two costs, $7845–$4680 = $3165, may be interpreted as the value of the additional 30 years of data.

Results are also obtained assuming other lengths of data period were available and are summarized in Table E2.22. The optimal cost appears to decrease with the increasing length of available data period. In the extreme case, when there are infinite years of data, no uncertainties

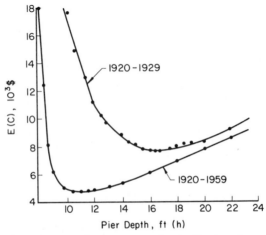

Figure E2.22 Annual total expected cost versus pier depth—based on 10 and 40 years of annual peak flow data on Rillito Creek, Tucson, Arizona (Davis and Dvoranchik, 1971).

should exist in λ and ζ. The corresponding value of the expected cost C under this condition of perfect information is estimated to be \$4349. Therefore, immediately after 1959, at most \$4680–\$4349 = \$331 will be gained if the statistical uncertainties in the parameters are eliminated.

Table E2.22 Optimal Depth and Cost of Pier Foundation

Data Period	1920–1924	1920–1929	1920–1939	1920–1959
Years	5	10	20	40
$\bar{x}$	3.7014	3.6670	3.6943	3.6840
s^2	0.0977	0.1314	0.0890	0.0977
C_{opt} (\$)	12,045	7845	5059	4680
h_{opt} (ft)	21.2	16.5	11.7	11.0

2.3 APPLICATIONS IN SAMPLING AND ESTIMATION

2.3.1 Bayes Point Estimator

In Chapter 8 of Vol. 1, we presented the Bayesian approach to estimation. This approach assumes that the parameters of a probability distribution are random variables. The uncertainty of a parameter is modeled by the corresponding prior or posterior distribution. If a point estimate of the parameter is desired, it can be formulated also as a problem of decision analysis. Consider the decision tree in Fig. 2.5 that depicts a situation in which the point estimator of the parameter θ is being selected. Depending on the particular choice of the estimator $\hat{\theta}$ and the actual value of the parameter θ, a prediction error will result, generally followed by a loss. By modeling this estimation process in the context of decision analysis, the *Bayes point estimator* may be determined such that the expected loss associated with the prediction error is minimized. Mathematically, if $\hat{\theta}$ is the estimator of a parameter whose actual value is governed by the distribution $f(\theta)$, then the expected loss resulting from the error in prediction is

$$L = \int_{-\infty}^{\infty} g(\theta, \hat{\theta}) f(\theta) \, d\theta \qquad (2.8)$$

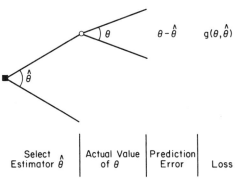

Figure 2.5 Selection of point estimator.

where $g(\theta, \hat{\theta})$ is the loss function. Generally, this loss function is expressed in terms of the prediction error, namely $(\theta - \hat{\theta})$. Since the Bayes estimator $\hat{\theta}$ minimizes the expected loss, it should satisfy the following equation:

$$\frac{dL}{d\hat{\theta}} = \frac{d}{d\hat{\theta}} \left[\int_{-\infty}^{\infty} g(\theta, \hat{\theta}) \, f(\theta) \, d\theta \right] = 0 \tag{2.9}$$

The Bayes estimators for some of the common loss functions are derived in the following examples.

EXAMPLE 2.23   (*Linear Loss Function*)

Suppose the loss function (see Fig. E2.23) is given by

$$g(\theta, \hat{\theta}) = \begin{cases} c(\hat{\theta} - \theta); & \theta \leq \hat{\theta} \\ c(\theta - \hat{\theta}); & \theta > \hat{\theta} \end{cases}$$

where c is a constant.

The expected loss is

$$L = \int_{-\infty}^{\hat{\theta}} c(\hat{\theta} - \theta) \, f(\theta) \, d\theta + \int_{\hat{\theta}}^{\infty} c(\theta - \hat{\theta}) \, f(\theta) \, d\theta$$

But

$$\frac{d[\int_{-\infty}^{\hat{\theta}} c(\hat{\theta} - \theta) \, f(\theta) \, d\theta]}{d\hat{\theta}} = \int_{-\infty}^{\hat{\theta}} cf(\theta) \, d\theta + c(\hat{\theta} - \theta) \, f(\hat{\theta}) \frac{d\hat{\theta}}{d\hat{\theta}}$$

$$= c \int_{-\infty}^{\hat{\theta}} f(\theta) \, d\theta$$

Similarly,

$$\frac{d[\int_{\hat{\theta}}^{\infty} c(\theta - \hat{\theta}) \, f(\theta) \, d\theta]}{d\hat{\theta}} = -c \int_{\hat{\theta}}^{\infty} f(\theta) \, d\theta$$

Applying Eq. 2.9 leads to

$$\frac{dL}{d\hat{\theta}} = c \int_{-\infty}^{\hat{\theta}} f(\theta) \, d\theta - c \int_{\hat{\theta}}^{\infty} f(\theta) \, d\theta = 0$$

or

$$\int_{-\infty}^{\hat{\theta}} f(\theta) \, d\theta - \int_{\hat{\theta}}^{\infty} f(\theta) \, d\theta = 0$$

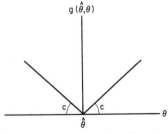

Figure E2.23   Linear loss function.

But

$$\int_{-\infty}^{\theta} f(\theta) \, d\theta + \int_{\theta}^{\infty} f(\theta) \, d\theta = 1$$

Hence,

$$\int_{-\infty}^{\theta} f(\theta) \, d\theta = \int_{\theta}^{\infty} f(\theta) \, d\theta = \tfrac{1}{2}$$

Therefore, in the case of a linear loss function, the Bayes estimator $\hat{\theta}$ is the median of θ.

EXAMPLE 2.24 (*Quadratic Loss Function*)

Suppose the loss function is

$$g(\theta, \hat{\theta}) = c(\theta - \hat{\theta})^2; \qquad -\infty < \theta < \infty$$

Applying Eqs. 2.8 and 2.9, we obtain

$$\frac{dL}{d\hat{\theta}} = \int_{-\infty}^{\infty} \frac{d[c(\theta - \hat{\theta})^2]}{d\hat{\theta}} f(\theta) \, d\theta$$

$$= \int_{-\infty}^{\infty} -2c(\theta - \hat{\theta}) f(\theta) \, d\theta$$

$$= -2c \int_{-\infty}^{\infty} \theta f(\theta) \, d\theta + 2c\hat{\theta} \int_{-\infty}^{\infty} f(\theta) \, d\theta$$

$$= -2cE(\theta) + 2c\hat{\theta} \equiv 0$$

Hence,

$$\hat{\theta} = E(\theta)$$

The Bayes estimator is the mean value of θ in the case of a quadratic loss function.

EXAMPLE 2.25 (*Constant Loss Function*)

Suppose the loss function is a constant over the entire range of θ, except that for the small region where $|\hat{\theta} - \theta| \leq \varepsilon/2$, there is no loss due to prediction error. Mathematically, the loss function as shown in Fig. E2.25 is given as follows

$$g(\theta, \hat{\theta}) = \begin{cases} c; & -\infty < \theta \leq \hat{\theta} - \dfrac{\varepsilon}{2} \\[2mm] 0; & \hat{\theta} - \dfrac{\varepsilon}{2} < \theta \leq \hat{\theta} + \dfrac{\varepsilon}{2} \\[2mm] c; & \hat{\theta} + \dfrac{\varepsilon}{2} < \theta < \infty \end{cases}$$

where ε is small.

The expected loss is

$$L = \int_{-\infty}^{\hat{\theta} - (\varepsilon/2)} cf(\theta) \, d\theta + \int_{\hat{\theta} + (\varepsilon/2)}^{\infty} cf(\theta) \, d\theta$$

$$= \int_{-\infty}^{\infty} cf(\theta) \, d\theta - \int_{\hat{\theta} - (\varepsilon/2)}^{\hat{\theta} + (\varepsilon/2)} cf(\theta) \, d\theta$$

$$\simeq c - c\varepsilon f(\hat{\theta})$$

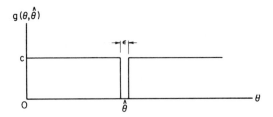

Figure E2.25   Constant loss function.

Applying Eq. 2.9, we obtain

$$\frac{dL}{d\hat{\theta}} = -c\varepsilon\frac{df(\hat{\theta})}{d\hat{\theta}} \equiv 0$$

or

$$\frac{df(\hat{\theta})}{d\hat{\theta}} = 0$$

Hence, the Bayes estimator for a constant loss function is the modal value of θ.

As shown in the above examples, the mean, median, mode, or perhaps other values (see Problem 2.20) can be used as the point estimator of a parameter, depending on the form of the loss function appropriate for the problem. Unlike the estimators in classical statistical estimation in which properties of unbiasedness, sufficiency, and efficiency are used to define the quality of an estimator, Bayes point estimators are determined on the basis of minimizing the loss associated with prediction errors. Physically, the Bayes point estimator is more meaningful for engineering purposes, and consistent with the decision maker's objectives.

2.3.2   Optimal Sample Size

One application of the preposterior analysis is in the determination of the optimal sample size in statistical sampling. In Chapters 5 and 8 of Vol. 1, we showed that the accuracy of the estimation of the statistical parameters of a random variable, such as the mean value and variance, increases with the sample size. A larger sample size, of course, involves a higher cost; naturally, the optimal sample size will involve a trade-off between accuracy and the cost of sampling.

Suppose X is a random variable with a parameter θ to be estimated from sampling observations. The decision model for determining the optimal sample size is shown in Fig. 2.6. Two phases of decision may be required. First, at B, a decision on the sample size n is required. Observe that $n = 0$ represents the case of no sampling needed. Then, a set of sample data $\{x\} = (x_1, x_2, \ldots, x_n)$ is observed; the appropriate estimator for θ is selected at C. The total loss for each path following the decision to sample generally depends on the sample size, the estimated and actual values of θ.

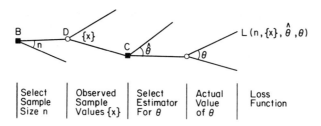

Select Sample Size n	Observed Sample Values {x}	Select Estimator For θ	Actual Value of θ	Loss Function

Figure 2.6   Decision tree for optimal sample size.

Decision analysis starts at the last node in which the expected loss for given n, $\{x\}$, and estimate $\hat{\theta}$ is calculated by weighing the loss over the posterior distribution of θ; that is,

$$E(L\,|\,n, \{x\}, \hat{\theta}) = \int_{-\infty}^{\infty} L(n, \{x\}, \hat{\theta}, \theta)\, f_{\theta}''(\theta)\, d\theta \qquad (2.10)$$

At node C, the optimal estimator $\hat{\theta}$ is the one that minimizes this expected loss function; thus,

$$\frac{dE(L\,|\,n, \{x\}, \hat{\theta})}{d\hat{\theta}} = 0 \qquad (2.11)$$

yields $\hat{\theta}_{\text{opt}}$. On this basis, the expected loss for a given sample size n (at D) is

$$E(L\,|\,n) = \int_{\{x\}} E[L\,|\,n, \{x\}, \hat{\theta}_{\text{opt}}]\, f_{\{X\}}(\{x\})\, d\{x\} \qquad (2.12)$$

The optimal sample size n_{opt} may then be evaluated from

$$\frac{dE(L\,|\,n)}{dn} = 0 \qquad (2.13)$$

As an illustration, consider the following loss function

$$L(n, \{x\}, \hat{\theta}, \mu) = c(\mu - \hat{\theta})^2 + kn \qquad (2.14)$$

where c and k are both constants. This function represents a quadratic loss because of an error in the estimation of μ and a linear sampling cost.

Applying Eq. 2.10, we obtain

$$E(L\,|\,n, \{x\}, \hat{\theta}) = \int_{-\infty}^{\infty} [c(\mu - \hat{\theta})^2 + kn]\, f_{\mu}''(\mu)\, d\mu$$

$$= cE_{\mu}''[(\mu - \hat{\theta})^2] + kn$$

$$= cE_{\mu}''[\{(\mu - \mu'') + (\mu'' - \hat{\theta})\}^2] + kn$$

$$= cE_{\mu}''[(\mu - \mu'')^2 + 2(\mu - \mu'')(\mu'' - \hat{\theta}) + (\mu'' - \hat{\theta})^2] + kn$$

$$= c\mathrm{Var}''(\mu) + c(\mu'' - \hat{\theta})^2 + kn \qquad (2.15)$$

where μ'' and $\text{Var}''(\mu)$ are the posterior mean and variance of μ, respectively. Applying Eq. 2.11 yields

$$\frac{dE(L|n, \{x\}, \hat{\theta})}{d\hat{\theta}} = 2c(\mu'' - \hat{\theta}) = 0$$

which gives

$$\hat{\theta}_{\text{opt}} = \mu'' \tag{2.16}$$

Hence, the optimal estimator is the posterior mean value of μ. However, the optimal estimator of μ is not always given by the posterior mean μ''. For example, if the loss function assumed in Eq. 2.14 was not symmetrical about μ, $\hat{\theta}_{\text{opt}}$ will not be equal to μ'' (see Problem 2.22).

For the loss function of Eq. 2.14, the corresponding expected loss at node C is

$$E[L|n, \{x\}, \hat{\theta}_{\text{opt}}] = c\{\text{Var}''(\mu) + [\mu'' - \hat{\theta}_{\text{opt}}]^2\} + kn$$

$$= c\,\text{Var}''(\mu) + kn \tag{2.17}$$

The posterior variance, $\text{Var}''(\mu)$, depends on the distribution of the underlying random variable (population). For the case in which the population is Gaussian with known σ, the posterior variance (see Section 8.4.2, Vol. 1) assuming random sampling is

$$\text{Var}''(\mu) = \frac{(\sigma')^2 \sigma^2/n}{(\sigma')^2 + \sigma^2/n} \tag{2.18}$$

where σ' is the prior standard deviation of μ. Therefore, in this case, the expected loss depends only on the sample size n, but not on any sample statistics of the observed values; that is,

$$E(L|n) = \int [c\,\text{Var}''(\mu) + kn]\, f_{\{x\}}(\{x\})\, d\{x\}$$

$$= c\,\text{Var}''(\mu) + kn \tag{2.19}$$

However, if the expected loss in Eq. 2.17 was also dependent on the posterior mean of μ, an integration over the distribution of the sample mean of the observed data would be required (see Example 2.27).

For a Gaussian population X with known standard deviation σ, if we substitute Eq. 2.18 into Eq. 2.19, the expected loss for given n is

$$E(L|n) = c\frac{(\sigma')^2 \sigma^2/n}{(\sigma')^2 + \sigma^2/n} + kn \tag{2.20}$$

Setting

$$\frac{dE(L|n)}{dn} = \frac{-c\sigma^2(\sigma')^2(\sigma')^2}{[\sigma^2 + n(\sigma')^2]^2} + k = 0$$

yields the optimal sample size

$$n_{\text{opt}} = \sqrt{\frac{c}{k}\sigma^2} - \frac{\sigma^2}{(\sigma')^2} \tag{2.21}$$

This expression is valid only if $\sqrt{c\sigma^2/k} > \sigma^2/(\sigma')^2$; otherwise, $n_{opt} = 0$ and no sampling should be exercised. In the case for which no prior information is available, σ' may be assumed to be ∞; hence, the optimal sample size becomes

$$n_{opt} = \sigma\sqrt{\frac{c}{k}} \qquad (2.22)$$

Extending the above concepts, Resendiz and Herrera (1969) studied a settlement problem for a foundation in which they determined the optimal design pressure (see Problem 2.23).

EXAMPLE 2.26

In a surveying project, suppose the loss due to error in a distance prediction is proportional to the square of the error (in inches). The cost per unit square of error is \$1200 and the cost for performing one measurement is \$10. The standard error in each measurement with the proposed device is 1 inch. Determine the optimal number of measurements for the following cases:

(a) No prior information is known on the distance.
(b) A prior information indicated that the distance is $N(1000, 0.4)$ inches.
(c) The prior distribution of the distance is $N(1000, 0.04)$ inches.

Solution

(a) The estimation of the actual distance is equivalent to the estimation of μ (see Section 5.2.3, Vol. 1). With Eq. 2.22, the optimal number of measurements with no prior information is

$$n_{opt} = 1.0\sqrt{\frac{1200}{10}} = 10.95 \simeq 11$$

(b) With the prior information on μ as $N(1000, 0.4)$ the optimal number of additional measurements (above and beyond the prior measurements) is determined from Eq. 2.21 as

$$n_{opt} = \sqrt{\frac{1200}{10}(1.0)^2 - \frac{1^2}{(0.4)^2}}$$

$$= 10.95 - 6.25$$

$$= 4.7$$

The expected loss for $n = 4$ and $n = 5$ can be computed with Eq. 2.20 giving respective losses of 157 and 156.5. Furthermore, observe that for the case of no sampling ($n = 0$), the expected loss is $c\,\mathrm{Var}'(u) = 1200(0.4)^2 = 192$. Hence, the optimal sample size is 5.

(c) If the prior information on μ is $N(1000, 0.04)$,

$$n_{opt} = \sqrt{\frac{1200}{10}(1.0)^2 - \frac{1^2}{(0.04)^2}}$$

$$= 11 - 625$$

$$= -614$$

which implies that realistically $n_{opt} = 0$. Hence, the best strategy in this case is not to make any additional measurements, but to estimate the distance using the available prior information giving an estimated distance of 1000 inches.

EXAMPLE 2.27

An environmental engineer obtains water samples from a stream to estimate its mean DO (dissolved oxygen) concentration. If DO in the stream is normally distributed with mean μ and standard deviation 0.5, determine the optimal sample size n assuming the following loss function:

$$L = 100(\mu - \hat{\theta})^2 - 5\mu^2 + n$$

where $\hat{\theta}$ is the estimator of the mean DO concentration; $100(\mu - \hat{\theta})^2$ represents the quadratic prediction loss; and $-5\mu^2$ denotes the benefit resulting from a higher mean DO concentration.

Solution

Substituting the given loss function in Eq. 2.10 and using the result of Eq. 2.15, we get

$$E(L|n, \{x\}, \hat{\theta}) = \int_{-\infty}^{\infty} [100(\mu - \hat{\theta})^2 - 5\mu^2 + n] f_{\mu}''(\mu)\, d\mu$$

$$= 100[\text{Var}''(\mu) + (\hat{\theta} - \mu'')^2] - 5E''(\mu^2) + n$$

$$= 95\, \text{Var}''(\mu) - 5\mu''^2 + 100(\hat{\theta} - \mu'')^2 + n$$

Differentiation with respect to $\hat{\theta}$ yields

$$200(\hat{\theta} - \mu'') = 0$$

Hence, $\hat{\theta}_{\text{opt}}$ is the posterior mean of μ. The corresponding minimum expected loss is

$$E[L|n, \{x\}, \hat{\theta}_{\text{opt}}] = 95\, \text{Var}''(\mu) - 5\mu''^2 + n$$

Since no prior information is available for estimating μ,

$$\mu'' = \bar{x}$$

and

$$\text{Var}''(\mu) = \frac{\sigma^2}{n} = \frac{0.25}{n}$$

Hence, the sample mean of the observed data is the only sample statistic needed for evaluating the optimal sample size. Equation 2.12 becomes

$$E(L|n) = \int_{-\infty}^{\infty} \left[95 \times \frac{0.25}{n} - 5\bar{x}^2 + n \right] f_{\bar{X}}(\bar{x})\, d\bar{x}$$

$$= \frac{23.75}{n} - 5E(\bar{X}^2) + n$$

$$= \frac{23.75}{n} - 5\left[\frac{0.25}{n} + E^2(\bar{X}) \right] + n$$

$$= \frac{22.5}{n} - 5\mu^2 + n$$

Differentiating $E(L|n)$ with respect to n, we obtain

$$n_{\text{opt}} = \sqrt{22.5} = 4.74$$

Since n must be an integer, observe the following:

$$\text{For } n = 4, \qquad E(L|n) = 9.63 - 5\mu^2$$

$$\text{For } n = 5, \qquad E(L|n) = 9.5 - 5\mu^2$$

The expected loss is less when $n = 5$; hence,

$$n_{\text{opt}} = 5$$

EXAMPLE 2.28

The occurrence of cracks along a highway pavement may be modeled as a Poisson process with a parameter m, which is the mean occurrence rate of cracks. From a previous inspection of similar pavements, the parameter m is estimated to have a mean of 2 per mile with a c.o.v. of 50%. Additional inspection may be conducted, at a cost of $20 per mile, to improve the accuracy of the estimated value of m. Suppose the loss due to error in the estimation of m is given by

$$L = 1000(m - \hat{\theta})^2$$

where m is the actual mean rate of cracks and $\hat{\theta}$ is the estimated value of m. Assume that the prior distribution of m is given by the conjugate distribution to the Poison random variable. Determine the optimal distance t (in miles) to be inspected that will minimize the expected total cost function.

Solution

From Table 8.1 of Vol. 1, we see that the conjugate concept requires a gamma distribution for the parameter m. Hence, the prior distribution of M will be of the form

$$f_M(m) = \frac{v'(v'm)^{k'-1}e^{-v'm}}{\Gamma(k')}$$

The parameters v' and k' can be evaluated from

$$E'(M) = \frac{k'}{v'} = 2$$

and

$$\text{Var}'(M) = \frac{k'}{v'^2} = (0.5 \times 2)^2 = 1$$

The solution of these equations gives $v' = 2$ and $k' = 4$. The decision model for this example is shown in Fig. E2.28.

At node D, the expected loss corresponding to each estimate $\hat{\theta}$ is given by Eq. 2.10 as

$$E(L|t, x, \hat{\theta}) = \int_0^\infty [1000\,(m - \hat{\theta})^2 + 20t]\,f_M''(m)\,dm$$

$$= 1000\,[\text{Var}''(M) + (m'' - \hat{\theta})^2] + 20t$$

Applying Eq. 2.11 again, we obtain $\hat{\theta}_{\text{opt}} = m''$. The corresponding optimal loss at node C becomes

$$E[L|t, x, \hat{\theta}_{\text{opt}}] = 1000\,\text{Var}''(M) + 20t$$

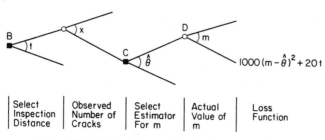

Select Inspection Distance	Observed Number of Cracks	Select Estimator For m	Actual Value of m	Loss Function

Figure E2.28 Decision tree for inspection distance.

in which the posterior variance of the parameter m may be obtained from Table 8.1 of Vol. 1 as

$$\text{Var}''(M) = \frac{k' + x}{(v' + t)^2} = \frac{4 + x}{(2 + t)^2}$$

where x is the number of observed cracks in distance t. Hence, from Eq. 2.12, the expected loss weighted by the distribution of X yields

$$E(L|t) = \int_0^\infty \left[1000 \frac{4 + x}{(2 + t)^2} + 20t \right] f_X(x) \, dx$$

$$= 1000 \frac{4 + E(X)}{(2 + t)^2} + 20t$$

$$= 1000 \frac{4 + E'(M)t}{(2 + t)^2} + 20t$$

$$= \frac{2000}{2 + t} + 20t$$

Setting the derivative to zero, we obtain

$$\frac{dE(L|t)}{dt} = -\frac{2000}{(2 + t)^2} + 20 = 0$$

$$(2 + t)^2 = 100$$

and

$$t_{\text{opt}} = 8 \text{ miles}$$

Hence, the optimal distance to be inspected for cracks is 8 miles.

2.4 ELEMENTARY CONCEPTS OF UTILITY THEORY

The desirability of an alternative may depend on several attributes such as monetary value, time factors, prestige, social acceptance, pleasure, and others. Ranking the feasible alternatives in a decision tree requires a scale for quantifying the degree of preference among the attributes. A well established scale of measure may exist for some of the attributes such as monetary value; however, there is no obvious way of quantifying the values for the majority of these attributes. The problem of value measurement is further complicated when the consequences in the decision tree require the evaluation of a combination of attributes. In order to establish a uniform scale for measuring the overall value of an alternative, the concept of *utility* may be introduced. Utility is defined as a true measure of value to the decision maker. Utility theory provides a framework whereby values may be measured, combined, and compared consistently with respect to a decision maker. Therefore, if the utility values of all the alternatives are available, the alternative with the highest utility value will be preferred.

2.4.1 Axioms of Utility Theory

Introduce the following symbols defining the degrees of preference between two events; namely,

$A > B$ denotes   "A is preferred to B"

$A \sim B$ denotes   "indifference between A and B"

$A \geqslant B$ denotes   "A is preferred at least as much as B"

The following are the important axioms in utility theory;

(i) *Orderability.* The states of preference between A and B can be arranged in order of increasing preference toward B as follows:

$$A \succ B, \qquad A \succeq B, \qquad A \sim B, \qquad A \preceq B, \qquad A \prec B$$

(ii) *Transitivity.* If $A \succ B$ and $B \succ C$, then $A \succ C$.

(iii) *Continuity.* If $A \succ B \succ C$, a value of p (between 0 and 1) exists such that a person is indifferent between (a) obtaining B definitely and (b) getting A or C with probabilities p and $(1 - p)$, respectively. This indifference may be presented schematically with the following lottery:

It is obvious from the lottery that if p is 0, getting C is certain, hence, B is clearly preferred. Whereas if p is 1, the lottery implies getting A for sure, and B is not preferred. Therefore, for some value of $0 < p < 1$, the condition of indifference between (a) and (b) will emerge. In mathematical terminology, B is the *certainty equivalent* of the latter lottery.

(iv) *Substitutability.* A lottery and its certainty equivalent are interchangeable without affecting preference. This is obvious from (iii) that shows the two are equally preferred.

(v) *Monotonicity.* If $A \succ B$, then

if and only if $p > p'$.

(vi) *Decomposability.* A series of branches may be replaced by a single branch as, for example,

Most of these axioms are obvious, yet they form the backbone of the field of utility theory.

2.4.2 Utility Function

An individual whose preference characteristics satisfy the utility axioms of Section 2.4.1 may have his preferences encoded in a utility function. A utility function quantifies the order of preferences. Mathematically, the function represents a mapping of the degree of preference into the real line, thus permitting preference to be expressed numerically.

Let the symbol $u(\cdot)$ denote the utility function of an event. The property of the utility function requires that

$$\text{If } A \succ B, \quad \text{then } u(A) > u(B);$$

$$\text{if } A \sim B, \quad \text{then } u(A) = u(B);$$

$$\text{if } A \succeq B, \quad \text{then } u(A) \geqslant u(B).$$

Another important property of the utility function is that the utility of a lottery is equal to the expected utility of its prizes. For example, the utility of the following lottery between A and C is

$$u \left(\begin{array}{c} \overset{p}{\diagup}\,A \\ \diagdown_{1-p}\,C \end{array} \right) = p\,u(A) + (1-p)\,u(C) \tag{2.23}$$

It can be demonstrated that a utility function is consistent with the axioms of utility theory. As an example, consider the axiom of monotonicity. For the case $A \succ B$, we have $u(A) > u(B)$, and hence, $u(A) - u(B)$ is a positive value. Furthermore, if $p > p'$, then

$$p[u(A) - u(B)] > p'[u(A) - u(B)]$$

Adding $u(B)$ to both sides of the inequality, we have

$$pu(A) + (1-p)u(B) > p'u(A) + (1-p')u(B)$$

or

$$u \left(\begin{array}{c} \overset{p}{\diagup}\,A \\ \diagdown_{1-p}\,B \end{array} \right) > u \left(\begin{array}{c} \overset{p'}{\diagup}\,A \\ \diagdown_{1-p'}\,B \end{array} \right)$$

or

$$\left(\begin{array}{c} \overset{p}{\diagup}\,A \\ \diagdown_{1-p}\,B \end{array} \right) \succ \left(\begin{array}{c} \overset{p'}{\diagup}\,A \\ \diagdown_{1-p'}\,B \end{array} \right)$$

which states the axiom of monotonicity.

If we apply the property of the utility function in Eq. 2.23 to the following preference statement,

$$\left(\begin{array}{c} \overset{1}{\diagup}\,B \end{array} \right) \sim \left(\begin{array}{c} \overset{p}{\diagup}\,A \\ \diagdown_{1-p}\,C \end{array} \right) \tag{2.24}$$

the utility of the lottery on the left side is

$$u(L_1) = 1 \times u(B) = u(B)$$

whereas the utility of the lottery on the right side is

$$u(L_2) = pu(A) + (1 - p)u(C)$$

Since lotteries L_1 and L_2 are indifferent, we have

$$u(L_1) = u(L_2)$$

or

$$u(B) = pu(A) + (1 - p)u(C) \tag{2.25}$$

In other words, for given utility values of A and C, the utility value of B can be determined from Eq. 2.25 by first establishing an indifference statement, Eq. 2.24, with an appropriate choice of p.

Linear Transformation of Utility Function   Another important property of the utility function is that the relative utility values will remain unchanged if the utility function undergoes a linear transformation. In other words, if a consistent set of utility values u are transformed to u' through the following

$$u'(E_i) = a + bu(E_i) \qquad i = 1, \ldots, n \tag{2.26}$$

where a is a constant and b is a positive constant, the new set of utility values $u'(E_i)$, $i = 1, \ldots, n$ will retain the decision maker's original preference. This property can be easily proven as follows.

Suppose that E_1, E_2, E_3 are in increasing order of preference with respective utility values $u(E_1)$, $u(E_2)$, and $u(E_3)$. From the continuity axiom of utility theory, a value of p exists such that

$$u(E_2) = pu(E_1) + (1 - p)u(E_3) \tag{2.27}$$

Multiply each term in Eq. 2.27 by b and add a to both sides of the equation; thus,

$$\begin{aligned} a + bu(E_2) &= pbu(E_1) + (1 - p)bu(E_3) + a \\ &= p[a + bu(E_1)] + (1 - p)[a + bu(E_3)] \end{aligned} \tag{2.28}$$

From Eq. 2.26, we define

$$u'(E_i) = a + bu(E_i) \qquad \text{for } i = 1, 2, 3$$

Hence, Eq. 2.28 becomes

$$u'(E_2) = pu'(E_1) + (1 - p)u'(E_3) \tag{2.29}$$

By comparing Eqs. 2.27 and 2.29, it is obvious that the same value of p is obtained at the indifference point between the lotteries involving E_1, E_2, and E_3. Therefore, Eq. 2.29 implies that the utility values $u'(E_i)$ are also a consistent set of utility assignments reflecting the decision maker's preference.

Since the values of a and b in the above transformation can be arbitrarily assigned, it follows that in determining a consistent set of utility values, the first two values can be arbitrarily assigned, as illustrated in the subsequent section. As long as all other utility values are consistently determined with respect to these two reference points, all the utility values will form a consistent set. In other words, utility values are not unique. In fact, absolute utility values are not necessary for decision making; relative utility values are sufficient.

EXAMPLE 2.29

Given the following utility function

$$u(x) = 10 + 10x + x^2; \qquad x \geq 0$$

determine the transformed utility function such that the utility values at $x = 0$ and $x = 10$ are, respectively, 0 and 1.

According to Eq. 2.26, a new utility function

$$u'(x) = a + bu(x)$$
$$= a + b(10 + 10x + x^2)$$

can be defined such that $u'(x)$ is still a valid utility function. Since a and b are constants to be determined such that

$$u'(0) = 0 = a + 10b$$

and

$$u'(10) = 1 = a + 210b$$

the solution to these simultaneous equations yields $a = -0.05$ and $b = 0.005$. Hence, the transformed utility function is

$$u'(x) = -0.05 + 0.005(10 + 10x + x^2)$$
$$= 0.05x + 0.005x^2$$

2.4.3   Determination of Utility Values

Suppose there are three events, say E_1, E_2, E_3, to which utilities values are to be assigned. First, arrange the events in decreasing order of preference; for example, $E_1 \succ E_2 \succ E_3$. It is common to assign a utility of 1.0 to the most preferred event and a zero to the least preferred event; that is, $u(E_1) = 1.0$ and $u(E_3) = 0$. This is similar to the assignment of 1.0 and 0 in probability scale to the sample space and the null set, respectively. To determine the utility of E_2, relative to those of E_1 and E_3, the decision maker is offered a choice between the following lotteries:

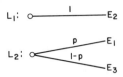

The value of p is adjusted until there is indifference between the two lotteries L_1 and L_2. An indirect way of obtaining the value of p is to use a probability wheel (Spetzler and Staël von Holstein, 1972). The probability wheel is a disk with two sectors, one blue and the other red, with a fixed pointer in the center of the disk. The disk is spun, finally stopping with the pointer either in the blue or red sector. Simple adjustments may be made to change the relative sizes of the two sectors, which in turn change the probabilities of the pointer landing on either sector as the disk stops spinning. In determining the value of p, the decision maker is asked whether he prefers L_1 or L_2. With L_1 he is certain to get E_2. In lottery L_2, the disk is spun. If the pointer ends up in the red sector, E_1 will occur; if it ends up in the blue sector, E_3 will occur instead. Starting with the absence of the red sector, the decision maker would definitely choose L_1 because E_3 will certainly occur in lottery L_2, and since $E_3 \prec E_2, u(L_1) > u(L_2)$. The red sector in the wheel is then increased until indifference is observed. At that point, the fraction of the red sector in the wheel is simply the probability p at which there is indifference between L_1 and L_2. Hence,

$$u(L_1) = u(L_2)$$

or

$$\begin{aligned} u(E_2) &= pu(E_1) + (1 - p)u(E_3) \\ &= p \times 1.0 + (1 - p) \times 0 \\ &= p \end{aligned}$$

The utility of E_2 is simply p in this case. This value of p is used only to determine $u(E_2)$; it is independent of the probabilities in the decision tree.

In a complex decision tree, a large number of events will require the assignment of utility values. The following procedure may be used to systematically determine the relative utility values of n events, namely, E_1 to E_n:

(1) Rank E_i in order of preference such that $E_1 \succ E_2 \succ \cdots \succ E_n$.
(2) Assign $u(E_1) = 1.0$ and $u(E_n) = 0$.
(3) Using the method suggested above for three events, determine $u(E_2)$ such that indifference is obtained between the following lotteries:

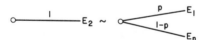

Observe that the two extreme events E_1 and E_n are taken as the reference points for the lottery.

(4) Repeat Step 3 $(n - 3)$ times with E_2 replaced each time by $E_3, \ldots, E_{n-1}$, respectively; of course, the value of p obtained each time will very likely be different.
(5) At this stage, a set of utilities $u(E_1), u(E_2), \ldots, u(E_n)$ has been determined. However, to provide cross checking on these values, repeat Step 3 using $u(E_1)$ and $u(E_{n-1})$ as the set of new reference points and determine a new utility value for E_2, namely $u'(E_2)$, such that indifference is achieved as follows:

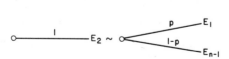

If the utility values are consistent, $u'(E_2)$ should be equal to $u(E_2)$.

(6) Repeat Step 5 $(n - 4)$ times with E_2 replaced each time by $E_3, \ldots, E_{n-2}$, respectively.

If any inconsistencies are found in the above procedure, repeat the process until all utility values agree satisfactorily. If time permits, steps 5 and 6 may be repeated with E_{n-1} replaced by E_{n-2} or E_{n-3}, respectively, in each cycle.

The ranking of preference may be difficult sometimes, especially if the E_i's are multiattributed; that is, they include several measures of effectiveness such as the combination of monetary value and pollution level. As an alternative, multi-attribute decision analysis (see Section 2.6) may be used.

EXAMPLE 2.30

In a decision analysis that includes various alternatives to overcome air pollution from sulfur dioxide (SO_2) and carbon monoxide (CO) in a large city, the utilities of the following four events are to be determined: (i) low CO, low SO_2, (ii) high CO, low SO_2; (iii) low CO, high SO_2, (iv) high CO, high SO_2 concentrations. It is obvious that the desirability of the events would vary inversely with the concentration of pollutants. Furthermore, the decision maker feels it is more urgent to reduce SO_2 than CO at this time. Let

$$E_1 = \{\text{low CO, low } SO_2\}$$
$$E_2 = \{\text{high CO, low } SO_2\}$$
$$E_3 = \{\text{low CO, high } SO_2\}$$
$$E_4 = \{\text{high CO, high } SO_2\}$$

Then, the preference order is

$$E_1 \succ E_2 \succ E_3 \succ E_4$$

In accordance with the procedure suggested above, assume

$$u(E_1) = 1.0, \qquad u(E_4) = 0.0$$

and start with the questioning process. Suppose the decision maker is indifferent between each of the following pairs of lottery, namely:

(lottery: E_2 with certainty $\sim$ lottery giving E_1 with probability 0.8 and E_4 with probability 0.2)

(lottery: E_3 with certainty $\sim$ lottery giving E_1 with probability 0.4 and E_4 with probability 0.6)

Then, from Eq. 2.25

$$u(E_2) = 0.8u(E_1) + 0.2u(E_4) = 0.8$$

$$u(E_3) = 0.4u(E_1) + 0.6u(E_4) = 0.4$$

In order to provide cross checks on these utility assignments, the decision maker is asked for the value of p that he would be indifferent between

(lottery: E_2 with certainty) and (lottery giving E_1 with probability p and E_3 with probability $1-p$)

Suppose the answer is $p = 0.5$. Then, the new estimate of the utility of E_2 is

$$u'(E_2) = 0.5u(E_1) + 0.5u(E_3)$$
$$= 0.5 \times 1 + 0.5 \times 0.4$$
$$= 0.7$$

This is not equal to the value of $u(E_2)$ determined earlier. Therefore, inconsistency exists and the decision maker is asked to reevaluate his assignment of probabilities. In the second round, after careful consideration, the decision maker is found to be indifferent between each of the following three pairs of lotteries:

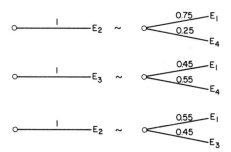

From the first two pairs, we compute

$$u(E_2) = 0.75 \qquad \text{and} \qquad u(E_3) = 0.45$$

Substituting these into the third pair, we obtain

$$u'(E_2) = 0.55 \times 1 + 0.45 \times 0.45$$
$$= 0.55 + 0.2025$$
$$= 0.7525$$

which is close to the value of $u(E_2) = 0.75$. Hence, the utility values of the four levels of air pollution would be:

$$u(\text{low CO, low SO}_2) = 1.0$$
$$u(\text{high CO, low SO}_2) = 0.75$$
$$u(\text{low CO, high SO}_2) = 0.45$$
$$u(\text{high CO, high SO}_2) = 0.0$$

2.4.4   Utility Function of Monetary Value

Consequences are often expressed in monetary terms. However, monetary value is not always a consistent measure of utility; that is, the preference order of the decision maker may depend on the magnitude or amount of money actually involved. Consider the following.

A decision maker is asked to choose between alternatives I and II from the following pair of lotteries:

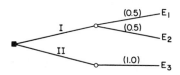

In lottery I, the outcome will be either E_1 or E_2 with equal likelihoods; whereas in lottery II, the only outcome is E_3 for sure. If the monetary values of the respective events are $d(E_1) = 10¢$, $d(E_2) = 0¢$, and $d(E_3) = 5¢$, alternatives I and II are indifferent to the decision maker. However, if $d(E_1) = \$1000$, $d(E_2) = \$0$, and $d(E_3) = \$500$, alternative II may be preferred as there is a sure gain of $500; whereas with alternative I there is a 50% chance of gaining nothing, and the chance of gaining an additional $500 may not be as appealing. Observe that the expected monetary values are the same for the two alternatives in all cases.

The above illustrates the fact that the order of preference for a decision maker may change with the amount of monetary value. Moreover, the preference order may also depend on the financial status of the decision maker relative to the monetary values of the alternatives; for example, for a millionaire, the above lotteries may be indifferent, even with $d(E_1) = \$1000$, $d(E_2) = \$0$, and $d(E_3) = \$500$. In general, therefore, a utility function involving monetary values for a specific decision maker must be established.

The construction of a utility function for money is similar to the method of determining utilities for general outcomes E_i as discussed in Section 2.4.3. The procedure is illustrated by the following example.

EXAMPLE 2.31

An engineer is performing a decision analysis involving consequences ranging from a loss of $20 to a gain of $100. His utility function of money over this range is to be determined.

To establish the utility scale, the utilities of $100 and $0 are assigned arbitrarily to be 1.0 and 0, respectively. Since the utility value of 0 is not assumed for the lowest dollar value involved, the utility function will take on negative values. The first pair of lotteries is presented to the decision maker as follows:

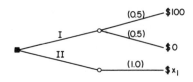

With lottery I, the engineer stands a 50–50 chance of gaining $100 or nothing; whereas he will get x_1 for sure with lottery II. At what value of x_1 will the two lotteries be indifferent?
Suppose $x_1 = 30$; then,

$$u(30) = 0.5u(100) + 0.5u(0) = 0.5$$

The second pair of lotteries presented to the decision maker is

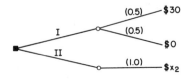

Now, suppose $x_2 = 15$; then,

$$u(15) = 0.5u(30) + 0.5u(0) = 0.25$$

Similarly, with a third pair of lotteries

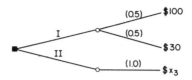

and with $x_3 = 55$,

$$u(55) = 0.5u(100) + 0.5u(30) = 0.5 + 0.25 = 0.75$$

In order to establish the utility function for negative monetary values, the decision maker is given the following pair of lotteries;

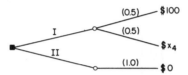

A value of x_4 is required to establish indifference. Suppose $x_4 = -25$; then,

$$u(0) = 0.5u(100) + 0.5u(-25)$$

and

$$u(-25) = 2 \times u(0) - u(100) = -1.0$$

At this stage, six points on the utility function for the range of monetary values from $-\$25$ to $\$100$ have been determined. Figure E2.31 graphically shows the utility function fitted through the six utility values.

In Example 2.31, the probabilities of each lottery are set at 0.5 and the decision maker is asked to determine the dollar values so that indifference is achieved. In earlier sections, the events E_i were fixed and the decision maker was asked to determine the probability p so that indifference is established. Clearly, therefore, there is more than one method of constructing the required utility function.

If the decision maker is a large firm or government organization, the utility function over a range of monetary values may be a straight line, for example, the dotted line in Fig. E2.31. Mathematically, if d stands for the dollar value, the utility function of money for a large firm may be given by the equation

$$u(d) = \alpha + \beta d \tag{2.30}$$

where α and β are constants. Inverting Eq. 2.30, we obtain

$$d = \frac{1}{\beta}[u(d) - \alpha]$$

$$= \alpha' + \beta' u(d) \tag{2.31}$$

where α' and β' are also constants. Since $u(d)$ is a utility function, applying the property of linear transformation, the dollar value d is also a consistent utility function. In other words, monetary value may be a valid utility function for a large firm.

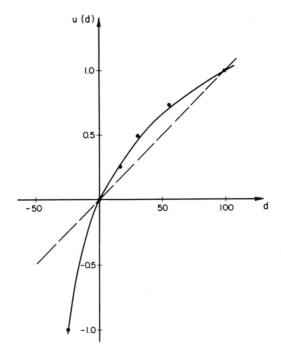

Figure E2.31   A utility function of money.

2.4.5   Utility Function of Other Variables

Using a similar approach, we can transform the value of other variables involved in a decision analysis, such as time, settlement, or the pollutant level to a utility scale. Mathematically, the construction of a utility function involves the transformation of the values of these continuous variables into respective consistent scales in the utility space, representing the decision maker's actual preference for values of the original variables.

As an illustration, consider the decision analysis of the development of the airport facilities in Mexico City (de Neufville and Keeney, 1972). One measure of effectiveness is the average access time to the proposed airport. Possible values for average access time ranged from 12 to 90 minutes. Since desirability increases with decreasing access time, utility values are arbitrarily assigned to these two extreme times as follows:

$$u(12) = 1.0 \quad \text{and} \quad u(90) = 0$$

After several pairs of indifferent lotteries are established, the utility function for average access time is plotted in Fig. 2.7.

If access time is the only measure of effectiveness, then this utility function alone will be sufficient for determining the best alternative. However, there are other measures of effectiveness such as cost, capacity, noise pollution, social disturbance, accident rates, etc. to be considered in the decision analysis. The methods for combining several utility functions into a single overall utility function for decision analysis is presented in Section 2.6.2.

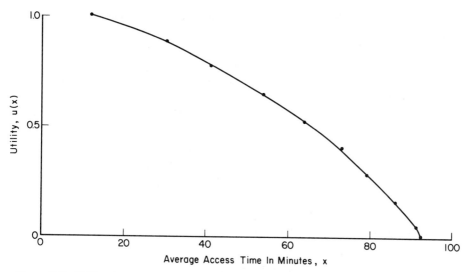

Figure 2.7 Utility function for average access time (from de Neufville and Keeney, 1972).

2.4.6   Maximum Expected Utility

The Maximum Expected Utility Criterion   It had been shown earlier that monetary value may not necessarily represent a true utility scale. Accordingly, the max EMV criterion may not always be the proper criterion for selecting alternatives that will reflect the decision maker's actual preference. A more general criterion for decision is as follows.

Once the utility of each consequence is known, the expected utility value of each alternative is given by

$$E(U_i) = \sum_j p_{ij} u_{ij} \qquad i = 1, \ldots, n \tag{2.32}$$

where u_{ij} is the utility of the jth consequence associated with alternative i and U_i is the utility of alternative i. Maximizing the expected utility, therefore, yields the *maximum expected utility value* criterion. The optimal alternative according to this max EUV criterion is the one whose expected utility value is

$$E(U_{\text{opt}}) = \max_i \left\{ \sum_j p_{ij} u_{ij} \right\} \tag{2.33}$$

In the case in which the consequences are expressed in monetary terms, d_{ij}, Eq. 2.32 becomes

$$E(U_i) = \sum_j p_{ij} u(d_{ij}) \qquad i = 1, \ldots, n \tag{2.34}$$

where $u(d_{ij})$ is the utility function of money established for the particular decision maker.

The procedures for decision analysis based on maximum expected utility value criterion will follow those described in Sections 2.2 and 2.3. The only modification is the replacement of d_{ij} by u_{ij} or $u(d_{ij})$ before taking the expectation at a chance node.

EXAMPLE 2.32

Consider the example discussed in Section 2.2.2 (see Fig. 2.2) in which the owner is deciding on whether or not to install an emergency power system. Suppose his utility function for money is

$$u(x) = -e^{-x/5000}; \qquad x < 0$$

The expected utility value of the first alternative (for installation) is thus

$$E(U_I) = 0.1(-e^{2000/5000}) + 0.9(-e^{2000/5000})$$
$$= 0.1(-e^{0.4}) + 0.9(-e^{0.4})$$
$$= -1.49$$

Similarly, for the no-installation alternative,

$$E(U_{II}) = 0.1(-e^{10,000/5000}) + 0.9(-e^{0/5000})$$
$$= 0.1(-e^2) + 0.9(-e^0)$$
$$= -1.64$$

According to the maximum EUV criterion, the owner should install the emergency power supply system. Then he can avoid the potential risk of losing $10,000, which is a considerable disutility to him. Although he would have chosen not to install the system had he been a maximum expected monetary value decision maker (see Example 2.6), his risk-aversive behavior as described by the exponential utility function causes him to select the more conservative alternative.

It may be worthwhile to observe that taking the expected value is only one method of combining the utility and probability values of all possible outcomes. For example, a conservative decision maker may subconsciously choose to assign more weights to those outcomes associated with the least utility. In doing so, he is overaccounting for his risk-aversive behavior, which should have already been properly accounted for in establishing his utility function. Since utility is a better measure of value to the decision maker than monetary values, the max EUV criterion should prove to be superior to the max EMV criterion. If the decision maker is consistently using the max EUV criterion in all his decisions, he will always be maximizing the true utility, and in the long run will be assured of obtaining the best value relative to decisions based on other criteria.

Value of Information For decision analysis based on the maximum expected utility value criterion, the value of information (e.g., from an experiment) may be obtained through a trial and error procedure. Assuming a specific value for the experiment cost, say x_1, we may calculate the expected utility of the experiment alternative $E(U_T)$ based on the total monetary value including the test cost. By repeating this calculation with several other cost values, a graph such as shown in Fig. 2.8 can be drawn. $E(U_{a^*})$ is the expected utility of the optimal alternative without the experiment. The point at which the $E(U_T)$ curve intersects $E(U_{a^*})$ gives the value of information, VI, accruable from the specific experiment. A similar procedure may be followed to determine the value of perfect information, $E(U_{PT})$, as indicated in Fig. 2.8.

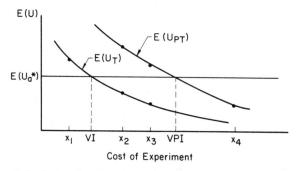

Figure 2.8 Determination of VI and VPI in expected utility analysis.

EXAMPLE 2.33

The value of a laboratory test and the value of perfect information for the waste treatment process design discussed earlier in Examples 2.14 and 2.15 can be reanalyzed using the maximum expected utility criterion. Suppose the utility function of monetary value is

$$u(x) = -e^{-x/50} \qquad x < 0$$

where x is the monetary value in $1000. Then, the expected utility value of design A is

$$\begin{aligned} E(U_A) &= 0.7u(0) + 0.3u(-100) \\ &= 0.7(-e^0) + 0.3(-e^2) \\ &= -2.92 \end{aligned}$$

whereas that for design B is

$$\begin{aligned} E(U_B) &= 0.7u(-50) + 0.3u(-20) \\ &= -2.35 \end{aligned}$$

Therefore, without the test, the engineer should select design B with $E(U_B) = -2.35$. To obtain the value of information from the proposed laboratory model test, several costs of the test are assumed, with the results for $E(U_T)$ summarized in Fig. E2.23. To illustrate the underlying calculations, consider a test cost of $10,000.

From the subtree of Fig. E2.14, we have

$$\begin{aligned} E(U_A|HR) &= 0.95u(-10) + 0.05u(-110) \\ &= -1.61 \end{aligned}$$

$$\begin{aligned} E(U_B|HR) &= 0.95u(-60) + 0.05u(-30) \\ &= -3.25 \end{aligned}$$

$$\begin{aligned} E(U_A|MR) &= 0.637u(-10) + 0.363u(-110) \\ &= -4.05 \end{aligned}$$

$$\begin{aligned} E(U_B|MR) &= 0.637u(-60) + 0.363u(-30) \\ &= -2.78 \end{aligned}$$

$$\begin{aligned} E(U_A|LR) &= 0.143u(-10) + 0.857u(-110) \\ &= -7.91 \end{aligned}$$

$$\begin{aligned} E(U_B|LR) &= 0.143u(-60) + 0.857u(-30) \\ &= -2.04 \end{aligned}$$

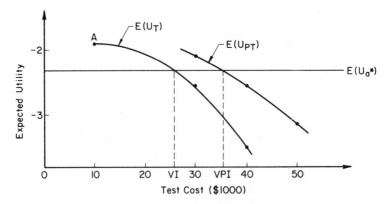

Figure E2.33   Value of information.

Hence, if the test indicates *HR*, design *A* is optimal; otherwise, design *B* will be optimal. The expected utility value of the test is

$$E(U_T) = 0.59(-1.61) + 0.165(-2.78) + 0.245(-2.04)$$
$$= -1.91$$

which is shown as point *A* in Fig. E2.33. Compare this value with the expected utility of -2.35 of the optimal alternative without the test. Since this $E(U_T)$ is higher than -2.35, a larger test cost is assumed in the next trial.

From Fig. E2.33, the VI and VPI are determined to be \$26,000 and \$35,000, respectively. These values far exceed those obtained earlier for EMV analysis (namely, \$15,140 and \$24,000, respectively). It appears that the risk-aversiveness of a decision maker increases the value of additional information. In other words, the decision maker will value any reduction in the level of uncertainty.

2.4.7   Common Types of Utility Functions

Most utility functions are convex such as that shown in Fig. E2.31; this means that the marginal increase in utility decreases with increasing value of the attribute. The preference behavior of a decision maker exhibited by a convex utility function is commonly referred to as "risk-aversiveness." Most people are risk-aversive to a certain degree; some may be more risk-aversive than others. The mathematical forms of utility functions commonly used to model such risk-aversive behavior would include the following.

1.   *Exponential Utility Function*

$$u(x) = a + be^{-\gamma x} \qquad \gamma \geq 0 \tag{2.35}$$

where the parameter γ is the measure of the degree of risk-aversion, and a and b are normalization constants. If the utility function is normalized such that $u(0) = 0$ and $u(1) = 1$, then the normalized exponential utility function becomes

$$u(x) = \frac{1}{1 - e^{-\gamma}} (1 - e^{-\gamma x}) \qquad \gamma \geq 0 \tag{2.36}$$

as γ increases, tne utility function becomes more convex (Fig. 2.9), indicating higher risk-aversion.

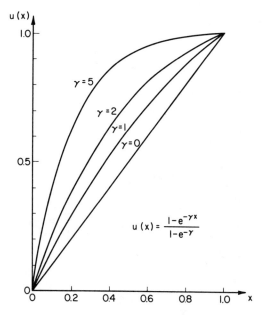

Figure 2.9 Exponential utility function.

2. Logarithmic Type

$$u(x) = a \ln (x + \beta) + b \qquad x + \beta > 0 \qquad (2.37)$$

where β is a parameter, generally corresponding to the amount of capital reserve of the decision maker; that is, as β increases, the decision maker has more utility to spare (such as money) and becomes less risk-aversive (see Fig. 2.10). A normalized logarithmic function may be as follows:

$$u(x) = \frac{1}{\ln \left(\dfrac{1 + \beta}{\beta} \right)} [\ln (x + \beta) - \ln \beta] \qquad x + \beta > 0 \qquad (2.38)$$

3. Quadratic Type

$$u(x) = a(x - \tfrac{1}{2}\alpha x^2) + b \qquad \alpha x \leq 1 \qquad (2.39)$$

where α is the parameter related to the degree of risk-aversion. A normalized quadratic utility function may be shown to be:

$$u(x) = \frac{1}{1 - \alpha/2} (x - \tfrac{1}{2}\alpha x^2) \qquad \alpha x \leq 1 \qquad (2.40)$$

The degree of risk-aversiveness of a decision maker is measured by the "risk-aversion coefficient"

$$r(x) = - \frac{u''(x)}{u'(x)} \qquad (2.41)$$

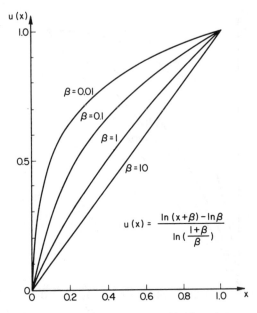

Figure 2.10   Logarithmic utility function.

where the prime denotes the derivative with respect to x. The coefficient measures the negative curvature of the utility function relative to the slope of the utility function at a given value of x. For the exponential utility function, the risk-aversiveness can be shown, using Eq. 2.36 in Eq. 2.41, to be a constant; that is,

$$r(x) = \gamma \qquad (2.42)$$

Applying Eq. 2.41 with Eqs. 2.38 and 2.40, we can show the coefficients of risk-aversion for the normalized logarithmic and quadratic utility functions to be, respectively,

$$r(x) = \frac{1}{x + \beta} \qquad (2.43)$$

and

$$r(x) = \frac{\alpha}{1 - \alpha x} \qquad (2.44)$$

which are both functions of x.

Observe that the coefficient of risk-aversion does not vary with the attribute in the case of the exponential utility function; whereas in the case of the logarithmic utility function the coefficient of risk aversion decreases with x, and in the case of the quadratic utility function the risk-aversion increases with x.

The utility function, of course, may also be concave upward; that is, the marginal increase in utility increases with increasing values of the attribute x. In such cases, the preference behavior of the decision maker is referred to as "risk-affinitive." The utility function of monetary value shown in Fig. 2.11 represents an example of this behavior. It is believed that this preference behavior is ordinarily not realistic.

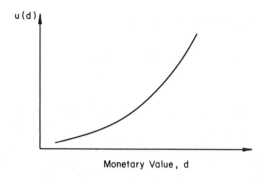

u(d)

Monetary Value, d

Figure 2.11   Utility function of money for a risk-affinitive person.

2.4.8   Sensitivity of Expected Utility to Form of Utility Function

Usually, it is difficult to ascertain what type of utility function is most appropriate; for example, whether it should be of the exponential or quadratic form. The correct choice of the form of the utility function, however, may not be very crucial, especially if the expected utility values are not sensitive to the form of the function.

In order to examine the sensitivity of the expected utility (associated with a given action) to the above three forms of utility functions, consider the simple case in which the possible outcomes from an action can be described by the value of a random variable X. In this case, the expected utility of a given action may be expressed as follows:

$$E(U) = \int_{-\infty}^{\infty} u(x)\, f_X(x)\, dx \tag{2.45}$$

where $f_X(x)$ is the probability density function of X and $u(x)$ is the utility function. Using the second-order approximation to evaluate the expected utility (see Chapter 4, Vol. 1) yields the result

$$E(U) \simeq u(\bar{x}) + \tfrac{1}{2}\, \mathrm{Var}(X) u''(\bar{x}) \tag{2.46}$$

where $\bar{x}$ and $\mathrm{Var}(X)$ are the mean and variance of the random variable X, and $u''(\bar{x})$ is the second derivative of the utility function evaluated at $\bar{x}$.

Applying Eq. 2.46 to the three types of utility functions described earlier, the second-order approximation of the expected utility becomes, respectively, as follows. For the exponential utility function,

$$E(U) \simeq \frac{1}{1 - e^{-\gamma}} \left[1 - e^{-\gamma\bar{x}} - \tfrac{1}{2}\gamma^2\, \mathrm{Var}(X) \cdot e^{-\gamma\bar{x}} \right] \tag{2.47}$$

For the logarithmic utility function,

$$E(U) \simeq \frac{1}{\ln\left(\dfrac{1+\beta}{\beta}\right)} \left[\ln(\bar{x} + \beta) - \ln \beta - \tfrac{1}{2}\, \mathrm{Var}(X) \cdot \frac{1}{(\bar{x} + \beta)^2} \right] \tag{2.48}$$

Finally, for the quadratic utility function,

$$E(U) \simeq \frac{1}{1 - \dfrac{\alpha}{2}} [\bar{x} - \tfrac{1}{2}\alpha\{\bar{x}^2 + \text{Var}(X)\}] \tag{2.49}$$

It has been shown (Ang and Tang, 1979) that the expected utility is relatively insensitive to the form of the utility function at a given level of risk-aversion, and that the expected utility does not change significantly over a wide range of risk-aversion coefficients. Hence, the exact form of the utility function may not be a crucial factor in the computation of an expected utility. Moreover, the risk-aversiveness coefficient in the utility function need not be very precise; that is, any error in the specification of the risk-aversiveness coefficient may not result in a significant difference in the calculated expected utility. In short, the problem of ascertaining an accurate utility function would not be crucial in the application of statistical decision analysis.

As indicated in Eq. 2.46, an approximate expected utility value may be computed on the basis of the mean and variance of the pertinent random variable. This would suggest that the entire probability density function may not be necessary in most decision analysis problems. In practice, the first two statistical moments could be all the information that may be available for a random variable; hence, Eq. 2.46 provides a convenient approximate formula for computing the expected utility of a given alternative.

2.5 ASSESSMENT OF PROBABILITY VALUES

2.5.1 Bases of Probability Estimation

There are basically three approaches to the estimation of probabilities. The first approach is based entirely on observed information. Probability could be estimated from the relative frequency of observed events. Moreover, a probability model may be fitted to a set of observed data. With the model, the probabilities of the events of interest can be computed. The validity of this approach requires a large set of observed data.

The second approach is to estimate probabilities of events or distributions of random variables solely on the basis of subjective judgment of the decision maker. This is necessary when direct observed data are not available; or when the decision maker believes that he has integrated all the relevant information in his mind, thus possessing a strong belief about the likelihood of the events of interest. The third approach is commonly referred to as the Bayesian approach in which the judgmental information of the decision maker is combined with the observed information for an updated estimation of the probabilities. Examples of these have been presented in Chapter 8 of Vol. 1. In both the second and the third approaches, the accuracy of the probability estimation will rely on how well the subjective probabilities can be realistically extracted from the decision maker. In the following sections, some of the methods for obtaining these subjective probabilities are presented.

Figure 2.12 A simple branch in a decision tree.

2.5.2 Subjective Probability

Consider a set of branches as shown in Fig. 2.12, where the probability p of outcome E_1 is to be estimated such that it will realistically represent the judgment of the decision maker.

Suppose the utilities of events E_1 and E_2 are already known, for example, through procedures presented in Section 2.4.3. If the decision maker is indifferent between the lottery in Fig. 2.12 and another alternative that guarantees an outcome of event E_3, then from Eq. 2.25,

$$u(E_3) = pu(E_1) + (1 - p)u(E_2)$$

or

$$p = \frac{u(E_3) - u(E_2)}{u(E_1) - u(E_2)} \tag{2.50}$$

which could be easily determined provided that the utility of event E_3 is also known.

EXAMPLE 2.34

A foundation engineer would like to estimate the probability of the presence of pockets of soft material in a soil stratum that will affect the amount of differential settlement in a structure. With the proposed design A, he expects that the structure will have only 0.5 inch of differential settlement if soft pockets are not present; whereas the presence of soft pockets will lead to a differential settlement of 2 inches. Suppose the following utility function is available:

$$u(x) = -3x^2$$

where x represents the differential settlement in inches.

The foundation engineer could be asked to choose between design A and a hypothetical design B at a comparable cost that will guarantee a differential settlement of x_0 inch, as shown in Fig. E2.34.

Let x_0 take on an initial value of 0.6 inch. If design B is preferred by the engineer, x_0 will be increased slightly, say 0.7 inch, and the engineer is asked again to state his preference between

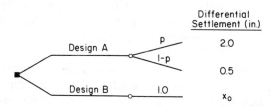

Figure E2.34 Decision tree for evaluation of probability p.

designs A and B. This procedure continues until an indifference condition is achieved between the two designs. Suppose $x_0 = 1.0$ inch at the indifference point. From Eq. 2.50,

$$p = \frac{u(1) - u(0.5)}{u(2) - u(0.5)}$$

$$= \frac{(-3 \times 1^2) - (-3 \times 0.5^2)}{(-3 \times 2^2) - (-3 \times 0.5^2)}$$

$$= 0.2$$

Hence, the foundation engineer's subjective probability that a soft pocket of soil is present at the site is 20%

2.5.3 Subjective Distribution of a Random Variable

Decision analysis may sometimes require the knowledge of the entire distribution of a random variable, including its mean and variance. In cases where observed data are not available, the distribution of the random variable may be established on the basis of subjective judgments. A common method for this purpose is the *fractile method*.

The decision maker would first estimate a value such that the random variable will be equally likely to exceed or fall below that value. This value is the 50 percentile value of the random variable, denoted as $x_{0.5}$. The decision maker would then concentrate on the upper range of the values. He may ask himself the question: "Suppose the value of the random variable is in this upper half range, what is the value at which the random variable will be equally likely to be above or below it?" This will give the value $x_{0.75}$; that is, the 75 percentile value of the random variable. Similarly, the procedure would be repeated for the lower half range to derive $x_{0.25}$.

As a consistency check, the decision maker may ask himself, "Is it equally likely that the random variable would be within the range $x_{0.25}$ to $x_{0.75}$ or outside the range?" If the answer is no, he should go back and revise his previous assignments on the percentile values until consistency is obtained. Similarly, he can continue to subdivide each of the four ranges, namely, $< x_{0.25}$, $x_{0.25}$ to $x_{0.5}$, $x_{0.5}$ to $x_{0.75}$, $> x_{0.75}$ into two equally likely intervals. Again, he would cross check if these eighth fractile values are consistent. Depending on how far out in each tail of the distribution function is required, this subdivision process can be repeated to yield low and very high percentile values.

The fractile values are then plotted to give the cumulative distribution of the random variable. As an illustration, Fig. 2.13 shows the cumulative distributions of the mean friction angle for a sand that has been extracted from two engineers using the fractile method. Curve B is extracted from an engineer who has more experience with this particular type of sand. Based on this cumulative distribution, the density function as well as the mean and variance of the mean friction angle can be computed. For a more detailed discussion and illustrations of the fractile method, readers may refer to Raiffa (1968) and Schlaifer (1969).

The fractile method could be very time-consuming and mentally strenuous for most engineers. Quite often the available information is limited to the upper and lower bounds of the random variable. In this case, a distribution may be assumed over the suggested range. For instance, a uniform distribution would be a convenient

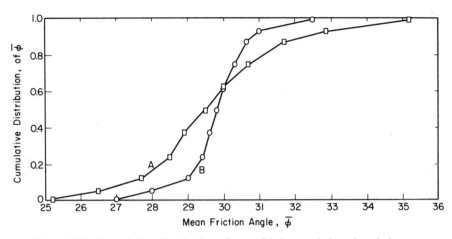

Figure 2.13 Cumulative distribution of mean friction angle based on judgment.

choice. Based on this subjectively assumed distribution, the mean and variance of the random variable can be estimated accordingly. The statistical properties of some commonly assumed distribution over the range $\{a, b\}$ are summarized in Table 6.1 of Chapter 6 that contains a detailed treatment of the analysis of uncertainties.

2.6 DECISIONS WITH MULTIPLE OBJECTIVES

With the increasing awareness for social relevance in engineering designs and planning, many engineering decisions involve conflicting interests of various groups. For example, suppose the mayor of a metropolitan city is faced with a decision on solid waste disposal for the city. The main objectives of the proposed waste disposal scheme might include the minimization of environmental pollution, the minimization of restrictions on future land use, the minimization of potential adverse political consequence, the minimization of systems costs to the city, and possibly others. Most of these objectives are nonmonetary, and methods to quantify them accurately and consistently in monetary terms may have serious limitations.

Decision analysis with multiple objectives requires the determination of multiattribute utility functions instead of a single utility function. These multiattribute functions would evaluate the joint utility value of several measures of effectiveness toward fulfilling the various objectives. However, manipulation of these multiattribute utility functions can be extremely complex, depending on the size of the problem and the degree of dependence among the various objectives. An in-depth treatment of decision with multiple objectives is found in Keeney and Raiffa (1976).

Churchman, Ackoff, and Arnoff (1957) suggested a simplified model for the treatment of multiobjective decision problems. A relatively simple and direct method is provided for evaluating a diverse set of objectives in terms of weighted relative utility values, and selecting the alternative that best balances the competing objectives. Stimson (1959) applied it to public health planning. The model has also been applied to urban planning by Davidson (1971) and Morris (1972), and to water resources development by Erskine and Shih (1972). The model may also have implications in engineering decision problems; the basic concepts of this model are presented in the next section.

2.6.1 Weighted Objective Decision Analysis

Consider a decision requiring the resolution of several objectives versus a number of feasible alternative engineering designs. The relative merits of importance of the various objectives may be different, and the likelihoods of the different alternatives achieving a stated objective may also be different. The information required for the weighted objective decision analysis can be summarized as shown in Table 2.1.

Table 2.1 Tableau for Weighted Objective Decision Analysis

Relative Weights	w_1 w_2 ...... w_j ...... w_n	
Objectives / Alternatives	O_1 O_2 ...... O_j ...... O_n	Overall relative utility
a_1	p_{11} p_{12} ...... p_{1j} ...... p_{1n}	$U_1 = \sum_j p_{1j} w_j$
a_2	p_{21} p_{22} ...... p_{2j} ...... p_{2n}	$U_2 = \sum_j p_{2j} w_j$
$\vdots$		$\vdots$
a_i	p_{i1} p_{i2} ...... p_{ij} ...... p_{in}	$U_i = \sum_j p_{ij} w_j$
$\vdots$		$\vdots$
a_m	p_{m1} p_{m2} ...... p_{mj} ...... p_{mn}	$U_m = \sum_j p_{mj} w_j$

a_i = The ith alternative, $i = 1$ to m.
O_j = The jth objective, $j = 1$ to n.
w_j = The relative weight of importance to objective O_j, $j = 1$ to n.
p_{ij} = The probability of attaining objective O_j through alternative a_i.
U_i = The overall relative utility of alternative a_i.

To implement the model, a set of relative weights has to be assigned to the various objectives. This involves a two-step ranking procedure as suggested by Churchman and Ackoff (1954, 1957). First, the n objectives are listed in decreasing order of importance; this is called *ordinal ranking*. In this step, a preference statement is also solicited from the decision maker with respect to combinations of objectives. For example, if O_1, O_2, O_3, O_4 represent four objectives in decreasing importance, then in addition to the preference statement

$$O_1 > O_2 > O_3 > O_4$$

the decision maker may also be asked if he prefers O_1 to a combination of O_2, O_3, and O_4; that is, whether or not

$$O_1 > O_2 + O_3 + O_4$$

Other similar preference statements may also be required.

Subsequently, a *cardinal ranking* of each of the objectives is established. The relative importance of each objective with respect to the other objectives is evaluated by assigning numerical weights to each objective. Starting with the most important objective, it is assigned an arbitrary weight, for example, 100. Next, the numerical weights are assigned to each of the other objectives. The set of initial weight assignment is then cross-checked for consistency with the preference statements established in the ordinal ranking. For example, if w_1, w_2, w_3, w_4 represent, respectively, the set of relative weights assigned to objectives O_1, O_2, O_3, O_4, then a consistency check would require

$$w_1 > w_2 + w_3 + w_4$$

if $O_1 \succ O_2 + O_3 + O_4$. This assumes implicitly that the overall relative weight of a combination of objectives is equal to the sum of the weights of the individual objectives. This assumption of *linearity* may not be valid in all decision problems. For instance, in a water supply planning project, if O_1 denotes the objective of reducing the demand for water and O_2 denotes the objective of increasing the water supply, then the importance of achieving both objectives would generally be less than the sum of the importances of the individual objectives since the problem of water shortage may be solved by either O_1 or O_2 and may not necessarily require the simultaneous achievements of O_1 and O_2. Hence, either w_1 or w_2 will be very close to the weight of the combination $w(O_1 + O_2)$. Because of diminishing return, $w(O_1 + O_2) \neq w_1 + w_2$. However, this "linearity" assumption does considerably simplify the decision analysis, and facilitates the quantification of the relative importance of each objective. As long as extreme care is exercised to insure that the objectives are mutually exclusive with respect to the contributions to the overall utility, linearity should yield useful results. For example, the problem previously mentioned could be avoided if the two objectives are combined and denoted as no water shortage.

In the event that inconsistency in the preference statement is found, the decision maker is asked to revise his preference statement or his assignment of relative weights or both, until all inconsistencies are eliminated.

The next step in the decision analysis includes a listing of feasible alternatives and the assignment of probabilities p_{ij}. Some of these probabilities can be computed based on probability models, whereas others may have to be estimated subjectively based on the decision maker's experience and judgment (see Section 2.5 for methods of assessing subjective probabilities).

Finally, the overall relative utility of each alternative is computed as

$$U_i = \sum_j p_{ij} w_j \qquad (2.51)$$

The optimal alternative is the one that has the maximum relative utility. Observe that the absolute numerical value of U_i is not important; the relative value of U_i is sufficient for the selection of the optimal alternative.

EXAMPLE 2.35 (excerpted from Davidson, 1971)

A planning group was asked to devise a general land use plan for a large section of a rural county in a southeastern state of the U.S.

There are four objectives that are of paramount importance in the planning, namely:

O_1:   *Provide for Regional Recreation.* This includes the provision of facilities for both weekend and day recreation, such as campsites, cabins, picnic areas, and hiking trails. The concept of a regional recreation area is within a 50 to 100-mile radius of the land under consideration.

O_2:   *Provide for Day Recreation.* This will primarily serve the local residents who are within short driving distances, and involves no overnight facilities.

O_3:   *Encourage Employment Opportunities.* The region is primarily a poor agricultural area with a minimum number of industrial concerns. Improving job opportunities for local residents is desired.

O_4:   *Conservation.* Because of the natural beauty and character of this mountain region, the planners felt that the conservation of the natural state is imperative.

Weighing of Objectives

In ordinal ranking, the following preference statements are obtained from the initial round of questioning.

$$O_4 > O_1 > O_2 > O_3$$

$$O_4 > O_1 + O_2 + O_3$$

$$O_1 < O_2 + O_3$$

In the cardinal ranking step, numerical weights are assigned to each of the four objectives. The initial assignment of relative weights is as follows:

$$w_1 = 50$$

$$w_2 = 40$$

$$w_3 = 20$$

$$w_4 = 100$$

These numbers imply that provisions for regional recreation, O_1, is only half as important as conservation, O_4; day recreation, O_2, is forty percent as important as O_4, and so on. The difference between achieving O_1 and O_2 is relatively small.

Next, the relative weights w_j are examined for consistency with the established preference statements; for example, the above weights yield

$$w_1 + w_2 + w_3 = 110$$

which implies that the combination of O_1, O_2, and O_3 is preferable to O_4. This is inconsistent with the second preference statement established in the ordinal ranking.

After discovering this inconsistency, the planners reconsidered both the preference statements and the assigned relative weights. Following extensive debate, the following set of preference statements and relative weights were determined:

$$O_4 > O_1 > O_2 > O_3$$

$$O_4 > O_1 + O_2 + O_3$$

$$O_1 > O_2 + O_3$$

and

$$w_1 = 50, \ w_2 = 25, \ w_3 = 10, \ w_4 = 100$$

which may be shown to contain no inconsistencies.

Establishment of Alternatives

Six alternative land use plans were proposed. Although each plan could be implemented, none of the plans can achieve most of the objectives.

a_1: *Full Range Recreation*, including picnic areas, cabins, campsites, beaches, boats, etc. The plan fully exploits the recreational potential of the area.

a_2: *Limited Recreation*, including a hunting and fishing preserve with limited cabins and campsites. This plan would retain much of the natural character of the area while providing limited service facilities for the users.

a_3: *Conservation Lockout*, development of a state or national forest with a number of hiking trails, but very limited service facilities.

a_4: *Agricultural*, setting aside the tillable land for agricultural use only and preserving the remainder in its natural state.

a_5: *Agricultural and Extractive Industry*, combining the farming aspect of a_4 with timber and other natural resource (if any) removed.

a_6: *Residential*, the development of a community of second homes or vacation homes. This would require the full development of the area in a traditional sense.

Efficiency Index

In this step, the relative probability of achieving the objectives with each alternative is estimated. For example, the conservation lockout alternative (a_3) will completely achieve conservation (O_4); thus, is assigned $p_{34} = 100\%$. The other alternatives will result in proportionately lesser degrees of conservation. The relative degree of achieving a given objective (e.g., O_j) with each alternative is given by p_{ij} as tabulated in Table E2.35.

Table E2.35 Decision Tableau for Land Use

Relative weights, w_j	50	25	10	100	
Objectives \ Alternatives	O_1 Regional Recreation	O_2 Day Recreation	O_3 Employment	O_4 Conservation	Overall Relative Utility
a_1 Full-range recreation	1.00	1.00	0.50	0.10	90.0
a_2 Limited recreation	0.70	0.80	0.25	0.80	137.5
a_3 Conservation lockout	0.05	0.10	0.05	1.00	105.5
a_4 Agriculture	0	0.10	0.25	0.50	55.0
a_5 Agriculture and extractive industry	0	0	0.70	0	7.0
a_6 Residential	0	0.10	0.40	0.10	16.5

Overall Relative Utility

The overall relative utility for each alternative is then computed according to Eq. 2.51 and summarized in the last column of Table E2.35. For example, the overall relative utility for the full-range-recreation alternative (a_1) is

$$U_1 = 1.0 \times 50 + 1.0 \times 25 + 0.5 \times 10 + 0.1 \times 100$$
$$= 90.0$$

A comparison of the overall relative utilities for the various alternatives indicates that alternative a_2, providing limited recreation, scores the maximum numerical utility values. Observe that although a_2 is the optimal alternative, it fails to provide the maximum possible contribution to any single objective. However, it does provide the best balance in meeting all objectives.

EXAMPLE 2.36 (excerpted from Erskine and Shih, 1972)

This is a case study of the water supply problem in San Angelo, Texas. Plans were developed to correct the problem of inadequate water supply. In 1972, the water supply in San Angelo is completely dependent on rainfall which has proved to be often inadequate. Moreover, there is a general lack of systematic resource planning and management.

Following a preliminary study, the feasible alternatives available for the future water resource development are proposed as follows:

a_1: Continuation of the present water management policies and complete reliance on rainfall.

a_2: Continuation of the present water management policies and development of adequate ground water resources in nearby McCulloch County.

a_3: Revision of the current water management policies to include rationing and rate adjustment programs, the development of a short-range water supply to smooth rainfall fluctuations, and the initiation of a waste water reuse program.

The major objectives to be achieved in this water resource planning consist of:

O_1: Meeting future water demand.

O_2: Improving water quality to meet minimum health requirements.

O_3: Minimizing annual cost.

O_4: Increasing recreational benefits such as water sports and lawn improvements.

O_5: Social acceptance, considering social preferences based on factors other than the above objectives.

Weighing of Objectives

In order to estimate the relative weights for the four objectives, questionnaires were sent out to experts in the fields of engineering and economic development, asking them to assign weights indicating the relative importance of the stated objectives. Using ordinal and cardinal rankings, each respondent reported his or her assigned weights for the five objectives together with a justification for such weight. These individual assessments were summarized and then circulated among all the experts involved. Each individual could revise his assessments at this point. The updated results were compiled and the mean weight for each objective computed and regarded as the consensus. After normalization, the relative weights representing the consensus opinion of the group on the five objectives were as follows:

$$w_1 = 44.9$$

$$w_2 = 23.8$$

$$w_3 = 15.8$$

$$w_4 = 11.1$$

$$w_5 = 4.4$$

This approach of subjective assessment is known as the Delphi Method (Dalkey, 1969, 1972).

Probability Assessments

In order to compute the probability of meeting future water demands, data on the annual runoff from rainfall and projection on the demand for water at the end of the planning period under each alternative are evaluated. From 30 years of observed runoff data for the Concho Rivers, and assuming that only 80% of the recorded runoff would be available for consumption because of evaporation, the annual net water supply may be shown to be lognormally distributed with a mean of 32,900 acre-ft and a coefficient of variation of 0.42. The planning period in this case is taken to be 30 years. At the year 2005 the projected overall water demand is estimated to be 22,000 acre-ft. Based on data compiled by Freese, Nichols, and Endress (1970, 1971), the estimated additional water supply under alternatives a_2 and a_3 are shown in Table E2.36a. Observe that the demand in the year 2005 is well-covered by the additional water supply alone under a_2; whereas for a_3, about $0.75 \times 22,000 - 8503 \simeq 8000$ acre-ft of water will still be needed from the rainfall. Hence, the probability of meeting the damand in the year 2005 is simply the probability that the net annual surface water available exceeds its demand for each alternative. These probabilities are computed to be 0.7865, 1.0, and 0.9995 for alternatives a_1, a_2, and a_3, respectively.

To evaluate each alternative with respect to the improvement of water quality, interviews with local political figures, city water department officials, and other experts were conducted. Based on the limited information and reports by Freer, Nichols and Endress (1970, 1971), ratings for each alternative with respect to the water quality were established, respectively, as 0.6, 0.9 and 0.8. These values indicate that the quality of surface water is judged inferior to that of underground water. The rating is equivalent, in a relative sense, to the probability of each alternative achieving the objective of good water quality. Similarly, ratings for additional recreational benefits are established as 0.0, 0.91, and 0.77, respectively, for the three alternatives.

The costs of the alternatives in most cases were estimated by informed officials. Based on an annual interest rate of 6% over a 30-year period, the annual cost figures for alternatives a_2 and a_3 are summarized in Table E2.36a. The additional annual overall costs for a_1, a_2, and a_3 are computed to be $0, $1,101,774, and $581,581, respectively. In order to convert this to numerical ratings, the alternative with the lowest annual cost is assigned a score of 1.0 for achieving the minimum cost objective, whereas the alternative with the highest annual cost scores a rating

Table E2.36a Cost and Water Supply Data (from Freese, Nichols, and Endress, 1970, 1971)

Alternative	Additional Source of Water	Additional Annual Cost	Additional Contribution to Water Supply
a_2	Ground water	$1,101,774	22,400 acre-ft/yr
a_3	Revised management	$0	25% of normal consumption
	Short-range supply	$94,540	2903 acre-ft/yr
	Waste water reuse	$487,041	5600 acre-ft/yr
	Total for a_3	$581,581	8503 acre-ft/yr + 25% reduction in water consumption

Table E2.36b   Decision Tableau for Water Resource Planning Example

Relative Weights	44.9	23.8	15.8	11.1	4.4	
Objectives	O_1 Meeting Future Demand	O_2 Water Quality	O_3 Annual Cost	O_4 Recreational Benefit	O_5 Social Acceptance	Overall Relative Utility
Alternatives						
a_1	0.7865	0.6	1.0	0.0	1.0	69.78
a_2	1.0	0.9	0.0	0.91	1.0	80.82
a_3	0.9995	0.8	0.472	0.77	0.0	79.92

of zero. The annual cost is assumed to vary linearly with a rating between these two points. Thus, a rating of 0.4721 for alternative a_3 is obtained.

Finally, with respect to the social acceptance objective, a_3 is the only alternative with a questionable social implication because it includes a waste water reuse program; therefore, the first two alternatives were rated 1.0 with the third 0.0

Table E2.36b summarizes the information for the decision analysis. The overall relative utility for each alternative is computed according to Eq. 2.51. For example, the overall relative utility for a_1 is

$$U_1 = 0.7865 \times 44.9 + 0.6 \times 23.8 + 1.0 \times 15.8 + 1.0 \times 4.4$$
$$= 69.78$$

The alternative with the highest relative utility value is a_2. This implies that the best course of action is to continue the present management policies and pursue the complete development of the ground water field in McCulloch County.

The weighted objective decision analysis provides a simplified method in the solution of multiattributed problems. The decision maker can directly assess the weights and probabilities as appropriate. In other words, he has complete knowledge of the entire process. A more rigorous approach to decision analysis with multiple objectives is presented in Section 2.6.3; however, this methodology lacks the brevity and directness of the weighted objective approach described above. In applying the weighted objective decision analysis, extreme care is required in identifying the objectives such that they do not overlap with each other. Furthermore, the limitations of the linearity assumption should be recognized.

2.6.2   Multiattribute Utility Approach

The previous sections show that if all the consequences in a decision model can be conveniently described by a single utility measure, the expected utility of an alternative, say a_i, is computed by

$$E(U_i) = \sum_j p_{ij} u_{ij} \tag{2.52a}$$

where u_{ij} is the jth consequence of the ith alternative, and p_{ij} is the corresponding probability. Alternatively, if the natural outcome is governed by the value of a continuous random variable X, the expected utility of a_i is given by

$$E(U_i) = \int_{-\infty}^{\infty} u(x) f_X^i(x) \, dx \tag{2.52b}$$

where $f_X^i(x)$ is the PDF of X, corresponding to alternative a_i.

This computation of the expected utility can be generalized to decision analysis in which the effectiveness is measured by several attributes, such as cost, time, noise pollution, and others. It is obvious that each of these attributes will require its respective units of measurement. For example dollars, minutes, and CNR are standard measures of cost, time, and noise level, respectively. The corresponding utility function and the associated probability density function will, therefore, be multidimensional. Hence,

$$E(U_i) = \int_{x_1} \cdots \int_{x_n} u(x_1, x_2, \ldots, x_n) f^i_{X_1, \ldots, X_n}(x_1, \ldots, x_n)\, dx_1 \cdots dx_n \quad (2.53)$$

where X_1 to X_n are random variables describing values of the n respective attributes associated with each alternative. Determining these joint utility and density functions requires the evaluation of the conditional utility and probability functions. This task can be cumbersome, especially when the number of attributes is large. Moreover, these functions may have to be developed entirely or largely on the basis of subjective judgments; interviews of experts to furnish descriptions of $u(x_1, \ldots, x_n)$ and $f^i_{X_1, \ldots, X_n}(x_1, \ldots, x_n)$, for all i, would generally be impractical. Appropriate assumptions have been proposed (Ting, 1971; Raiffa, 1969; Fishburn, 1970; Keeney, 1972; Keeney and Raiffa, 1976) to simplify the determination of these functions. Some of these assumptions are as follows.

Statistical Independence If the random variables X_i's, $i = 1$ to n, are statistically independent, the joint density function

$$f^i_{X_1, \ldots, X_n}(x_1, \ldots, x_n) = f^i_{X_1}(x_1) \cdots f^i_{X_n}(x_n) \quad (2.54)$$

where $f^i_{X_j}(x_j)$ is the marginal density function of X_j corresponding to alternative a_i. Hence, the assessment of an n-dimensional function is simplified to that of n one-dimensional functions.

Preferential Independence Trade-offs between any two attributes are governed by the unique indifference curve between these two attributes regardless of the values of other attributes.

Utility Independence The relative utility of X_i remains the same regardless of the values of other X_j's. In other words, the utility of each of the X_i's can be separately determined.

The assumptions of preferential independence and utility independence together imply that the joint utility function may be expressed as a function of the marginal (single attribute) utility function, namely

$$u(x_1, \ldots, x_n) = g[u_1(x_1), \ldots, u_n(x_n)] \quad (2.55)$$

After the marginal utility functions have been obtained, the function g may be determined by scaling $u_i(x_i)$ with respect to other utility functions such that they are consistent with each other. The procedure is described in the following.

For an n-dimensional utility function whose variables are mutually preferential and utility independent, the joint utility function $u(x_1, x_2, \ldots, x_n)$ satisfies the following equation (Kenney and Raiffa, 1976)

$$1 + ku(x_1, \ldots, x_n) = \prod_{i=1}^{n} [1 + kk_i u_i(x_i)] \qquad (2.56)$$

where k and k_i ($i = 1$ to n) are constants to be evaluated; $u_i(x_i)$ is the marginal utility function that has been scaled such that $u_i(^*x_i) = 0$ and $u_i(x_i^*) = 1$; $^*x_i$ and x_i^* are defined as the values of x_i that give the minimum and maximum values of $u_i(x_i)$, respectively.

It is obvious that the outcomes $(^*x_1, ^*x_2, \ldots, ^*x_n)$ and $(x_1^*, x_2^*, \ldots, x_n^*)$ are, respectively, the least and most desirable ones. To set up the relative utility scale, $u(^*x_1, ^*x_2, \ldots, ^*x_n)$ and $u(x_1^*, x_2^*, \ldots, x_n^*)$ are assigned the reference values of 0 and 1, respectively. Then, from Eq. 2.56, the utility function with x_i at the most desirable state and all other attributes at the least desirable states is given by

$$1 + ku(^*x_1, \ldots, x_i^*, \ldots, ^*x_n)$$
$$= [1 + kk_1 u_1(^*x_1)] \ldots [1 + kk_i u_i(x_i^*)] \ldots [1 + kk_n u_n(^*x_n)]$$
$$= 1 \ldots (1 + kk_i) \ldots 1 = 1 + kk_i$$

Thus,

$$k_i = u(^*x_1, \ldots, x_i^*, \ldots, ^*x_n) \qquad i = 1, \ldots, n \qquad (2.57)$$

Moreover, by substituting $u(x_1^*, \ldots, x_n^*)$ into Eq. 2.56, we obtain

$$1 + ku(x_1^*, \ldots, x_n^*) = \prod_{i=1}^{n} [1 + kk_i u_i(x_i^*)]$$

or

$$1 + k = \prod_{i=1}^{n} (1 + kk_i) \qquad (2.58)$$

Hence, the constants k_i and k can be determined from Eqs. 2.57 and 2.58 once the utility value $u(^*x_1, \ldots, x_i^*, \ldots, ^*x_n)$ is known for all i.

Consider the two-dimensional case; assign $u(^*x_1, ^*x_2) = 0$ and $u(x_1^*, x_2^*) = 1.0$. The value of $u(x_1^*, ^*x_2)$ can be determined from a pair of indifferent lotteries as shown in Fig. 2.14. Suppose the decision maker is indifferent between alternatives I and II at probability p_1; then,

$$u(x_1^*, ^*x_2) = p_1 u(x_1^*, x_2^*) + (1 - p_1) u(^*x_1, ^*x_2)$$
$$= p_1$$

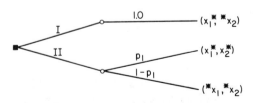

Figure 2.14 Indifferent lotteries.

Similarly, the value of $u(*x_1, x_2^*)$ may also be determined by replacing $(x_1^*, *x_2)$ with $(*x_1, x_2^*)$ in alternative 1 above. Again, it may be shown that $u(*x_1, x_2^*)$ will be equal to another probability p_2.

Hence, from Eq. 2.57,

$$k_1 = p_1$$
$$k_2 = p_2$$

and from Eq. 2.58,

$$1 + k = (1 + kk_1)(1 + kk_2)$$

or

$$k = \frac{1 - k_1 - k_2}{k_1 k_2} = \frac{1 - p_1 - p_2}{p_1 p_2}$$

Eq. 2.56, therefore, becomes

$$1 + ku(x_1, x_2) = [1 + kk_1 u_1(x_1)][1 + kk_2 u_2(x_2)]$$

or

$$
\begin{aligned}
u(x_1, x_2) &= k_1 u_1(x_1) + k_2 u_2(x_2) + kk_1 k_2 u_1(x_1)u_2(x_2) \\
&= p_1 u_1(x_1) + p_2 u_2(x_2) + (1 - p_1 - p_2)u_1(x_1)u_2(x_2) \quad (2.59)
\end{aligned}
$$

This procedure may be extended to derive the joint utility function for the n-dimensional case. It can be shown that the joint function will be a sum of various products of marginal utility functions. Again, the constants k_i ($i = 1$ to n) may be evaluated from n pairs of indifferent lotteries.

EXAMPLE 2.37

As the scheduled completion date of a construction project approaches, a contractor has to decide whether he should rent a set of modern equipment to speed up the construction. With the present equipment, the job may be completed within 2 to 5 weeks (see Fig. E2.37a)

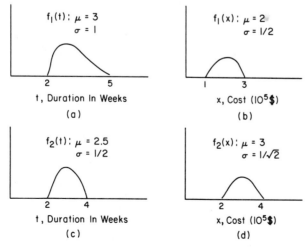

Figure E2.37 Distributions of durations and costs of projects.

costing from \$100 to 300 thousand (Fig. E2.37*b*). However, if the contractor rents modern equipment, the time of completion will range between 2 and 4 weeks (Fig. E2.37*c*); whereas the cost will be increased to between \$200 and 400 thousand (Fig. E2.37*d*).

After interviewing the contractor, a decision analyst determined that the following conditions are appropriate:

(i) The cost and job duration for each alternative are statistically independent.
(ii) The contractor's preference on cost and job duration are preferential- and utility-independent.
(iii) The marginal utility functions on cost x and job duration t are

$$u_x(x) = -x; \qquad x \text{ in } 10^5 \text{ dollars}$$

$$u_t(t) = \frac{1}{21}(25 - t^2); \qquad t \text{ in weeks}$$

(iv) The contractor is indifferent in the following pairs of lotteries:

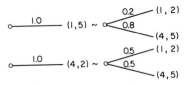

where, for example (1, 5) denotes the event that the job costs \$100 thousand and requires 5 weeks.

(a) Assume that $u(1, 2) = 1.0$ and $u(4, 5) = 0$; determine the joint utility function $u(x, t)$.
(b) Should the contractor rent the additional equipment to speed up the job?

Solution

(a) The marginal utility functions should be checked first if they are scaled from zero to one. The most and least desirable costs in this example are \$100 and 400 thousand. Since

$$u_x(1) = -1$$

$$u_x(4) = -4$$

the given $u_x(x)$ should be modified. Applying a linear transformation, define

$$u_x'(x) = a + bu_x(x)$$

The conditions to be satisfied are:

$$u_x'(1) = a + b(-1) \equiv 1$$

$$u_x'(4) = a + b(-4) \equiv 0$$

The solution to these simultaneous equations gives $a = \frac{4}{3}$ and $b = \frac{1}{3}$; thus,

$$u_x(x) = \frac{4}{3} - \frac{x}{3}$$

Similarly, the most and least desirable durations of the project are, respectively, 2 and 5 weeks. Since

$$u_t(2) = \tfrac{1}{21}(25 - 4) = 1$$

$$u_t(5) = \tfrac{1}{21}(25 - 25) = 0$$

the given utility function is scaled properly. From the given pairs of indifferent lotteries,

$$u(x^*, {}^*t) = u(1, 5) = 0.2u(1, 2) + 0.8u(4, 5)$$
$$= 0.2$$

Similarly, $u({}^*x, t^*) = u(4, 2) = 0.5$. Hence, substituting $p_1 = 0.2$ and $p_2 = 0.5$ in Eq. 2.59, the joint utility function is

$$u(x, t) = 0.2u_x(x) + 0.5u_t(t) + 0.3u_x(x)u_t(t)$$

$$= 0.2\left(\frac{4}{3} - \frac{x}{3}\right) + \frac{0.5}{21}(25 - t^2) + \frac{0.3}{21}\left(\frac{4}{3} - \frac{x}{3}\right)(25 - t^2)$$

$$= 1.338 - 0.186x - 0.043t^2 + 0.005xt^2; \qquad 1 \le x \le 4, 2 \le t \le 5$$

(b) The expected utility of not renting additional equipment (alternative a_1) can be determined from Eq. 2.53, in which x and t are statistically independent.

$$E(U_1) = \int_{-\infty}^{\infty} \int_{-\infty}^{\infty} u(x, t) f_{X,T}^1(x, t) \, dx \, dt$$

$$= \int_{2}^{5} \int_{1}^{3} (1.338 - 0.186x - 0.043t^2 + 0.005xt^2) \, f_X^1(x) \, f_T^1(t) \, dx \, dt$$

Integrating term by term, we obtain

$$E(U_1) = 1.338 - 0.186\mu_{X1} - 0.043(\mu_{T1}^2 + \sigma_{T1}^2) + 0.005\mu_{X1}(\mu_{T1}^2 + \sigma_{T1}^2)$$
$$= 1.338 - 0.186(2) - 0.043(3^2 + 1^2) + 0.005(2)(3^2 + 1^2)$$
$$= 0.636$$

Similarly, for the other alternative a_2, that is, renting the modern equipment,

$$E(U_2) = 1.338 - 0.186\mu_{X2} - 0.043(\mu_{T2}^2 + \sigma_{T2}^2) + 0.005\mu_{X2}(\mu_{T2}^2 + \sigma_{T2}^2)$$
$$= 1.338 - 0.186(3) - 0.043(2.5^2 + 0.5^2) + 0.005(3)(2.5^2 + 0.5^2)$$
$$= 0.598$$

Since $E(U_1) > E(U_2)$, the contractor should not rent the additional equipment.

EXAMPLE 2.38

Two alternative designs of an offshore platform are proposed. The equivalent annual construction cost of designs A and B are \$5 and \$6 million, respectively. Waves are the major loading on the platform. The distribution of the annual maximum wave height H is lognormal with a median of 20m and c.o.v. of 20%. Design A will suffer some damage if H exceeds 26m; moreover, if H exceeds 30m, it will collapse. On the other hand, design B will be damaged when

Table E2.38 Statistics of Consequences

	Design A				Design B			
	Monetary loss[a]		Casualty		Monetary loss[a]		Casualty	
	Mean	c.o.v.	Mean	c.o.v.	Mean	c.o.v.	Mean	c.o.v.
Collapse	50	0.3	10	0.5	55	0.4	5	0.5
Damage	10	0.3	1	0.5	15	0.4	0	0

[a] In addition to initial construction costs.

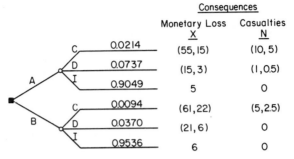

Figure E2.38   Decision tree for platform design.

H exceeds 28m and will collapse when H exceeds 32m. The respective consequences associated with the damage and collapse states of the two proposed designs are summarized in Table E2.38.

The joint utility function of x and n has been established as

$$u(x, n) = 1 - 0.004x - 0.005n^2$$

where x is the total monetary cost (in $ million), and n is the number of casualties. Determine the optimal design alternative.

The decision tree is shown in Fig. E2.38, where the consequences X and N for each path are described in terms of the respective means and standard deviations (μ, σ). The probability of collapse (C) or damage (D) or intact (I) associated with each proposed design can be computed from the given statistical distribution on wave heights. For example, for platform design A,

$$P(D) = P(26 < H < 30)$$

$$= \Phi\left(\frac{\ln 30 - \ln 20}{0.2}\right) - \Phi\left(\frac{\ln 26 - \ln 20}{0.2}\right)$$

$$= 0.0737$$

The expected utility of platform design A if a collapse does occur is

$$\begin{aligned}
E(U_A|C) &= E[1 - 0.004X_{AC} - 0.005N_{AC}^2] \\
&= 1 - 0.004E(X_{AC}) - 0.005E(N_{AC}^2) \\
&= 1 - 0.004(55) - 0.005(10^2 + 5^2) \\
&= 0.155
\end{aligned}$$

where X_{AC} and N_{AC} are the random variables describing the monetary loss and casualty associated with path AC. Similarly,

$$E(U_A|D) = 1 - 0.004(15) - 0.005(1^2 + 0.5^2)$$
$$= 0.934$$

$$E(U_A|I) = 1 - 0.004(5) - 0.005(0)$$
$$= 0.980$$

Hence, the expected utility value associated with design A is

$$E(U_A) = 0.155(0.0214) + 0.934(0.0737) + 0.980(0.9049)$$
$$= 0.959$$

The expected utility of design B can be similarly calculated, yielding

$$E(U_B) = 0.970$$

According to the above analysis, platform design B is preferred.

EXAMPLE 2.39 (excerpted from de Neufville and Keeney, 1972)

In 1970, the Mexican government was faced with the problem of deciding how the airport facilities of Mexico City could be developed to insure adequate service for the region for the next 30 years.

The decision required the following:

(1) The location of the airport relative to the city.
(2) The extent of the services.
(3) The timing of the several stages in the development of the airport.

The Alternative Sites

Because of severe environmental constraints, there are only two sites suitable for a large, international airport in the Mexico City metropolitan area. One is the Texcoco site of the existing airport. The site is close to the city; however, it has been surrounded on three sides by mixed residential and commercial developments, and thus, further expansion is severely limited. This site is also on top of a thick alluvium that has caused considerable differential settlement problems in the past. The second site is in Zumpango. The Zumpango site is farther from the city, has more room for expansion, and is on higher and firmer ground. Problems of differential settlement, such as those experienced at the Texcoco site, are not expected here.

Objectives and Measures of Effectiveness

The interest groups in the planning of the airport consist of the government as *operator*, the *users* of the air facilities, and the interested public as *nonusers*. To evaluate the potential consequences of the different alternatives, it is necessary to study the impacts of the airport sites on these interest groups. Six major objectives were identified as appropriate, as follows:

(1) Minimize total construction and maintenance costs.
(2) Provide adequate capacity to meet the projected air traffic demands.
(3) Minimize access time to the airport.
(4) Maximize the safety of the system (airplane, airport, and its surroundings).
(5) Minimize social disruption caused by the construction and operation of new airport facilities.
(6) Minimize the effects of noise pollution caused by air traffic.

Although there is some overlap among the different interest groups, the first two objectives are primarily of concern to the government as operator, the third is important to the users, and the last two objectives are primarily nonusers' concerns.

The measures of effectiveness, X_i, associated with each of these six objectives were defined as follows:

X_1 = Total cost in millions of pesos.

X_2 = Hourly capacity in terms of the number of aircraft operations per hour.

X_3 = Average access time in minutes, weighted by the number of travellers from each zone of Mexico City.

X_4 = Expected number of people seriously injured or killed per aircraft accident.

X_5 = Number of people displaced by airport expansion.

X_6 = Number of people subjected to a high noise level, specified to be ≥ 90 CNR.

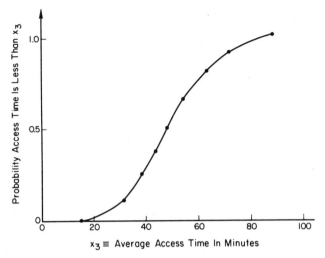

Figure E2.39   Probability distribution for average access time.

Probability Distributions

Because of inherent randomness and uncertainties in prediction, the six measures of effectiveness cited above cannot be accurately known for each alternative. It is, therefore, necessary to derive the joint PDF for the six measures of effectiveness, denoted as $f^i_{X_1, \ldots, X_6}(x_1, \ldots, x_6)$ for alternative a_i.

A first step toward simplifying the assessment of the joint density function is to examine if the X_i's are statistically independent. The size of the airport and levels of services would obviously affect both construction costs and the amount of noise generated; thus X_1 and X_6 are likely to be correlated. For simplicity, however, the X_i's were assumed to be mutually statistically independent in the study. The development of the marginal density functions for each alternative was performed by the staff of the Ministry of Public Works, the Department of Airports, and the Center for Computation and Statistics. The fractile method discussed in Section 2.5.3 was used. In most cases, data were available to validate and confirm the judgmental probability measures. Hence, the derived density functions represent the judgment of a group of people familiar with the problem and the impacts of the different alternatives. As an example, Fig. E2.39 shows the CDF of X_3 if the Texcoco site is selected.

Utility Functions

To simplify the determination of the multiattribute utility function $u(x_1, \ldots, x_6)$, preferential and utility independence were assumed. Discussions with the people responsible for the final decision show that these assumptions appear reasonable.

The six marginal utility functions $u_i(x_i)$, $i = 1$ to 6, were first formulated. As an example of these results, the utility function for mean access time is presented in Fig. 2.8 of Section 2.4.5. The procedure summarized in Eqs. 2.56 through 2.58 is then used to combine these marginal utility functions to obtain the multiattribute functions $u(x_1, \ldots, x_6)$.

Change of Effectiveness with Time

Since a duration of 30 years is involved, the measure of effectiveness associated with each objective for each alternative may not be constant throughout this period, or the achievement of the objective may occur at different points in time. It was necessary to determine a common basis for comparison. For the monetary cost in the first objective this was done using a discounted present value method. For all but one of the other measures of effectiveness, average values over the years seemed appropriate. Specifically, the average access time, the expected number of casualties per accident in the 30-year period, the total number of people displaced, and the average of people subjected to high noise level were used, respectively, for X_3, X_4, X_5, X_6. The capacity, X_2, was divided into three subattributes: $(X_2)_{75}$, $(X_2)_{85}$, $(X_2)_{95}$, each denoting the capacity at 1975, 1985, 1995, respectively.

Results

The expected utilities for each of the feasible alternatives were obtained using a computer program developed for this purpose. The results show that two general types of developmental strategies appeared to be superior to all others; these were:

(1) "Phased Development at Zumpango," that is, a gradual development of this site over the next 30 years, with some activities remaining at Texcoco.
(2) "All at Zumpango," implying that all airport activities should eventually be moved to Zumpango, by 1985 at the latest.

EXAMPLE 2.40   (excerpted from Keeney and Wood, 1977)

The multiattribute utility approach has been used for planning the water resources development of the Tisza River Basin, covering an area of 130,000 km², of which 30,000 km² is in Hungary, as shown in Fig. E2.40. The main river runs from the Carpathian mountains and after flowing through the flat Hungarian Plain, joins the Danube River in Yugoslavia. Five alternative system plans for the Tisza River basin were proposed for further consideration, which may be described as follows:

Figure E2.40   Schematic map of the existing water resources system.

System 1

A multipurpose canal-reservoir system to transfer water between the Tisza and Danube river. The water would be transferred all year round from the Danube by a gravity canal in the flatland area and by a pumped canal-reservoir system in the Börzsöny-Cserhát mountains.

System 2

A pumped reservoir system is developed in the northeastern part of the region, which is supplied from the Tisza River. In this system, the natural supply of water is available only 4 to 5 months per year. It would use all of the natural supply of this part of the Tisza River basin but not all of the available storage capacity.

System 3

A reservoir system will be developed on the flatland part of the region, consisting of 2 to 4 m-deep reservoirs, using water from the Tisza River. Only a limited volume of 5.5 km^3 may be used with such a system, and its capacity to regulate is limited to only 10 km^3/y.

System 4

Develop a mountain reservoir system in the upper Tisza River basin, which involves building reservoirs in the upper Tisza River basin outside of Hungary. For this system to function properly, extensive and long-term international cooperation would be necessary.

System 5

Involves a ground water storage system, mainly in the flatland part to the east of the region. The system would be a conjunctive groundwater system utilizing the Tisza River water for both supply and groundwater exchange. The maximum yield of the aquifer is a limiting factor.

Objectives and Attributes

The basic purpose of the water resources planning is to develop, over a 55-year period, the natural supply of water resources by comprehensive runoff regulation, in terms of both quantity and quality over space and time. The plans call for significant development with growth in irrigated agriculture, industry, and hydroelectric power generation, as well as an implementation of new technologies. The Hungarian National Water Authority planners have established the following goals:

   (i)   To satisfy the demand for the quantity and quality of water.
   (ii)   To provide protection at least against the 50-year flood.
   (iii)   To develop the efficient use and reuse of water.
   (iv)   To be able to drain an area of 15,000 km^2 in 10 days on the average.
   (v)   To minimize the resources needed for the implementation and operation of the system; the resources include water, energy, land and forest, capital and manpower.
   (vi)   To minimize environmental impact caused by the construction of the reservoir, canal networks, and groundwater pumping facilities.
   (vii)   To provide flexibility in the system to accommodate the future development of the region and to cope with the uncertainties inherent in forecasting, economics, and technologies.

Twelve attributes have been identified for evaluating the alternative systems, namely:

X_1 = Cost in 10^9 forints/yr

X_2 = Water shortage in percent of demand

X_3 = Water quality in a scale of 1 to 100

X_4 = Energy reuse factor in terms of the ratio of energy produced to the energy used

X_5 = Recreation in a scale of 1 to 100

X_6 = Flood protection in terms of the mean recurrence interval

X_7 = Land and forest use in thousand hectares

X_8 = Social impact in a scale of 1 to 100

X_9 = Environmental impact in a scale of 1 to 100

X_{10} = International cooperation in a scale of 1 to 100

X_{11} = Development possibility in a scale of 1 to 100

X_{12} = Flexibility in a scale of 1 to 100

Utility Functions

The assumptions of preferential independence and utility independence are verified as follows.
For example, it was found that $(x_1 = 80, x_2 = 60)$ was indifferent to $(x_1 = 110, x_2 = 30)$ regardless of the fixed levels of other attributes x_3 to x_{12}. The same condition was observed for other pairs (x_1, x_2) that were found to be indifferent. Thus, it seemed appropriate to assume that X_1, X_2 is preferentially independent of $\{X_3, \ldots, X_{12}\}$. In a similar manner each pair of attributes was verified to be preferentially independent of the other 10.
In establishing the utility function of X_1, the levels of all the eleven other attributes were first fixed; the certainty equivalent of a 50–50 chance lottery with $x_1 = 80$ and $x_1 = 110$ was then obtained as 98. It was found that the certainty equivalent of 98 did not change when only the levels of the other attributes x_2 to x_{12} were varied. Next, the certainty equivalent for the 50–50 chance lottery with $x_1 = 80$ and $x_1 = 98$ was assessed to be 91 and this also did not depend on the levels of the attributes x_2 through x_{12}. Hence, one felt justified in assuming X_1 is utility-independent of $\{X_2, \ldots, X_{12}\}$. The analyses of the other attributes established that they too were utility-independent.
According to Eq. 2.57, the joint utility function $u(x_1, x_2, \ldots, x_{12})$ is given by

$$1 + ku(x_1, \ldots, x_{12}) = \prod_{i=1}^{12} [1 + kk_i u_i(x_i)]$$

where $u_i(x_i)$ is the normalized marginal utility function of X_i; k and k_i, $i = 1$ to 12 are scaling constants. Table E2.40a summarizes this information obtained from extensive interview procedures.

Table E2.40a Summary of Information for Evaluating Joint Utility Function

Attribute X_i	Worst Value $^*x_i$	Best Value x_i^*	Utility Function $u_i(x_i)$	k_i
X_1	110	80	$1.784 - 0.088e^{0.027x_1}$	0.150
X_2	60	0	$1.611 - 0.611e^{0.016x_2}$	0.243
X_3	0	100	$1.198 - 1.198e^{-0.019x_3}$	0.189
X_4	0	1	$-0.784 + 0.784e^{0.822x_4}$	0.090
X_5	0	100	$-0.476 + 0.476e^{0.011x_5}$	0.132
X_6	40	500	$-1.08 - 1.354e^{-0.0056x_6}$	0.200
X_7	100	50	$2 - 0.02x_7$	0.090
X_8	0	100	$1.198 - 1.198e^{-0.018x_8}$	0.165
X_9	0	100	$1.784 - 1.784e^{-0.00822x_9}$	0.132
X_{10}	0	100	$1.198 - 1.198e^{-0.018x_{10}}$	0.189
X_{11}	0	100	$1.198 - 1.198e^{-0.018x_{11}}$	0.034
X_{12}	0	100	$0.01x_{12}$	0.034
				$k = -0.715$

Evaluation of Utility of Alternatives

The levels of attributes associated with each alternative system are assessed and summarized in Table E2.40b. The uncertainties associated with these attribute levels are assumed to be negligible. By substituting these attribute levels into the joint utility function, the utility corresponding to each of the five proposed systems are determined as:

$$U_1 = 0.832$$

$$U_2 = 0.821$$

$$U_3 = 0.503$$

$$U_4 = 0.648$$

$$U_5 = 0.521$$

indicating that systems 1 and 2 are superior to the others. Hence, systems 1 and 2 should be subject to further investigation. Nevertheless, based on the current utility values, system 1 should be preferred to system 2. In order to determine how much system 1 is better than system 2, the cost of system 1 in Table E2.40b may be increased with all other attribute levels remaining fixed, until the utility of system 1 equals the current utility value of system 2; namely, 0.821. This occurred at $x_1 = 104.2 \times 10^9$ forints/yr; so system 1 is better than system 2 by $(104.2 - 99.6) \times 10^9 = 4.6 \times 10^9$ forints/yr.

Table E2.40b.   Attribute Levels for Alternative Systems

	Alternative Systems				
Attribute	1	2	3	4	5
X_1	99.6	85.7	101.1	95.1	101.8
X_2	4	19	50	50	50
X_3	80	60	20	80	40
X_4	0.7	0.5	0.01	0.1	0.01
X_5	80	60	40	20	20
X_6	100	200	67	200	500
X_7	90	80	80	60	70
X_8	80	80	60	40	40
X_9	80	60	20	60	40
X_{10}	80	60	40	20	40
X_{11}	80	60	40	20	40
X_{12}	80	80	20	40	20

PROBLEMS

2.1 Figure P2.1 shows a network of roads linking five towns: A, B, C, D, and E. The respective travel times between the towns are independently normally distributed with means and standard deviations shown in parentheses in Fig. P2.1. For example, the travel time between A and C is $N(5, 2)$ in hours. A bus company is planning a route going from A to E with intermediate stops at the other three towns. The sequence of service is not important. Assume that the time spent at each station is negligible. Determine the best route if the objective is to:
   (a) Minimize the expected travel time.
   (b) Maximize the probability of completing the entire route from A to E within 25 hours.
   (c) Repeat part (b) for 35 hours.

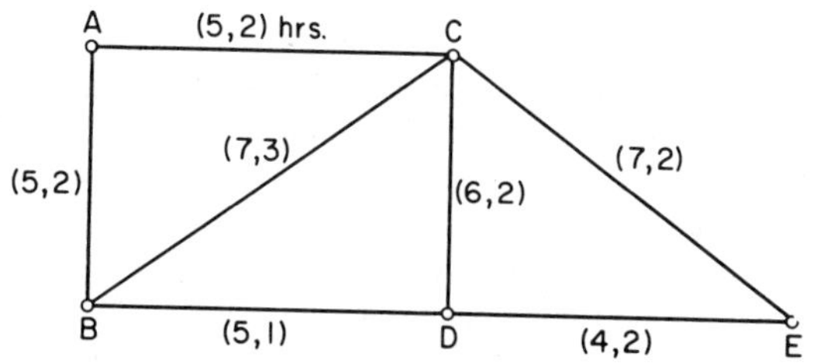

Figure P2.1   Road network

If one of the roads was to be improved so that the mean and standard deviation of the travel time for that road is reduced by 10%, which road should the bus company recommend if the objective is (a)? Repeat this if the objective is (c).

2.2   A highway intersection will be closed for the sanitary department to improve the sewer line crossing the intersection. The project will require at least 10 days but no more than 30 days, and will most likely take 16 days. A service-station owner on the corner of the intersection estimates that his loss as a function of the project duration is as follows:

Project Duration T (days)	Loss
$T \leq 15$	negligible
$15 < T \leq 20$	$4000
$20 < T \leq 25$	$10,000
$25 < T \leq 30$	$20,000

Assume that the distribution of the project time T is triangular; determine the expected loss to the service station owner arising from the closing of the intersection.

2.3   The construction of a causeway is proposed over a long stretch of swampy areas. The causeway is a system of girders resting on piers. As the span length of the girder increases, the construction cost per foot of girder decreases, whereas the cost of pier construction increases substantially. For the following costs,

$$\text{Cost per pier} = 2x^2$$

$$\text{Cost per ft of girder} = \frac{2000}{\sqrt{x}}$$

where x is the span length of each girder, determine the optimal span length.

2.4   The performance function of an engineering system is given by

$$G = X_1 + X_2 + X_3 - \alpha D$$

where the X_i's are independent normal variates $N(0, 2)$, $N(2, 1)$ and $N(20, 3)$, respectively; D is the present design system capacity; and α is an additional safety factor that could be imposed to achieve a higher reliability. Failure of the system is denoted by the event $(G > 0)$. Assume that $D = 30$; the cost of the present design is C_o; the cost of a system with safety factor α is αC_o; and the cost in the event of system failure is C_F. Derive an expression for the optimum value of α in terms of C_F/C_o, such that the total expected cost is minimized. Determine α_{opt} for $C_F/C_o = 10^3$ and 10^6.

$$Hint: \quad \frac{d}{dx}\{\Phi[g(x)]\} = \frac{d}{dg(x)}\{\Phi[g(x)]\}\frac{dg(x)}{dx} = \phi[g(x)] \cdot \frac{dg(x)}{dx}$$

where $\Phi(\ )$ and $\phi(\ )$ are the CDF and PDF of the standard normal variate.

2.5   Construct the new decision tree for Example 2.4 if the following alternatives are also included:

(i)   A field test of the embankment system can be constructed. The amount of seepage recorded during the test period is classified into three categories: low, medium, and high. If low seepage is indicated, the actual flow will be Q_1 with 95% probability; whereas, if high seepage is indicated, the actual flow will be Q_2 with 80%

probability. If medium seepage is indicated, Q_1 and Q_2 will be equally likely. Assume that the field test is equally likely to indicate any one of the three seepage levels. The total cost of the field test program is 3 units.

(ii) A perfect test is available, which will cost 10 units, that will give perfect information on the actual flow.

2.6 A housing project is proposed for low-income citizens. From feasibility studies, three alternative sizes of the housing project were suggested for further study, namely:

a_1: 1200 units

a_2: 1500 units

a_3: 1000 units in the first phase with an optional second phase of 500 units

The actual demand for housing, however, is uncertain. From the results of a survey study, the probabilities of the various levels of demand were estimated as follows:

Demand	Probability
Less than 1000 units	0.2
1000–1100	0.1
1100–1200	0.2
1200–1300	0.3
1300–1400	0.1
1400–1500	0.05
>1500	0.05

After a consideration of the socio-economic factors, the following monetary consequences are inferred, in terms of $ thousand.

(i) For alternatives a_1 and a_2, if the demand does not exceed the number of housing units provided, the expected losses are 100 and 200, respectively; otherwise, if the demand is not met, the expected losses are 400 in both cases.

(ii) For alternative a_3, if the demand does not exceed the capacity provided in the first phase, the expected loss is 100; if the demand exceeds 1000 units but is less than 1500 units, the expected loss is 200 associated with the inconvenience caused by delay; whereas, if the demand exceeds 1500, the expected loss is 400.

Draw the decision tree and determine the optimal alternative based on the maximum EMV criterion.

2.7 A contractor is faced with the opportunity of bidding on two projects, A and B. His alternatives are to submit bids for both projects, or for either project, or none at all. Based on his experience and considering the expected number of bidders for the jobs, he estimates that his chances of winning the bids for A and B are, respectively, 0.2 and 0.1. The costs for preparing the bids will be $1000 for the two jobs, and $600 for one job. The estimated profit on job A will be $4000, whereas that for job B will be $6000. If he fails to win any bid, the contractor will lose $400; this includes the cost of employees and equipment left idle. Assume that the probabilities of winning projects A and B are statistically independent. Construct the decision tree and determine the best alternative based on the maximum EMV criterion.

2.8 In order to protect an $8 million dam from overtopping in the event of extreme floods, two alternative spillway designs are proposed; system A costing $3.5 million and system B costing $3 million. The spillway system may fail because of exceptionally large flood inflow during the expected life of the dam and/or incorrect assessment of the spillway capacity. If the spillway system fails to perform satisfactorily, the dam could be destroyed; however, because of the location of the dam, a failure will not cause other losses except the replacement cost of the dam and the spillway system.

With the available information, the lifetime failure probabilities for spillway systems A and B are estimated to be 0.15 and 0.3, respectively.

The uncertainty in the capacity of the proposed spillway systems may be reduced by performing a model test at a cost of $0.3 million to examine the performance under large flood condition. In the judgment of the engineer, there is a 0.7 probability that the test will indicate that system B is preferable, and a 0.3 probability that system A is preferable. If the test indicates a preference for system A, system B may still be built but the failure probability will be increased to 0.6, whereas system A will have a failure probability of 0.2. On the other hand, if the test indicates that system B is preferred, its failure probability will be 0.1, whereas system A will have a corresponding failure probability of 0.05.

(a) Draw a decision tree and include all of the relevant alternatives, outcomes, probabilities, and monetary consequences.

(b) Determine whether a model test of the dam and spillway should be conducted before building the actual spillway. Assume that decisions will be based on EMV.

2.9 A contractor is deciding between submitting a bid for a highway project or a dam project. Because of time and manpower limitations, he cannot afford to submit bids for both projects. The estimated bid for the highway job is $1.5 million; whereas, that for the dam is $3.0 million. The cost for preparing a bid is about 2% of the bid price. Suppose the contractor estimated his gross profit is 10% of the bid price. After a careful study of the potential bidders for the two jobs, he estimates from his experience that there is a 70% chance of winning the highway job, but only a 50–50 chance of winning a bid on the dam project. If the contractor does not acquire any job, he will incur an additional loss of $50,000 due to the idling of machinery and permanent employees.

(a) Construct the decision tree.

(b) Determine the best alternative based on the following decision criteria:

(i) mini-max (minimize maximum loss);

(ii) maxi-max (maximize maximum gain);

(iii) maximum expected monetary value.

2.10 The following data are taken from a study of Whitman, et al. (1975) in which earthquake damages are evaluated.

(i) For buildings designed according to the Uniform Building Code Zone 3 requirement, the damage probabilities after an earthquake of given Modified Mercalli Intensity are as follows:

| Damage State | \multicolumn{6}{c}{Modified Mercalli Intensity} |
	V	VI	VII	VIII	IX	X
None O	1.0	0.57	0.25	0	0	0
Light L	0	0.43	0.50	0.25	0	0
Moderate M	0	0	0.25	0.53	0.20	0
Heavy H	0	0	0	0.21	0.52	0
Total T	0	0	0	0.01	0.23	0.80
Collapse C	0	0	0	0	0.05	0.20

(ii) The losses, *expressed as the percentage of the replacement cost of the building,* for each damage state are tabulated below:

Damage State	Damage Repair Cost	Incident Cost	Human Costs
None O	0	0	0
Light L	0.3	0.3	0
Moderate M	5	0.4	0.6
Heavy H	30	2	7
Total T	100	3	30
Collapse C	100	3	600

The total loss for each damage state is the sum of the damage repair cost, incident cost, and human cost.

(iii)   For a building founded on soft soil in a given region, the probability that the largest annual earthquake will be of MM intensity V, VI, VII, and VIII, respectively, are estimated to be 0.08, 0.0155, 0.0043, and 0.0002, respectively.

(a)   Determine the annual expected total loss from the earthquake of a building designed according to the UBC Zone 3 requirement.

(b)   For a building whose expected service life is 50 years, determine the present value of the total lifetime expected earthquake loss. Assume a net discount rate of 2% per annum, accounting for future interest rate and inflation rate.

(c)   An alternative design is to satisfy the UBC Zone 2 requirement whose damage probabilities are given below.

Damage State	Modified Mercalli Intensity					
	V	VI	VII	VIII	IX	X
O	1.0	0.47	0.20	0	0	0
L	0	0.53	0.50	0.1	0	0
M	0	0	0.29	0.53	0	0
H	0	0	0.01	0.31	0	0
T	0	0	0	0.05	0.8	0.6
C	0	0	0	0.01	0.2	0.4

Furthermore, suppose for a given type of building, the initial cost for the UBC Zone 3 requirement is 1% higher than that for the UBC Zone 2 requirement. Which code will minimize the total expected cost (initial construction plus expected earthquake losses) for this type of building with an expected service life of 50 years?

(d)   Would the decision in part (c) change if the site is more seismically active? The corresponding probabilities for the annual largest earthquake are 0.3, 0.1, 0.02, and 0.0001 for MM intensity VI, VII, VIII, and IX, respectively.

2.11   In Example 2.5, suppose the reliability of the Tharp and Louisiana models are expressed by the conditional probabilities given below:

	Conditional Probabilities of $P(H'_A H'_F \mid H_A H_F)$, etc.							
	Tharp				Louisiana			
	$H'_A H'_F$	$H'_A L'_F$	$L'_A H'_F$	$L'_A L'_F$	$H'_A H'_F$	$H'_A L'_F$	$L'_A H'_F$	$L'_A L'_F$
$H_A H_F$	0.4	0.3	0.2	0.1	0.25	0.25	0.25	0.25
$H_A L_F$	0.2	0.5	0.2	0.1	0.15	0.35	0.25	0.25
$L_A H_F$	0.2	0.2	0.5	0.1	0.2	0.2	0.4	0.2
$L_A L_F$	0.15	0.15	0.25	0.45	0.1	0.1	0.3	0.5

The Louisiana model will cost $10,000 less to perform than the Tharp model. Determine the optimal decision. Assume that the prediction models apply only to the improved road scheme, whereas the probabilities associated with the present (no improvement) road design have been well established from records of past performance as listed in Fig. E2.5.

2.12 An engineer is driving his car to attend an important meeting in a nearby town. The meeting is scheduled to begin in another 90 minutes. He judges that if he stays within the legal speed limit, his travel time will be uniformly distributed over 2 ± 0.6 hours. However, if he decides to overspeed, say by x mph, his travel time will decrease and be uniformly distributed over $(2 - x/20) \pm 0.6$ hours. In this case, however, he would be running the risk of being ticketed by a traffic officer; this probability is given by $x/25$ for $0 \le x \le 25$. Suppose that the loss associated with the delay in attending the meeting is equal to the amount of the delay (in hours). If the engineer is ticketed, 0.3 hour needs to be added to his travel time; in addition, the loss resulting from the traffic ticket is $x/30$, which depends also on the overspeed. Assume there is no other loss. Formulate the decision problem and determine his optimal overspeed.

2.13 An engineer is selecting the pavement material for the landing strip of a new airport. There are two kinds of pavement materials available: namely, good quality material (G) that costs \$0.4 million and regular quality material (R) costing \$0.28 million. Both of these materials will last for 10 years with proper maintenance. The annual maintenance cost will depend on the quality of the pavement material as well as the volume of traffic, as follows:

Material Quality	Traffic	Annual Maintenance Cost
Good	High	\$0.02 million
Good	Low	0.01 million
Regular	High	0.04 million
Regular	Low	0.02 million

In the planner's judgment, there is a 30% probability that in the next 10 years the air traffic volume will be high. Assume that the decision will be based on the maximum EMV criterion and the interest rate is 6% per annum.
   (a) Without any additional information, which pavement material should the engineer select?
   (b) A research group is willing to conduct a study to predict the traffic volume for a price of \$20,000. However, the prediction from this research group is not perfectly reliable. If the actual future traffic is high (H), the group will predict high traffic (H_o) with probability 0.9 and low traffic (L_o) with probability 0.1, that is, $P(H_o|H) = 0.9$ and $P(L_o|H) = 0.1$. Similarly, $P(L_o|L) = 0.9$ and $P(H_o|L) = 0.1$. Should such a study be performed before the selection of the pavement material?
   (c) What is the value of perfect information on the actual volume of future traffic?

2.14 The present waste treatment facility at a Midwestern city consists of a trickling filter operating at a cost of \$5.40 per capita per year. The filter is packed with particles so that a large surface area is provided for the oxidation and purification process as the waste water is sprinkled from the top and allowed to percolate through the particles. The efficiency of this device is equally likely to be between 60 and 80%. There is, however, public pressure for meeting higher stream standards. Based on the prevailing technology and after a feasibility study, the following alternative waste treatment processes were suggested:

   A: Enlarging the existing trickling filter at an additional cost of \$2.10 per capita/year.
   B: The output from the existing trickling filter will be treated chemically and the wastes allowed to precipitate slowly into a lagoon and removed. The cost of this scheme depends partly on the cost of acquiring the lagoon. Based on the judgment of the planner, the additional costs for this scheme will be either \$4.10 or \$5.60 per capita/year with probabilities of $\frac{2}{3}$ and $\frac{1}{3}$, respectively.
The efficiency of scheme A will depend on the workmanship of the expansion of the trickling filter; assume that it is 70% likely that the workmanship will be good. The

probabilities associated with the levels of efficiency of each alternate scheme are summarized as follows:

		Probabilities		
		Improvement Scheme *A*		
Efficiency Level (%)	Present Device	Good Workmanship	Bad Workmanship	Improvement Scheme *B*
60–70	0.5	0.2	0.3	0.1
70–80	0.5	0.7	0.7	0.2
80–90	0	0.1	0	0.5
90–100	0	0	0	0.2

The classification of water quality and its estimated loss from social and other considerations are tabulated as follows:

Efficiency (% Removal of Pollutants)	Water Quality	Estimated Monetary Loss per Capita/Year
<70	poor	$10
70–90	good	$2
90–100	excellent	$0

Suppose you were consulted by the planner of the pollution control board to perform a decision analysis on whether or not to improve the waste treatment.

(a) Draw the decision tree and include all the relevant alternatives, outcomes, probabilities, and monetary consequences.

(b) Determine the best alternative assuming EMV decision criterion.

(c) Suppose that alternative *A* is chosen. The planner is given the option whereby the quality of the workmanship may be tested by a week's trial performance. Because of the limited trial time, it can only be estimated that if the workmanship is good, the performance will be satisfactory with a probability of 80%, whereas if the workmanship is bad, the performance will be satisfactory with a probability of only 10%. Suppose that the job will be accepted if there are satisfactory performances during the trial week; otherwise, the contractor will have to lower the cost by an amount equivalent to $1.00 per capita/year due to poor workmanship. Draw the decision tree for this alternative including all the outcomes and their probabilities. At most, how much money (equivalent to dollars per capita/year) should the planner pay for pursuing this option.

2.15 A contractor has to decide whether he should order new machinery for assembling precast structural elements. Each of the new machines is subject to breakdown, repair, and maintenance. During each breakdown, the repair time is estimated to be 5 days. Assume that the time between breakdowns for each machine is T, which is exponentially distributed with mean λ regardless of the number of previous breakdowns. The contractor estimates that his long-term benefit will be a function of the number of pieces of equipment that are operating on an average day, as shown in the following table:

Number of Operating Machines (on an Average Day)	Benefits (in thousand $)
0	0
1	120
2	180

After studying the recommendation of the manufacturer, he assumes that λ is either 15 or 25 days with relative likelihoods of 1 to 3.

(a)   If the contractor has the option of ordering 0, 1, or 2 pieces of equipment, which alternative should he choose if the cost of the first piece of equipment is $60,000, and any additional equipment may be acquired at a 5% discount?

(b)   If the contractor is given a free trial period, say 20 days, on a piece of the equipment, would the additional information from the 20-day trial period influence his decision or improve his expected utility from (a)? Assume that there are only two possible outcomes in the trial period: the equipment will or will not break-down.

(c)   What is the value of perfect information in this case?

2.16   Two alternative designs, namely A and B, are considered in a project. There are two modes of failure associated with each design. Let α_{A1}, α_{A2} be the performance indicators of design A against the two modes of failure. Similarly α_{B1}, α_{B2} are the performance indicators of design B against the corresponding two modes of failure. The performance indicators are all statistically independent random variables as follows:

$$\alpha_{A1} - N(1.5, 0.3) \qquad \alpha_{A2} - N(1.6, 0.4)$$

$$\alpha_{B1} - N(1.6, 0.4) \qquad \alpha_{B2} - N(1.4, 0.28)$$

If the performance indicator is less than 1, the failure of the corresponding mode will occur; that is, failure $= \{\alpha < 1\}$. Suppose the initial costs of design A and design B are, respectively, 1.0 and 0.8 $ million. Moreover, failure in modes 1 and 2 will incur additional costs of $5 million and $2 million, respectively. Assume that a design can be subject to both modes of failure and all costs are additive.

(a)   Construct the decision tree and determine the optimal design based on *minimum expected total cost* criterion.

(b)   Suppose design B can be load-tested with respect to its first mode of failure. The survival of a specific load test will imply that α_{B1} will be at least 0.9, and all the other α's will remain unchanged. On the other hand, if design B does not survive the load test, it will be removed from further consideration; assume $0.5 million is lost and design A will be adopted. Determine the value of this load test.

(c)   In part (a), if the failure costs associated with mode 2 depend on the actual values of α_{A2} and α_{B2} such that

$$C(\alpha_{A2}) = 5(1 - \alpha_{A2}) + 5(1 - \alpha_{A2})^2 = 5\alpha_{A2}^2 - 15\alpha_{A2} + 10; \qquad \alpha_{A2} \le 1$$

$$C(\alpha_{B2}) = 5(1 - \alpha_{B2}) + 5(1 - \alpha_{B2})^2 = 5\alpha_{B2}^2 - 15\alpha_{B2} + 10; \qquad \alpha_{B2} \le 1$$

In other words, as α_{A2} or α_{B2} is significantly below unity, larger damages will be expected compared to the case in which α_{A2} or α_{B2} is close to unity.

Construct the decision tree and determine the expected total cost associated with design A. Numerical integration may be needed.

2.17   In a project involving tunnel construction through rock presented in Einstein et al (1978), an engineer may select one of three construction strategies: $S1$, $S2$, or $S3$. The cost of construction depends on the strategy adopted *and* the geology of the tunnel, which could be classified into three conditions: namely good (G), fair (F), and poor (P). The construction cost is given by the following cost matrix, in terms of dollars per foot of construction.

		Tunnel Geology		
		G	F	P
Construction	S_1	200	500	1500
strategy	S_2	300	600	1000
	S_3	500	600	800

Furthermore, the prior probabilities of the tunnel geology conditions at a proposed site are 0.2, 0.6 and 0.2 for G, F, and P, respectively.

(a) If no further exploration is conducted, which construction strategy should be adopted at the proposed site?

(b) The engineer could perform soil exploration programs to gather additional information about the tunnel geology. The reliability of one such program is summarized by the following reliability matrix:

		Exploration Result		
		E_G	E_F	E_P
Geologic	G	0.6	0.2	0.2
state	F	0.3	0.6	0.1
	P	0.2	0.3	0.5

For example, if the actual geologic state is good (G), the exploration will indicate good geology (E_G) with probability 0.6, that is, $P(E_G|G) = 0.6$. Draw the decision tree including this soil exploration program as an alternative. Determine the value (in terms of dollars per foot of construction) of this exploration program.

(c) Suppose a more elaborate soil exploration program is available with the following reliability matrix:

		Exploration Result		
		E_G	E_F	E_P
Geologic	G	0.8	0.1	0.1
state	F	0.15	0.8	0.05
	P	0.1	0.2	0.7

How much more is this program worth than the one in part (b)?

(d) What is the maximum value (in terms of dollars per foot of construction) that the engineer should spend in verifying the actual geologic condition of the site?

2.18 A geotechnical engineer has to decide between two foundation designs for a given site, which is believed to contain a lens of soft clay. The size of the lens could be discretized into large L (width 20 to 30 ft), medium M (width 10 to 20 ft), and small S (width 0 to 10 ft). Design A basically assumes that there is indeed a large lens of soft clay, whereas design B assumes that the size of the soft clay lens, if any, is insignificant. The expected cost, which includes initial cost and expected failure costs, for each design and lens size combination is given by the following cost matrix (in units of $ million):

		Size of Soft Clay Lens		
		L	M	S
Design	A	4.0	3.8	3.6
	B	4.8	3.8	3.0

Suppose the engineer believes that the soft clay lens is equally likely among L, M, and S based on preliminary information.

(a) Which is the better design based on the maximum EMV criterion?

(b) Suppose four borings spaced at 60 ft could be driven at the site to verify the existence of soft clay lens. Probabilistic analyses show that the probabilities of such soil exploration programs encountering a weak spot are 0.99, 0.44, and 0.02 for lens sizes L, M, and S, respectively. Draw the decision tree and determine if such borings should be driven before adopting the foundation design. Assume that the boring program will cost $0.1 million.

(c) The geotechnical éngineer could perform a more extensive soil exploration program instead. How much, at most, should the geotechnical engineer spend for verifying the size of soft clay lens?

2.19 Consider Problem 2.4 again and

(a) Suppose X_1 represents the model error, whose uncertainty level could be reduced by half, that is, to $N(0, 1)$, by additional research. What is the maximum amount that should be spent on such research? Assume $C_F/C_o = 10^3$ and express your answer in terms of C_o.

(b) What is the value of perfect information on X_1? Assume $C_F/C_o = 10^3$ and express your answer in terms of C_o.

2.20 For most engineering variables, such as material strengths, the penalty associated with underestimating the actual value of the variable, say θ, is generally less than the loss associated with overestimation. Suppose the loss function is given by

$$g(\theta, \hat{\theta}) = \begin{cases} C_1(\hat{\theta} - \theta); & \theta \le \hat{\theta} \\ C_2(\theta - \hat{\theta}); & \theta > \hat{\theta} \end{cases}$$

where $\hat{\theta}$ is the estimated value; $C_1 > C_2$ and both are constants in the respective linear loss functions. Determine the optimal Bayes' estimator.

2.21 For a medium-size town (population 50,000 to 100,000) the daily per capita demand for water may be modeled by a normal distribution $N(\mu, \sigma)$. Data on water demand (per capita per day) collected for four such towns are as follows:

$$120, 110, 128, 112 \quad \text{(in gallons/day, gpd)}$$

obtaining: sample mean = 117.5 gpd; sample variance = 71.7 gpd^2. Assume that σ is known, with $\sigma = 8.47$ gpd.

(a) Data on the water demand for larger towns are also available. Based on this information, it is believed that μ will be 120 ± 12 gpd with 95% probability. Determine the distribution of μ based on the combined information.

(b) Additional data on water consumption from other towns of similar size can be obtained at a cost of approximately $2100 per town. Assume that the error in the prediction of μ will introduce an expected loss of (in dollars)

$$L = 2000\sigma_\mu''^2$$

where $\sigma_\mu''^2$ is the posterior variance of μ. Determine the optimal number of additional towns that should be surveyed for data on water consumption. Determine the corresponding expected total loss.

2.22 (a) Determine the optimum sample size in the estimation of the mean value μ if the loss function is

$$L(n, \hat{\mu}, \mu) = c(\mu - \hat{\mu})^2 + b(\mu - \hat{\mu}) + kn$$

where $\hat{\mu}$ is the estimate chosen for μ, k is the unit sample cost, and n is the sample size, under the following conditions:

(i) There is no prior information except that obtained from the sample observations.

(ii) The prior distribution of μ is $N(\mu', \sigma')$.

Assume that the basic random variable is normally distributed whose standard derivation σ is known.

(b) Repeat part (a) under condition (ii) if the loss function is

$$L(n, \hat{\mu}, \mu) = c(\mu - \hat{\mu})^2 + b\mu + kn$$

2.23 A foundation engineer has to determine the design pressure for a rigid foundation in order to perform a trade-off study between construction cost, soil exploration cost, and the penalty for excessive settlement. The compression index X is measured from soil samples during soil testing. Let X be a normal random variable with mean μ and standard deviation σ. The parameter μ depends on the soil encountered, which will be estimated from laboratory measurements; whereas, σ may be assumed to be a known constant. Settlement of the foundation, Y, has a mean and variance given by

$$\mu_Y = bq\mu_\mu$$

$$\sigma_Y^2 = a^2 q^2 \sigma_\mu^2$$

where a and b are foundation dimensions; q is the design pressure; μ_μ and σ_μ are, respectively, the mean and standard deviation of μ. Suppose the total load of the structure is Q (per unit foundation area). Because of excavation to depth d, the net design pressure of the foundation is

$$q = Q - \gamma d$$

where γ is the density of the soil excavated. The following overall cost function is assumed:

$$L = C_1 + C_2 d + C_3 Y^2 + C_4 n + C_5$$

where

C_1 = Cost of construction, excluding foundation cost.

C_2 = Additional cost per meter of excavation depth.

C_3 = Loss per (meter of settlement)^2.

C_4 = Unit cost for sampling and testing.

C_5 = Cost of setting up the testing program.

n = Number of tests for determining compression index.

Suppose a prior distribution of μ is available and is normal $N(\mu', \sigma')$. Decisions are to be based on minimizing the expected overall cost.
  (a) Draw the decision tree including the sampling alternative and the selection of a design pressure.
  (b) Determine q_{opt} for given sample mean $\bar{x}$ and sample size n.
  (c) For the case of no sampling, determine q_{opt} and the corresponding expected overall cost.

2.24 A home-owner is building a house on the outskirts of a town and is considering the alternatives between drilling his own well on a one-acre property and subscribing to the city water. It costs $500 to drill a well and $200 to complete the necessary facilities if water is found in adequate quantity. The probability of striking a water-bearing stratum is estimated to be 0.4 in the vicinity. However, if the first well is dry, it may be assumed that any other well on the property will yield the same result. The city requires an immediate payment of $500 to make a connection. The present worth of water payments to the city is estimated at $1000, whereas, the present worth of operation and maintenance on a successful well is only $300. Answer the following questions, and draw the corresponding decision trees wherever appropriate.
  (a) What should the engineer do if he is an EMV decision maker?
  (b) An ultrasonic device to detect the presence of water can be rented. Although the device will indicate "wet" if there is a water-bearing stratum, it may fail to indicate "dry" in the absence of a water-bearing stratum. In this latter case, the device will read "wet" 30% of the time. How much should the engineer pay to rent the ultrasonic device to obtain additional information on the presence of water beneath his property?
  (c) At most, how much should perfect information be worth (i.e., certain knowledge about whether or not a water-bearing stratum is present)?
  (d) Suppose the probability of striking a water-bearing stratum in the vicinity can only be estimated within the range 0.4 ± 0.1. Will the engineer still insist on his optimal decisions previously obtained in (a)? Justify.

(e) Would the decision in part (a) change if the utility function of money for the home-owner is

$$u(x) = -\exp\left(\frac{-x}{1000}\right); \quad x \le 0$$

where x is the total present monetary value in dollars? Is the utility function risk-aversive or risk-affinitive?

2.25 A contractor may order ready-mix concrete from plant A or plant B at \$820 and \$900, respectively. Although the concrete mix from plant A is cheaper, it is of poorer quality; the chance that it may not pass the strength test is 10%. The probability that concrete mix from plant B will fail the strength test is only 5%. In the event that a batch of concrete fails the strength test, the contractor will lose an additional \$2000, in the form of penalty and delay.
  (a) Draw the decision tree.
  (b) Because of keen competition, plant A offers an option to the contractor. The plant will guarantee that the concrete mix has the required quality, but the cost will be 20% higher.
Plant B is also offering an option for an additional cost of 6% on the price of the concrete. Instead of placing a guarantee on its products, plant B will replace any concrete mix that had been proved to be under-strength. However, because of a delay in the project, the contractor will still lose \$540 (but not \$2000) in addition to the cost of material from plant B under this alternate plan. Draw the new decision tree including these options.
  (c) Determine the best alternative if the decision is based on expected monetary value (EMV).
  (d) If the decision is based on expected utility value (EUV), determine the best alternative for each of the following utility functions.
     (i) $u(d) = 0.2d$ where d = monetary value
     (ii) $u(d) = -0.0001d^2$; for $d < 0$

2.26 The utility function of monetary value for the contractor in Problem 2.9 is given as shown in Fig. P2.26.

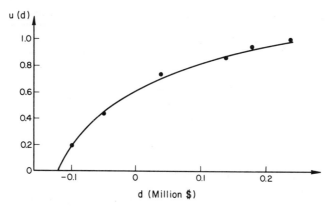

Figure P2.26 Utility function of money.

Determine the best alternative for the contractor on the basis of the maximum expected utility value criterion.

2.27 If the utility function for monetary value is exponential, that is, $u(x) = 1 - e^{-0.5x}$, determine the expected utility of a project whose monetary return is normally distributed with a mean of \$10,000 and c.o.v. of 30%. Observe that x is in units of \$ thousand.

2.28 Two modes of public transportation are available between two cities 150 miles apart. A passenger could either ride the train or take a plane. The travel time for each transportation mode is randomly distributed according to the weather condition as follows:

| | | Travel time (hours) | |
Mode	Weather	Mean	Standard Deviation
Train	Any	3	0.5
Plane	Good	1	0.2
	Bad	6	2.5

During the winter months, the probability of bad weather on a given day of travel is 0.2. Suppose only travel time is considered in the choice of a specific mode of transportation. Furthermore, because of limited seat capacities in both modes of transportation, a passenger would have to decide on his travel plan at least one week ahead.

(a) Should train travel be planned if the decision criterion is to minimize expected travel time?

(b) Suppose the utility function for time is

$$u(t) = -10t^2$$

Which mode should be chosen if one wants to maximize expected utility?

(c) What is the value of perfect information on the weather condition on the day of travel, in terms of time units? Assume the same criterion as in (b).

2.29 A manufacturer is deciding between two alternatives:

A: Devoting the entire firm to the production of a conventional product, whose volume of sale units is described by a normal variate $N(100, 10)$; the profit is $70 for each unit sold.

B: Devoting the entire firm to the production of a new product, whose volume of sale will depend on public acceptance; if the product is well accepted, the manufacturer is 90% confident that he will net a profit of $10,000, with the remaining 10% chance of zero profit. Whereas, if the product does not meet public approval, the manufacturer is certain to make zero profit.

With his five years of experience in the manufacturing business, the manufacturer believes that the probability of the new product meeting public acceptance is 80%.

(a) What should be the decision on the basis of EMV analysis?

(b) Suppose the utility function of money is

$$u(y) = \left(\frac{y}{10,000}\right)^2 \qquad y \geq 0$$

where y is the profit in dollars. What should he decide on the basis of EUV analysis?

(c) Suppose a market research firm is offering to conduct a market survey, guaranteeing a reliable indication of the public reaction to the new product, at a price of $1600. Should the manufacturer obtain such information? Assume EMV analysis.

(d) The manufacturer doubts the validity of the assumption of 80% public acceptance probability for the new product.

  (i) What should be the value of this probability be before he changes his mind about the original decision made in part (a)?

  (ii) What should the value of this probability be before the manufacturer reverses his choice of part (c)?

2.30 The traffic in the existing highway between towns A and B is approaching the full capacity of the highway. The local government is considering alternatives to expand the traffic capacity between the two towns, for the next ten years. The alternatives considered are:

(I)   *Expand the Existing Highway.* The existing highway is a 4-lane road with a capacity of 50 cars per minute. Expanding it to a 6-lane road will provide an additional capacity of 30 cars per minute. The cost required for this expansion is estimated to be \$3 million for construction and \$3 million for displacing and relocating the existing business and buildings adjacent to the existing roadway.

(II) *Construct a New Highway.* A new highway is proposed that will go through a presently sparsely populated area such that negligible cost will be incurred in displacing the existing buildings to make way for the new road. It is expected that the new highway will increase the land value in the adjacent area, which is estimated to bring \$2 million of revenue to the local government. The proposed highway will provide a capacity of 50 cars per minute, at a construction cost of \$10 million.

(III) *Rapid-Transit System.* This alternative is a response to reduce the amount of air pollution caused by autos, which is becoming a problem in this region. The cost of constructing the rapid-transit system is estimated to be \$8 million. In addition, the loss resulting from the displacement of some of the existing structures will be \$1 million.

The local government will subsidize the operation of the transit system at a total cost of \$2 million. As far as the number of potential passengers are concerned, it is expected that 50% of the future traffic demand between the two towns would be handled by the transit system. The desired objectives for each alternative are as follows:

(O_1): *To meet future traffic demand.* This is measured by the probability that an alternative is able to handle the future peak hour traffic demand.

(O_2): *To minimize total cost.* This is measured by the following utility function:

$$u = e^{-0.1x} \qquad x > 0$$

where x is the total cost in \$ million associated with each alternative.

(O_3): *To minimize air pollution.* The possible effects on air pollution from each alternative have been studied. The ratings stating how well each alternative can achieve this objective have been subjectively determined as 0.0, 0.5, and 1.0 for alternatives I, II, III, respectively. In other words, expanding the present roadway will increase the amount of air pollution the most, whereas the transit system will result in the maximum reduction of air pollution.

It is assumed that the peak hour traffic demand in the next ten years is a normal random variable with a mean of 80 cars per minute and a coefficient of variation of 20%.

(a)   Suppose the relative importance of achieving objectives O_1, O_2, and O_3 is 5:2:3. Construct the decision tableau and determine the best alternative using the *weighted objective decision approach.*

(b)   What additional alternatives and other relevant objectives do you think should be included in the decision analysis?

2.31 In a multiattribute utility analysis, suppose the following joint utility function has been established:

$$u(x, y) = 2 + x + xy + x^2 y^2$$

(i)   Determine the expected utility of an alternative in which X and Y are statistically independent normal random variables $N(2, 1)$ and $N(1, 0.5)$, respectively.

(ii) Suppose one of the following improvements can be made to part (i):

I   Increasing μ_X by 50%

II   Increasing μ_Y by 100%

whereas σ_X and σ_Y remain the same as before. Which improvement alternative is better?

2.32 A foundation engineer is deciding between two types of foundation schemes for a proposed building. The first scheme is called *flotation* involving extensive excavation and will cost $200,000 if normal conditions exist. However, there is a 40% probability that an excessive heave and drainage problem may occur; in such events an additional cost of $50,000 will be required. The second alternative is a rigid base foundation at shallow depth, costing $150,000. There is a thick layer of normally consolidated clay beneath the ground surface; consequently, settlement may be encountered with the second alternative. Depending on whether the clay has high or low compressibility, the settlement is expected to be as follows:

Compressibility	Settlement	Probability
High	2 in.	0.4
	4 in.	0.6
Low	2 in.	0.8
	4 in.	0.2

The engineer believes that the clay is equally likely to have high or low compressibility. Assume that the settlement is negligible with the flotation scheme. From a study of the utility of money and settlement, the following joint utility function is assumed to be appropriate:

$$u(d, s) = -0.1d - s^2$$

where

$$d = \text{monetary value in thousand dollars}$$

$$s = \text{settlement in inches}$$

(a) Draw the decision tree and determine the best foundation scheme with the present information.

(b) Suppose a test exists that will yield perfect information about the compressibility of the clay at the site. At most how much should be invested for such informations?

(c) Suppose there is an indirect test that will furnish the engineer with additional information on the compressibility of the clay stratum. If the compressibility is actually high, the probability that the test will indicate high compressibility is 90%; whereas if the actual compressibility is low, this probability is only 20%. Draw the decision tree for this alternative. Assume that the indirect test will indicate either high or low compressibility only. Should the engineer proceed with the indirect test if it costs $5000?

2.33 The desirability of a proposed design is evaluated in terms of both cost x (in units of million dollars) and reliability r. Suppose the following marginal utility functions are given:

$$u(r) = 4r^2; \qquad 0.5 \leq r \leq 1$$

$$u(x) = e^{-x}; \qquad 1 \leq x \leq 2$$

where the above ranges of r and x cover the possible values of r and x associated with all the designs proposed. Moreover, the decision maker is found to be indifferent in the following pairs of lotteries where (1, 2) denotes the event that the reliability of a design is 1 and the design costs $2 million. Assume that r and x are statistically independent; also assume preferential- and utility-independence.

(a) Determine the joint utility function $u(r, x)$.

(b) Suppose design A has a reliability of 0.99 and its cost is uniformly distributed between 1.5 and 2.0 million dollars. Determine its expected utility.

(c) Suppose the reliability of another design B has a mean of 0.8 and a c.o.v. of 0.1, whereas its cost is uniformly distributed between 1.0 and 1.5 million dollars. Determine its expected utility. Which design is better?

(d) What should the reliability of design A be before the decision maker is indifferent between designs A and B?

2.34 Consider a contractor evaluating various new business markets; the attributes influencing his decisions include the markup (x_1) in percent return, contract size (x_2) in millions of dollars, and the regulatory environment (x_3) that is defined by a subjective scale. Suppose the marginal utility functions of each of these attributes have been established (see Ibbs and Crandall, 1982) as follows:

$$u_1(x_1) = 0.8 \ln x_1 - 1.11; \qquad 4 \le x_1 \le 14$$
$$u_2(x_2) = x_2 - 0.1x_2^2 - 1.48; \quad 1.8 \le x_2 \le 4.55$$
$$u_3(x_3) = e^{-0.4x_3}; \qquad\qquad 0 \le x_3$$

and the joint utility function for the contractor is given as

$$
\begin{aligned}
u(x_1, x_2, x_3) = {} & 0.35u_1(x_1) + 0.25u_2(x_2) \\
& + 0.25u_3(x_3) + 0.0525u_1(x_1)u_2(x_2) \\
& + 0.0525u_1(x_1)u_3(x_3) + 0.0375u_2(x_2)u_3(x_3) \\
& + 0.007u_1(x_1)u_2(x_2)u_3(x_3)
\end{aligned}
$$

(a) For a construction project whose attributes have the following statistical information:

Attribute	Mean	Standard Deviation
$\ln X_1$	1.66	0.4
X_2	3.1	0.3
X_3	2.25	1.0

and X_3 follows a gamma distribution, determine the expected utility of the project.

(b) Suppose two other alternatives, namely A_2 and A_3, are available to the contractor. For A_2, the mean and standard deviation of $\ln X_1$ are 1.66 and 0.2, the statistics of X_2 and X_3 are the same as those for A_1; whereas for A_3, the mean and standard deviation of $\ln X_1$ are 2.17 and 0.23, the mean of X_3 is 4.0, and all other statistics of X_2 and X_3 are the same as those for the previous alternative A_1. Determine which among the three alternatives is optimal.

3. Markov, Queueing, and Availability Models

3.1 THE MARKOV CHAIN

3.1.1 Introduction

The state of a system invariably changes with respect to some parameter, for example, time or space. The transition from one state to another as a function of the parameter, or its corresponding *transition probability*, may generally depend on the prior states. However, if the transition probability depends only on the current state, the process of change may be modeled with the *Markov process*. If the state space is a countable or finite set, the process is called a *Markov chain*. Moreover, if change can occur only at discrete points of the parameter, for example, at discrete instants of time, the process is a *discrete parameter Markov chain*. If the transition probability is independent of the state of the system, the process reduces to the Poisson process (see Chapter 3, Vol. I).

As an example, after an earthquake a structure may be damaged to various degrees. If the damages are not repaired, the structure could be subject to even further damage during the next earthquake. It is obvious that a structure is more likely to sustain heavy damage in the next earthquake if it had prior damage from earlier earthquakes. Conceptually, the damage may be classified into several *damage states*, such as no damage, light damage, moderate damage, heavy damage, and total collapse. The structure either remains in the same state or moves to another state after each earthquake according to the transitional probabilities between states. These probabilities, therefore, will determine the likelihood of the structure being in a particular damage state after a number of earthquake occurrences or at some future time. Of interest here is perhaps the average number of earthquake occurrences that will bring a structure to collapse, or the average time a structure could remain undamaged from earthquakes.

3.1.2 The Basic Model

Consider a system with m possible states, namely $1, 2, \ldots, m$, and changes in state can occur only at discretized values of the parameter; for example, at times $t_1, t_2, \ldots, t_n$. Let X_{n+1} denote the state of the system at t_{n+1}. In general, the probability of a future state of the system may depend on its entire history; that is, its conditional probability is

$$P(X_{n+1} = i \mid X_0 = x_0, X_1 = x_1, \ldots, X_n = x_n) \tag{3.1}$$

where $X_0 = x_0, \ldots, X_n = x_n$ represent all previous states of the system. If the future state is governed solely by the present state of the system, that is, the conditional probability, Eq. 3.1, is

$$P(X_{n+1} = i | X_0 = x_0, \ldots, X_n = x_n) = P(X_{n+1} = i | X_n = x_n) \qquad (3.2)$$

then the process is a *Markov chain*. For a discrete parameter Markov chain, the transitional probability from state i at time t_m to state j at time t_n may be denoted by

$$p_{i,j}(m, n) = P(X_n = j | X_m = i); \qquad n > m \qquad (3.3)$$

The Markov chain is *homogeneous* if $p_{i,j}(m, n)$ depends only on the difference $t_n - t_m$; in this case, we define

$$p_{i,j}(k) = P(X_k = j | X_0 = i) = P(X_{k+s} = j | X_s = i) \qquad s \geq 0$$

as the k-step transition probability function. Physically, this represents the conditional probability that a homogeneous Markov chain will go from state i to state j after k time stages. This probability can be determined from the one-step transition probabilities, namely $p_{i,j}(1)$ or simply $p_{i,j}$, between all pairs of states in the system. These transition probabilities can be summarized in a matrix for a system with m states, called the *transition probability matrix*

$$\mathbf{P} = \begin{bmatrix} p_{1,1} & p_{1,2} & \cdots & p_{1,m} \\ p_{2,1} & p_{2,2} & \cdots & p_{2,m} \\ \vdots & \vdots & & \vdots \\ p_{m,1} & p_{m,2} & \cdots & p_{m,m} \end{bmatrix} \qquad (3.4)$$

As the states of a system are mutually exclusive and collectively exhaustive after each transition, the probabilities in each row add up to 1.0. For a homogeneous discrete Markov chain, the probabilities of the initial states are the only other information needed to define the model behavior at any future time.

3.1.3 State Probabilities

The probabilities of the respective initial states of a system may be denoted by a row matrix

$$\mathbf{P}(0) = [p_1(0), p_2(0) \ldots p_m(0)]$$

where $p_i(0)$ is the probability that the system is initially at state i. In the special case for which the initial state of the system is known, for example, at state i, then $p_i(0) = 1.0$ and all other elements in the row matrix $\mathbf{P}(0)$ are zero. After one transition, the probability that the system is in state j is given by the theorem of total probability as

$$p_j(1) = P(X_1 = j) = \sum_i P(X_0 = i) P(X_1 = j | X_0 = i)$$

Hence,

$$p_j(1) = \sum_i p_i(0) p_{i,j} \qquad (3.5)$$

In matrix notation, the single stage state probabilities become

$$\mathbf{P}(1) = \mathbf{P}(0)\mathbf{P} \tag{3.6}$$

which is also a row matrix.

Similarly, the probability that the system is in state j after two transitions is given by

$$p_j(2) = \sum_k P(X_1 = k)P(X_2 = j|X_1 = k)$$

$$= \sum_k p_k(1)p_{k,j} \tag{3.7}$$

or in matrix notation

$$\mathbf{P}(2) = \mathbf{P}(1)\mathbf{P} = \mathbf{P}(0)\mathbf{P}\mathbf{P} = \mathbf{P}(0)\mathbf{P}^2 \tag{3.8}$$

Therefore, by induction, it can be shown that the n-stage state probability matrix is given by

$$\mathbf{P}(n) = \mathbf{P}(n-1)\mathbf{P} = \mathbf{P}(n-2)\mathbf{P}\mathbf{P} = \cdots = \mathbf{P}(0)\mathbf{P}^n \tag{3.9}$$

Chapman–Kolmogorov Equations   The state probabilities after n stages of transition can be determined alternatively by considering the fact that when the process goes from state i to state j in n time stages, the process must be in some intermediate state k after exactly r stages ($r < n$). Hence, on the basis of the theorem of total probability, the transition from state i to state j in n stages is given by:

$$p_{i,j}(n) = P(X_n = j|X_0 = i)$$

$$= \sum_k P(X_n = j|X_0 = i, X_r = k)P(X_r = k|X_0 = i)$$

$$= \sum_k p_{k,j}(n-r)p_{i,k}(r); \quad 0 < r < n \tag{3.10}$$

Equation 3.10 is known as the *Chapman–Kolmogorov Equation*. It is valid between any two states i and j, and any $r < n$. In the case in which $n = 2$, and $r = 1$, Eq. 3.10 becomes

$$p_{i,j}(2) = \sum_k p_{k,j}(1)p_{i,k}(1)$$

$$= \sum_k p_{i,k}p_{k,j} \tag{3.11}$$

which is the product of the ith row by the jth column of the one-stage transition probability matrix.

EXAMPLE 3.1

The transition of wet and dry days in a town is modeled as a homogeneous Markov chain with the following transition probability matrix

$$
\begin{array}{cc}
 & \begin{array}{cc} \text{Dry} & \text{Wet} \end{array} \\
\begin{array}{c} \text{Dry} \\ \text{Wet} \end{array} & \begin{bmatrix} 0.8 & 0.2 \\ 0.5 & 0.5 \end{bmatrix}
\end{array}
$$

For example, the probability of going from a dry day to a wet day is 0.2. Let the dry and wet days be denoted as states 1 and 2, respectively.

(a)   If it is dry today, the probability that it will be dry 2 days from now may be computed by first determining $\mathbf{P}(2)$ using Eq. 3.8.

$$\mathbf{P}(2) = \mathbf{P}(0)\mathbf{P}^2$$

$$= [1 \quad 0]\begin{bmatrix} 0.8 & 0.2 \\ 0.5 & 0.5 \end{bmatrix}\begin{bmatrix} 0.8 & 0.2 \\ 0.5 & 0.5 \end{bmatrix}$$

$$= [0.8 \quad 0.2]\begin{bmatrix} 0.8 & 0.2 \\ 0.5 & 0.5 \end{bmatrix}$$

$$= [0.74 \quad 0.26]$$

Hence, it is 74% probable that the day after tomorrow will be dry. Alternatively, Eq. 3.11 may be applied to yield

$$p_{1,1}(2) = \sum_k p_{1,k}p_{k,1}$$
$$= p_{1,1}p_{1,1} + p_{1,2}p_{2,1}$$
$$= 0.8 \times 0.8 + 0.2 \times 0.5 = 0.74$$

Similarly for five days from now,

$$\mathbf{P}(5) = \mathbf{P}(0)\mathbf{P}^5$$

$$= [1.0 \quad 0]\begin{bmatrix} 0.8 & 0.2 \\ 0.5 & 0.5 \end{bmatrix}^5$$

$$= [0.715 \quad 0.285]$$

Hence, the probability of dry weather 5 days from now is 0.715.

(b)   However, if today is wet instead of dry, then the corresponding probabilities are:

$$\mathbf{P}(2) = [0 \quad 1]\begin{bmatrix} 0.8 & 0.2 \\ 0.5 & 0.5 \end{bmatrix}^2$$

$$= [0.65 \quad 0.35]$$

$$\mathbf{P}(5) = [0 \quad 1]\begin{bmatrix} 0.8 & 0.2 \\ 0.5 & 0.5 \end{bmatrix}^5$$

$$= [0.713 \quad 0.287]$$

In this case the probability that 2 days from now it will be dry is less than that in part (a); however, the probability that it will be dry 5 days from now is practically the same as that in part (a). Table E3.1 summarizes the state probabilities at subsequent time stages. Row 1 lists

Table E3.1   Successive State Probabilities with Two Different Initial States

n	1	2	3	4	5	6	7	...
$p_{1,1}(n)$	0.8	0.74	0.722	0.717	0.715	0.714	0.714	...
$p_{2,1}(n)$	0.5	0.65	0.695	0.709	0.713	0.714	0.714	...

the results starting with a dry day, whereas row 2 indicates the results starting with a wet day. Observe that when the number of stages becomes large, the state probabilities appear to become independent of the initial states.

EXAMPLE 3.2

The amount of stored water in a reservoir may be idealized into three states, namely: full, half-full, and empty. Because of the probabilistic nature of the inflowing water to the reservoir as well as the outflow from the reservoir to meet uncertain demands for water, the amount of water storage may shift from one state to another between seasons. Suppose the transition probabilities from one state to another are as indicated in Fig. E3.2. Denote empty, half-full,

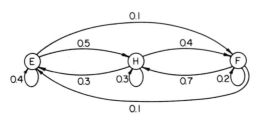

Figure E3.2   Transition probabilities between states of reservoir storage.

and full storage with the states 1, 2, and 3, respectively. The corresponding transition probability matrix is as follows:

$$\mathbf{P} = \begin{bmatrix} 0.4 & 0.5 & 0.1 \\ 0.3 & 0.3 & 0.4 \\ 0.1 & 0.7 & 0.2 \end{bmatrix}$$

Suppose two weeks before the end of a season, it is predicted that there is 80% probability the reservoir will be full at the beginning of the next season; if it is not full, the reservoir will equally likely be empty or half-full. What is the probability that the reservoir will be full at the end of the next season?

The initial state probabilities are: $\mathbf{P}(0) = [0.1 \quad 0.1 \quad 0.8]$; therefore, after 1 season, the state probabilities will become

$$\mathbf{P}(1) = [0.1 \quad 0.1 \quad 0.8] \begin{bmatrix} 0.4 & 0.5 & 0.1 \\ 0.3 & 0.3 & 0.4 \\ 0.1 & 0.7 & 0.2 \end{bmatrix}$$

$$= [0.15 \quad 0.64 \quad 0.21]$$

The probability that the reservoir will be full at the end of the following season is only 21%.

EXAMPLE 3.3

The record of hurricane occurrences at Mustang Island, Texas (1818–1970) is shown in Fig. E3.3.

A homogeneous Markov chain is assumed to describe the number of hurricane occurrences in a year. Suppose states 0, 1, 2 denote, respectively, the occurrence of 0, 1, and 2 hurricanes in a year.

(a)   Determine the transition probability matrix.

In estimating the transitional probabilities, we have to examine all pairs of consecutive years. From the record of observations, there are 152 pairs of consecutive years, out of which 122 are preceded by years with no hurricane. Among these 122 pairs, 97 were followed by another

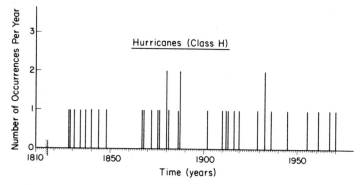

Figure E3.3 Hurricane occurrences on Mustang Island, Texas (1818–1970) (from Russell and Schuëller, 1971).

year of no hurricane; 23 were followed by a year with one hurricane and the remaining 2 by a year with two hurricanes. On the basis of these statistics, the following probabilities may be estimated:

$$p_{0,0} = \tfrac{97}{122} = 0.7951; \quad p_{0,1} = \tfrac{23}{122} = 0.1885; \quad p_{0,2} = \tfrac{2}{122} = 0.0164$$

Similarly,

$$p_{1,0} = \tfrac{22}{27} = 0.8148; \quad p_{1,1} = \tfrac{4}{27} = 0.1482; \quad p_{1,2} = \tfrac{1}{27} = 0.0370$$

$$p_{2,0} = \tfrac{2}{3} = 0.6667; \quad p_{2,1} = \tfrac{1}{3} = 0.3333; \quad p_{2,2} = \tfrac{0}{3} = 0.0000$$

Hence, the single-stage transition probability matrix is

$$\mathbf{P} = \begin{bmatrix} 0.7951 & 0.1885 & 0.0164 \\ 0.8148 & 0.1482 & 0.0370 \\ 0.6667 & 0.3333 & 0.0000 \end{bmatrix}$$

(b)   If the above model was used to predict hurricane frequencies in the beginning of 1971, what is the probability that there will be one hurricane in 1973 on Mustang Island?

The state probabilities after 3 transitions are required. With the initial state as 1 (i.e., there was 1 hurricane in 1970), the state probability matrix after 3 years is

$$\mathbf{P}(3) = \mathbf{P}(0)\mathbf{P}^3$$

$$= \begin{bmatrix} 0 & 1 & 0 \end{bmatrix} \begin{bmatrix} 0.7951 & 0.1885 & 0.0164 \\ 0.8148 & 0.1482 & 0.0370 \\ 0.6667 & 0.3333 & 0.0000 \end{bmatrix}^3$$

$$= \begin{bmatrix} 0.7964 & 0.1836 & 0.0200 \end{bmatrix}$$

Hence, the probability of one hurricane in 1973 is about 18%.

(c)   Suppose the engineer in (b) would also like to estimate the probability that there will be hurricanes in 1971, 1972, and 1973, but not in 1974. Since both states 1 and 2 imply occurrences of hurricanes, considerable simplification in computation can be achieved in this case by combining these two states. Redefine the year without a hurricane as state 0 and the year with hurricane(s) as state 1. Hence, the transition probabilities are

$$p_{0,0} = \tfrac{97}{122} = 0.7951 \qquad p_{0,1} = \tfrac{25}{122} = 0.2049$$

$$p_{1,0} = \tfrac{24}{30} = 0.8000 \qquad p_{1,1} = \tfrac{6}{30} = 0.2000$$

The probability that each of the years 1971, 1972, and 1973 would be subjected to at least one hurricane (but not in 1974) at Mustang Island, is

$$p = p_{1,1}^3 p_{1,0}$$
$$= (0.2)^3(0.8)$$
$$= 0.0064$$

3.1.4 Steady State Probabilities

In Example 3.1, we saw that the state probabilities starting with two different initial states approach one another as the number of transition stages increases. In fact, for a class of Markov chain possessing the *ergodic property*, the state probabilities will converge to a set of *steady-state probabilities* p^*, which are independent of the initial states. Therefore, at steady-state condition,

$$\mathbf{P}(n + 1) = \mathbf{P}(n) = \mathbf{P}^* \tag{3.12}$$

Also from Eq. 3.9,

$$\mathbf{P}(n + 1) = \mathbf{P}(n)\mathbf{P} \tag{3.13}$$

Hence,

$$\mathbf{P}^* = \mathbf{P}^*\mathbf{P}$$

For a Markov chain with m states, this matrix equation represents a set of simultaneous equations as follows:

$$[p_1^* \quad \cdots \quad p_m^*] = [p_1^* \quad \cdots \quad p_m^*] \begin{bmatrix} p_{1,1} & \cdots & p_{1,m} \\ \vdots & & \vdots \\ p_{m,1} & \cdots & p_{m,m} \end{bmatrix}$$

or

$$\begin{aligned} p_1^* &= p_{1,1}p_1^* + p_{2,1}p_2^* + \cdots + p_{m,1}p_m^* \\ &\vdots \\ p_m^* &= p_{1,m}p_1^* + p_{2,m}p_2^* + \cdots + p_{m,m}p_m^* \end{aligned} \tag{3.14}$$

where p_i^* is the probability in state i when a steady-state condition is achieved. Adding both sides of Eq. 3.14, we obtain the identity

$$p_1^* + \cdots + p_m^* = p_1^* + \cdots + p_m^*$$

Hence, Eq. 3.14 contains one degree of freedom. The required constraint to obtain $p_1^*, p_2^*, \ldots, p_m^*$ is

$$p_1^* + p_2^* + \cdots + p_m^* = 1.0 \tag{3.15}$$

EXAMPLE 3.4

For the town in Example 3.1, what is the probability that the day one year from now will be dry?

According to Eq. 3.9, the probability matrix after 365 transitions is

$$\mathbf{P}(365) = \mathbf{P}(0)\mathbf{P}^{365}$$

This obviously requires considerable computations. Alternatively, it is reasonable to assume that the state probabilities will have reached steady-state. Thus, it is sufficient to compute $\mathbf{P}^*$ instead. From the given transition probability matrix, the first equation in Eq. 3.14 yields

$$p_1^* = 0.8p_1^* + 0.5p_2^*$$

whereas Eq. 3.15 gives

$$p_1^* + p_2^* = 1.0$$

The solution of these two simultaneous equations yields

$$p_1^* = 0.714$$

$$p_2^* = 0.286$$

Therefore, the probability that it will be a dry day one year from today is 0.714. This result may be observed also in Table E3.1, which shows that the state probabilities have reached steady-state by the sixth day.

EXAMPLE 3.5

The bus in a mass-transit system is operating on a continuous route with intermediate stops. Suppose the arrival of a bus at an intermediate stop is classified into three states, namely:

1—Early arrival.

2—On-time arrival.

3—Late arrival.

The state of arrival at an intermediate stop may be assumed to depend on the state of arrival at the previous stop, governed by the transitional probability matrix as follows:

$$\mathbf{P} = \begin{bmatrix} 0.5 & 0.4 & 0.1 \\ 0.1 & 0.6 & 0.3 \\ 0.1 & 0.2 & 0.7 \end{bmatrix}$$

Over a long period of operation, what fraction of the intermediate stops can be expected to be late?

From the solution to the following set of simultaneous equations

$$p_1^* = 0.5p_1^* + 0.1p_2^* + 0.1p_3^*$$

$$p_2^* = 0.4p_1^* + 0.6p_2^* + 0.2p_3^*$$

$$p_1^* + p_2^* + p_3^* = 1.0$$

we obtain the steady-state probabilities $p_1^* = 0.167$, $p_2^* = 0.389$, and $p_3^* = 0.444$. Therefore, we can expect the bus to be late at 44.4% of its intermediate stops.

EXAMPLE 3.6

Hundreds of prestressed tendons are evenly spaced along a circular circumferential concrete wall of a nuclear reactor containment structure. After a period of operation, these tendons may become corroded and/or subject to loss of prestress. The occurrences of these phenomena between adjacent tendons may not be independent. An inspection of a sequence of 10 adjacent tendons in a structure revealed the following:

Tendon	Corroded	Loss of Prestress
1	no	no
2	no	no
3	yes	no
4	yes	yes
5	no	yes
6	no	no
7	no	no
8	no	no
9	no	yes
10	no	yes

Assume a Markov model for occurrences of corroded tendons and also tendons with loss of prestress. What percentage of the tendons are expected to be corroded in this structure?

The transition probability matrix that governs the occurrence of corroded tendons is first determined from the given information. Observe that out of 7 tendons (1, 2, 5, 6, 7, 8, and 9) that are not corroded, 6 of them are followed by another noncorroded tendon; tendon number 2 is the only one out of the seven noncorroded tendons that is not followed by another noncorroded tendon. Suppose states 1 and 2 denote noncorroded and corroded tendons, respectively. The transitional probabilities $p_{1,1}$ and $p_{1,2}$ are estimated to be $\frac{6}{7}$ and $\frac{1}{7}$, respectively. Similarly, for the two corroded tendons (3 and 4), only one is followed by a corroded tendon. Hence, $p_{2,1}$ and $p_{2,2}$ are both $\frac{1}{2}$. The transitional probability matrix is

$$\mathbf{P} = \begin{bmatrix} \frac{6}{7} & \frac{1}{7} \\ \frac{1}{2} & \frac{1}{2} \end{bmatrix}$$

The steady-state probabilities are then evaluated from the following simultaneous equations:

$$p_1^* = \tfrac{6}{7}p_1^* + \tfrac{1}{2}p_2^*$$

$$p_1^* + p_2^* = 1$$

yielding $p_2^* = 0.222$. In other words, 22% of the tendons are expected to be corroded in the structure.

Alternatively, it appears that the probability of finding a corroded tendon could have been estimated simply by the fact that only two out of the ten tendons observed were corroded. This would yield 0.2 as the fraction of corroded tendons in the structure; this latter result, of course, is based on the assumption of random sampling. In a case in which dependent behavior is exhibited among adjacent tendons, observing ten tendons in a sequence would not constitute a set of independent sample selected from the entire population of tendons in the structure. Therefore, estimates based on the Markov model should give better results.

EXAMPLE 3.7

Suppose that sand and clay layers are the two basic soil materials that constitute the stratum at a given site. Assume that the layers may be idealized as horizontal. Consider a 1-foot layer as the unit distance of transition, that is, within each 1-foot layer (e.g., 0–1′ or 2–3′, etc.), it is either all sand or all clay. An example of a soil profile is shown in Fig. E3.7. Suppose that a homogeneous Markov chain could be used to model the transition between clay and sand

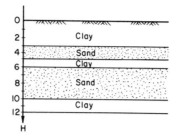

Figure E3.7   A soil profile.

for each foot interval over a depth of 20 feet, and the transition probability matrix has been estimated based on boring logs as follows:

$$\begin{array}{c}\text{Clay}\quad\text{Sand}\\ \begin{array}{c}\text{Clay}\\ \text{Sand}\end{array}\begin{bmatrix}0.8 & 0.2\\ 0.5 & 0.5\end{bmatrix}\end{array}$$

(a)   If the surface layer (0–1′) is clay,
(i)   What is the probability that sand is found in $2 < H < 3$? in $4 < H < 5$?
(ii)   Suppose that sand is found in $2 < H < 3$. What is the probability that sand is also found in $4 < H < 5$?
(iii)   Suppose a layer of at least 2 feet of sand is required for proper foundation support within the first 4 feet of the soil stratum. What is the probability that such conditions will be met for the site?
(iv)   What is the probability that $18 < H < 20$ consists of sand only?

(b)   What is the expected cumulative thickness of sand over a 20-foot stratum?

Solution

(a)(i)   The probability that sand is found in $\{2 < H < 3\}$ can be determined from the 2-stage probability vector, with the initial state being clay.

$$\mathbf{P}(2) = \begin{bmatrix}1 & 0\end{bmatrix}\begin{bmatrix}0.8 & 0.2\\ 0.5 & 0.5\end{bmatrix}^2 = \begin{bmatrix}0.74 & 0.26\end{bmatrix}$$

Hence,

$$P(S_{2-3}) = 0.26$$

Similarly,

$$\mathbf{P}(4) = \begin{bmatrix}1 & 0\end{bmatrix}\begin{bmatrix}0.8 & 0.2\\ 0.5 & 0.5\end{bmatrix}^4 = \begin{bmatrix}0.72 & 0.28\end{bmatrix}$$

Hence,

$$P(S_{4-5}) = 0.28$$

(ii)   If sand is found in $(2 < H < 3)$, this information supersedes that for $(0 < H < 1)$; the initial state will be sand and the probability that sand is found in $(4 < H < 5)$ is determined from the 2-stage probability vector.

$$\mathbf{P}(2) = \begin{bmatrix}0 & 1\end{bmatrix}\begin{bmatrix}0.8 & 0.2\\ 0.5 & 0.5\end{bmatrix}^2 = \begin{bmatrix}0.65 & 0.35\end{bmatrix}$$

Hence,

$$P(S_{4-5}) = 0.35$$

(iii)   The combinations of stages for the first four feet of soil that will provide at least 2 feet of sand may be identified as follows:

$$C - S - S - S/C$$
$$C - C - S - S$$

where C and S denote clay and sand, respectively. Hence, the probability of proper foundation support is

$$P(C - S - S) + P(C - C - S - S) = p_{cs}p_{ss} + p_{cc}p_{cs}p_{ss}$$
$$= 0.2 \times 0.5 + 0.8 \times 0.2 \times 0.5$$
$$= 0.18$$

(iv)   Since the depths of $(18 < H < 19)$ are far from the surface, the states may be assumed to be independent of the initial state; the probabilities, therefore, are expected to approach the steady-state probabilities. The solution to the following simultaneous equations

$$0.8p_C^* + 0.5p_S^* = p_C^*   \quad \text{and} \quad   p_C^* + p_S^* = 1$$

yields $p_S^* = \frac{2}{7}$. Hence,

$$P(S_{18-20}) = p_S^* p_{ss} = \frac{2}{7} \times 0.5 = \frac{1}{7}$$

(b)   Since the steady-state probability of finding sand in a 1-foot layer is $\frac{2}{7}$ from (iv) above, the fraction of sand over a stratum is expected to also be $\frac{2}{7}$. The expected cumulative thickness of sand over a 20-foot stratum is thus $\frac{2}{7} \times 20 = 5.7$ feet.

3.1.5   First Passage Probabilities

In addition to the n-stage and steady-state probabilities, another quantity of interest associated with a Markov chain is the number of transitions before a system goes from state i to state j for the first time. For instance, in Example 3.7 the foundation engineer may be interested in knowing if sand would be encountered within the first ten feet of pile driven through the stratum, whereas a contractor may be interested in the expected length of a wet spell that could seriously hamper the progress of a construction job. The probabilities of these events may be computed from the *first passage probabilities*. Define $f_{i,j}(n)$ as the probability that a system starting from state i will be in state j *for the first time* after n transitions.

Consider first a two-state Markov chain (with states 1 and 2) as shown in Fig. 3.1 with the corresponding transition probabilities. The first passage probabilities may be evaluated as follows:

$$f_{1,1}(1) = p_{1,1}$$
$$f_{1,1}(n) = p_{1,2}\,p_{2,2}^{n-2}\,p_{2,1}   \qquad n \geq 2$$
$$f_{1,2}(n) = p_{1,1}^{n-1}\,p_{1,2}$$
$$f_{2,1}(n) = p_{2,2}^{n-1}\,p_{2,1}   \tag{3.16}$$
$$f_{2,2}(1) = p_{2,2}$$
$$f_{2,2}(n) = p_{2,1}\,p_{1,1}^{n-2}\,p_{1,2}   \qquad n \geq 2$$

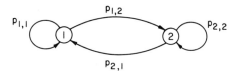

Figure 3.1   A two-state Markov chain.

For example, the event that the system starting at state 1 will return to state 1 for the first time after n transitions (for $n \geq 2$) requires that the system starting at state 1 switches to state 2 in the first transition, remains in state 2 during the next $n - 2$ transitions, and then switches back to state 1 at the last transition.

When more than two states are involved in a system, the direct determination of the first passage probabilities $f_{i,j}(n)$ using the procedure illustrated above could become unmanageable. Alternatively, the first passage probabilities may be evaluated from the n-stage probabilities through a step-by-step procedure as follows:

$$f_{i,j}(1) = p_{i,j}(1) = p_{i,j}$$

$$f_{i,j}(2) = p_{i,j}(2) - f_{i,j}(1)p_{j,j}$$

$$f_{i,j}(3) = p_{i,j}(3) - f_{i,j}(1)p_{j,j}(2) - f_{i,j}(2)p_{j,j} \qquad (3.17)$$

$$\vdots$$

$$f_{i,j}(n) = p_{i,j}(n) - f_{i,j}(1)p_{j,j}(n-1) - \cdots - f_{i,j}(n-1)p_{j,j}$$

It is obvious that the first passage probability in one transition ($n = 1$) is the same as the one-step transitional probability. For $n = 2$, the two-stage transitional probability $p_{i,j}(2)$ includes the probability of visiting state j immediately after the first transition and remaining in state j during the second transition. Hence, the probability of this event should be subtracted from $p_{i,j}(2)$ to obtain the first passage probability in exactly two transitions. The rest of Eq. 3.17 is also based on similar reasoning. Through a recursive procedure and applications of Eq. 3.17 the first passage probabilities between any two states and at any stage can be computed accordingly.

Once these first passage probabilities have been determined, they may be used to evaluate various probabilities and quantities of interest, including the following:

(1) The probability that a system, initially at state i, will visit state j (at least once) within m transitions is

$$q_{i,j}(m) = \sum_{n=1}^{m} f_{i,j}(n) \qquad (3.18)$$

The event of interest is the same as the event that the first passage from i to j occurs within m transitions.

(2) The probability that a system starting at state i will eventually reach state j is thus given by

$$q_{i,j}(\infty) = \sum_{n=1}^{\infty} f_{i,j}(n) = f_{i,j} \qquad (3.19)$$

(3)   The probability that a system starting in state i will eventually return to state i is

$$q_{i,i}(\infty) = \sum_{n=1}^{\infty} f_{i,i}(n) = f_{i,i} \qquad (3.20)$$

(4)   The mean recurrence time for state i is the weighted average period from the time the system leaves state i until it subsequently returns to state i for the first time. Hence,

$$t_i = E(\text{recurrence time}) = \sum_{n=1}^{\infty} n f_{i,i}(n) \qquad (3.21)$$

Furthermore, the expected length of occupation in state i is the weighted average period until the system first leaves state i. Hence the expected occupation time is

$$l_i = \sum_{n=1}^{\infty} n p_{i,i}^{n-1}(1 - p_{i,i}) = \frac{1}{1 - p_{i,i}} \qquad (3.22)$$

For illustrations, consider the two-state Markov chain in Fig. 3.1. Applying Eqs. 3.18 and 3.16, we obtain

$$q_{1,1}(m) = \sum_{n=1}^{m} f_{1,1}(n)$$

$$= p_{1,1} + \sum_{n=2}^{m} p_{1,2} p_{2,2}^{m-2} p_{2,1}$$

$$= p_{1,1} + p_{1,2} p_{2,1} \frac{1 - p_{2,2}^{m-1}}{1 - p_{2,2}}$$

Since $1 - p_{2,2} = p_{2,1}$, and $p_{1,1} + p_{1,2} = 1$, it can be shown that

$$q_{1,1}(m) = 1 - p_{1,2} p_{2,2}^{m-1} \qquad (3.23)$$

Similarly,

$$q_{1,2}(m) = \sum_{n=1}^{m} f_{1,2}(n)$$

$$= \sum_{n=1}^{m} p_{1,1}^{n-1} p_{1,2}$$

$$= p_{1,2} \frac{1 - p_{1,1}^{m}}{1 - p_{1,1}}$$

$$= 1 - p_{1,1}^{m} \qquad (3.24)$$

$$q_{2,1}(m) = 1 - p_{2,2}^{m} \qquad (3.25)$$

$$q_{2,2}(m) = 1 - p_{2,1} p_{1,1}^{m-1} \qquad (3.26)$$

For the average occupation time, Eq. 3.22 yields

$$l_1 = \frac{1}{1 - p_{1,1}} = \frac{1}{p_{1,2}} \tag{3.27}$$

Similarly,

$$l_2 = \frac{1}{p_{2,1}} \tag{3.28}$$

From Eq. 3.21, the average recurrence time

$$t_1 = \sum_{n=1}^{\infty} n f_{1,1}(n)$$

$$= p_{1,1} + \sum_{n=2}^{\infty} n p_{1,2} p_{2,2}^{n-2} p_{2,1}$$

$$= p_{1,1} + p_{1,2} p_{2,1} \left[\sum_{i=1}^{\infty} (i+1) p_{2,2}^{i-1} \right]$$

$$= p_{1,1} + p_{1,2} p_{2,1} \left[\frac{1}{(1 - p_{2,2})^2} + \frac{1}{1 - p_{2,2}} \right]$$

$$= 1 + \frac{p_{1,2}}{p_{2,1}} \tag{3.29}$$

Similarly,

$$t_2 = 1 + \frac{p_{2,1}}{p_{1,2}} \tag{3.30}$$

3.1.6 Recurrent, Transient, and Absorbing States

If $f_{i,i}$ in Eq. 3.20 is equal to unity, state i is known as a *recurrent state*; that is, a system starting at state i will surely return to state i eventually. On the other hand, if $f_{i,i} < 1$, it implies that a system starting at state i will have a finite probability of never returning; in such a case, state i is then called a *transient (or nonrecurrent) state*. In the case of a two-state Markov chain, $f_{1,1}$ and $f_{2,2}$ can be determined by substituting $m = \infty$ in Eqs. 3.23 and 3.26. If the one-step transitional probabilities, namely $p_{i,j} < 1.0$ $(i = 1, 2; j = 1, 2)$, then $f_{1,1} = f_{2,2} = 1.0$, indicating that states 1 and 2 are recurrent states. On the other hand, if $p_{1,1}$ is 1.0 and $p_{2,1} > 0$, then $f_{2,2} = q_{2,2}(\infty) = 1 - p_{2,1} p_{1,1}^{\infty} = 1 - p_{2,1} < 1.0$, indicating that state 2 may never be reached again. State 2, therefore, is an example of a transient state.

When there are many states in a Markov chain, the calculation of the first passage probabilities using Eq. 3.17 could be cumbersome. If all the states in the chain are recurrent, the mean first passage time from one state to another, or the mean recurrence time of any state, may be computed without these first passage probabilities; such alternative methods are presented later in this section.

EXAMPLE 3.8

For the probability matrix given in Example 3.1, determine the following:

(a) Suppose it is wet today. What is the probability that it will be wet again for the first time the day after tomorrow?

(b) Suppose it is wet today. What is the probability that there will be a dry day within the next five days?

(c) What is the expected length of a dry spell? A wet spell?

(d) If it is wet today, what is the expected length of time (in days) until the next wet day?

Solution

Let states 1 and 2 denote dry and wet days, respectively.

(a) From Eq. 3.16, the probability that it will be wet for the first time two days from now is

$$f_{2,2}(2) = p_{2,1}p_{1,2}$$
$$= 0.5 \times 0.2$$
$$= 0.1$$

(b) From Eq. 3.25, the probability there will be a dry day within the next five days is given by

$$q_{2,1}(5) = 1 - p_{2,2}^5$$
$$= 1 - (0.5)^5$$
$$= 0.96875$$

(c) From Eqs. 3.27 and 3.28, the average lengths of dry and wet spells are, respectively:

$$l_1 = \frac{1}{p_{1,2}} = \frac{1}{0.2} = 5 \text{ days}$$

and

$$l_2 = \frac{1}{p_{2,1}} = \frac{1}{0.5} = 2 \text{ days}$$

(d) From Eq. 3.30, the expected recurrence time of wet days is

$$t_2 = 1 + \frac{p_{2,1}}{p_{1,2}} = 1 + \frac{0.5}{0.2} = 3.5 \text{ days}$$

EXAMPLE 3.9

A quality control program is set up to inspect the quality of aggregates that are produced daily from a rock core. Suppose the daily aggregate quality is divided into two classes, namely, satisfactory (S) and unsatisfactory (U). It is found that the probability of a given quality of aggregates in a given day depends on the quality of the product for the previous day, in accordance with the following transition probability matrix:

$$\begin{array}{cc} & \begin{array}{cc} S & \quad U \end{array} \\ \begin{array}{c} S \\ U \end{array} & \begin{bmatrix} 0.9 & 0.1 \\ 0.4 & 0.6 \end{bmatrix} \end{array}$$

(a) If the aggregate is of satisfactory quality today, what is the probability that it will become unsatisfactory within the next week?

(b) What is the average period that satisfactory quality aggregates will be produced? That unsatisfactory quality aggregates will be produced?

Solution

Let states 1 and 2 denote satisfactory and unsatisfactory aggregate quality, respectively.

(a) From Eq. 3.24,

$$q_{1,2}(7) = 1 - p_{1,1}^7$$
$$= 1 - (0.9)^7$$
$$= 0.522$$

(b) The expected period of satisfactory quality aggregate is given by Eq. 3.27 as

$$l_1 = \frac{1}{p_{1,2}} = 10 \text{ days}$$

whereas from Eq. 3.28, the expected period of unsatisfactory quality aggregate is

$$l_2 = \frac{1}{p_{2,1}} = 2.5 \text{ days}$$

EXAMPLE 3.10

Consider Example 3.7 again; the expected depth of excavation (from the surface) until sand is encountered is given by

$$l_c = \frac{1}{p_{cs}} = \frac{1}{0.2} = 5 \text{ feet}$$

whereas the expected thickness of a sand layer is

$$l_s = \frac{1}{p_{sc}} = \frac{1}{0.5} = 2 \text{ feet}$$

EXAMPLE 3.11

For the nuclear reactor containment structure in Example 3.6, the size (number of tendons) of an average group of corroded tendons may be determined as the expected occupation time of the corroded state; thus,

$$l_2 = \frac{1}{p_{2,1}} = 2$$

Suppose an inspection plan calls for the inspection of a group of tendons in sequence. If the first tendon inspected is free of corrosion and loss of prestress, what is the average number of tendons inspected before a defective tendon (corroded and/or loss of prestress) will be discovered?

In this case, the probability matrix in Example 3.6 will not be applicable. A new two-state transition probability matrix must be constructed with state 1 representing defective tendon (corroded and/or loss of prestress) and state 2 representing nondefective tendon; thus, with the data summarized in Example 3.6

$$\mathbf{P} = \begin{bmatrix} \frac{3}{4} & \frac{1}{4} \\ \frac{2}{5} & \frac{3}{5} \end{bmatrix}$$

With the first tendon inspected being nondefective, the average number of tendons inspected until a defective tendon is found is equivalent to determining the average group size of nondefective tendons, which is given by

$$l_2 = \frac{1}{p_{2,1}} = \frac{1}{\frac{2}{5}} = 2.5$$

Absorbing States   A state j in the Markov chain is an *absorbing state* if its one-step transition probability is $p_{j,j} = 1.0$. An absorbing state, therefore, is defined as a state that cannot reach any other states in the system except itself.

Consider the deflection at mid-span of a wooden beam subject to time-varying loads. If the performance of the wooden beam at a particular time is described by the states (level) of deflection, the beam system will switch from one state to another as the load on the beam varies. However, there is a maximum limit to which the beam may deflect, beyond which the beam may collapse, and the system will not be able to move back to its previous state again. The state corresponding to the maximum deflection limit in this case is an absorbing state. Another example is the bankruptcy state of a construction firm. Once bankruptcy is declared, no other financial states of the firm can be recovered, as the firm ceases operation.

If an absorbing state exists in a Markov chain, the system will eventually reach and be trapped in that absorbing state. A Markov chain may contain more than one absorbing state. In this case, only one of the absorbing states will ever be visited. Of interest is the probability that a system will be absorbed in a particular absorbing state b for a given initial state j. This absorption probability is equal to the probability of ever getting to state b from state j, namely $f_{j,b}$. Applying the theorem of total probability, we obtain

$$f_{j,b} = P(\text{ever getting to } b | X_0 = j)$$

$$= \sum_{i=1}^{r} P(\text{ever getting to } b | X_0 = j, X_1 = i) P(X_1 = i | X_0 = j)$$

$$= \sum_{i=1}^{r} P(\text{ever getting to } b | \text{initial state at } i) p_{j,i}$$

$$= \sum_{i=1}^{r} f_{i,b} p_{j,i} \qquad \begin{array}{l} j = 1, \ldots, r \\ j \neq \text{absorbing states} \end{array} \qquad (3.31)$$

where r is the number of states in the system. When the initial state is b, it is already absorbed in b; hence, clearly $f_{b,b} = 1.0$; whereas when the initial state is at some other absorbing state, say a, then it will never be absorbed at b. Hence, $f_{a,b} = 0$. For all other initial states, a set of simultaneous equations may be written from Eq. 3.31, whose solution is the absorption probability, $f_{j,b}$. The computation of absorption probabilities are demonstrated in Examples 3.14 and 3.15.

Another quantity of interest is the mean time before absorption if the system starts from a given initial state. Let m_j be the mean time of absorption if the initial state is j; then,

$$m_j = E[\text{absorption time} | X_0 = j]$$

$$= \sum_{i=1}^{r} E[\text{absorption time} | X_0 = j, X_1 = i] P(X_1 = i | X_0 = j)$$

$$= \sum_{i=1}^{r} \{1 + E[\text{absorption time} | \text{initial state } i]\} p_{j,i}$$

$$= 1 + \sum_{i=1}^{r} m_i p_{j,i} \qquad \begin{array}{l} j = 1, \ldots, r \\ j \neq \text{absorbing states} \end{array} \qquad (3.32)$$

Equation 3.32 represents a set of simultaneous linear equations whereby the mean absorption time m_j's can be determined in terms of the transition probabilities. Observe that if state i is an absorbing state, the system is already in an absorbing state; in such a case, $m_i = 0$.

EXAMPLE 3.12

A pile is to be driven into a soil stratum until rock is reached. Consider each foot of pile-driving as a unit distance of transition between states, consisting of sand (state 1), clay (state 2), and rock (state 3). In other words, it is implicitly assumed that within each 1-foot layer, the soil is either sand or clay or rock. Although the pile may penetrate sand and clay layers, pile-driving will stop once rock is encountered; state 3, therefore, is an absorbing state. The Markov chain model is shown in Fig. E3.12. Let m_1 and m_2 denote the mean depths of pile-driving until rock is encountered if the surface of the soil stratum starts with sand and clay, respectively.

Applying Eq. 3.32, we have

$$m_1 = 1 + p_{1,1}m_1 + p_{1,2}m_2$$

$$m_2 = 1 + p_{2,1}m_1 + p_{2,2}m_2$$

Solving these two simultaneous equations, we obtain

$$m_1 = \frac{1 - p_{2,2} + p_{1,2}}{(1 - p_{1,1})(1 - p_{2,2}) - p_{1,2}p_{2,1}}$$

$$m_2 = \frac{1 - p_{1,1} + p_{2,1}}{(1 - p_{1,1})(1 - p_{2,2}) - p_{1,2}p_{2,1}}$$

Suppose

$$p_{1,1} = 0.2; p_{1,2} = 0.74; p_{1,3} = 0.06; p_{2,1} = 0.1; p_{2,2} = 0.88; p_{2,3} = 0.02$$

Then,

$$m_1 = \frac{1 - 0.88 + 0.74}{(1 - 0.2)(1 - 0.88) - 0.74 \times 0.1} = 39.1$$

$$m_2 = \frac{1 - 0.2 + 0.1}{(1 - 0.2)(1 - 0.88) - 0.74 \times 0.1} = 40.9$$

Therefore, a pile starting penetration in sand will take 39.1 ft, on the average, before hitting rock, whereas a pile starting penetration in clay will require an average of 40.9 ft.

Mean First Passage Time for Recurrent States   For a Markov chain in which all the states are recurrent, the mean first passage time from one state to another can be conveniently computed by applying the concept of absorbing states.

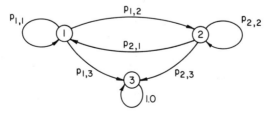

Figure E3.12   Markov chain with an absorbing state.

Let $m_{j,k}$ denote the mean first passage time from state j to k. If a new Markov chain is constructed from the original Markov chain by artificially making k an absorbing state, the mean absorption time m_j in the new Markov chain is equivalent to the mean first passage time $m_{j,k}$ of the original Markov chain, and Eq. 3.32 may be used to compute the required probability. The construction of the new Markov chain requires the definition of a new transition probability matrix $\mathbf{P}' = [p'_{i,j}]$ as follows:

$$p'_{i,j} = 1 \qquad \text{if } i = k, j = k$$
$$p'_{i,j} = 0 \qquad \text{if } i = k, j \neq k \qquad (3.33)$$
$$p'_{i,j} = p_{i,j} \qquad \text{if } i \neq k$$

The resulting Markov chain consists of a single absorbing state k with all other states becoming nonrecurrent. The behavior of the new Markov chain before absorption is the same as that of the original Markov chain before state k is visited for the first time.

EXAMPLE 3.13

For the nuclear reactor containment structure in Example 3.6, suppose the first tendon inspected is free of corrosion and has no loss of prestress. Determine the average number of tendons inspected before a corroded tendon (with no loss of prestress) is discovered.

Four states may be identified to characterize each tendon; namely:

State 1: Not corroded, no loss of prestress.
State 2: Not corroded, loss of prestress.
State 3: Corroded, no loss of prestress.
State 4: Corroded, loss of prestress.

Based on the data for the ten consecutive tendons inspected, a transition probability matrix can be established as follows:

$$\mathbf{P} = \begin{bmatrix} \frac{3}{5} & \frac{1}{5} & \frac{1}{5} & 0 \\ \frac{1}{2} & \frac{1}{2} & 0 & 0 \\ 0 & 0 & 0 & 1 \\ 0 & 1 & 0 & 0 \end{bmatrix}$$

Starting with a tendon in state 1, we see that the mean number of tendons inspected before a corroded tendon with no loss of prestress is discovered is equivalent to determining the mean first passage time from state 1 to state 3. From Fig. E3.13a it may be observed that all four states of the system are recurrent. Hence, if state 3 is an absorbing state (see Fig. E3.13b), the mean absorption time for the modified Markov chain gives the mean first passage time for the original chain. The modified transition probability matrix is

$$\mathbf{P}' = \begin{bmatrix} \frac{3}{5} & \frac{1}{5} & \frac{1}{5} & 0 \\ \frac{1}{2} & \frac{1}{2} & 0 & 0 \\ 0 & 0 & 1 & 0 \\ 0 & 1 & 0 & 0 \end{bmatrix}$$

Applying Eq. 3.32, we obtain the following set of simultaneous equations:

$$m_1 = 1 + \tfrac{3}{5}m_1 + \tfrac{1}{5}m_2$$
$$m_2 = 1 + \tfrac{1}{2}m_1 + \tfrac{1}{2}m_2$$

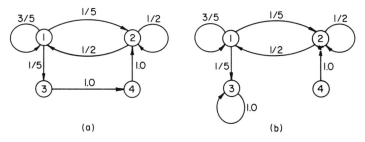

Figure E3.13   Transitional probabilities.

which yields $m_1 = 7$ and $m_2 = 9$. Since the initial state is 1, it will require on the average the inspection of seven tendons before a corroded tendon with no loss of prestress is discovered.

Mean Recurrence Time of Recurrent States   The mean recurrence time associated with state j in a Markov chain may be computed by Eq. 3.21. However, simplification is possible if all the states in the Markov chain are recurrent states. It requires first determining the steady-state probabilities $p_1^*, \ldots, p_k^*$. Observe that the mean recurrence time will be inversely proportional to the corresponding steady-state probability (see Parzen, 1962). Hence,

$$t_j = \frac{1}{p_j^*} \qquad j = 1, \ldots, k \tag{3.34}$$

As a verification, consider a two-state Markov chain. Solving Eqs. 3.14 and 3.15 gives the steady-state probabilities.

$$p_1^* = \frac{p_{2,1}}{p_{1,2} + p_{2,1}} \quad \text{and} \quad p_2^* = \frac{p_{1,2}}{p_{1,2} + p_{2,1}}$$

Hence, Eq. 3.34 yields

$$t_1 = \frac{1}{p_1^*} = 1 + \frac{p_{1,2}}{p_{2,1}}$$

which is the same as Eq. 3.29.

EXAMPLE 3.14

The daily BOD concentration of a stream has been measured over a long period of time and may be divided into four levels, namely: negligible, low, medium, and high. By representing these four concentration levels as states 1, 2, 3, and 4, respectively, a homogeneous Markov chain is constructed with the following transition probability matrix:

$$\mathbf{P} = \begin{bmatrix} 0.5 & 0.4 & 0.1 & 0 \\ 0.2 & 0.5 & 0.2 & 0.1 \\ 0.05 & 0.25 & 0.6 & 0.1 \\ 0 & 0.2 & 0.4 & 0.4 \end{bmatrix}$$

If the BOD concentration in a stream is high, it implies a polluted state.

(a) Determine the steady-state probabilities.

(b) Suppose the stream starts out with negligible BOD concentration. How long will it take, on the average, to reach a polluted state? Repeat this with the stream starting out with a low BOD concentration.

(c) How long will it take, on the average, for a polluted stream to become one with negligible or low BOD concentration?

(d) For the stream starting with medium BOD concentration, what is the probability that it will become one with negligible BOD concentration (excellent water quality) before becoming polluted (i.e., state 4).

(e) Determine the mean recurrence time of the polluted state, that is, the mean time until the stream becomes polluted again.

Solution

(a) If we apply Eqs. 3.14 and 3.15, the set of four simultaneous equations required for the solution of the steady-state probabilities are:

$$p_1^* = 0.5p_1^* + 0.2p_2^* + 0.05p_3^*$$

$$p_2^* = 0.4p_1^* + 0.5p_2^* + 0.25p_3^* + 0.2p_4^*$$

$$p_4^* = 0.1p_2^* + 0.1p_3^* + 0.4p_4^*$$

$$p_1^* + p_2^* + p_3^* + p_4^* = 1.0$$

The solution to these equations yields $p_1^* = 0.1788$, $p_2^* = 0.3613$, $p_3^* = 0.3426$, $p_4^* = 0.1173$.

(b) The first passage time to state 4 is required; hence, state 4 is made into an absorbing state in the modified Markov chain. The corresponding transition probabilities become

$$\mathbf{P'} = \begin{bmatrix} 0.5 & 0.4 & 0.1 & 0 \\ 0.2 & 0.5 & 0.2 & 0.1 \\ 0.05 & 0.25 & 0.6 & 0.1 \\ 0 & 0 & 0 & 1.0 \end{bmatrix}$$

With Eq. 3.32, the mean absorption times m_1 (corresponding to the initial state 1) and m_2 (corresponding to initial state 2) can be determined from the solution to the following equations:

$$m_1 = 1 + 0.5m_1 + 0.4m_2 + 0.1m_3$$

$$m_2 = 1 + 0.2m_1 + 0.5m_2 + 0.2m_3$$

$$m_3 = 1 + 0.05m_1 + 0.25m_2 + 0.6m_3$$

yielding $m_1 = 14.8$, $m_2 = 12.9$, and $m_3 = 12.4$. Therefore, it takes, on the average, approximately 15 days for a stream with negligible BOD concentration to become polluted; whereas it takes an average of about 13 days for a stream with low BOD concentration to become polluted.

(c) In this case, both states 1 and 2 must be converted to absorbing states in the new Markov chain. The transition probability matrix becomes

$$\mathbf{P'} = \begin{bmatrix} 1 & 0 & 0 & 0 \\ 0 & 1 & 0 & 0 \\ 0.05 & 0.25 & 0.6 & 0.1 \\ 0 & 0.2 & 0.4 & 0.4 \end{bmatrix}$$

The set of equations to be solved is

$$m_3 = 1 + 0.6m_3 + 0.1m_4$$

$$m_4 = 1 + 0.4m_3 + 0.4m_4$$

The solution yields $m_3 = 3.5$ and $m_4 = 4$. Hence, the mean time for a polluted stream to become one containing negligible or low BOD concentration is 4 days.

(d)   To obtain the desired probabilities requires converting states 1 and 4 into absorbing states. Hence, the transition probability matrix becomes

$$P' = \begin{bmatrix} 1 & 0 & 0 & 0 \\ 0.2 & 0.5 & 0.2 & 0.1 \\ 0.05 & 0.25 & 0.6 & 0.1 \\ 0 & 0 & 0 & 1 \end{bmatrix}$$

Applying Eq. 3.31 and observing that $f_{1,1} = 1$ and $f_{4,1} = 0$, we obtain

$$f_{2,1} = p_{2,1} + p_{2,2}f_{2,1} + p_{2,3}f_{3,1}$$
$$f_{3,1} = p_{3,1} + p_{3,2}f_{2,1} + p_{3,3}f_{3,1}$$

or

$$f_{2,1} = 0.2 + 0.5f_{2,1} + 0.2f_{3,1}$$
$$f_{3,1} = 0.05 + 0.25f_{2,1} + 0.6f_{3,1}$$

The solution yields $f_{2,1} = 0.6$ and $f_{3,1} = 0.5$. By similar analyses, we obtain $f_{3,4} = 0.5$. (Observe that $f_{3,1} + f_{3,4} = 1.0$ as required.) Therefore, a stream starting with medium BOD concentration is equally likely to turn into one with excellent water quality or become polluted.

(e)   Applying Eq. 3.34, we obtain

$$t_4 = \frac{1}{p_4^*} = \frac{1}{0.1173} = 8.53$$

Therefore, on the average, it will take 8.53 days for a polluted stream to become polluted again.

EXAMPLE 3.15

The amount of storage in a reservoir may be discretized into 0, 1, 2, or 3 units of water. Empty and full reservoirs correspond to states 0 and 3 units, respectively. Suppose the amount of storage between two successive months is governed by a homogeneous Markov chain with the following transition probability matrix:

$$P = \begin{bmatrix} 0.4 & 0.3 & 0.3 & 0 \\ 0.1 & 0.4 & 0.4 & 0.1 \\ 0 & 0.2 & 0.2 & 0.6 \\ 0 & 0 & 0.2 & 0.8 \end{bmatrix}$$

If the reservoir contains 1 unit of storage in a given month,

(a)   What is the mean time until the reservoir will be full?

(b)   What is the probability that the reservoir will be empty before it ever becomes full?

Solution

(a)   Make state 3 an absorbing state, with the corresponding transition probability matrix as follows:

$$P' = \begin{bmatrix} 0.4 & 0.3 & 0.3 & 0 \\ 0.1 & 0.4 & 0.4 & 0.1 \\ 0 & 0.2 & 0.2 & 0.6 \\ 0 & 0 & 0 & 1.0 \end{bmatrix}$$

Applying Eq. 3.32 for $j = 0$, we obtain

$$m_0 = 1 + \sum_{i=0}^{3} m_i p_{0,i} = 1 + 0.4m_0 + 0.3m_1 + 0.3m_2$$

Similarly,

$$m_1 = 1 + \sum_{i=0}^{3} m_i p_{1,i} = 1 + 0.1m_0 + 0.4m_1 + 0.4m_2$$

and

$$m_2 = 1 + \sum_{i=0}^{3} m_i p_{2,i} = 1 + 0.2m_1 + 0.2m_2$$

After simplification, these become

$$0.6m_0 - 0.3m_1 - 0.3m_2 = 1$$
$$-0.1m_0 + 0.6m_1 - 0.4m_2 = 1$$
$$-0.2m_1 + 0.8m_2 = 1$$

whose solution yields $m_1 = 3.95$. In other words, with a present storage of 1 unit, we can expect almost 4 months before the reservoir becomes full.

(b) In this case, both states 0 and 3 must be converted into absorbing states. The corresponding transition probability matrix becomes

$$\begin{bmatrix} 1 & 0 & 0 & 0 \\ 0.1 & 0.4 & 0.4 & 0.1 \\ 0 & 0.2 & 0.2 & 0.6 \\ 0 & 0 & 0 & 1 \end{bmatrix}$$

Applying Eq. 3.31 for $j = 1$ and $b = 0$, we have

$$f_{1,0} = \sum_{i=0}^{2} f_{i,0} p_{1,i} = f_{0,0} p_{1,0} + f_{1,0} p_{1,1} + f_{2,0} p_{1,2}$$

$$= 1 \times 0.1 + f_{1,0} \times 0.4 + f_{2,0} \times 0.4$$

Similarly,

$$f_{2,0} = \sum_{i=0}^{2} f_{i,0} p_{2,i} = f_{0,0} p_{2,0} + f_{1,0} p_{2,1} + f_{2,0} p_{2,2}$$

$$= 1 \times 0 + f_{1,0} \times 0.2 + f_{2,0} \times 0.2$$

which can be reduced to the following:

$$0.6f_{1,0} - 0.4f_{2,0} = 0.1$$

and

$$-0.2f_{1,0} + 0.8f_{2,0} = 0$$

The solutions are $f_{1,0} = 0.2$ and $f_{2,0} = 0.05$. In other words, therefore, for a reservoir with an initial storage of 1 unit, there is a 20% probability that it will become empty before it ever becomes full.

EXAMPLE 3.16

The degree of damage to an existing building following an earthquake of MM scale $\geq V$ may be classified into several damage states as follows (Whitman et al., 1975):

State	Damage Level
1	No damage
2	Light damage
3	Moderate damage
4	Heavy damage
5	Total damage or collapse

Suppose the additional damage that may be incurred in a subsequent earthquake depends on the initial state of the structure prior to the next earthquake, as defined by the following transition probability matrix.

$$\mathbf{P} = \begin{bmatrix} 0.8 & 0.15 & 0.045 & 0.005 & 0 \\ 0 & 0.6 & 0.3 & 0.07 & 0.03 \\ 0 & 0 & 0.5 & 0.4 & 0.1 \\ 0 & 0 & 0 & 0.3 & 0.7 \\ 0 & 0 & 0 & 0 & 1 \end{bmatrix}$$

If building damages are not repaired,

(a) Determine the probabilities of the building being in the various damage states after three earthquakes.

(b) On the average, how many earthquakes can a building withstand before it incurs at least moderate damage (i.e., state 3)?

Solution

(a) For a building that is initially in an undamaged state (state 1), the state probability matrix following three earthquakes is

$$\mathbf{P} = \begin{bmatrix} 1 & 0 & 0 & 0 & 0 \end{bmatrix} \begin{bmatrix} 0.8 & 0.15 & 0.045 & 0.005 & 0 \\ 0 & 0.6 & 0.3 & 0.07 & 0.03 \\ 0 & 0 & 0.5 & 0.4 & 0.1 \\ 0 & 0 & 0 & 0.3 & 0.7 \\ 0 & 0 & 0 & 0 & 1 \end{bmatrix}^3$$

$$= \begin{bmatrix} 0.512 & 0.222 & 0.144 & 0.069 & 0.053 \end{bmatrix}$$

(b) To determine the mean number of earthquakes necessary to inflict at least moderate damage to a building, states 3, 4, and 5 may be combined into a single absorbing state 3′; the corresponding transitional probability matrix becomes

$$\mathbf{P}' = \begin{bmatrix} 0.8 & 0.15 & 0.05 \\ 0 & 0.6 & 0.4 \\ 0 & 0 & 1 \end{bmatrix}$$

The mean time of transition from state 1 to state 3′ can be evaluated by applying Eq. 3.32

$$m_1 = 1 + 0.8m_1 + 0.15m_2$$
$$m_2 = 1 + 0.6m_2$$

from which $m_1 = 6.88$ and $m_2 = 2.5$. In other words, it will require, on the average, seven earthquakes before the building suffers at least moderate damage. On the other hand, if the building already had light damage from earlier quakes, a mean number of 2.5 additional earthquakes will be required before the building suffers at least moderate damage.

3.1.7 Random Time Between Stages

The time required for each stage of transition in a Markov chain model may be a random variable itself. In such a case, the number of transitions in a given time period is not deterministic. Therefore, the probability that the system is in state i at time t, namely $p_i(t)$, will depend on both the number of stages n that have taken place and the state probabilities after n stages. Applying the theorem of total probability, we obtain

$$p_i(t) = \sum_{n=0}^{\infty} p_i(n)p_N(n;t) \tag{3.35}$$

where $p_i(n)$ is the n-stage probability of the system at state i, calculated according to Section 3.1.3, and $p_N(n;t)$ is the probability mass function of the number of stages N within time t.

The event that the number of stages N exceeds n in time t is the same as the event that the total time taken for the $(n+1)$ stages of transition is less than or equal to t. Hence, the cumulative distribution function of N

$$F_N(n) = 1 - P(N > n) = 1 - P(T_1 + T_2 + \cdots + T_{n+1} \le t) \tag{3.36}$$

where $T_i, i = 1$ to $n+1$, is the time taken for the ith stage of transition.

Let S_{n+1} denote the sum of $n+1$ transition times, namely $T_1 + T_2 + \cdots + T_{n+1}$. Equation 3.36 becomes

$$F_N(n) = 1 - F_{S_{n+1}}(t) \tag{3.37}$$

where $F_{S_{n+1}}(t)$ is the CDF of S_{n+1}. The PMF

$$
\begin{aligned}
p_N(n;t) &= F_N(n) - F_N(n-1) \\
&= F_{S_n}(t) - F_{S_{n+1}}(t)
\end{aligned}
\tag{3.38}
$$

Suppose the transition times T_j's are independent and identically normally distributed as $N(\mu, \sigma)$. Then, S_n will also be normally distributed as $N(n\mu, \sqrt{n}\sigma)$. Hence,

$$p_N(n;t) = \Phi\left(\frac{t - n\mu}{\sqrt{n}\sigma}\right) - \Phi\left[\frac{t - (n+1)\mu}{\sqrt{n+1}\sigma}\right] \tag{3.39}$$

On the other hand, if the transition times are exponentially distributed, as in the case of the Poisson process for stage transition occurrences, $p_N(n;t)$ in Eq. 3.35 could be calculated simply as

$$p_N(n;t) = \frac{e^{-vt}(vt)^n}{n!} \tag{3.40}$$

where v is the mean rate of transition.

EXAMPLE 3.17

For the earthquake damage in Example 3.16, suppose severe earthquakes occur at the site according to a Poisson process with a mean occurrence rate of once in ten years.

(a)   Determine the probability that a building will suffer at least moderate damage over a period of 20 years.

(b)   Repeat part (a) if the building is restored to its original condition after each earthquake.

(c)   How would the answer to part (a) change if the time between severe earthquake occurrences is normally distributed with a mean of 10 years and a standard deviation of $\sqrt{10}$ years?

Solution

(a)   The probability that the building will suffer at least moderate damage ($D \geq 3$) over a period of 20 years depends on the number of earthquakes that may occur during this period. Applying Eqs. 3.35 and 3.40, we obtain

$$P(D \geq 3) = \sum_{n=0}^{\infty} P(D \geq 3|n) \frac{e^{-2} 2^n}{n!}$$

where $P(D \geq 3|n)$ is the probability of the building in state $3'$ after n earthquakes; this is the n-stage probability of state $3'$, which can be calculated from the modified transition probability matrix of part (b) in Example 3.16. The following table summarizes the required calculations:

n	$P(N = n) = \dfrac{e^{-2} 2^n}{n!}$	$P(D \geq 3|n)$	$P(D \geq 3|n)P(N = n)$
0	0.1353	0	0
1	0.2707	0.050	0.0135
2	0.2707	0.150	0.0406
3	0.1804	0.266	0.0480
4	0.0902	0.380	0.0343
5	0.0361	0.485	0.0175
6	0.0120	0.576	0.0069
7	0.0034	0.654	0.0022
8	0.0009	0.719	0.0006
9	0.0002	0.773	0.0002
10	0.00004	0.817	0.0000
			$\overline{0.1638}$

Therefore, there is a 16.4% probability that the building will suffer at least moderate damage during a 20-year period.

(b)   If the building is carefully inspected after each earthquake and then completely restored to its initial condition before the occurrence of the next earthquake, the probability that it will suffer at least moderate damage in the next earthquake is simply $p_{13'} = 0.05$. Then, over a 20-year period, the probability of a building suffering at least moderate damage is

$$P(D \geq 3) = 1 - P(D < 3) = 1 - \sum_{n=0}^{\infty} P(D < 3|n)P(N = n)$$

$$= 1 - \sum_{n=0}^{\infty} (1 - 0.05)^n \frac{e^{-2} 2^n}{n!}$$

$$= 1 - \frac{e^{-2}}{e^{-2 \times 0.95}}$$

$$= 0.0952$$

which is less than the probability of part (a). Obviously, therefore, inspection and repair of a building after each earthquake will increase its safety.

(c)   The probabilities of n earthquakes in 20 years are reevaluated according to Eq. 3.39 for $t = 20$ years as

$$P(N = n) = \Phi\left(\frac{20 - 10n}{\sqrt{n}\sqrt{10}}\right) - \Phi\left(\frac{20 - 10(n + 1)}{\sqrt{n + 1}\sqrt{10}}\right)$$

whose results are summarized as follows:

n	$P(N = n)$	$P(D \geq 3 \mid n)$	$P(D \geq 3 \mid n)P(N = n)$
0	0.000785	0	0
1	0.4992	0.050	0.0250
2	0.466	0.150	0.0699
3	0.0332	0.266	0.0088
4	0.000774	0.380	0.0003
5	0.000011	0.485	0.0000
			0.1040

Therefore, there is only 10.4% probability that the building will suffer at least moderate damage during a 20-year period, which is less than the probability of 16.4% obtained for the Poisson occurrence model in part (a).

3.1.8   Continuous Parameter Homogeneous Markov Chain

In the previous sections, the transition between states has been assumed to occur in discrete stages; for example, from year to year, from earthquake to earthquake, or from bus stop to bus stop. In some physical systems, the transitions may occur at any value of the parameter, such as at any time. For instance, the queue length at a toll station may change at any time. In a continuous parameter Markov chain, the time of transition is described by a continuous variable with $\{X(t), t \geq 0\}$ denoting the state of the system at time t. For a *homogeneous Markov chain*, the transition probability of the system from state j to state k in a time interval t is given by the conditional probability

$$p_{j,k}(t) = P[X(t + s) = k \mid X(s) = j] = P[X(t) = k \mid X(0) = j]$$

Instead of the one-step transition probability as in the discrete case, the likelihood of changes between states is expressed in terms of the *transition intensity*. The transition intensity is simply the rate of changing from one state to another per unit time. Let $q_{k,j}$ be the intensity of transition from state k to j, whereas q_k is the intensity of passage out of state k. Hence, in a small time interval Δt, the probability of a system going from state k to state j is approximately given by

$$p_{k,j}(\Delta t) \simeq q_{k,j} \cdot (\Delta t); \qquad k \neq j \tag{3.41a}$$

and the probability that a system remains at state k is

$$p_{k,k}(\Delta t) = 1 - q_k \cdot (\Delta t) \tag{3.41b}$$

The Chapman-Kolomogorov equations are also applicable to the continuous parameter Markov process. In a Markov chain with m states, the transition probability between states i and j in time t is

$$p_{i,j}(t) = P[X(t) = j \,|\, X(0) = i]$$

$$= \sum_{k=1}^{m} P[X(t) = j \,|\, X(0) = i, X(r) = k] P[X(r) = k \,|\, X(0) = i]$$

$$= \sum_{k=1}^{m} p_{k,j}(t - r) p_{i,k}(r)$$

where $0 < r < t$. Similarly, the probability that the process will be at state j at time $(t + \Delta t)$ is

$$p_j(t + \Delta t) = \sum_{k=1}^{m} p_k(t) p_{k,j}(\Delta t) \tag{3.42}$$

For small Δt Eqs. 3.41a and 3.41b may be substituted into Eq. 3.42 to yield

$$p_j(t + \Delta t) = \sum_{k \neq j} p_k(t) q_{k,j}(\Delta t) + p_j(t)(1 - q_j \cdot \Delta t)$$

or

$$\frac{p_j(t + \Delta t) - p_j(t)}{\Delta t} = \sum_{k \neq j} p_k(t) q_{k,j} - p_j(t) q_j$$

Taking the limit as $\Delta t \to 0$, we have

$$\frac{dp_j(t)}{dt} = \sum_{k \neq j} p_k(t) q_{k,j} - p_j(t) q_j; \qquad j = 1, \dots, m \tag{3.43}$$

Equation 3.43 provides a set of differential equations that can be used to determine the state probability at a future time t for any given probabilities of the initial state, provided that the transition intensities between all pairs of states in the system are known.

On the other hand, when t is very large, the steady-state condition on the state probabilities may have been achieved, as in the discrete case of Section 3.1.4. Thus,

$$\frac{dp_j(t)}{dt} = 0$$

and

$$p_j(t) = p_j^*$$

for $j = 1, \dots, m$. Equation 3.43 then becomes

$$p_j^* q_j = \sum_{k \neq j} p_k^* q_{k,j}; \qquad j = 1, \dots, m \tag{3.44}$$

Again with Eq. 3.15, the steady-state probabilities of a continuous parameter Markov process can be determined.

As an example, for a two-state Markov chain, substituting $j = 1$ in Eq. 3.44 we have

$$p_1^* q_1 = p_2^* q_{2,1}$$

Since q_1 is the intensity of passage out of state 1, it will be equal to $q_{1,2}$ in this case. Furthermore, since

$$p_1^* + p_2^* = 1$$

the steady-state probabilities can be shown to be

$$p_1^* = \frac{q_{2,1}}{q_{1,2} + q_{2,1}} \tag{3.45}$$

and

$$p_2^* = \frac{q_{1,2}}{q_{1,2} + q_{2,1}} \tag{3.46}$$

Section 3.2 describes an application of the continuous parameter Markov chain to queueing problems in engineering. Availability problems, presented in Section 3.3, further discuss applications of the continuous parameter Markov chain with emphasis on the transient behavior of a system.

3.2 QUEUEING MODELS

3.2.1 Introduction

A class of physical problems involves the arrivals of "calling units" that must be served by "service units." Depending on the arrival rate versus the service rate, a queue of calling units may be formed. The study of the queue length as a function of the arrival and service rates requires *queueing models*.

Queueing models have been applied to predict traffic flow characteristics (Haight, 1963), in the study of the efficiency of operations in construction projects (Shaffer, 1967), the determination of optimal construction crew size (Griffis, 1968), and in the study of waste treatment processes (Thatcher, 1968). A common feature in these problems is a waiting line, or queue, that may consist of vehicles, delivery trucks, shovels, construction workers, or solid waste loads.

A typical example of a queueing problem is the toll booth operation at a turnpike exit. The arrivals of vehicles constitute a random process, and the service time for each vehicle is a random variable. The system may have been designed such that the average service capacity of the toll booth can adequately handle the long-term average traffic demand; however, the instantaneous arrival rate of vehicles at the toll booth may exceed the service capacity of the booth resulting, temporarily, in a queue of vehicles. Through appropriate queueing models of the toll booth operation, information such as the expected length of the waiting line and the mean waiting time for a vehicle may be derived.

3.2.2 Poisson Process Arrivals and Departures

A basic queueing system may be schematically represented as in Fig. 3.2, with m service units that can operate simultaneously. The length of the queue is defined as the total number of calling units in the system including those being served. In

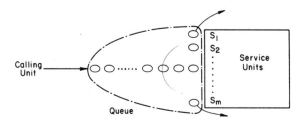

Figure 3.2 A queueing system.

general, the mean rates of arrival and service will depend on the queue length. Suppose λ_n is the mean arrival rate of the calling units, and μ_n is the corresponding mean service rate of the system when the queue length is n. Consider a system with queue length n at time $t + \Delta t$. During an incremental time interval, Δt, between t and $t + \Delta t$, one of the following events must occur:

(i)   Exactly one arrival.
(ii) Exactly one departure.
(iii) No arrival or departure.
(iv) Multiple arrivals and/or departures.

If we assume that the arrivals and departures are respectively Poisson processes, the probability of multiple arrivals or departures will be of higher order in Δt; hence, as $\Delta t \to 0$, the probability of event (iv) will be negligible relative to those of the first three events. In other words, the queue length at time t can only be one of three mutually exclusive possibilities, namely: $n - 1, n$, and $n + 1$. Thus, by virtue of the theorem of total probability, the probability that the queue length X is n at time $(t + \Delta t)$ is given by

$$
\begin{aligned}
P[X(t + \Delta t) = n] &= P[X(t + \Delta t) = n | X(t) = n - 1]P[X(t) = n - 1] \\
&\quad + P[X(t + \Delta t) = n | X(t) = n]P[X(t) = n] \\
&\quad + P[X(t + \Delta t) = n | X(t) = n + 1]P[X(t) = n + 1]
\end{aligned}
$$

The conditional probabilities correspond, respectively, to the events of exactly one arrival at $X = n - 1$, no arrival or departure at $X = n$, and exactly one departure at $X = n + 1$, within the time interval Δt. In terms of transitional probabilities and with the notation $p_n(t) = P[X(t) = n]$, the above equation becomes

$$
p_n(t + \Delta t) = p_{n-1,n}(\Delta t)p_{n-1}(t) + p_{n,n}(\Delta t)p_n(t) + p_{n+1,n}(\Delta t)p_{n+1}(t)
$$

in which the transition probabilities can be shown to be

$$
\begin{aligned}
p_{n-1,n}(\Delta t) &= \lambda_{n-1}\Delta t \\
p_{n,n}(\Delta t) &= 1 - \lambda_n \Delta t - \mu_n \Delta t \\
p_{n+1,n}(\Delta t) &= \mu_{n+1}\Delta t
\end{aligned}
$$

Hence,

$$
p_n(t + \Delta t) = \lambda_{n-1}\Delta t p_{n-1}(t) + (1 - \lambda_n \Delta t - \mu_n \Delta t)p_n(t) + \mu_{n+1}\Delta t p_{n+1}(t)
$$

Rearranging the above yields

$$\frac{p_n(t + \Delta t) - p_n(t)}{\Delta t} = \lambda_{n-1}p_{n-1}(t) + \mu_{n+1}p_{n+1}(t) - (\lambda_n + \mu_n)p_n(t)$$

Therefore, as $\Delta t \to 0$, the following differential equation is obtained:

$$\frac{dp_n(t)}{dt} = \lambda_{n-1}p_{n-1}(t) + \mu_{n+1}p_{n+1}(t) - (\lambda_n + \mu_n)p_n(t); \qquad \text{for } n = 1, 2, \ldots \quad (3.47)$$

When $n = 0$, the state X_{-1} for the system at time t is an impossible event; hence, $p_{-1}(t) = 0$. Moreover, at state X_0, that is, queue length equals zero, the service rate is obviously zero. Hence,

$$\frac{dp_0(t)}{dt} = \mu_1 p_1(t) - \lambda_0 p_0(t) \qquad (3.48)$$

Equations 3.47 and 3.48 provide a system of differential equations for determining $p_n(t)$. The solution is mathematically involved. Fortunately, for most practical engineering problems, the queueing system may attain a steady-state condition in a relatively short time. For such a long-term system performance, the steady-state probabilities may be adequate for engineering planning purposes. At the steady-state condition,

$$\frac{dp_n(t)}{dt} = 0, \qquad \text{for } n = 0, 1, 2, \ldots \qquad (3.49)$$

and the steady-state probabilities p^*'s will be independent of time. Thus, Eqs. 3.47 and 3.48 become

$$0 = \lambda_{n-1}p_{n-1}^* + \mu_{n+1}p_{n+1}^* - (\lambda_n + \mu_n)p_n^*; \qquad \text{for } n > 0 \qquad (3.50)$$

$$0 = \mu_1 p_1^* - \lambda_0 p_0^*; \qquad \text{for } n = 0 \qquad (3.51)$$

The recursive solutions of Eqs. 3.50 and 3.51 would yield

$$p_1^* = \frac{\lambda_0}{\mu_1} p_0^*$$

$$p_2^* = \frac{(\lambda_1 + \mu_1)p_1^*}{\mu_2} - \frac{\lambda_0 p_0^*}{\mu_2} = \frac{\lambda_0 \lambda_1}{\mu_1 \mu_2} p_0^*$$

$$p_n^* = \frac{\lambda_0 \cdots \lambda_{n-1}}{\mu_1 \cdots \mu_n} p_0^* = \frac{\prod\limits_{i=0}^{n-1} \lambda_i}{\prod\limits_{i=1}^{n} \mu_i} p_0^* \qquad (3.52)$$

In addition, of course,

$$\sum_{n=0}^{\infty} p_n^* = 1.0$$

or

$$p_0^* + \sum_{n=1}^{\infty} \frac{\prod_{i=0}^{n-1} \lambda_i}{\prod_{i=1}^{n} \mu_i} p_0^* = 1.0$$

from which

$$p_0^* = \frac{1}{1 + \sum_{n=1}^{\infty} \frac{\prod_{i=0}^{n-1} \lambda_i}{\prod_{i=1}^{n} \mu_i}} \qquad (3.53)$$

The steady-state probabilities p_n^* can be determined from Eqs. 3.52 and 3.53 for given values of λ_i and μ_i, $i = 1, 2, \ldots, n$. The values of λ_i and μ_i generally depend on the number of available servers and calling units. The expected queue length is, therefore, given by

$$L = \sum_{n=1}^{\infty} n p_n^* \qquad (3.54)$$

Queueing Model with One Server In the simplest case when there is only one server with an unlimited number of calling units, both the arrival rate and service rate will be constant regardless of the queue length (except for the obvious case $\mu_0 = 0$); that is,

$$\mu_i = \mu; \qquad i \geq 1$$

and

$$\lambda_i = \lambda; \qquad i \geq 0$$

Then,

$$\prod_{i=0}^{n-1} \lambda_i = \lambda^n$$

and

$$\prod_{i=1}^{n} \mu_i = \mu^n$$

For the case $\lambda/\mu < 1.0$, Eq. 3.53 becomes

$$p_0^* = \frac{1}{1 + \sum_{n=1}^{\infty} \left(\frac{\lambda}{\mu}\right)^n}$$

$$= \frac{1}{1 + \frac{\lambda/\mu}{1 - \lambda/\mu}}$$

$$= 1 - \frac{\lambda}{\mu}$$

Substituting this into Eq. 3.52, we obtain

$$p_n^* = \left(\frac{\lambda}{\mu}\right)^n \left(1 - \frac{\lambda}{\mu}\right); \qquad n = 0, 1, \ldots, \text{ and } \frac{\lambda}{\mu} < 1.0 \qquad (3.55)$$

The corresponding queue length, Eq. 3.54, is

$$
\begin{aligned}
L &= \sum_{n=1}^{\infty} n \left(\frac{\lambda}{\mu}\right)^n \left(1 - \frac{\lambda}{\mu}\right) \\
&= \left(1 - \frac{\lambda}{\mu}\right) \frac{\lambda}{\mu} \sum_{n=1}^{\infty} n \left(\frac{\lambda}{\mu}\right)^{n-1} \\
&= \left(1 - \frac{\lambda}{\mu}\right) \frac{\lambda}{\mu} \frac{1}{\left(1 - \frac{\lambda}{\mu}\right)^2} \\
&= \frac{\lambda}{\mu - \lambda}; \qquad \text{for } \frac{\lambda}{\mu} < 1.0 \qquad (3.56)
\end{aligned}
$$

As the rate of arrivals approaches the rate at which the queue is served, that is, $\lambda \to \mu$, the queue would become very long (i.e. $L \to \infty$).

EXAMPLE 3.18

During rush hours, the mean rate of arriving planes at an airport is estimated to be 40 per hour. The airport is equipped with two traffic controllers who can independently process the landings. Each controller takes approximately two minutes, on the average, to process one landing. For a plane arriving at this airport during rush hours, what is the chance that it will be immediately processed for landing?

The mean arrival rate will not be affected by the number of planes in the system. If there are two or less planes in the queueing system, the mean process rate will be proportional to the number of planes in the system, whereas if there are more than two planes in the system, both of the controllers will be busy and the process rate is limited to 60 per hour. Thus,

$$\lambda_0 = \lambda_1 = \cdots = \lambda_\infty = \lambda = 40$$

$$\mu_0 = 0; \mu_1 = \mu = 30; \mu_2 = \mu_3 = \cdots = \mu_\infty = 2\mu = 60$$

Substituting these into Eq. 3.53, we have

$$
\begin{aligned}
p_0^* &= \left[1 + \frac{\lambda}{\mu} + \frac{\lambda}{\mu}\left(\frac{\lambda}{2\mu}\right) + \frac{\lambda}{\mu}\left(\frac{\lambda}{2\mu}\right)^2 + \cdots\right]^{-1} \\
&= \left[1 + \left(\frac{\lambda}{\mu}\right) \frac{1}{1 - (\lambda/2\mu)}\right]^{-1} \\
&= \frac{2\mu - \lambda}{2\mu + \lambda} \\
&= 0.2
\end{aligned}
$$

and from Eq. 3.52

$$p_1^* = \frac{\lambda_0}{\mu_1} p_0^* = \frac{40}{30} \times 0.2 = 0.267$$

The event that the arriving plane will be immediately processed for loading is equivalent to the event that there is zero or one plane in the queue. Hence, the pertinent probability is

$$p_0^* + p_1^* = 0.20 + 0.267 = 0.467$$

The expected number of planes ahead of an arriving plane may be computed, using Eq. 3.54, as

$$L = 1 \cdot \frac{\lambda}{\mu} p_0^* + 2 \cdot \frac{\lambda}{\mu} \left(\frac{\lambda}{2\mu}\right) p_0^* + 3 \cdot \frac{\lambda}{\mu} \left(\frac{\lambda}{2\mu}\right)^2 p_0^* + \cdots$$

$$= \frac{\lambda}{\mu} p_0^* \frac{1}{\left(1 - \frac{\lambda}{2\mu}\right)^2}$$

$$= \frac{40}{30} \times 0.2 \times \frac{1}{\left(1 - \frac{40}{60}\right)^2}$$

$$= 2.403$$

that is, on the average, the pilot of an arriving plane can expect 2 or 3 planes ahead of him in the queue.

EXAMPLE 3.19 (Excerpted from Griffis, 1968)

An earth excavation operation using a single shovel is schematically represented in Fig. E3.19.
A typical cycle of the excavation operation consists of :

(i)   Truck is loaded by the shovel at the production unit.
(ii) Travel to the fill area.
(iii) Dump the truck's load.
(iv) Travel back to the production unit.
(v)   Wait in the queue until it is the truck's turn to be reloaded.

The contractor would like to estimate the expected production per unit time for a single shovel and a fleet of five trucks. First, this requires determining the fraction of time the shovel is idle in the present mode of operation. This is p_0^*, which is the steady-state probability that no trucks are in the queue. The efficiency of the shovel is $(1 - p_0^*)$, and the expected production Y (or amount of excavation in cu yd) in a time interval T is

$$Y = (1 - p_0^*)T\mu C$$

where μ is the rate of shoveling in terms of the number of truck loads and C is the capacity of each truck load in cu yd.

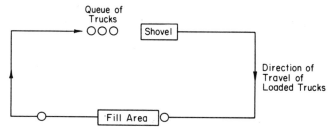

Figure E3.19 Schematic diagram of earth excavation operation.

In the present case, of course, the number of trucks in the queue ranges from 0 to 5. Hence, the mean rate of arrivals of trucks in the queue depends on the number of trucks not in the queuing system. Assume that the mean arrival rate is proportional to the number of trucks not in the queueing system; namely,

$$\lambda_n = (5 - n)\lambda; \qquad n = 0, 1, \ldots, 5$$

where λ is the mean arrival rate of a particular truck. Since there is only one shovel, the mean rate of departure for loaded trucks would be constant, equal to μ when one or more trucks are present in the queue; otherwise, the shovel is not operating and the mean rate of service is zero. Hence,

$$\mu_0 = 0, \mu_1 = \mu_2 = \mu_3 = \mu_4 = \mu_5 = \mu$$

With these values of λ_n and μ_n, Eq. 3.53 yields

$$p_0^* = \left[1 + 5\left(\frac{\lambda}{\mu}\right) + 5 \cdot 4\left(\frac{\lambda}{\mu}\right)^2 + 5 \cdot 4 \cdot 3\left(\frac{\lambda}{\mu}\right)^3 + 5 \cdot 4 \cdot 3 \cdot 2\left(\frac{\lambda}{\mu}\right)^4 + 5 \cdot 4 \cdot 3 \cdot 2 \cdot 1\left(\frac{\lambda}{\mu}\right)^5\right]^{-1}$$

Suppose the engineer estimates $\lambda = 7.5$ arrivals per hour, and $\mu = 30$ truck loads per hour. The steady-state probability that no trucks are at the shovel, therefore, is

$$p_0^* = \frac{1}{1 + 1.25 + 1.25 + 0.9375 + 0.4688 + 0.1172} = 0.199$$

which implies that the shovel is idle about 20% of the time. The expected production per hour for the present excavation operation (with 5 trucks, each with a capacity of 8 cu yd) is

$$Y = (1 - 0.199) \times 1 \times 30 \times 8 = 192 \text{ cu yd}$$

The contractor may also be concerned with whether the present operation is optimal; that is, should he order additional trucks or transfer some of the trucks to other operations. Intuitively, as the truck fleet expands, the shovel is less likely to become idle, and thus, the efficiency of the shovel is increased. However, this also means an increased cost for hiring additional trucks and drivers. Suppose from past cost records, the contractor determines that each truck with a driver costs $14.80 per hour, whereas the power shovel, operator, and oiler cost $35.60 per hour. Then, for a fleet of k trucks, the expected total cost per unit production is

$$E[TC(k)] = \frac{\text{Total cost per hour}}{\text{Expected production per hour}}$$

$$= \frac{14.8k + 35.6}{Y(k)}$$

Since

$$\lambda_i = (k - i)\lambda; \qquad i = 0, 1, \ldots, k$$
$$= 0; \qquad\qquad i > k$$

and

$$\mu_i = \mu; \qquad i = 1, 2, \ldots, k$$

the steady-state probability p_0^* associated with a fleet of k trucks can be derived from Eq. 3.53 as

$$p_0^*(k) = \left[1 + \sum_{n=1}^{k} \frac{k(k-1)\cdots(k-n+1)\lambda^n}{\mu^n}\right]^{-1}$$

$$= \left[\sum_{n=0}^{k} \frac{k!}{(k-n)!}\left(\frac{\lambda}{\mu}\right)^n\right]^{-1}$$

Hence, the expected production per hour with a fleet of k trucks is

$$Y(k) = [1 - p_0^*(k)] \times 1 \times 30 \times 8$$

Table E3.19 Summary of Calculations for Expected Total Cost

Fleet Size k	$p_0^*(k)$	$Y(k)$	$E[TC(k)]$
1	0.800	48.0	1.05
2	0.615	92.5	0.705
3	0.451	132.0	0.599
4	0.311	165.0	0.568
5	0.199	192.0	0.567
6	0.117	211.5	0.581
7	0.063	224.5	0.618

Table E3.19 summarizes the calculations of the expected total cost for various fleet sizes; the results are based on constant values of λ and μ. It may be observed that the present fleet of 5 trucks is, in fact, the optimal fleet size for the excavation operation.

EXAMPLE 3.20

A municipal agency is studying the operation of a waste-treatment process. The solid waste arrives in truck loads at an average rate of 10 loads per hour. Because of the random size of each truck load and the randomness in the treatment process, the processing time for each truck load is assumed to be exponentially distributed with a mean of 12 minutes. The total cost for operating and maintaining a processing unit is $80 per hour. On the other hand, any backlog of the solid waste will cause inconvenience and pollution problems, involving a loss estimated to be $100 per truck load per hour on the queue (excluding those loads being processed). Determine the optimal number of processing units such that the expected total cost and loss will be minimum.

Solution

The expected total cost is given by

$$T = 80m + 100N$$

where m is the number of solid waste treatment processing units and N is the net expected number of loaded trucks waiting to be processed (this will depend on m).

For a treatment system with m processing units, the mean service rate per hour is

$$\mu_n = n\mu = 5n; \qquad \text{for } n \le m$$

$$= m\mu = 5m; \qquad \text{for } n \ge m$$

If we assume that the arrivals of trucks follow a homogeneous Poisson process, $\lambda_n = \lambda = 10$ per hour for all n. Therefore, from Eq. 3.53

$$p_0^* = \left[1 + \sum_{n=1}^{m} \frac{\lambda^n}{\prod_{i=1}^{n} i\mu} + \sum_{n=m+1}^{\infty} \left(\frac{\lambda}{\mu}\right)^n \frac{1}{m!} \frac{1}{m^{n-m}} \right]^{-1}$$

$$= \left[1 + \sum_{n=1}^{m} \left(\frac{\lambda}{\mu}\right)^n \frac{1}{n!} + \frac{m^m}{m!} \left(\frac{\lambda}{m\mu}\right)^{m+1} \Big/ \left(1 - \frac{\lambda}{m\mu}\right) \right]^{-1}$$

$$= \left[1 + \sum_{n=1}^{m} \frac{2^n}{n!} + \frac{m^m}{m!} \left(\frac{2}{m}\right)^{m+1} \Big/ \left(1 - \frac{2}{m}\right) \right]^{-1}; \qquad \text{for } m > 2$$

which can be determined for given m. Subsequently, p_n^* can be determined using Eq. 3.52.

For the case $m = 3$,

$$p_0^* = \left[1 + 2 + 2 + \frac{4}{3} + \frac{3^3}{3!} \left(\frac{2}{3}\right)^4 \Big/ \left(\frac{1}{3}\right) \right]^{-1} = 0.111$$

$$p_1^* = \frac{10}{5} p_0^* = 0.222$$

$$p_2^* = \frac{10}{10} p_1^* = 0.222$$

$$p_3^* = \frac{10}{15} p_2^* = 0.148$$

$$p_4^* = \frac{10}{15} p_3^* = 0.099$$

$$p_5^* = \frac{10}{15} p_4^* = 0.066$$

When there are only three or fewer trucks in the queue, none of these trucks will need to wait; hence, no pollution cost will be incurred. If there are more than three trucks in the queue, all the trucks except the first three will give rise to pollution. Hence,

$$N = \sum_{i=4}^{\infty} p_i^*(i - 3) = 0.099 \times 1 + 0.066 \times 2 + \cdots = 0.9$$

The total expected cost is

$$T = 80 \times 3 + 100 \times 0.9$$
$$= \$330 \text{ per hour}$$

Table E3.20 summarizes the numerical calculations for other values of m. When $m = 1$

Table E3.20 Summary of Calculations for Expected Total Cost

m	p_0^*	N	$80m$	$100N$	T
1	—	∞	80	∞	∞
2	—	∞	160	∞	∞
3	0.111	0.9	240	90	$330
4	0.130	0.18	320	18	$338
5	0.134	0.27	400	27	$427

and 2, the incoming rate exceeds the service rate; thus, the queue is expected to become infinitely long. From the total costs shown in Table E3.20, the minimum-cost number of treatment units would be three, with an expected total cost of $330 per hour.

EXAMPLE 3.21

In a given region, storms occur as a Poisson process with a mean arrival rate $\lambda = 0.1$ per day. For each occurrence, the duration of the storm follows an exponential distribution with a mean of 1 day. Assume that the next storm will not occur until the previous storm has ended. Determine the probability of stormy weather in the region in a given day.

The problem may be formulated as a queueing problem with Poisson arrivals and exponential service time, in which the queue length is limited to 1 or 0. Hence,

$$\lambda_0 = 0.1, \quad \mu_1 = 1$$

Substituting these into Eq. 3.53, we obtain

$$p_0^* = \frac{1}{1 + \dfrac{\lambda_0}{\mu_1}} = \frac{\mu_1}{\mu_1 + \lambda_0} = \frac{1}{1.1}$$

Therefore, on a given day, the probability of stormy weather in the region is

$$p_1^* = 1 - \frac{1}{1.1} = 0.09$$

3.2.3 Poisson Arrivals and Arbitrarily Distributed Service Time

The results derived in Section 3.2.2 are valid only if the arrivals of the calling units follow a Poisson process and the distribution of the service time is exponential. The assumption of exponential service time may be reasonable for routine operations involving a single stage in which the service time is generally short, with only occasional longer service time, as in the case of a toll booth where changing big bills may occasionally be necessary. However, if the service consists of several operations, the total service time would not be exponentially distributed even though the individual operation times are exponentially distributed. In general, therefore, the service time may have to be some other distribution, although the arrivals may reasonably be Poisson.

Consider the instances at which departures from a service system take place, which will be referred to as *departure points*. The corresponding queue length (not including the one just departed) is defined as the *state of the system* at that departure point. If the system is at state i at a preceding departure point, the event that it will be at state j at the present departure point is equivalent to the event that $(j - i + 1)$ arrivals have occurred during the last service period. For a Poisson arrival process with mean rate λ, the probability of n arrivals within a service duration x is

$$k_n = \frac{e^{-\lambda x}(\lambda x)^n}{n!} \tag{3.57}$$

Since the service time is random, this probability becomes

$$k_n = \int_0^\infty \frac{e^{-\lambda x}(\lambda x)^n}{n!} f_X(x)\, dx \tag{3.58}$$

where $f_X(x)$ is the PDF of the service time. Hence, the transition probability between states i and j for two successive departure points is

$$p_{i,j} = k_{j-i+1}; \quad j - i + 1 \geq 0, \quad i > 0 \tag{3.59}$$

For the case in which the queue length is zero at the previous departure point, that is, $i = 0$, the transition to state j will involve j additional arrivals (immediately after the next arrival), during the period required to process the next arrival. Hence,

$$p_{0,j} = k_j \tag{3.60}$$

and $p_{i,j} = 0$ in all other cases, that is, $j - i + 1 < 0$ and $i > 0$.

From Eq. 3.14, the one-stage transition probabilities are related to the steady-state probabilities as

$$p_j^* = \sum_{i=0}^{\infty} p_{i,j} p_i^*; \quad j = 0, 1, 2, \ldots \tag{3.61}$$

Substituting Eqs. 3.59 and 3.60 into Eq. 3.61, we obtain

$$\begin{aligned} p_j^* &= \sum_{i=1}^{\infty} p_{i,j} p_i^* + p_0^* k_j \\ &= \sum_{i=1}^{j+1} k_{j-i+1} p_i^* + p_0^* k_j; \quad j = 0, 1, \ldots \end{aligned} \tag{3.62}$$

or

$$\begin{aligned} p_0^* &= p_0^* k_0 + p_1^* k_0 \\ p_1^* &= p_0^* k_1 + p_1^* k_1 + p_2^* k_0 \\ &\vdots \\ p_n^* &= p_0^* k_n + p_1^* k_n + p_2^* k_{n-1} + \cdots + p_{n+1}^* k_0 \end{aligned} \tag{3.63}$$

The steady-state probabilities are obtained as follows.

Observe that the moment-generating function (see p. 96, Vol. I) of the steady-state probabilities in Eq. 3.62 is

$$\begin{aligned} G_P(s) &= \sum_{j=0}^{\infty} p_j^* e^{sj} \\ &= \sum_{j=0}^{\infty} \sum_{i=1}^{j+1} k_{j-i+1} p_i^* e^{sj} + \sum_{j=0}^{\infty} p_0^* k_j e^{sj} \\ &= \sum_{i=1}^{\infty} \sum_{j=i-1}^{\infty} k_{j-i+1} p_i^* e^{sj} + p_0^* \sum_{j=0}^{\infty} k_j e^{sj} \\ &= \sum_{i=1}^{\infty} \sum_{l=0}^{\infty} k_l p_i^* e^{s(l+i-1)} + p_0^* G_n(s) \end{aligned}$$

where $l = j - i + 1$ and $G_n(s)$ is the moment-generating function of n, the number of arrivals during a service period, whose probabilities are k_n as defined in Eq. 3.58. Hence,

$$\begin{aligned} G_P(s) &= e^{-s} \sum_{i=1}^{\infty} p_i^* e^{si} \sum_{l=0}^{\infty} k_l e^{sl} + p_0^* G_n(s) \\ &= e^{-s} [G_P(s) - p_0^*] G_n(s) + p_0^* G_n(s) \end{aligned}$$

Therefore,

$$G_P(s) = \frac{p_0^* G_n(s)[1 - e^{-s}]}{1 - e^{-s} G_n(s)}$$

According to L'Hopital's rule,

$$\lim_{s \to 0} G_P(s) = \lim_{s \to 0} \frac{p_0^*[1 - e^{-s}]G_n'(s) + p_0^* G_n(s)e^{-s}}{-e^{-s}G_n'(s) + e^{-s}G_n(s)}$$

$$= \frac{p_0^* G_n(0)}{G_n(0) - G_n'(0)} \tag{3.64}$$

Observe that

$$G_n(0) = \sum_{j=0}^{\infty} k_j = 1.0$$

and

$$E(n) = G_n'(0) = \sum_{j=0}^{\infty} jk_j$$

$$= \sum_{j=0}^{\infty} \int_0^{\infty} j \frac{e^{-\lambda x}(\lambda x)^j}{j!} f_X(x) \, dx$$

$$= \int_0^{\infty} \lambda x \left(\sum_{j=1}^{\infty} \frac{e^{-\lambda x}(\lambda x)^{j-1}}{(j-1)!} \right) f_X(x) \, dx$$

$$= \frac{\lambda}{\mu} \tag{3.65}$$

where $1/\mu$ is the mean service time. Moreover, $G_P(0) = 1.0$. Hence, Eq. 3.64 yields

$$p_0^* = 1 - \frac{\lambda}{\mu} \tag{3.66}$$

The probabilities $p_j^*, j \geq 1$ can then be determined recursively from Eq. 3.63.

The expected queue length at the steady-state condition may be determined as follows.

Let N and N' denote the queue lengths at two successive departure points. For $N > 0$,

$$N' = N + n - 1 \tag{3.67}$$

where n is the number of arrivals within one service period, whose probability mass function is given in Eq. 3.58.

For $N = 0$, the queue length at the next departure point equals the number of additional arrivals immediately after the next arrival, that is, during the service period of the next arrival. Hence, for $N = 0$,

$$N' = n \tag{3.68}$$

By introducing the Kronecker delta function $\delta(-)$, we can combine Eqs. 3.67 and 3.68 into the following equation:

$$N' = N + n - 1 + \delta(N) \tag{3.69}$$

Squaring Eq. 3.69 and observing that $[\delta(N)]^2 = \delta(N)$ yields

$$N'^2 = N^2 + n^2 + 1 - \delta(N) + 2Nn - 2N - 2n + 2n\delta(N)$$

Taking the expected values on both sides and observing that N and n are statistically independent, we obtain

$$E(N'^2) = E(N^2) + E(n^2) + 1 - E[\delta(N)] + 2E(N)E(n)$$
$$- 2E(N) - 2E(n) + 2E(n)E[\delta(N)] \qquad (3.70)$$

At the steady-state condition, $E(N'^2) = E(N^2)$ and $E[\delta(N)] = p_0^* = 1 - \rho$ where $\rho = \lambda/\mu$ from Eq. 3.65; moreover, with $E(n) = \rho$, Eq. 3.70 becomes

$$2E(N)(1 - \rho) = E(n^2) + \rho - 2\rho + 2\rho(1 - \rho)$$
$$= E[n(n - 1)] + 2\rho(1 - \rho)$$

Hence,

$$E(N) = \rho + \frac{E[n(n - 1)]}{2(1 - \rho)}$$

Applying Eq. 3.58, we obtain

$$E[n(n - 1)] = \sum_{j=0}^{\infty} j(j - 1) \int_0^{\infty} \frac{e^{-\lambda x}(\lambda x)^j}{j!} f_X(x)\, dx$$
$$= \int_0^{\infty} (\lambda x)^2 f_X(x)\, dx \sum_{j=2}^{\infty} \frac{e^{-\lambda x}(\lambda x)^{j-2}}{(j - 2)!}$$
$$= \lambda^2 \mathrm{Var}(X) + \rho^2$$

Hence, the steady-state mean queue length is

$$L = E(N) = \rho + \frac{\lambda^2 \mathrm{Var}(X) + \rho^2}{2(1 - \rho)} \qquad (3.71)$$

EXAMPLE 3.22

A ready-mix concrete supply firm delivers concrete on a per order basis. Suppose orders are received according to a Poisson process with a mean rate of one per hour. Because of variations in the destination and volumes of deliveries, the time required to process and deliver each order is described by a gamma distribution with a mean of 40 minutes and a c.o.v. of 50%. Determine the pertinent probabilities and the expected queue length at the steady-state condition.

For a gamma distribution with parameters v and k, the PDF is

$$f_X(x) = \frac{v(vx)^{k-1} e^{-vx}}{\Gamma(k)}$$

where

$$E(X) = \frac{k}{v}$$

$$\delta_X = \frac{1}{\sqrt{k}}$$

From the given data,

$$\delta_X = 0.5 = \frac{1}{\sqrt{k}}$$

Hence, $k = 4$, and $v = k/E(X) = 4/\frac{2}{3} = 6$. Moreover, $\lambda = 1$; hence, $\rho = \lambda/\mu = \frac{2}{3}$. Therefore, according to Eq. 3.66, the probability of zero queue length at steady-state is

$$p_0^* = 1 - \tfrac{2}{3} = \tfrac{1}{3}$$

and from Eq. 3.71 the steady-state mean queue length is

$$L = \frac{2}{3} + \frac{1^2(0.5 \times \frac{2}{3})^2 + (\frac{2}{3})^2}{2(1 - \frac{2}{3})} = 1.5$$

To obtain the steady-state probabilities for other queue lengths, values of k_n are first determined. With Eq. 3.58

$$k_n = \int_0^\infty \frac{e^{-\lambda x}(\lambda x)^n}{n!} \frac{v(vx)^{k-1}e^{-vx}}{\Gamma(k)} \, dx$$

$$= \int_0^\infty \frac{e^{-(\lambda+v)x}x^{n+k-1}(v+\lambda)^{n+k}}{\Gamma(n+k)} \, dx \cdot \frac{\lambda^n v^k \Gamma(n+k)}{(v+\lambda)^{n+k}\Gamma(k)n!}$$

$$= \frac{\lambda^n v^k}{(v+\lambda)^{n+k}} \frac{\Gamma(n+k)}{\Gamma(k)n!}$$

Substituting the values of λ, k, and v, we have

$$k_n = \frac{1^n 6^4}{(6+1)^{n+4}} \frac{\Gamma(n+4)}{\Gamma(4)n!}$$

$$= \frac{0.09}{7^n} \frac{(n+3)!}{n!}$$

or, specifically,

$$k_0 = 0.54;\, k_1 = 0.30;\, k_2 = 0.110;\, k_3 = 0.032;\, k_4 = 0.008;$$

$$k_5 = 0.002;\, k_6 = k_7 = \cdots k_\infty = 0.0$$

With the first equation of Eq. 3.63; that is,

$$p_0^* = p_0^* k_0 + p_1^* k_0$$

we have

$$\tfrac{1}{3} = \tfrac{1}{3} \times 0.54 + p_1^* \times 0.54$$

Thus,

$$p_1^* = 0.284$$

Similarly, the other steady-state probabilities may be obtained recursively with Eq. 3.63, yielding

$$p_2^* = 0.174;\, p_3^* = 0.097;\, p_4^* = 0.052;\, p_5^* = 0.027;\, p_6^* = 0.013;\, p_7^* = 0.006, \text{etc.}$$

EXAMPLE 3.23

In Example 3.22, suppose it takes exactly 40 minutes to process an order. What would be the queue length probabilities and the expected queue length at the steady-state condition? From Eq. 3.66, the probability of zero queue length is

$$p_0^* = 1 - \tfrac{2}{3} = \tfrac{1}{3}$$

which is the same as that for Example 3.22. From Eq. 3.71, since $Var(X) = 0$, the mean queue length is

$$L = \frac{2}{3} + \frac{(\frac{2}{3})^2}{2(1 - \frac{2}{3})} = 1.33$$

Hence, the steady-state queue length is shorter if the uncertainty in the service time is removed.

Since k_n is the probability of n arrivals during one service period, this is simply the probability of n occurrences in a 40-minute period. Hence, for Poisson arrivals,

$$k_0 = \frac{e^{-1 \times 2/3}(1 \times \frac{2}{3})^0}{0!} = 0.513$$

Similarly $k_1 = 0.342$; $k_2 = 0.114$; $k_3 = 0.025$; $k_4 = 0.004$ and $k_5 = 0.001$. Recursively applying Eq. 3.63, we obtain

$$p_1^* = 0.316; p_2^* = 0.183; p_3^* = 0.090; p_4^* = 0.043; p_5^* = 0.023, \text{etc.}$$

Comparing these probabilities with those in Example 3.22, we can see that the probabilities of long queue lengths (e.g., exceeding two) are smaller here than those of Example 3.22.

3.2.4 Distribution of Waiting Time

The total time spent by a calling unit in a queue is defined as the *waiting time T*. This includes the waiting time before and during service. Under steady-state condition, the CDF of T may be determined using the theorem of total probability as

$$P(T \le t) = \sum_{n=0}^{\infty} P(T \le t \mid N = n)p_n^* \tag{3.72}$$

where $N = n$ is the event that the queue length is n at steady-state.

Consider a queueing system with a single server. Suppose that the arrivals constitute a Poisson process with mean rate λ and the service time is exponentially distributed with mean $1/\mu$. Given that there is no queue in the system, the CDF of the waiting time for an arriving unit is simply the probability that its service time X does not exceed t; that is,

$$P(T \le t \mid N = 0) = P(X \le t) = \int_0^t \mu e^{-\mu x} \, dx$$

If there is already one unit in the system, the waiting time T will be the sum of its own service time and the additional time required to serve the unit already in the system. Because of the memoryless characteristic of the exponential distribution, this additional service time required for the previous caller is also exponentially distributed with mean $1/\mu$.

If we extend this to the case of $N = n - 1$, that is, $n - 1$ units in the system, the waiting time will be the sum of n identically distributed exponential random variables, which can be shown to follow a gamma distribution (see, e.g., Hoel et al., 1971) as

$$f_X(x) = \frac{\mu^n x^{n-1} e^{-\mu x}}{(n - 1)!}$$

if the service times are assumed to be statistically independent. Hence,

$$P(T \leq t \,|\, N = n - 1) = \int_0^t \frac{\mu^n x^{n-1} e^{-\mu x}}{(n - 1)!} \, dx; \qquad \text{for } n = 1, 2, \ldots$$

and Eq. 3.72 becomes

$$P(T \leq t) = \sum_{n=1}^{\infty} p^*_{n-1} \int_0^t \frac{\mu^n x^{n-1} e^{-\mu x}}{(n - 1)!} \, dx$$

Using the steady-state probabilities of Eq. 3.55, we obtain

$$P(T \leq t) = (1 - \rho) \sum_{n=1}^{\infty} \rho^{n-1} \int_0^t \frac{\mu^n x^{n-1} e^{-\mu x}}{(n - 1)!} \, dx \tag{3.73}$$

where $\rho = \lambda/\mu$.

Differentiating Eq. 3.73 with respect to t yields the PDF of T

$$f_T(t) = (1 - \rho) \sum_{n=1}^{\infty} \rho^{n-1} \left[\frac{\mu^n t^{n-1} e^{-\mu t}}{(n - 1)!} \right]$$

$$= \mu(1 - \rho) e^{-\mu t} \sum_{n=1}^{\infty} \frac{(\mu \rho t)^{n-1}}{(n - 1)!}$$

$$= \mu(1 - \rho) e^{-\mu t(1 - \rho)} \tag{3.74}$$

which is an exponential distribution with parameter $\mu(1 - \rho)$. The corresponding mean waiting time, therefore, is

$$E(T) = \frac{1}{\mu(1 - \rho)} \tag{3.75}$$

For queueing systems with more than one server, and/or where the service time is not exponentially distributed, the determination of the distribution of the waiting time would be more involved. For a steady-state queueing process, Little (1961) has shown that under most conditions, the mean queue length is

$$L = \lambda E(T) \tag{3.76}$$

where $E(T)$ is the expected waiting time, and λ is the mean arrival rate (assumed to be constant). Equation 3.76 provides a simple formula to compute the expected waiting time where L and λ are known.

In particular, if the arrivals are Poisson, with a single server whose service time is exponentially distributed, the expected waiting time is given by Eq. 3.75. Therefore, Eq. 3.76 yields

$$L = \lambda E(T) = \frac{\lambda}{\mu(1 - \rho)} = \frac{\lambda}{\mu - \lambda}$$

this is the same result as Eq. 3.56.

EXAMPLE 3.24

In Example 3.18, suppose that the two independent controllers are combined into one group of controllers that can process landings at the rate of 50 per hour. Assume the processing time of each landing is exponentially distributed. What is the probability that an arriving plane will have to wait more than 2 minutes before landing?

Using the PDF in Eq. 3.74, we obtain

$$P\left(T > \frac{2}{60}\right) = \int_{2/60}^{\infty} 50\left(1 - \frac{40}{50}\right)e^{-50(1-40/50)t}\, dt$$

$$= 0.717$$

The expected waiting time (including landing) is given by Eq. 3.75 as

$$E(T) = \frac{1}{50(1 - \frac{40}{50})}$$

$$= 0.1 \text{ hour or 6 minutes}$$

For more complicated queueing problems including those involving multiple servers, finite queues, or priority-discipline queueing models, the corresponding steady-state probabilities, expected queue lengths, and waiting times are available; for example, see Saaty (1961), Hillier and Lieberman (1967).

3.3   AVAILABILITY PROBLEMS

3.3.1   Introduction

Many engineering systems can be classified into two states; for example, *operating* and *nonoperating, safe* and *unsafe* states, or *satisfactory* and *unsatisfactory* states. The quality of a stream may or may not be in a polluted state; a section of a highway pavement may or may not contain severe cracks; equipment may or may not be operating; and a structural member may or may not be damaged. In many cases, a system in a *failed* or *nonoperating* state may be returned to an *operating* state through proper repair, whereas in time an operating state may eventually become a failed state. The time between breakdowns as well as the time required for repair are generally random variables. Moreover, in practice, the performance of a system may be monitored only at regular intervals. Hence, a *failed* state may not be discovered immediately and thus not corrected until the next scheduled inspection.

The *availability* of a system may be defined as the probability that a system is in the *operating* state. Consequently, the system availability probability at a given time and its average availability over a long period of time would be of interest to engineers and managers.

3.3.2   Continuous Inspection–Repair Process

The performance of engineering systems may be continuously monitored. If a system is discovered to be in the *failed* state, repair is immediately undertaken to restore it to the *operating* state. Suppose the time until the breakdown of a system is X, with PDF $f_X(x)$; then, the rate at which an operating state would become a failed state at time t is given by the hazard rate (or function), defined as the conditional probability density that the system will fail at time t given that it has survived through time t. Namely,

$$\lambda(t) = \frac{f_X(t)}{P(X \geq t)} = \frac{f_X(t)}{\int_t^{\infty} f_X(x)\, dx} \tag{3.77}$$

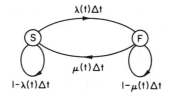

Figure 3.3   Transition probabilities between states S and F.

Similarly, if $f_Y(y)$ is the PDF of the repair time, the corresponding rate at which a *failed* state under repair will revert to an *operating* state is

$$\mu(t) = \frac{f_Y(t)}{\int_t^\infty f_Y(y)\,dy} \tag{3.78}$$

The transition between states for the system at time t and the corresponding transition probabilities are depicted in Fig. 3.3. S and F correspond, respectively, to the *operating* and *failed* states. In a small time interval, Δt, there is a probability $\lambda(t)\,\Delta t$ that the system will go from state S to state F; the probability that the system will remain in state S is, therefore, $1 - \lambda(t)\,\Delta t$. Conversely, the transition from state F to state S involves the probabilities $\mu(t)\,\Delta t$ and $1 - \mu(t)\,\Delta t$, as shown in Fig. 3.3. Observe that this is an example of a continuous-parameter Markov chain, as discussed in Section 3.1.8. If $p_S(t)$ and $p_F(t)$ denote the probabilities of the system in states S and F, respectively, at time t, the probability that the system will be in state F at time $(t + \Delta t)$ is

$$p_F(t + \Delta t) = p_S(t) \cdot \lambda(t)\,\Delta t + p_F(t)[1 - \mu(t)\,\Delta t] \tag{3.79}$$

Since $p_S(t) = 1 - p_F(t)$, Eq. 3.79 becomes

$$p_F(t + \Delta t) - p_F(t) = \lambda(t)\,\Delta t - p_F(t)\lambda(t)\,\Delta t - p_F(t)\mu(t)\,\Delta t$$

As $\Delta t \to 0$, this yields the differential equation,

$$\frac{dp_F(t)}{dt} + p_F(t)[\lambda(t) + \mu(t)] = \lambda(t)$$

whose solution is

$$p_F(t) = \exp\left\{ - \int_0^t [\lambda(t) + \mu(t)]\,dt \right\} \int_0^t \exp\left\{ \int_0^t [\lambda(t) + \mu(t)]\,dt \right\} \cdot \lambda(t)\,dt$$

$$+ c \exp\left\{ - \int_0^t [\lambda(t) + \mu(t)]\,dt \right\} \tag{3.80}$$

The constant $c = 0$ if the system starts from state S at $t = 0$. Hence,

$$p_F(t) = \exp[-Q(t)] \int_0^t \exp[Q(t)]\lambda(t)\,dt \tag{3.81}$$

where $Q(t) = \int_0^t [\lambda(t) + \mu(t)]\,dt$. The probability that the system will be in the *failed* state at a given time t may then be determined from Eqs. 3.81, 3.77, and 3.78 for given density functions $f_X(x)$ and $f_Y(y)$.

The instantaneous availability at a given time is given by the probability that the system is in the operating state at time t; that is,

$$A(t) = p_S(t) = 1 - p_F(t) \tag{3.82}$$

whereas the average availability through time t is

$$\bar{A}(t) = \frac{1}{t} \int_0^t A(t)\, dt \tag{3.83}$$

Exponential Breakdown and Repair Time Consider the case in which X and Y are both exponentially distributed with means $1/\lambda$ and $1/\mu$, respectively.

For the first cycle of operations, the breakdown time X ranges from 0 to ∞. Hence,

$$f_X(x) = \lambda e^{-\lambda x}; \qquad 0 \le x < \infty$$

Applying Eq. 3.77, we obtain

$$\lambda(t) = \frac{\lambda e^{-\lambda t}}{\int_t^\infty \lambda e^{-\lambda x}\, dx} = \lambda$$

Suppose the breakdown time in the first cycle is τ; the repair time Y would range from τ to ∞, which is a shifted exponential distribution with PDF

$$f_Y(y) = \mu e^{-\mu(y-\tau)}; \qquad \tau \le y < \infty$$

then applying Eq. 3.78, we obtain

$$\mu(t) = \frac{\mu e^{-\mu(t-\tau)}}{\int_t^\infty \mu e^{-\mu(y-\tau)}\, dy} = \mu$$

Therefore, regardless of the time at which the first breakdown occurs, the mean repair rate is constant. It can be shown for subsequent cycles of operation that $\lambda(t)$ and $\mu(t)$ will be equal to λ and μ. Hence,

$$Q(t) = \int_0^t (\lambda + \mu)\, dt = (\lambda + \mu)t \tag{3.84}$$

From Eq. 3.81,

$$p_F(t) = e^{-(\lambda+\mu)t} \int_0^t e^{(\lambda+\mu)t} \lambda\, dt$$

$$= \frac{\lambda}{\lambda + \mu} [1 - e^{-(\lambda+\mu)t}] \tag{3.85}$$

The instantaneous availability, Eq. 3.82, is then

$$A(t) = 1 - \frac{\lambda}{\lambda + \mu} [1 - e^{-(\lambda+\mu)t}] \tag{3.86}$$

whereas the average availability, according to Eq. 3.83, is

$$A(t) = \frac{1}{t} \int_0^t \left\{ 1 - \frac{\lambda}{\lambda + \mu} [1 - e^{-(\lambda + \mu)t}] \right\} dt$$

$$= \frac{\mu}{\lambda + \mu} + \frac{\lambda}{(\lambda + \mu)^2} \frac{1 - e^{-(\lambda + \mu)t}}{t} \qquad (3.87)$$

As $t \to \infty$, the availability reaches a steady-state value; namely,

$$A(\infty) = 1 - \frac{\lambda}{\lambda + \mu} [1 - 0]$$

$$= \frac{\mu}{\lambda + \mu} \qquad (3.88)$$

Observe that this is the same as $\bar{A}(\infty)$ of Eq. 3.87.

Eqs. 3.85 to 3.88 were derived for a system starting from an operating state S. If a system starts from a failed state, similar expressions for availability may be derived accordingly.

EXAMPLE 3.25

An engineering consulting firm has its own computer. From past experiences, it was observed that the average length of trouble-free service for the computer is 20 days. During a breakdown it takes, on the average, two days to restore the computer to its operating condition. Assume that the time between breakdowns and the repair time are both exponentially distributed. If the computer is in operating condition now, determine:

(a) The probability that it will not break down within the next month (30 days).
(b) The probability that it will be down on the day one week (7 days) from now; and one month (30 days) from now.
(c) The proportion of time that the computer will be in operating condition over a long period of time.
(d) The average availability within the first week (7 days), the first month (3 days), and the first year (365 days).

Solution

The parameters λ and μ are simply the reciprocals of the mean breakdown and repair times, which are, respectively, 0.05 and 0.5.

(a) The probability that a breakdown will not occur within the next 30 days is

$$P(X > 30) = 1 - \int_0^{30} \frac{1}{20} e^{-t/20} dt = 0.223$$

(b) From Eq. 3.85, the probabilities that it will be down on the 7th and 30th day are, respectively,

$$p_F(7) = \frac{0.05}{0.05 + 0.5} [1 - e^{-(0.05 + 0.5)7}]$$

$$= 0.091(1 - 0.021) = 0.089$$

and

$$p_F(30) = 0.091[1 - e^{-(0.55)30}] = 0.091$$

(c)   At steady-state, $p_F(\infty) = 0.091$; hence, $p_S(\infty) = 1 - 0.091 \simeq 0.91$, that is, the computer will be operating approximately 91 % of the time.

(d)   The average availability within the next week is calculated from Eq. 3.87 as

$$\bar{A}(7) = \frac{0.5}{0.55} + \frac{0.05}{(0.55)^2} \frac{1 - e^{-0.55 \times 7}}{7} = 0.933$$

whereas the average availability within the next month is

$$\bar{A}(30) = 0.91 + \frac{0.05}{(0.55)^2} \frac{1 - e^{-0.55 \times 30}}{30} = 0.9155$$

and within the next year

$$\bar{A}(365) = 0.91 + \frac{0.05}{(0.55)^2} \frac{1 - e^{-0.55 \times 365}}{365} = 0.9105$$

Observe that the average availability (as the time period increases) approaches the long-term availability of 0.91.

EXAMPLE 3.26

The rate of excavating a tunnel under normal conditions is 10 m/day. However, for each large boulder encountered during a tunneling operation, a delay will be incurred, which is exponentially distributed with a mean of 1 day. Suppose the occurrence of large boulders along the proposed tunnel is a Poisson process with a mean distance between boulders of 40 m. If no boulder exists at the starting point of the tunnel excavation,

(a)   What is the proportion of time during the first week of tunneling that normal conditions will prevail?

(b)   If the entire project requires three months (90 days) to complete, what is the average number of days that the excavation can operate under normal conditions?

Solution

(a)   This problem may be modeled as an availability problem. Suppose availability is defined here as the probability of excavation under normal condition. Since the mean distance between large boulders is 40 m and the rate of tunneling in the absence of boulders is 10 m/day, the mean duration of normal excavation condition is 4 days whereas the mean duration of repair (i.e., delay caused by the occurrence of boulders) is 1 day. Since exponential distributions are assumed for these two durations, Eq. 3.87 may be applied with $\lambda = \frac{1}{4}$ and $\mu = 1$. Hence, during the first week of tunnel excavation, the proportion of time that it will remain under normal conditions is given by

$$\bar{A}(7) = \frac{1}{(1 + \frac{1}{4})} + \frac{\frac{1}{4}}{(1 + \frac{1}{4})^2} \frac{1 - e^{-7(1 + 1/4)}}{7} = 0.82$$

(b)   Over a period of 90 days, the average availability may be approximated by Eq. 3.88 as

$$A(\infty) = \frac{1}{1 + \frac{1}{4}} = 0.8$$

In other words, about 80 % of the 90 days, or 72 days, can be expected to be under normal conditions.

EXAMPLE 3.27

A pipeline is susceptible to corrosion depending on the environmental conditions. Suppose the corrosive parts of the pipeline appear as in Fig. E3.27 where C and $\bar{C}$ denote the corroded

Figure E3.27   Corroded sections in a pipeline.

and noncorroded sections along the pipeline. The length of a corroded section is exponentially distributed with a mean of 100 ft, whereas each noncorroded section is also exponentially distributed with a mean length of 400 ft.

(a)   If no corrosion is observed at a given location,
 (i)   What is the probability that it will be corroded 200 ft away?
 (ii)   What is the expected total length of the corroded sections within the next 200 ft?

(b)   Suppose a 100-foot section of the pipeline is selected at random for inspection. What is the probability that no corrosion will be found?
(c)   Instead of inspecting 1 section of 100 feet, ten sections, each 10 ft long, are randomly selected along the pipeline for inspection. What is the probability of finding no corrosion?

Solution

(a)   This problem is similar to the availability problem, in which the noncorroded section $\bar{C}$ is the satisfactory state S and the corroded section C is the failure state F. The mean rates λ and μ are, respectively, $\frac{1}{400}$ and $\frac{1}{100}$. Given that there was no corrosion, state $\bar{C}$, the probability that corrosion will be discovered 200 ft away is given by Eq. 3.85 as

$$p_C(200) = \frac{\frac{1}{400}}{\frac{1}{400}+\frac{1}{100}}\{1 - e^{-200[(1/400)+(1/100)]}\}$$

$$= 0.184$$

Over the next 200 ft of the pipeline, the fraction of noncorroded pipe is simply the average availability over 200 ft; from Eq. 3.87, this is

$$\bar{A}(200) = \frac{\frac{1}{100}}{\frac{1}{400}+\frac{1}{100}} + \frac{\frac{1}{400}}{\left(\frac{1}{400}+\frac{1}{100}\right)^2}\frac{1 - e^{-200[(1/400)+(1/100)]}}{200} = 0.873$$

Thus, the fraction of corroded pipe is $1 - 0.873 = 0.127$ and the expected total length of corroded pipes over the next 200 ft is $0.127 \times 200 = 25.4$ ft.
(b)   For a 100-foot section of the pipeline selected at random, the event that the entire section is corrosion-free requires that both the starting point of the section is not corroded, and no corrosion is found within the next 100 ft. The probability of the first event is simply the steady-state probability of a noncorroded state, namely,

$$p_{\bar{C}}(\infty) = A(\infty) = \frac{\frac{1}{100}}{\frac{1}{400}+\frac{1}{100}} = 0.8$$

The event of no corrosion within 100 ft, given that the pipe starts out at state $\bar{C}$, is equivalent to the event that the length of the noncorroded state exceeds 100 ft. Hence, the probability of a corrosion-free 100-foot section is

$$P(\bar{C}_{100}) = p_{\bar{c}}(\infty)P(L_{\bar{c}} > 100)$$

$$= 0.8 \int_{100}^{\infty} \frac{1}{400} e^{-x/400}\, dx = 0.623$$

(c)   Similarly for each 10-foot section picked at random, the probability that it is entirely corrosion-free is

$$P(\bar{C}_{10}) = p_{\bar{c}}(\infty)P(L_{\bar{c}} > 10)$$

$$= 0.8 \int_{10}^{\infty} \frac{1}{400} e^{-x/400}\, dx = 0.780$$

If we assume that the ten 10-foot sections are statistically independent, the probability that all ten sections are corrosion-free is

$$P(\bar{C}) = (0.780)^{10} = 0.084$$

which is significantly smaller than the probability of part (b).

EXAMPLE 3.28

Delays in subsurface investigations of a construction site may be caused by: bad weather, equipment breakdown, operator absences, and standby resulting from bad job organization. Suppose each one of these four classes of delay occurs according to independent Poisson processes with mean rates of 0.124, 0.129, 0.069, and 0.105 per day, respectively. Moreover, the duration of each delay (regardless of class) follows an exponential distribution with a mean duration of 0.4 per day.

(a)   If a subsurface investigation job requires five days of work, what is the probability that no delay of any kind will occur?
(b)   For a normal job, what fraction of the time is expected to be wasted because of delays?
(c)   Suppose the subsurface investigation is presently proceeding smoothly. What is the probability that the operation will be suspended due to delay exactly one day from now?
(d)   Suppose exactly two delays occur during the subsurface investigation. Determine the probability that the total delay will not exceed half a day. Assume the durations of the two delays are statistically independent.

Solution

(a)   The combined occurrences of all delays follow a Poisson process (see Chapter 4, Vol. I) with a mean rate

$$v_T = 0.124 + 0.129 + 0.069 + 0.105 = 0.427.$$

The probability of no delay over a five-day period is, therefore,

$$P(N_T = 0) = \frac{(0.427 \times 5)^0 e^{-0.427 \times 5}}{0!} = 0.118$$

(b)   For a normal job, without any information on the initial condition, the fraction of time wasted in delays is given by the average unavailability at the steady-state condition. Hence, applying Eq. 3.88, we obtain

$$\bar{A}(\infty) = \frac{0.427}{0.427 + \dfrac{1}{0.4}} = 0.146$$

(c)   If there is no delay at present, the probability of work being suspended one day from now is given by the unavailability

$$1 - A(1) = \frac{0.427}{0.427 + 2.5}(1 - e^{-0.427-2.5}) = 0.138$$

(d)   The total delay X is the sum of the two independent identically distributed exponential random variables, each with a mean of 0.4 day. The result is a gamma distribution with PDF

$$f_X(x) = 2.5^2 e^{-2.5x}$$

Therefore, the probability that the total delay is less than half a day is

$$P(X < 0.5) = \int_0^{0.5} 6.25e^{-2.5x}\,dx = 0.355$$

Availability with Nonexponential Breakdown and Repair Times   Thus far, all time intervals between breakdowns and all repair times are assumed to be exponentially distributed. Consequently the rates of transition between states are constant and, thus, independent of time t. In general, $\lambda(t)$ and $\mu(t)$ will be functions of t. In the general case, the probabilities between states at any time and at the steady-state condition may be derived using transform analysis as follows.

Suppose $p_{SF}(t)$ denotes the probability that a system starting from state S at time zero will be in state F after time t; $f_X(x)$ denotes the PDF of the breakdown time. By conditioning on the time of the first breakdown, namely x, we obtain

$$p_{SF}(t) = \int_0^t p_{FF}(t - x)f_X(x)\,dx \tag{3.89}$$

where $p_{FF}(t - x)$ is the probability that the system starting in state F will also be in state F after time $(t - x)$. Define the *Laplace transform* of the PDFs of the breakdown time X and the repair time Y, respectively, as

$$G_X(s) = \int_0^\infty e^{-sx}f_X(x)\,dx$$

and

$$G_Y(s) = \int_0^\infty e^{-sy}f_Y(y)\,dy$$

Then, the Laplace transform of Eq. 3.89 is

$$P_{SF}(s) = \int_0^\infty e^{-st}p_{SF}(t)\,dt = G_X(s)P_{FF}(s) \tag{3.90}$$

Similarly, the probability $p_{FF}(t)$ can be expressed as the sum of the probabilities of two mutually exclusive events; namely,

(i)   Repair is not completed by time t *and* the system is therefore in state F at time t.

(ii)   Repair is completed within time t *and* the system moves from state S to state F in time $(t - y)$.

Hence,

$$p_{FF}(t) = \left[1 - \int_0^t f_Y(y)\, dy\right] + \int_0^t p_{SF}(t - y) f_Y(y)\, dy$$

or, in terms of the Laplace transform

$$P_{FF}(s) = \frac{1}{s} - \frac{G_Y(s)}{s} + G_Y(s) P_{SF}(s) \tag{3.91}$$

Combining Eqs. 3.90 and 3.91 yields

$$P_{SF}(s) = \frac{1}{s} \frac{G_X(s)[1 - G_Y(s)]}{1 - G_X(s) G_Y(s)} \tag{3.92}$$

$P_{SF}(s)$ and $P_{FF}(s)$ may be inverted to obtain the transition probabilities between states at any time instant. For the steady-state solution, observe that expanding e^{-sx} in a Taylor series,

$$G_X(s) = 1 - s\mu_X + \mathbf{o}(s^2)$$

where μ_X is the mean breakdown time and $\mathbf{o}(s^2)$ denotes all terms of order s^2. Similarly,

$$G_Y(s) = 1 - s\mu_Y + \mathbf{o}(s^2)$$

From Eq. 3.92 as $s \to 0$, we have

$$\lim_{s \to 0} s P_{SF}(s) = \frac{\mu_Y}{\mu_X + \mu_Y}$$

Alternatively, using integration by parts, we can show that

$$s P_{SF}(s) = \int_0^\infty s e^{-st} p_{SF}(t)\, dt$$

$$= p_{SF}(0) + \int_0^\infty e^{-st} p'_{SF}(t)\, dt$$

where $p'_{SF}(t)$ is the first derivative of $p_{SF}(t)$ with respect to t. In the limit, as $s \to 0$,

$$\lim_{s \to 0} s P_{SF}(s) = p_{SF}(0) + [p_{SF}(\infty) - p_{SF}(0)]$$

$$= p_{SF}(\infty)$$

Therefore, the steady-state probability of the system in state F is

$$p_F^* = p_{SF}(\infty) = \frac{\mu_Y}{\mu_X + \mu_Y}$$

and the long-term average availability is

$$p_S^* = 1 - p_F^* = \frac{\mu_X}{\mu_X + \mu_Y} \tag{3.93}$$

which is the same as that obtained by Eq. 3.88 for exponential breakdown and repair times (observing in such a case $\mu_X = 1/\lambda$ and $\mu_Y = 1/\mu$). However, the result in Eq. 3.93 is more general as X and Y can have any distributions. If the repair time is a constant, τ_r, instead of a random variable, Eq. 3.93 is still applicable; the long-term availability becomes

$$p_S^* = \frac{\mu_X}{\mu_X + \tau_r} \tag{3.94}$$

EXAMPLE 3.29

Suppose an environmental protection agency requires that storm water must pass through a treatment process before it is disposed into a stream. A pump is used to convey the storm water from a sewer pipe to a treatment plant. The time between breakdowns of the pump follows a Weibull distribution with a mean of 500 hours and a c.o.v. of 20%, and the distribution of the repair time is $N(10, 2)$ hours. Assume that each repair will restore the pump to its new condition. The probability that the pump is operating normally at any given time is, according to Eq. 3.93,

$$p_S^* = \frac{500}{500 + 10} = 0.98$$

EXAMPLE 3.30

An old lighthouse tower is deemed to be unsafe during hurricanes. It is required that all human occupants be removed from the tower during such events. Suppose the frequency of hurricanes at the lighthouse location is 20 times a year (365 days) on the average, and the duration of each hurricane is lognormally distributed with a mean of two days and a c.o.v. of 30%. What is the mean percentage of time that the lighthouse is out of service?

There are two states describing the service condition of the lighthouse tower; namely, service (S) and no service (F). The mean duration of each occurrence of state F is two days. On the average, state F occurs 20 times in a year; therefore, the mean time between two successive occurrences of F is $\frac{365}{20} = 18.25$ days. Hence, after the tower is restored to service, the mean time until it is out of service again is $18.25 - 2 = 16.25$ days. The probability that the tower is out of service at a particular time is calculated from Eq. 3.93 as

$$p_F^* = 1 - \frac{16.25}{2 + 16.25} = 0.11$$

The percentage of time that the tower is out of service is thus 11%.

EXAMPLE 3.31

Suppose the breakdown time (in years) X for a system follows a gamma distribution; that is,

$$f_X(x) = 4xe^{-2x}; \quad x \geq 0$$

and the repair time Y is exponentially distributed with a mean of 0.1 year. Determine the probability that the system initially at a satisfactory state will be in the failed state after six months.

The Laplace transform of X is

$$G_X(s) = \int_0^\infty e^{-sx} 4x e^{-2x}\, dx$$

$$= \frac{4}{(s+2)^2}$$

whereas the Laplace transform of Y is

$$G_Y(s) = \int_0^\infty e^{-sy} 10 e^{-10y}\, dy$$

$$= \frac{10}{s+10}$$

Substituting these expressions in Eq. 3.92, we have

$$P_{SF}(s) = \frac{1}{s}\frac{\dfrac{4}{(s+2)^2}\left[1 - \dfrac{10}{s+10}\right]}{1 - \dfrac{4}{(s+2)^2}\dfrac{10}{s+10}}$$

$$= \frac{4}{s(s^2 + 14s + 44)}$$

$$= \frac{4}{s(s+4.76)(s+9.24)}$$

whose inverse transform can be shown to be

$$p_{SF}(t) = \frac{-4}{(-4.76)(4.76 - 9.24)(9.24)}[4.76 - 9.24 + 9.24e^{-4.76t} - 4.76e^{-9.24t}]$$

$$= 0.0909 - 0.1876e^{-4.76t} + 0.0966e^{-9.24t}$$

Therefore, the probability that the system will be at the failed state after six months, or $t = 0.5$ year, is

$$p_{SF}(0.5) = 0.0909 - 0.1876e^{-2.38} + 0.0966e^{-4.62}$$
$$= 0.0745$$

3.3.3 Inspection and Repairs at Regular Intervals

Seldom is the performance of a system monitored continuously; instead, most systems are inspected and subsequently repaired (if necessary) at regular intervals. Therefore, a failed state may not be discovered immediately, or if discovered it may not be repaired until the next scheduled maintenance period.

Consider a system that is subjected to a regular inspection-repair at a time interval of τ. At the beginning of the interval, the system is in a satisfactory state, S; during the interval, the system may change to a failed state, F. If the time between breakdowns of the system is X, with PDF $f_X(x)$, the probability that the system will fail during the interval τ is

$$p_F(\tau) = P(X < \tau) = \int_0^\tau f_X(x)\, dx \qquad (3.95)$$

Within the interval τ, two mutually exclusive events may prevail; the system may or may not break down. If there is no breakdown within an interval τ, the system is operating satisfactory over the entire period τ and hence the availability over the period τ is 1.0; on the other hand, if a breakdown occurs, the fraction of operating time within period τ is X'/τ, where X' denotes the time at which breakdown occurred. Hence, the mean availability of the system over the period τ is

$$\bar{A} = \frac{X'}{\tau} \cdot p_F(\tau) + 1 \cdot p_S(\tau) \tag{3.96}$$

where $p_S(\tau) = 1 - p_F(\tau)$. Since X' is the breakdown time within the interval τ, its value will range from zero to τ and its density function is the conditional PDF of X given $X < \tau$; that is,

$$f_{X'}(x) = f_X(x \,|\, X < \tau)$$

$$= \frac{f_X(x)}{p_F(\tau)}; \quad 0 \le x \le \tau$$

The expected value of X', then, is

$$E(X') = \frac{1}{p_F(\tau)} \int_0^\tau x f_X(x)\, dx$$

$$= \frac{1}{p_F(\tau)} \left[x F_X(x) \Big|_0^\tau - \int_0^\tau F_X(x)\, dx \right]$$

$$= \frac{1}{p_F(\tau)} \left[\tau F_X(\tau) - \int_0^\tau F_X(x)\, dx \right] \tag{3.97}$$

The expected average availability of the system is obtained by taking the expected value of Eq. 3.96. Hence,

$$E(\bar{A}) = \frac{E(X')}{\tau} p_F(\tau) + p_S(\tau)$$

$$= \frac{1}{\tau} \left[\tau F_X(\tau) - \int_0^\tau F_X(x)\, dx \right] + [1 - F_X(\tau)]$$

$$= 1 - \frac{1}{\tau} \int_0^\tau F_X(x)\, dx \tag{3.98}$$

where $F_X(x)$ is the CDF of the breakdown time X. Therefore, for a system that is regularly inspected at interval τ, and any breakdown is restored to its initial or new operating condition, the mean fraction of time that the system is operating satisfactorily is given by Eq. 3.98. This assumes, of course, that the duration of the inspection-repair process is negligible. However, a more realistic assumption may be a duration of τ_r for each scheduled inspection and repair. In this case, the net availability would be

$$A' = \frac{\tau}{\tau + \tau_r} \bar{A} \tag{3.99}$$

and

$$E(\bar{A}') = \frac{\tau}{\tau + \tau_r} E(\bar{A})$$

$$= \frac{\tau}{\tau + \tau_r} - \frac{1}{\tau + \tau_r} \int_0^\tau F_X(x)\,dx \tag{3.100}$$

From Eq. 3.96, the variance of the average availability is

$$\text{Var}(\bar{A}) = \left[\frac{p_F(\tau)}{\tau}\right]^2 \text{Var}(X')$$

$$= \left[\frac{p_F(\tau)}{\tau}\right]^2 [E(X'^2) - E^2(X')] \tag{3.101}$$

where

$$E(X'^2) = \frac{1}{p_F(\tau)} \int_0^\tau x^2 f_X(x)\,dx$$

$$= \frac{1}{p_F(\tau)} \left[x^2 F_X(x) \Big|_0^\tau - \int_0^\tau 2x F_X(x)\,dx \right]$$

$$= \frac{\tau^2}{p_F(\tau)} F_X(\tau) - \frac{2}{p_F(\tau)} \int_0^\tau x F_X(x)\,dx \tag{3.102}$$

Combining Eqs. 3.97, 3.101, and 3.102, and observing $p_F(\tau) = F_X(\tau)$, we have

$$\text{Var}(\bar{A}) = -\frac{2F_X(\tau)}{\tau^2} \int_0^\tau x F_X(x)\,dx$$

$$+ \frac{2}{\tau} F_X(\tau) \int_0^\tau F_X(x)\,dx - \frac{1}{\tau^2} \left[\int_0^\tau F_X(x)\,dx \right]^2 \tag{3.103}$$

Also, from Eq. 3.99,

$$\text{Var}(\bar{A}') = \left(\frac{\tau}{\tau + \tau_r}\right)^2 \text{Var}(\bar{A}) \tag{3.104}$$

On the other hand, if a significant portion of the inspection-repair duration is devoted to repair, the net average availability defined in Eq. 3.99 will be conservative, since it assumes a system operating satisfactorily at the end of the period will also be idle for a duration τ_r, though, in reality, practically no time would be spent on such systems. Consider the case in which the inspection time is negligible, and the repair time is of duration τ_r. A system that breaks down within a period τ is only available for a duration X' over that period and in addition will be idle for a period τ_r. Effectively, the average availability for a system breaking down by inspection time is $X'/(\tau + \tau_r)$. Hence, the average availability for the system would be

$$\bar{A}' = \frac{X'}{\tau + \tau_r} p_F(\tau) + 1 \cdot p_s(\tau) \tag{3.105}$$

Applying Eq. 3.97 we can show that the expected average availability, in this case, becomes

$$E(\bar{A}') = \frac{p_F(\tau)}{\tau + \tau_r} E(X') + p_s(\tau)$$

$$= \frac{1}{\tau + \tau_r} \left[\tau F_X(\tau) - \int_0^\tau F_X(x)\, dx \right] + 1 - F_X(\tau)$$

$$= 1 - \frac{\tau_r}{\tau + \tau_r} F_X(\tau) - \frac{1}{\tau + \tau_r} \int_0^\tau F_X(x)\, dx \qquad (3.106)$$

It can be shown that the variance of the average availability will be the same as in Eq. 3.104.

EXAMPLE 3.32

The pile-driving equipment in a construction firm is scheduled semi-annually for periodic maintenance. The time between breakdowns for this equipment follows an exponential distribution with a mean of 12 months. Assume that repairs of the equipment are per ormed only at the regularly scheduled time and the duration of repair is negligible; the equipment is restored to a brand-new condition after each inspection and repair. Determine:

(a) The probability that the equipment will break down before the scheduled maintenance time.
(b) The mean and variance of the average availability of the pile-driving equipment over an interval of six months.

Solution

(a) The breakdown time (in months) has the following density function:

$$f_X(x) = \tfrac{1}{12} e^{-x/12}; \qquad 0 \le x < \infty$$

Substituting this in Eq. 3.95, we obtain

$$p_F(6) = \int_0^6 \frac{1}{12} e^{-x/12}\, dx$$

$$= 1 - e^{-1/2} = 0.393$$

(b) The CDF of the breakdown time is

$$F_X(x) = \int_0^x \frac{1}{12} e^{-x/12}\, dx = 1 - e^{-x/12}$$

Applying Eq. 3.98, we get the mean availability

$$E(\bar{A}) = 1 - \frac{1}{6} \int_0^6 [1 - e^{-x/12}]\, dx = 0.787$$

From Eq. 3.103,

$$\mathrm{Var}(\bar{A}) = -\frac{2F_X(6)}{36} \int_0^6 x F_X(x)\, dx + \frac{2}{6} F_X(6) \int_0^6 F_X(x)\, dx - \frac{1}{6^2} \left[\int_0^6 F_X(x)\, dx \right]^2$$

Since,

$$F_X(6) = 1 - e^{-6/12} = 0.393$$

$$\int_0^6 F_X(x)\,dx = \int_0^6 (1 - e^{-x/12})\,dx = 1.278$$

$$\int_0^6 xF_X(x)\,dx = \int_0^6 x(1 - e^{-x/12})\,dx = 5.011$$

we obtain

$$\mathrm{Var}(\bar{A}) = -\frac{2}{36}(0.394)(5.011) + \frac{0.394}{3}(1.278) - \frac{1.278^2}{36} = 0.0128$$

EXAMPLE 3.33

Flaws in a submersible welded structure generally tend to increase in size during normal operation. The satisfactory performance of the submersible requires that flaws in the welds should not exceed a critical depth. Suppose the time of operation x until a critical flaw depth is reached follows a Weibull distribution

$$f_X(x) = \frac{\beta x^{\beta-1}}{\lambda^\beta}\exp\left[-\left(\frac{x}{\lambda}\right)^\beta\right]; \qquad 0 \le x < \infty$$

with parameters $\lambda = 26.5$ and $\beta = 4$ (these correspond approximately to a mean of 24 months and a c.o.v. of 28%). After nine months of operation, the submersible is brought back to the yard where it is inspected and repaired for any deterioration and damage. Assume this process takes a month and the submersible is restored to a completely new condition before the next expedition. Determine the expected fraction of time over its entire life span that the submersible structure is operating under a satisfactory condition.

Solution

With the given parameters, the Weibull density function for time until the development of critical flaw depth is

$$f_X(x) = \frac{4x^3}{26.5^4}\exp\left[-\left(\frac{x}{26.5}\right)^4\right], \qquad 0 \le x < \infty$$

and the corresponding cumulative distribution is

$$F_X(x) = \int_0^x \frac{4x^3}{26.5^4}\exp\left[-\left(\frac{x}{26.5}\right)^4\right]dx = 1 - \exp\left[-\left(\frac{x}{26.5}\right)^4\right]$$

From Eq. 3.100, the average fraction of time that the submersible is operating in a satisfactory condition is

$$E(\bar{A}') = \frac{9}{9+1} - \frac{1}{9+1}\int_0^9 \left\{1 - \exp\left[-\left(\frac{x}{26.5}\right)^4\right]\right\}dx$$

$$= \frac{9}{10} - \frac{1}{10}(0.0243) = 0.898$$

where numerical integration was used.

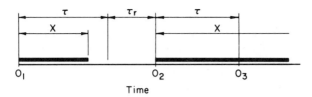

Figure E3.34a   Time history of dam operations.

EXAMPLE 3.34   (*Dam Inspection Program*)

Suppose the physical conditions of a dam during operation can be classified as either satisfactory (state S) or unsatisfactory (state F). Operating a dam under undesirable conditions could lead to disastrous consequences; hence, it is desirable to frequently inspect the dam in order to minimize the likelihood of operating under state F. Define availability of the system as the event that the dam is operating under satisfactory conditions. When state F occurs or when repair work is being done on the dam after an inspection, the dam is unavailable. The operating condition of a dam can be schematically shown in Fig. E3.34a as a function of time. The dam starts out in state S at O_1. It is scheduled for inspection at a time interval τ after O_1. During inspection, the policy is that if state S is encountered, no repair work except routine maintenance needs to be performed and the system is assumed to have a fresh start for another cycle of operation. On the other hand, if state F is discovered, dam operation is immediately suspended and repair undertaken. After a repair duration of τ_r, the system is assumed to have a fresh start for another cycle of operation. In a given cycle of operation, a dam may or may not encounter state F depending on its breakdown characteristics. Let X be the breakdown time which is random. If X is less than τ, as in the first cycle of operation in Fig. E3.34a, a period of τ_r will follow after inspection; whereas if X exceeds τ such as during the second cycle of operation of Fig. E3.34a, the system immediately begins its third cycle of operation without losing any time. During a cycle of operation, two mutually exclusive but collectively exhaustive events may occur:

(1)   No breakdown occurs and the system is 100% available during the cycle.
(2)   A breakdown occurs, hence, the system is available only during a fraction of the time, namely $X'/(\tau + \tau_r)$, where X' is the breakdown time for a cycle.

Since the probabilities of these two events are $p_S(\tau)$ and $p_F(\tau)$, respectively, the average availability of the system is

$$\bar{A}' = 1 \cdot p_S(\tau) + \frac{X'}{\tau + \tau_r} p_F(\tau)$$

and the expected average availability, according to Eq. 3.103, is

$$E(\bar{A}') = 1 - \frac{\tau_r}{\tau + \tau_r} F_X(\tau) - \frac{1}{\tau + \tau_r} \int_0^\tau F_X(x)\, dx$$

where $F_X(x)$ is the CDF of the breakdown time X.

The results derived so far rest on the assumption that if a dam is operating in state F at inspection time, that condition is always detected during inspection. The same perfect detectability also applies to the condition of state S. However, for a complex system such as a dam operating in an unsatisfactory state, it is difficult to achieve perfect detection. An improved model would allow a certain probability that state F will not be detected during inspection. Moreover, if state F is not detected during inspection, the physical condition of the dam will generally worsen. Thus, the chance of state F being detected during subsequent inspections would be equal or higher than at the previous inspection even though the same level of inspection is performed. Three different models of the detectability p are investigated as follows.

Case 1

Perfect second inspection is assumed; namely, state F missed at a previous inspection will be detected with certainty at a subsequent inspection. In this case, during a cycle of operation, three mutually exclusive and collectively exhaustive events are possible:

(1) No breakdown occurs, and the system is 100% available during the cycle.

(2) A breakdown occurs, and state F is detected at inspection; availability is thus $X'/(\tau + \tau_r)$.

(3) A breakdown occurs but state F was not detected at the first inspection; state F remained in effect during the subsequent cycle of operation and is detected at the subsequent inspection; the availability is thus $X'/(2\tau + \tau_r)$.

It can be shown that the average availability would be

$$\bar{A}' = 1 \cdot p_S(\tau) + \frac{X'}{\tau + \tau_r} p \cdot p_F(\tau) + \frac{X'}{2\tau + \tau_r} (1 - p) \cdot p_F(\tau)$$

where p is the detectability of state F at the first inspection. Also, the expected average availability can be shown to be

$$E(\bar{A}') = [1 - F_X(\tau)] + \left[\frac{p}{\tau + \tau_r} + \frac{1 - p}{2\tau + \tau_r} \right] \left[\tau F_X(\tau) - \int_0^\tau F_X(x) \, dx \right]$$

Observe that four factors determine the expected average availability, namely:

(1) The quality of the dam as described by the cumulative distribution of breakdown time $F_X(x)$.

(2) The level or quality of inspection as defined by the detectability p. A better inspection procedure will be associated with a higher value of p.

(3) The frequency of inspection as defined by the time between inspection τ.

(4) The time τ_r required to repair and restore the dam to satisfactory condition. Since τ_r is generally small relative to τ, it has been assumed to be deterministic for simplicity.

If the breakdown time follows an exponential distribution with mean time μ; that is,

$$F_X(x) = 1 - e^{-x/\mu}$$

then,

$$\int_0^\tau F_X(x) \, dx = \int_0^\tau (1 - e^{-x/\mu}) \, dx = \tau - e^{-\tau/\mu}(\tau + \mu)$$

The expected average availability becomes

$$E(\bar{A}') = e^{-\tau/\mu} + \left[1 - e^{-\tau/\mu}\left(\frac{\tau}{\mu} + 1 \right) \right] \left[\frac{p}{\tau/\mu + \tau_r/\mu} + \frac{1 - p}{2\tau/\mu + \tau_r/\mu} \right]$$

which is a function of the dimensionless parameters p, τ/μ and τ_r/μ. Figure E3.34b shows the variation of $E(\bar{A}')$ with τ/μ and p for the case $\tau_r/\mu = 0.05$. The curve for $p = 1$ denotes the case of perfect inspection. Clearly, as the inspection interval increases relative to the mean breakdown time, the chance that the system will break down increases, and thus the expected availability decreases. As the detectability decreases, the reduction in the expected availability becomes more dramatic. For example, for $p = 0.5$, $\mu = 10$ years, and $\tau = 2$ years, the expected availability is 87.5%; if the inspection interval is lengthened to 3 years, the expected availability

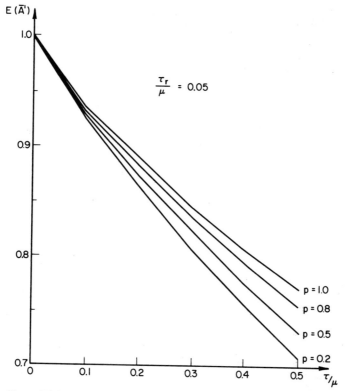

Figure E3.34b Availability versus τ/μ (perfect second inspection).

reduces to 82.3%. However, if the level of inspection is increased (with $p = 0.8$) to compensate for the longer inspection interval, the expected availability increases to 83.6%.

Case 2

The detectability p remains constant regardless of the history of inspection. Furthermore, the events of detecting state F between inspections are assumed to be statistically independent. In this case, the average availability is

$$\bar{A}' = 1 \cdot p_S(\tau) + p_F(\tau)X'\left[\frac{p}{\tau + \tau_r} + \frac{p(1-p)}{2\tau + \tau_r} + \frac{p(1-p)^2}{3\tau + \tau_r} + \cdots\right]$$

$$= p_S(\tau) + p_F(\tau)X'p\left[\sum_{n=1}^{\infty}\frac{(1-p)^{n-1}}{n\tau + \tau_r}\right]$$

For an exponentially distributed breakdown time, the expected average availability can be shown to be

$$E(\bar{A}') = e^{-\tau/\mu} + \left[1 - e^{-\tau/\mu}\left(\frac{\tau}{\mu} + 1\right)\right]p\sum_{n=1}^{\infty}\frac{(1-p)^{n-1}}{n\tau/\mu + \tau_r/\mu}$$

again expressed in terms of p, τ/μ and τ_r/μ. Numerical results for the case $\tau_r/\mu = 0.05$ are shown in Fig. E3.34c. If we compare these with those for *Case 1* (Fig. E3.34b), the expected availability is slightly lower.

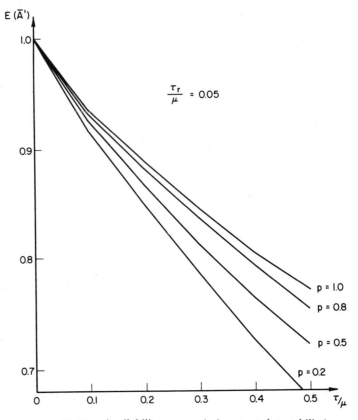

Figure E3.34c   Availability versus τ/μ (constant detectability).

Case 3

The detectability at each inspection increases with the number of cycles of operation. A possible detectability function is $p^{1/n}$. For example, the detectability is p during the first inspection, but increases to $\sqrt{p}$ at the second inspection. The average availability would be

$$\bar{A}' = 1 \cdot p_S(\tau) + p_F(\tau)\left[\frac{X'}{\tau + \tau_r}p + \sum_{n=2}^{\infty}\frac{X'}{n\tau + \tau_r}p^{1/n}\left\{\prod_{i=1}^{n-1}(1 - p^{1/i})\right\}\right]$$

For an exponentially distributed breakdown time, the expected availability can be shown to be

$$E(\bar{A}') = e^{-\tau/\mu} + \left[1 - e^{-\tau/\mu}\left(\frac{\tau}{\mu} + 1\right)\right]\left[\frac{p}{\tau/\mu + \tau_r/\mu} + \sum_{n=2}^{\infty}\frac{p^{1/n}}{n\tau/\mu + \tau_r/\mu}\left\{\prod_{i=1}^{n-1}(1 - p^{1/i})\right\}\right]$$

The expected availability in this case will be between those of *Cases 1* and *2*.

Optimal Inspection Interval

Obviously, frequent and detailed inspections will improve the operating condition and availability of a design. A trade-off between the increasing inspection cost and the economic loss resulting from dam unavailability may be used to determine the optimal inspection interval.

Suppose C_I is the cost of performing a dam inspection at a detectability p_o (e.g., $p_o = 0.2$). Assume further that the cost of a more detailed inspection (with detectability p) increases with the ratio p/p_o. If τ is the inspection interval, then the equivalent annual cost of inspection is $(C_I/\tau)(p/p_o)$. Suppose that the annual cost of operating the dam under unsatisfactory conditions or allowing the dam to remain idle during repairs is denoted by C_F. Then, the expected total cost per year can be expressed as

$$C_T = [1 - E(\bar{A}')]C_F + \frac{C_I}{\tau}\frac{p}{p_o}$$

where $E(\bar{A}')$ is the fraction of time the dam is operating satisfactorily; this is, of course, a function of p, τ, μ, and τ_r. Dividing both sides by C_I, we obtain

$$\frac{C_T}{C_I} = [1 - E(\bar{A}')]\frac{C_F}{C_I} + \frac{1}{\tau}\frac{p}{p_o}$$

The optimal inspection interval may be obtained by minimizing C_T/C_I with respect to τ for a given p, τ_r, and μ. Numerical results are shown in Fig. E3.34d, for $p_o = 0.2$, $\tau_r = 0.05$, and $C_F/C_I = 10$, and for different detectability functions. For example, curve D denotes a case with $\mu = 20$, $p = 0.8$, and a perfect second inspection; the optimal inspection interval in this case is four years. For the same type of dam, with a lower inspection level, say $p = 0.5$, the optimal inspection interval is three years (curve F), indicating that more frequent inspection is necessary. For a dam whose mean breakdown time is only 10 years (i.e., half of that in curve D), the optimal inspection interval is 3.2 years (curve B). In other words, a dam of poorer quality requires more frequent inspection. It is also interesting to observe that the optimal inspection interval can be sensitive to the detectability function.

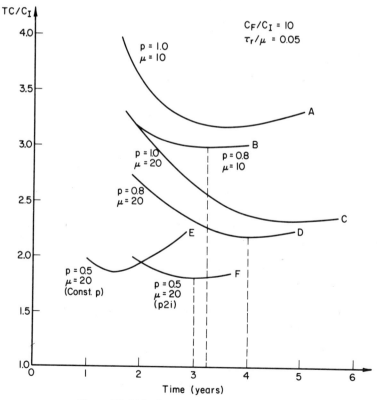

Figure E3.34d Optimal inspection interval.

3.3.4 Deteriorating Systems

In the previous sections a system is assumed to be restored to its new condition after each inspection and repair. However, this condition is generally not achieved. In practice, repairs are usually not performed if the system undergoes only normal deterioration but has not yet broken down. Moreover, if breakdown does occur during an operating period τ, the routine repair program may often fail to bring the system back to its original condition. Conceptually, a *deteriorating system* may be defined as one that is not restored to new condition after each inspection and repair; instead the inspection-repair process only transforms the failed state to an operating state. The system had deteriorated in the meanwhile, and the breakdown time is always measured with respect to the initial starting point of the system instead of the beginning of each operating cycle. An example of this is the temporary patching of the isolated damaged spots of a highway pavement; it is obvious that the pavement sections are not restored to their new condition unless the entire highway pavement is replaced.

Consider an inspection interval of τ. The conditional density function of the breakdown time in the jth interval, given that a breakdown occurs in that interval, is

$$f_{X_j}(x) = f_X(x|\text{breakdown in } j\text{th period})$$

$$= \frac{f_X(x)}{\int_{(j-1)\tau}^{j\tau} f_X(x)\,dx}; \qquad (j-1)\tau \le x \le j\tau \qquad (3.107)$$

where $f_X(x)$ is the PDF of the breakdown time of the system. The average availability in the jth interval is defined as

$$\bar{A}_j = \frac{X_j' - (j-1)\tau}{\tau} \frac{P[(j-1)\tau < X < j\tau]}{P[X > (j-1)\tau]} + 1 \cdot \frac{P(X > j\tau)}{P[X > (j-1)\tau]} \qquad (3.108)$$

where X_j' is the breakdown time of a system that is known to break down at the jth interval. The probability $P[X > (j-1)\tau]$ is needed in the denominator of Eq. 3.108 because in the jth time interval the system behaves as if it has survived the first $(j-1)$ intervals. On the basis of Eq. 3.108, the expected average availability in the jth interval can be shown to be

$$E(\bar{A}_j) = \frac{1}{1 - F_X[(j-1)\tau]} \left[1 - \frac{1}{\tau} \int_{(j-1)\tau}^{j\tau} F_X(x)\,dx \right] \qquad (3.109)$$

For n periods of operation, the overall expected availability assuming negligible repair time is

$$E(\bar{A}) = \frac{1}{n} \sum_{j=1}^{n} E(\bar{A}_j) \qquad (3.110)$$

The mean number of breakdowns over n periods of operation is the sum of the expected number of breakdowns in each of the n periods, which is simply the sum of the probabilities of breakdowns during each of the n periods; thus,

$$E(N) = \sum_{j=1}^{n} p_F(j\tau)$$

$$= \sum_{j=1}^{n} \frac{F_X(j\tau) - F_X[(j-1)\tau]}{1 - F_X[(j-1)\tau]} \qquad (3.111)$$

EXAMPLE 3.35

A stream is inspected daily for its DO concentration. If the DO concentration falls below a desired level at any location, oxygen is injected into the stream, sufficient only to temporarily restore the stream to a satisfactory level of *DO* concentration. The time required to restore the water quality may be assumed to be negligible. Suppose the time X for this stream to fall below the desirable *DO* concentration is given by the following density function (special Erlangian distribution)

$$f_X(x) = \frac{x}{\lambda^2} \exp\left(-\frac{x}{\lambda}\right); \qquad 0 < x < \infty$$

with a mean value of 4 days (96 hours), which yields $\lambda = 2$. The average availability in this case represents the proportion of time in which the DO concentration in the stream is satisfactory. Determine the following:

(a) The expected average availability on days 1, 2, 5, 10, and 20.
(b) The overall expected availability in the first 20 days of operation.
(c) The expected number of days in which the DO concentration will fall below the desired level during the first 20 days of operation.
(d) The mean availability over a long period of time.

Solution

(a) The CDF of X is

$$F_X(x) = \int_0^x \frac{x}{4} \exp\left(\frac{-x}{2}\right) dx$$

$$= 1 - \left(1 + \frac{x}{2}\right) \exp\left(-\frac{x}{2}\right); \qquad 0 \le x < \infty$$

From Eq. 3.109, with $\tau = 1$, the expected average availability on the jth day is

$$E(\bar{A}_j) = \frac{1}{1 - F_X(j-1)} \left[1 - \int_{j-1}^j F_X(x)\, dx\right]$$

$$= \frac{1}{\left(1 + \dfrac{j-1}{2}\right)\exp\left(\dfrac{1-j}{2}\right)} \left[1 - \int_{j-1}^j \left\{1 - \left(1 + \frac{x}{2}\right)\exp\left(\frac{-x}{2}\right)\right\} dx\right]$$

$$= \frac{1}{\left(\dfrac{j+1}{2}\right)\exp\left(\dfrac{1-j}{2}\right)} \left[(j+3)\exp\frac{(1-j)}{2} - (4+j)\exp\left(-\frac{j}{2}\right)\right]$$

$$= \frac{2}{j+1} [(j+3) - (4+j)e^{-1/2}]$$

$$= \frac{2}{j+1} [(j+1)(1 - e^{-1/2}) + (2 - 3e^{-1/2})]$$

$$= 0.787 + \frac{0.361}{j+1}$$

Hence,

$$E(\bar{A}_1) = 0.968$$

$$E(\bar{A}_2) = 0.907$$

$$E(\bar{A}_5) = 0.847$$

$$E(\bar{A}_{10}) = 0.820$$

$$E(\bar{A}_{20}) = 0.804$$

(b) The overall mean availability over the first 20 days is, from Eq. 3.110,

$$E(\bar{A}) = \frac{1}{20} \sum_{j=1}^{20} \left(0.787 + \frac{0.361}{j+1}\right) = 0.835$$

(c) Applying Eq. 3.111, we obtain

$$
\begin{aligned}
E(N) &= \sum_{j=1}^{20} \frac{F_x(j) - F_x(j-1)}{1 - F_x(j-1)} \\
&= \sum_{j=1}^{20} \frac{\left[1 - \left(1+\frac{j}{2}\right)\exp\left(-\frac{j}{2}\right)\right] - \left[1 - \left(1+\frac{j-1}{2}\right)\exp\left(-\frac{j-1}{2}\right)\right]}{1 - \left[1 - \left(1+\frac{j-1}{2}\right)\exp\left(-\frac{j-1}{2}\right)\right]} \\
&= \sum_{j=1}^{20} \left(1 - \frac{j+2}{j+1} e^{-1/2}\right) = 6.266
\end{aligned}
$$

Hence, it is expected that in approximately six out of the first 20 days, the *DO* in the stream will fall below the desired level.

(d) Over a long period of time, the mean availability is determined from Eq. 3.110, as $n \to \infty$. Hence,

$$E(\bar{A}) = \lim_{n \to \infty} \frac{1}{n} \sum_{j=1}^{n} \left(0.787 + \frac{0.361}{j+1}\right)$$

$$= 0.787$$

At the steady-state condition, the stream will have adequate DO concentration about 79% of the time.

PROBLEMS

3.1 Each spring a county is subject to flooding caused by the melting of snow. The record on flooding indicates the following:

1966—No flood	1973—Flood
1967—No flood	1974—No flood
1968—No flood	1975—No flood
1969—Flood	1976—No flood
1970—Flood	1977—Flood
1971—No flood	1978—Flood
1972—No flood	

The annual flood condition may be modeled with a two-state Markov chain.
(a) Determine the transition probability matrix.
(b) What is the probability that there will be a flood in the year 1980? In both 1980 and 1981?

(c) What is the probability that there will be a flood in the year 2000.

(d) What is the expected time between floods?

(e) What is the expected length of a sequence of consecutive floods?

3.2 The record of earthquake occurrence in the Greater San Francisco Bay area from 1905 to 1971 shows that two large-magnitude (exceeding 6.5 in the Richter scale) earthquakes have occurred: one in 1906 with a magnitude of 8.3, and the other in 1911 with a magnitude of 6.6. A homogeneous two-state Markov chain may be constructed as follows:

State 1: A year in which a large-magnitude earthquake occurs.
State 2: A year without any large-magnitude earthquake.

Based on the above information, answer the following:

(a) Determine the transition probability matrix between years with and without a large-magnitude earthquake.

(b) Determine the probability that there will be large-magnitude earthquake(s) in the year 2020.

(c) If no large-magnitude earthquake is observed in 1980 in the San Francisco Bay Area,

   (i) What is the probability that a large-magnitude earthquake will occur in 1985?

   (ii) What is the probability that the next large-magnitude earthquake will occur in 1985?

   (iii) What is the probability that the next large-magnitude earthquake will occur within the next five years?

(d) What is the mean recurrence time of a large-magnitude earthquake?

3.3 Suppose the bus in Example 3.5 arrives on time at the station.

(a) How many more stops, on the average, must the bus make before it is late? Repeat this problem if the bus arrives early at the station.

(b) What is the probability that there will be early arrivals before late arrivals?

(c) How many stops, on the average, will it take until the bus again arrives on time?

3.4 Suppose one percent of the wooden ties along a railroad are defective. The defective ties are not randomly located. There is a 10% probability that a defective tie is followed by another defective tie. Assume a homogeneous Markov chain model for the occurrence of good (state 1) and defective (state 2) ties.

(a) Determine the one-step transitional probability matrix based on the above information.

(b) What is the mean size (in terms of the number of ties) of a group of defective ties?

(c) If a group of five ties in sequence is selected at random, what is the probability that they will all be defective?

3.5 In Example 3.6, suppose four tendons are selected to be inspected for corrosion, and acceptance requires all four tendons to be corrosion-free. Which plan below would yield a higher probability of acceptance?

(a) The four tendons are at random locations (assume the tendons selected are far apart).

(b) The four tendons are adjacent to each other.

3.6 The quality of a concrete mix is evaluated each day and classified into one of the three categories: (1) below standard; (2) meets the standard; (3) exceeds the standard. The record of performance of a specific supplier reveals the following transitional information about the concrete qualities in two successive days.

$$
\begin{array}{c}
 & \begin{array}{ccc} 1 & \quad 2 & \quad 3 \end{array} \\
\begin{array}{c} 1 \\ 2 \\ 3 \end{array} &
\left[\begin{array}{ccc}
5 & 3 & 2 \\
6 & 15 & 9 \\
1 & 2 & 7
\end{array}\right]
\end{array}
$$

For example, there are nine recorded cases in which the quality goes from "meeting standard" to "exceeding standard" on the following day. Assume a homogeneous Markov model for the transitional characteristics of the concrete quality.

(a)   Determine the transitional probability matrix.
(b)   What is the percentage of time that the concrete quality meets the standard?
(c)   How many days in a row, on the average, would the concrete quality remain in the substandard level?
(d)   How many days would it take, on the average, to go from "substandard quality" to "meeting standard quality"?
(e)   Suppose the quality in each of the last five days met the standard. What is the probability that the quality will be substandard for the next two days?
(f)   Suppose the supplier is subject to a penalty of $200 for each day of substandard concrete mix supplied. What is the penalty that this supplier expects to pay over a month (30 days)?
(g)   To reduce the testing schedule, the quality of the concrete will only be spot-checked on the 1st and 15th of each month *without the supplier's knowledge.* According to this testing program, what is the probability that the quality will be "meeting standard" for the next 3 months? Would this probability be different if the quality is spot-checked on the 1st and 2nd of each month instead?
(h)   Repeat parts (d) and (e) if the phrase "meeting standard" is changed to "meeting or exceeding standard."

3.7   A Markov chain model was suggested to study the traffic pattern *between* various zones of the John Hancock Center in Chicago (Engi and Solbey, 1974). Seven zones are identified and denoted by the states of the Markov chain; namely,

States	Zone
1	Observation deck
2	Apartment
3	Grocery–restaurant
4	Office
5	Shops
6	Lower restaurant
7	Outside of building

The report for one day of observation at the center reveals the following trip distribution matrix. For instance, 1000 trips to the shops were observed coming from the observation deck.

	1	2	3	4	5	6	7
1	—	0	0	0	1000	150	1750
2	0	—	420	10	160	60	1160
3	0	420	—	170	0	0	0
4	0	10	170	—	160	1920	4200
5	950	160	0	160	—	50	1375
6	50	60	0	1920	50	—	320
7	1900	1160	0	4200	1325	220	—

(a)   Determine the transition probability matrix.
(b)   Determine the steady-state probabilities. What do the steady-state probabilities mean in this problem?
(c)   What is the mean number of intermediate stops for a trip from the shops to the office?

3.8 Suppose the annual winter weather in a midwestern town can be classified as good, normal, or bad, with the following transition probabilities:

$$
\begin{array}{c c c c}
 & G & N & B \\
G & \begin{bmatrix} 0.5 & 0.4 & 0.1 \\ N & 0.1 & 0.8 & 0.1 \\ B & 0.1 & 0.6 & 0.3 \end{bmatrix}
\end{array}
$$

For example, if the current year has a good winter, there is a 50% probability that the succeeding winter will also be good. Assume a homogeneous Markov chain model.
   (a) For an 80-year period, what is the mean number of years with bad winters in this town?
   (b) What is the expected length of a sequence of bad winters?
   (c) If a bad winter has just passed, what is the mean number of years until the next bad winter?
   (d) If a bad winter has just passed, what is the mean number of years until the next good winter?
   (e) Consider the year 2020 AD. What is the probability that both the year 2020 and 2022 will have bad winters?

3.9 In Example 3.15, answer the following additional questions:
   (a) Are all four states recurrent? Justify.
   (b) If the reservoir contains one unit of storage this month:
       (i)   What is the probability that it will be full within the next two months?
       (ii) What is the mean time until the reservoir will be full?
       (iii) What is the probability that the reservoir will be full the 20th, 40th, and 41st months from now.
       (iv) The probability that the reservoir will be full before it becomes empty is 0.8; verify this by applying Eq. 3.31 with $b = 3$.

3.10 Consider a row of footings as shown below:

The settlement of each footing can be classified into three levels as follows:

N: Normal.
B: Below Normal.
A: Above Normal.

The settlements of footings can be modeled as a Markov chain. Prior inspections of a sequence of similar footings reveal the following transition probability matrix:

$$
\begin{array}{c c c c}
 & N & B & A \\
N & \begin{bmatrix} \frac{1}{2} & \frac{1}{3} & \frac{1}{6} \\ B & \frac{1}{5} & \frac{3}{5} & \frac{1}{5} \\ A & \frac{1}{3} & \frac{1}{6} & \frac{1}{2} \end{bmatrix}
\end{array}
$$

For example, if a footing picked at random has normal settlement, the probability that the settlement of the footing to the right will be below normal settlement is $p_{NB} = \frac{1}{3}$
   (a) Suppose there are 100 footings in the sequence. How many of these will have abnormal (i.e., above or below normal) settlements?
   (b) What is the mean size of a group of consecutive footings with above normal settlement?

(c) During the inspection of a sequence of footings, if the first footing inspected has normal settlement:
(i) What is the expected number of additional footings that would need to be inspected before a footing with above normal settlement is detected?
(ii) What is the probability of detecting a footing with above normal settlement prior to detecting one with below normal settlement?
(d) Suppose excessive differential settlement will occur if the footing with above normal settlement is adjacent to one with below normal settlement. What fraction of the total number of pairs of adjacent footings will have excessive differential settlement?

3.11 The clay stratum supporting an offshore platform can be weakened by the cyclic loading effects of wave action during severe storms. Suppose the stiffness of the clay stratum is divided into five states with state 1 denoting the initial state with unweakened stiffness, and state 5 the failed state with negligible stiffness. The transition probabilities between states after a severe storm are as follows:

$$
\mathbf{P} = \begin{bmatrix}
0.9 & 0.07 & 0.026 & 0.0035 & 0.0005 \\
0 & 0.8 & 0.15 & 0.04 & 0.01 \\
0 & 0 & 0.7 & 0.2 & 0.1 \\
0 & 0 & 0 & 0.7 & 0.3 \\
0 & 0 & 0 & 0 & 1
\end{bmatrix}
$$

Observe that as the stiffness decreases, there is a progressive deterioration of stiffness. Assume that the clay will not recover its stiffness once it is weakened.
(a) Determine the mean number of severe storms before the stiffness of the clay stratum becomes negligible.
(b) Suppose severe storms occur as a Poisson process with a mean occurrence rate of 0.1 per year. What is the probability that the clay will fail during an expected service period of 20 years.
(c) After a severe storm, the stiffness of the weakened clay may improve with time due to the dissipation of excess pore pressure. An alternative assumption is that the clay will always be at the initial state before all severe storms. In this case, what is the probability of failure in the clay stratum during a 20-year period?

3.12 In Example 3.17, repeat the calculation for the probability of *at least heavy damage* over a period of 20 years.

3.13 A parking space is provided for customers of a local post office. Suppose customers arrive according to a Poisson process with a mean rate of 10 per hour; the time taken by each customer in the post office is exponentially distributed with a mean of five minutes. Assume that a customer will wait for a parking space if necessary.
(a) What percentage of the time would parking spaces be available to incoming customers?
(b) How would the answer to part (a) change if an additional parking space is provided? Consider two cases separately:
(i) The post office can only handle one customer at a time.
(ii) The post office can serve more than one customer simultaneously.
(c) Repeat part (a) if the service time is not exponentially distributed but has a coefficient of variation of 60%, and a mean of five minutes.

3.14 Consider a garage with capacity for four trucks. Suppose trucks arrive at the garage according to a Poisson process with a mean rate of one per day. The amount of time a truck spends in the garage is exponentially distributed with a mean of two days. If an arriving truck finds the garage already fully occupied, it will go elsewhere.
(a) What fraction of the time is the garage fully occupied?
(b) Suppose each truck weighs 50 tons. What is the average load applied to the garage's foundation over a long time period?

(c) If the weight of each truck is normally distributed $N(50, 10)$ tons, and the weights between trucks are statistically independent, what is the probability the total load from the trucks in the garage exceeds 200 tons at a given time?

3.15 A left-turn bay is provided at a street intersection. The time until a gap long enough for a safe left turn follows an exponential distribution with a mean of 20 sec. Suppose the left-turn traffic arrives according to a Poisson process at a mean rate of two per minute.
   (a) What should the size of the left-turn bay (in terms of the number of car spaces) be such that it will be adequate 96 % of the time?
   (b) What is the expected waiting time before one can make a left turn at this intersection?
   (c) What is the probability that the waiting time will exceed one minute?

3.16 Figure P3.16 represents a multiserver operation where R parallel shovels are involved in a queueing system of hauling trucks.

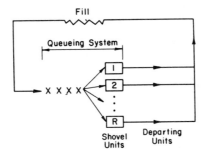

Figure P3.16 Queueing system of hauling trucks with R parallel shovels.

Suppose that the arrival of a particular hauling truck (not in the queue) follows a Poisson process with mean rate λ, and the service time of each shovel follows an exponential distribution with a mean shoveling time of $1/\mu$. Trucks are served on a first-come-first-serve basis.
   (a) For a fleet of k trucks, derive the expressions for the probabilities of the queue length at the steady-state condition.
   (b) Determine the production per hour if four shovels are serving seven trucks. The average cycle time of each shovel is 1.5 minutes and the average traveling time for each truck is four minutes. The capacity of each truck is 16 cu yd.
   (c) Determine the optimum number of trucks required to work with three shovels, such that the cost per cu yd of earth moved is minimized. The average traveling time for each truck is six minutes and the average loading time is three minutes. A shovel costs $100 per hour including operator and oiler, whereas a truck costs $40 per hour including a driver. The capacity of each truck is 16 cu yd.

3.17 Determine the optimum number of pusher tractors needed to push the load of eight scrapers, if the average cycle time of a scraper is 11 minutes and the average cycle time of a pusher is three minutes. The capacity of each scraper is 28 cu yds. The rental cost of each scraper is $34 per hour; the corresponding rental cost of each pusher is $30.

3.18 A solid waste treatment plant processes wastes at an average rate of 10 minutes per truck load with a c.o.v. of 40 %. Suppose the truckloads arrive according to a Poisson process with a mean arrival rate of four per hour. Determine:
   (a) The proportion of time in which the treatment plant is idle.
   (b) The mean queue length at the steady-state condition.
   (c) The steady-state probabilities for queue lengths of 1, 2, and 3 truck loads if the service time follows a gamma distribution.

3.19   In studying the feasibility of a personal rapid transit (PRT) system, a parallel bay station, as shown in Fig. P3.19, is suggested. This design enables each vehicle to leave the station as soon as it is ready. However, no spaces are provided for waiting if a customer arrives and finds that all bays are in use. Suppose that heavy traffic is expected and the successive arrivals will be regulated to T seconds apart. The service time required for each user is exponentially distributed with a mean of μ sec. The number of parallel bays in the station is m.

Consider a Markov process in which the time stages are discretized at instances just before each arrival, called "arrival points." The state of the Markov chain is defined as the number of vehicles in service at the bays at the instant before an arrival.
  (a)   For a vehicle sitting at the bay at an arrival point, what is the probability that it completes service in an interval of time T?
  (b)   What is the proportion of arriving customers who will be rejected if $m = 2$, $T = 10$ sec, and $\mu = 10$ sec? [*Hint*: First determine the transitional probability matrix.]
  (c)   How would the answer in part (b) change if the interarrival time is not a constant, but the arrival process follows a Poisson distribution with the same mean rate of arrival of $\frac{1}{10}$ per sec?
  (d)   Repeat part (b) if the number of bays is increased to four.
  (e)   Develop a general expression for the transition probability $p_{r,q}$ which is the probability that q bays will be occupied at the next arrival point when r bays are occupied at this arrival point.

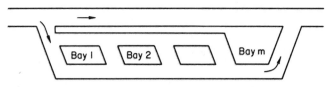

Figure P3.19   A four parallel-bay station.

3.20   The dissolved oxygen concentration in a nearby stream fluctuates between satisfactory and unsatisfactory levels. The record shows that the average lengths of satisfactory and unsatisfactory periods are seven and two days, respectively; also, the actual duration in each period (satisfactory or unsatisfactory) is exponentially distributed.
  (a)   Determine the percentage of time that the stream will have satisfactory dissolved oxygen concentration.
  (b)   Suppose the dissolved oxygen concentration in the stream is presently satisfactory.
     (i)   What is the probability that it will be unsatisfactory on the fifth day from now.
     (ii) What is the expected fraction of time within the next two weeks that the DO concentration will be satisfactory?
  (c)   Repeat part (a) if the time of each period (satisfactory or unsatisfactory) is normally distributed, with a c.o.v. of 30%.

3.21   The repair record of a bus company with a fleet of 10 buses shows that the rate of breakdown per bus is one per month (30 days) on the average, and the time required to repair a disabled bus is normally distributed with a mean of four days and a standard deviation of one day. If the average cost of breakdown for each bus is $600 a day, determine the long-run average monthly cost arising from the breakdown of buses.

3.22   The performance of an automatic pore pressure recorder is inspected at monthly intervals (30 days). If the recorder has already broken down by inspection time, it will be repaired and restored to its new condition with negligible repair time. Suppose the time

duration between breakdowns follows an exponential distribution with a mean of six months.
  (a) Determine the average number of repairs in one year.
  (b) Determine the mean and variance of the average availability of the automatic pore pressure recorder over a long period of operation.
  (c) How would the mean and variance of the average availability change if inspection is scheduled for every two months?

3.23 Consider the pile-driving equipment in Example 3.32. Suppose there are two alternative schedules for maintenance; namely, at a 6-month interval (a_1) or at a 4-month interval (a_2). The inspection-maintenance-repair time, τ_r, is assumed to be 0.5 months, regardless of the operating condition of the equipment. The utility of each alternative is a function of the average availability $\bar{A}'$ and the cost of maintenance C, or

$$U = \bar{A}' + \bar{A}'^2 - C$$

where C for alternative a_1 is 0.2 and that for alternative a_2 is 20% higher, that is, 0.24.
  Determine which alternative is better if the decision criterion is to maximize the expected utility.

3.24 Consider Example 3.26 again.
  (a) If the construction crew had developed a systematic method of removing boulders such that the delay during each encounter of a boulder took only half a day, what would be the expected duration of the tunnel excavation project?
  (b) Suppose the delay for each boulder encountered is independently normally distributed with a mean of one day and a c.o.v. of 20%. What is the probability that the excavation of a 50m tunnel section can be completed within a week?

3.25 The pavement of a landing strip at an airport has just been completed. According to past experiences with similar pavements, and taking into account the traffic and environmental conditions, the time (in years) until cracks appear in the pavement follows a Weibull distribution as follows:

$$f_X(x) = \frac{\beta x^{\beta-1}}{\lambda^\beta} \exp\left[-\left(\frac{x}{\lambda}\right)^\beta\right]; \quad x \geq 0$$

where

$$\lambda = 4 \quad \text{and} \quad \beta = 2$$

The inspection of cracks in the pavement is scheduled for every 12 months. If cracks are found during inspection, they will be repaired so that local conditions are restored to a level consistent with the rest of the pavement. The availability in this case denotes the proportion of time in which the pavement is free of cracks.
  (a) Determine the expected proportion of time in which the pavement is free of cracks during the first year of operation.
  (b) Determine the overall mean fraction of time in which the pavement is free of cracks during the first three years of operation.
  (c) During the next three years, how often will repair work be required at scheduled inspection times?

4. Statistics of Extremes

4.1 INTRODUCTION

An important class of probability problems are those involving the extreme values of random variables; that is, the "largest" and/or "smallest" values of a random variable. Statistically, these pertain to the maximum and minimum values from a set of observations. Conceivably, if the set of observations (samples of size n) were repeated, other maximum and minimum values will be obtained; thus, the possible largest and smallest values comprise populations of their own. Therefore, these extreme values may also be modeled as random variables with respective probability distributions. Such distributions and their associated parameters have special characteristics that are unique to the extreme values. These special concepts and unique characteristics have significance in engineering, and their understanding is important for their proper applications.

The subject of the statistics of the largest and smallest values is a subtopic within the broader subject of *order statistics* (e.g., Wilks, 1948), that is, the statistics of the kth value from the top as well as the kth value from the bottom.

A comprehensive treatment of the statistical theory of extreme values is available in the work of Gumbel (1958) and examples of applications to physical problems have also been discussed (e.g., Gumbel, 1954). The principal elements of the statistical theory of extreme values are highlighted herein, with emphasis on their significance and potential applications to engineering problems.

4.1.1. Engineering Significance of Extreme Values

Extreme values from observational data are of special importance to many engineering applications. For example, in the consideration of structural safety, the high loads and low structural resistances are the values most relevant to the assurance of safety or reliability of a structure; in flood control programs, the maximum flood of a river (e.g., in a year) is the most significant flow of concern, whereas in irrigation work and water supply programs, the minimum flow would be of the most relevance. Therefore, in dealing with problems involving such extreme conditions, the maximas or minimas from sets of observations are the only pertinent data. The statistics and probabilities associated with such extreme values, therefore, are the information of special importance.

The prediction of future conditions is often required in engineering planning and design, and may involve the prediction of the largest (or smallest) value. Extrapolation from previously observed extreme value data, therefore, is invariably necessary; for this purpose, the asymptotic theory of statistical extremes often provides a powerful basis for developing the required engineering information.

4.1.2 Objective and Coverage

The study of extreme value statistics is an established field of mathematical statistics. However, the significance of this mathematical theory to practical engineering problems is not widely recognized. Gumbel (1954) described the potential applications in a number of areas, whereas practical results derived on the basis of the statistical extremes have been obtained, for example, by Press (1950), Chow (1951), and Ochi (1973).

The practically relevant aspects of extreme value statistics may not immediately be transparent from the available mathematical treatises on the subject; on the other hand, practical results are often presented presuming the necessary mathematical or statistical basis. It is our intention in this chapter to present the statistical and/or probability bases of extreme values in terms that are accessible to or can be more easily understood by engineers and students of engineering. The concepts involved will be illustrated extensively with examples to clarify the underlying mathematical principles.

Specific problems in several areas of engineering applications are also discussed, and where appropriate (e.g., in parameter estimation) actual data are used to illustrate the theoretical methods.

4.2 PROBABILITY DISTRIBUTION OF EXTREMES

4.2.1 Exact Distributions

The largest and smallest values from samples of size n are also random variables and therefore have probability distributions of their own. These distributions can be expected to be related to the distribution of the *initial variate* (or population).

Let X be the initial random variable with known initial distribution function $F_X(x)$. Consider samples of size n taken from the population (sample space) of X; each sample will be a set of observations $(x_1, x_2, \ldots, x_n)$ representing respectively the first, second, $\ldots$, and nth observed values. Since every observed value is unpredictable prior to actual observation, we may assume that each observation is the value of a random variable, and the set of observations $(x_1, x_2, \ldots, x_n)$ is a realization of the *sample random variables* $(X_1, X_2, \ldots, X_n)$. Therefore, in speaking of the extreme values from a sample of size n, we are considering the maximum and minimum of $(X_1, X_2, \ldots, X_n)$; that is, the random variables

$$Y_n = \max(X_1, X_2, \ldots, X_n) \tag{4.1a}$$

and

$$Y_1 = \min(X_1, X_2, \ldots, X_n) \tag{4.1b}$$

The probability distributions of these extreme values, that is, of Y_n and Y_1, can be developed in the context of the transformation of probability distribution (see Chapter 4, Vol. I), specifically for the functions Y_n and Y_1 of Eq. 4.1.

In other words, the largest value and the smallest value from samples of size n taken from a population X are also random variables whose probability distributions may be derived from that of the initial variate X.

We observe that if Y_n, the largest among $(X_1, X_2, \ldots, X_n)$, is less than some value y, then all the other sample random variables must necessarily also be less than y. For mathematical simplicity and consistent with random sampling theory, assume that $X_1, X_2, \ldots, X_n$ are statistically independent and identically distributed as the initial variate X; that is,

$$F_{X_1}(x) = F_{X_2}(x) = \cdots = F_{X_n}(x) = F_X(x)$$

On these bases, the distribution function of Y_n is

$$
\begin{aligned}
F_{Y_n}(y) &\equiv P(Y_n \leq y) \\
&= P(X_1 \leq y, X_2 \leq y, \ldots, X_n \leq y) \\
&= [F_X(y)]^n
\end{aligned}
\tag{4.2}
$$

The corresponding density function for Y_n, therefore, is

$$
\begin{aligned}
f_{Y_n}(y) &= \frac{\partial F_{Y_n}(y)}{\partial y} \\
&= n[F_X(y)]^{n-1} f_X(y)
\end{aligned}
\tag{4.3}
$$

From Eq. 4.2, we see that for a given y the probability $[F_X(y)]^n$ decreases with n; this means that the functions $F_{Y_n}(y)$ and $f_{Y_n}(y)$ will shift to the right with increasing values of n, as illustrated in Fig. 4.1 for an exponential initial distribution $f_X(x) = e^{-x}; x \geq 0$.

The distribution function for Y_1, the smallest value in a sample of size n, can be similarly derived. In this case, we observe that if Y_1, the smallest among $(X_1, X_2, \ldots, X_n)$ is larger than y, then all the other values in the same sample must be larger than y. Hence, the "survival function" (i.e., the complement of the distribution function) is

$$
\begin{aligned}
1 - F_{Y_1}(y) &\equiv P(Y_1 > y) \\
&= P(X_1 > y, X_2 > y, \ldots, X_n > y) \\
&= [1 - F_X(y)]^n
\end{aligned}
$$

The distribution function of Y_1, therefore, is

$$F_{Y_1}(y) = 1 - [1 - F_X(y)]^n \tag{4.4}$$

and the corresponding density function becomes

$$f_{Y_1}(y) = n[1 - F_X(y)]^{n-1} f_X(y) \tag{4.5}$$

In the case of the smallest value from samples of size n, the density function $f_{Y_1}(y)$ and distribution function $F_{Y_1}(y)$ will shift to the left with increasing sample size n.

Equations 4.2 through 4.5 are the exact probability distributions of the extremes from samples of size n taken from a population X. Clearly, these distributions depend on the initial distribution $F_X(x)$ of the population and also on the sample size n. The distributions of Y_n and Y_1 are generally difficult to obtain or derive in analytic form. Initial variates with the exponential distribution are one of the exceptions.

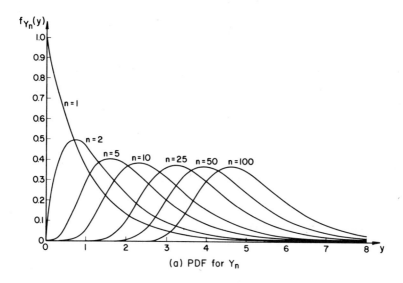

(a) PDF for Y_n

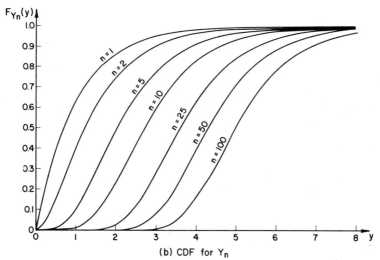

(b) CDF for Y_n

Figure 4.1 PDF and CDF of the largest value from an exponential initial variate.

EXAMPLE 4.1

Consider the initial variate X having the exponential density function as follows:

$$f_X(x) = \lambda e^{-\lambda x}; \qquad x \geq 0$$

The corresponding distribution function is

$$F_X(x) = 1 - e^{-\lambda x}$$

Therefore, the largest value from samples of size n will have the distribution function

$$F_{Y_n}(y) = (1 - e^{-\lambda y})^n; \qquad y \geq 0$$

and the corresponding density function (PDF) is

$$f_{Y_n}(y) = \lambda n(1 - e^{-\lambda y})^{n-1}e^{-\lambda y}$$

Graphically, the above functions $f_{Y_n}(y)$ and $F_{Y_n}(y)$ would appear (for $\lambda = 1.0$) as shown in Figs. 4.1a and 4.1b for different values of n. Clearly, both functions shift to the right as n increases.

If the above $F_{Y_n}(y)$ is expanded in series with the binomial theorem,

$$(1 - e^{-\lambda y})^n = 1 - ne^{-\lambda y} + \frac{n(n-1)}{2!}e^{-2\lambda y} - \cdots$$

For large n, the above series approaches the double exponential function $e^{-ne^{-\lambda y}}$. Hence, for large n the CDF of the largest value from an exponential population approaches

$$F_{Y_n}(y) = e^{-ne^{-\lambda y}}; \qquad y \geq 0$$

which is a double exponential function.

The CDF of the smallest value from the above initial distribution is, according to Eq. 4.4,

$$F_{Y_1}(y) = 1 - [1 - (1 - e^{-\lambda y})]^n$$
$$= 1 - e^{-n\lambda y}; \qquad y \geq 0$$

The corresponding PDF is

$$f_{Y_1}(y) = n\lambda e^{-n\lambda y}; \qquad y \geq 0$$

It should be observed that the distributions of the largest value and of the smallest value from an exponential initial variate are of different forms. For the largest value, Y_n, the CDF approaches the double exponential form, whereas for the smallest value, Y_1, the corresponding CDF is of the single exponential form.

EXAMPLE 4.2

Consider an initial variate with

$$f_X(x) = \frac{1}{x^2}; \qquad x \geq 1$$
$$= 0; \qquad x < 1$$

Then,

$$F_X(x) = 1 - \frac{1}{x}; \qquad x \geq 1$$

Equation 4.2 then yields the CDF for the largest values from samples of size n

$$F_{Y_n}(y) = \left(1 - \frac{1}{y}\right)^n$$

The PDF for Y_n then is

$$f_{Y_n}(y) = \frac{n}{y^2}\left(1 - \frac{1}{y}\right)^{n-1}; \qquad y \geq 1$$
$$= 0; \qquad y < 1$$

These PDF and CDF are illustrated graphically in Fig. E4.2 for different values of n.

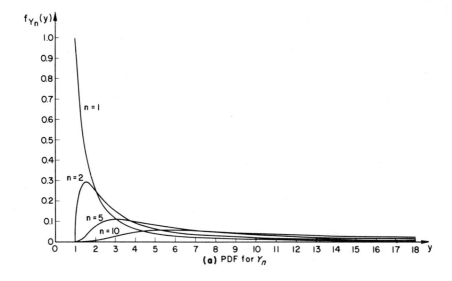

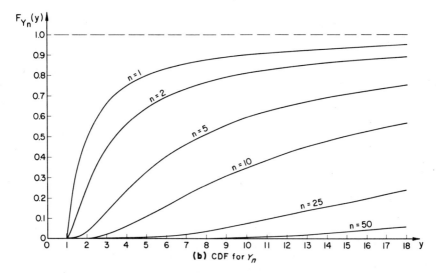

Figure E4.2 PDF and CDF of the largest value from a polynomial initial distribution.

The corresponding CDF and PDF for the smallest value from this initial variate are as follows:

$$F_{Y_1}(y) = 1 - \left(\frac{1}{y}\right)^n$$

and

$$f_{Y_1}(y) = \frac{n}{y^2}\left(\frac{1}{y}\right)^{n-1}; \qquad y \geq 1$$

$$= 0; \qquad\qquad y < 1$$

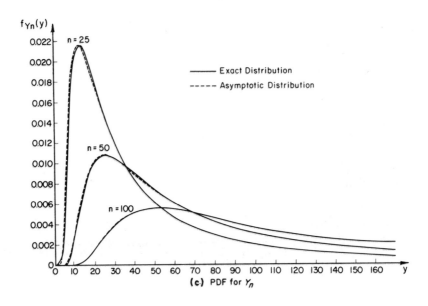

$f_{Y_n}(y)$

Exact Distribution
----- Asymptotic Distribution

(c) PDF for Y_n

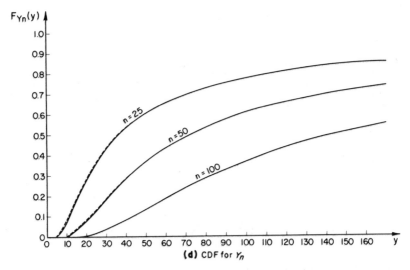

$F_{Y_n}(y)$

(d) CDF for Y_n

Figure E4.2   PDF and CDF of the largest value from a polynomial initial distribution.

EXAMPLE 4.3

For an initial variate with the uniform distribution between 0 and 1, that is,

$$f_X(x) = 1; \qquad 0 \le x \le 1$$
$$= 0; \qquad \text{elsewhere}$$

and

$$F_X(x) = x$$

the distribution of its largest value is

$$F_{Y_n}(y) = y^n; \qquad 0 \le y \le 1$$

and

$$f_{Y_n}(y) = ny^{n-1}; \qquad 0 \le y \le 1$$
$$= 0; \qquad \text{elsewhere}$$

Its smallest value will have CDF

$$F_{Y_1}(y) = 1 - (1 - y)^n; \qquad 0 \le y \le 1$$

and

$$f_{Y_1}(y) = n(1 - y)^{n-1}; \qquad 0 \le y \le 1$$
$$= 0; \qquad \text{elsewhere}$$

The PDF and CDF for the largest value from the above uniformly distributed variate X are shown in Fig. E4.3.

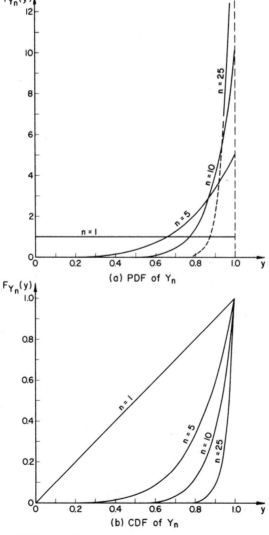

(a) PDF of Y_n

(b) CDF of Y_n

Figure E4.3 PDF and CDF of the largest value from a uniform initial variate.

EXAMPLE 4.4

Consider the initial variate X having the standard normal distribution, with density function

$$f_X(x) = \frac{1}{\sqrt{2\pi}} e^{-(1/2)x^2}$$

The corresponding cumulative distribution function is

$$F_X(x) = \frac{1}{\sqrt{2\pi}} \int_{-\infty}^{x} e^{-(1/2)z^2} dz = \Phi(x)$$

The largest value from samples of size n, therefore, will have a CDF given as follows:

$$F_{Y_n}(y) = \left(\frac{1}{\sqrt{2\pi}} \int_{-\infty}^{y} e^{-(1/2)z^2} dz \right)^n = [\Phi(y)]^n$$

and the corresponding PDF

$$f_{Y_n}(y) = \frac{n}{\sqrt{2\pi}} [\Phi(y)]^{n-1} e^{-(1/2)y^2}$$

In this case, the distribution function cannot be derived analytically; the evaluation of the cumulative probabilities will require numerical integration.

The smallest values from samples of size n from the same population, according to Eq. 4.4, will have corresponding CDF and PDF as follows:

$$F_{Y_1}(y) = 1 - \left[1 - \frac{1}{\sqrt{2\pi}} \int_{-\infty}^{y} e^{-(1/2)z^2} dz \right]^n = 1 - [1 - \Phi(y)]^n$$

and

$$f_{Y_1}(y) = \frac{n}{\sqrt{2\pi}} [1 - \Phi(y)]^{n-1} e^{-1/2y^2}$$

Again, numerical integration will be required to evaluate the corresponding cumulative probabilities.

Through numerical integration, the cumulative distribution function (CDF) for increasing values of n are obtained for $F_{Y_n}(y)$ as shown graphically in Fig. E4.4b; the corresponding PDFs are shown in Fig. E4.4a.

4.2.2   Asymptotic Distributions

Observation of Eqs. 4.2 through 4.5 may lead one to inquire about the characteristics of the distributions of the extremes as n becomes large. In particular, are there limiting or asymptotic forms of $F_{Y_n}(y)$ and $f_{Y_n}(y)$ as $n \to \infty$; similarly, for $F_{Y_1}(y)$ and $f_{Y_1}(y)$ as $n \to \infty$? This question has been the subject of study by early statisticians; for example, Fisher and Tippett (1927), Gnedenko (1943), and has come to be known as the *asymptotic theory of statistical extremes*.

The usefulness of extreme value statistics is greatly enhanced by the asymptotic theory of extremal distributions. This is the analytic theory concerned with the limiting forms of $F_{Y_n}(y)$ and $F_{Y_1}(y)$, or $f_{Y_n}(y)$ and $f_{Y_1}(y)$, as n becomes large or increases without bound. That is, as $n \to \infty$, $F_{Y_n}(y)$ and/or $F_{Y_1}(y)$ may converge (in distribution) to a particular functional form. It is of interest and practical significance that the asymptotic form of an extremal distribution does not depend

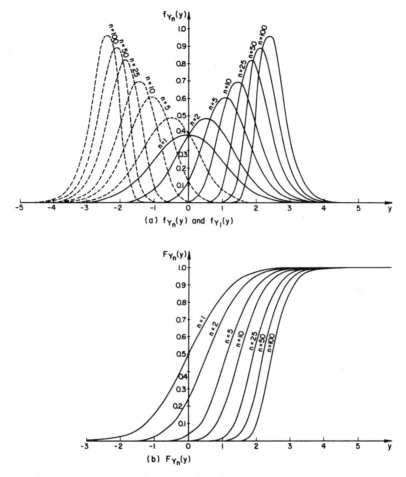

Figure E4.4 PDFs and CDFs of Y_n and Y_1 from a standard normal initial variate.

on the exact form of the initial distribution; rather, it depends largely on the tail behavior of the initial distribution in the direction of the extreme. The central portion of the initial distribution has little influence on the asymptotic form of the extremal distribution; the extremal parameters, however, will depend on the form of the initial distribution.

Derivation of Asymptotic Extremal Distributions; Cramer's Method The analytical derivation of the appropriate asymptotic extremal distribution (i.e., as $n \to \infty$), given the distribution of an initial variate, is facilitated by the method of Cramer (1946).

Consider first the extremal distribution for the largest value from an initial variate X. Following Cramer (1946), we define the transformed random variable

$$\xi_n = n[1 - F_X(Y_n)] \tag{4.6}$$

Then,

$$F_{\xi_n}(\xi) = P(\xi_n \leq \xi)$$

$$= P\{n[1 - F_X(Y_n)] \leq \xi\}$$

$$= P\left[F_X(Y_n) \geq 1 - \frac{\xi}{n}\right]$$

$$= P\left[Y_n \geq F_X^{-1}\left(1 - \frac{\xi}{n}\right)\right]$$

$$= 1 - F_{Y_n}\left[F_X^{-1}\left(1 - \frac{\xi}{n}\right)\right]$$

$$= 1 - \left\{F_X\left[F_X^{-1}\left(1 - \frac{\xi}{n}\right)\right]\right\}^n$$

$$= 1 - \left(1 - \frac{\xi}{n}\right)^n$$

As $n \to \infty$, we have

$$\lim_{n \to \infty} n \ln\left(1 - \frac{\xi}{n}\right) = \lim_{n \to \infty} \frac{\ln\left(1 - \frac{\xi}{n}\right)}{1/n} = -\xi$$

Therefore, the term $(1 - \xi/n)^n$ becomes

$$\lim_{n \to \infty} \left(1 - \frac{\xi}{n}\right)^n = e^{-\xi}$$

hence, as $n \to \infty$,

$$F_{\xi_n}(\xi) = 1 - e^{-\xi} \tag{4.7}$$

The corresponding asymptotic PDF is

$$f_{\xi_n}(\xi) = e^{-\xi} \tag{4.8}$$

Inverting Eq. 4.6 for Y_n, we have

$$Y_n = F_X^{-1}\left(1 - \frac{\xi_n}{n}\right) \tag{4.9}$$

Therefore, for large n, the asymptotic distribution of Y_n may be determined from that of ξ_n as given in Eqs. 4.7 and 4.8, based on the functional relationship of Eq. 4.9. In this regard, we observe that in Eq. 4.6, ξ_n decreases as Y_n increases; therefore,

$$P(Y_n \leq y) = P[\xi_n > g(y)]$$

where $g(y)$ is the expression on the right side of Eq. 4.6. Hence, the CDF of Y_n may be obtained from that of ξ_n as

$$F_{Y_n}(y) = 1 - F_{\xi_n}[g(y)]$$
$$= \exp[-g(y)] \tag{4.10}$$

The derivative of Eq. 4.10 then yields the PDF,

$$f_{Y_n}(y) = -\frac{dg(y)}{dy}\exp[-g(y)] \tag{4.11}$$

EXAMPLE 4.5

Consider again the initial variate of Example 4.1, with the exponential PDF

$$f_X(x) = \lambda e^{-\lambda x}; \qquad x \geq 0$$

and CDF

$$F_X(x) = 1 - e^{-\lambda x}$$

In this case, Eq. 4.6 is

$$\xi_n = n[1 - (1 - e^{-\lambda Y_n})]$$
$$= ne^{-\lambda Y_n}$$

Then, according to Eq. 4.10, the CDF of the largest value from the above X is, for large n,

$$F_{Y_n}(y) = \exp(-ne^{-\lambda y})$$

and its PDF is

$$f_{Y_n}(y) = n\lambda e^{-\lambda y}\exp(-ne^{-\lambda y})$$

Observe that these were also the results obtained earlier in Example 4.1 for large n. For $\lambda = 1.0$, the above CDF for Y_n becomes

$$F_{Y_n}(y) = e^{-ne^{-y}}$$

The asymptotic distribution, PDF and CDF, of Y_n derived above are shown in Fig. E4.5 for increasing values of n. The corresponding exact PDF and CDF of Y_n, as obtained earlier in Example 4.1, are also shown in Fig. E4.5 for different n. The comparison of the two sets of curves serves to demonstrate the "convergence" of the extremal distribution, as n increases, to certain asymptotic forms.

EXAMPLE 4.6

Consider the initial variate X with

$$f_X(x) = \frac{k}{x^{k+1}}; \qquad x \geq 1$$
$$= 0; \qquad x < 1$$

Its CDF is

$$F_X(x) = 1 - \frac{1}{x^k}; \qquad x \geq 1$$

Then, Eq. 4.6 yields

$$\xi_n = n\left[1 - \left(1 - \frac{1}{Y_n^k}\right)\right]$$
$$= \frac{n}{Y_n^k}$$

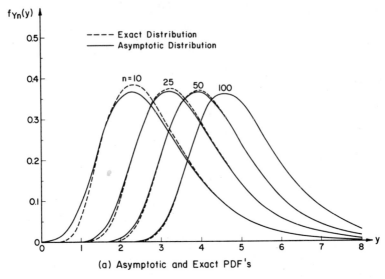

(a) Asymptotic and Exact PDF's

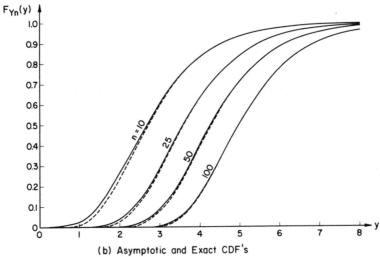

(b) Asymptotic and Exact CDF's

Figure E4.5 Exact and asymptotic distributions of the largest value Y from an exponential initial distribution.

The CDF of Y_n, therefore, is

$$F_{Y_n}(y) = \exp(-n/y^k); \qquad y \geq 1$$

and the corresponding PDF is

$$f_{Y_n}(y) = \frac{nk}{y^{k+1}} \exp(-n/y^k); \qquad y \geq 1$$

In this case, the CDF is of the form e^{-b/y^k}.

EXAMPLE 4.7

Next, consider an initial random variable with a finite upper bound. For this purpose, consider X with a uniform PDF between 0 and a; that is,

$$f_X(x) = \frac{1}{a}; \quad 0 \le x \le a$$

$$= 0; \quad \text{elsewhere}$$

The corresponding CDF is

$$F_X(x) = \frac{x}{a}; \quad 0 \le x \le a$$

In this case, the transformed variate, Eq. 4.6, would be

$$\xi_n = n\left(1 - \frac{Y_n}{a}\right)$$

Then, with Eq. 4.10, the CDF of Y_n is

$$F_{Y_n}(y) = \exp\left[-\frac{n}{a}(a - y)\right]; \quad 0 \le y \le a$$

and

$$f_{Y_n}(y) = \frac{n}{a}\exp\left[-\frac{n}{a}(a - y)\right]; \quad 0 \le y \le a$$

Observe that the CDF, in this case, is of the exponential form $e^{-c(w - y)^k}$, where w is the upper-bound value of the initial variate X.

EXAMPLE 4.8

Finally, consider an initial variate with the standard normal PDF,

$$f_X(x) = \frac{1}{\sqrt{2\pi}} e^{-(1/2)x^2}; \quad -\infty < x < \infty$$

The CDF is

$$F_X(x) = \frac{1}{\sqrt{2\pi}} \int_{-\infty}^{x} e^{-(1/2)z^2}\, dz$$

In this case, the transformed variate of Eq. 4.6 is

$$\xi_n = \frac{n}{\sqrt{2\pi}} \int_{Y_n}^{\infty} e^{-(1/2)z^2}\, dz$$

Integrating by parts, with

$$u = \frac{1}{z} \quad \text{and} \quad dv = ze^{-(1/2)z^2}\, dz$$

we obtain

$$\xi_n = \frac{n}{\sqrt{2\pi}} \frac{1}{Y_n} e^{-(1/2)Y_n^2}[1 + o(1/Y_n^2)]$$

where $o(1/Y_n^2)$ consists of the terms of order $1/Y_n^2$. Cramer (1946) gives the following asymptotic solution for Y_n as $n \to \infty$:

$$Y_n = \sqrt{2 \ln n} - \frac{\ln \ln n + \ln 4\pi}{2\sqrt{2 \ln n}} - \frac{\ln \xi_n}{\sqrt{2 \ln n}}$$

Denoting,

$$u_n = \sqrt{2 \ln n} - \frac{\ln \ln n + \ln 4\pi}{2\sqrt{2 \ln n}}$$

and

$$\alpha_n = \sqrt{2 \ln n}$$

we obtain

$$Y_n = u_n - \frac{1}{\alpha_n} \ln \xi_n$$

From this the transformed variate becomes

$$\xi_n = e^{-\alpha_n(Y_n - u_n)}$$

Therefore, the CDF of Y_n is

$$F_{Y_n}(y) = \exp[-e^{-\alpha_n(y - u_n)}]$$

and the corresponding PDF is

$$f_{Y_n}(y) = \alpha_n e^{-\alpha_n(y - u_n)} \exp[-e^{-\alpha_n(y - u_n)}]$$

Observe that this is also of the double exponential form. The convergence of the distribution of Y_n to the *Type I asymptotic distribution*, as $n \to \infty$, (see Sect. 4.3.2) may be observed graphically in Fig. E4.8.

The above results for the standard normal distribution $N(0, 1)$ may be extended for the extremal distribution of an initial variate with a general Gaussian distribution $N(\mu, \sigma)$. From Chapter 3, Vol. I, we note that if X is $N(\mu, \sigma)$, then $(X - \mu)/\sigma$ is $N(0, 1)$; therefore, the asymptotic distribution of the largest value of $(X - \mu)/\sigma$ will have the double exponential CDF and the parameters u_n and α_n obtained above.

Now, suppose Y_n' is the largest from the initial Gaussian variate X; then,

$$Y_n = \frac{Y_n' - \mu}{\sigma}$$

is the largest from $(X - \mu)/\sigma$. The CDF for Y_n', therefore, is

$$F_{Y_n'}(y') = F_{Y_n}\left(\frac{y' - \mu}{\sigma}\right)$$

$$= \exp\{-e^{-\alpha_n[(y' - \mu)/\sigma - u_n]}\}$$

$$= \exp[-e^{(-\alpha_n/\sigma)(y' - \mu - \sigma u_n)}]$$

Hence, the CDF of Y_n' is of the same double exponential form as Y_n, with the parameters

$$u_n' = \sigma u_n + \mu$$

$$\alpha_n' = \alpha_n/\sigma$$

Similar to Eq. 4.6, we define a transformed variable for the smallest value from an initial variate X as follows:

$$\xi_1 = n F_X(Y_1) \tag{4.12}$$

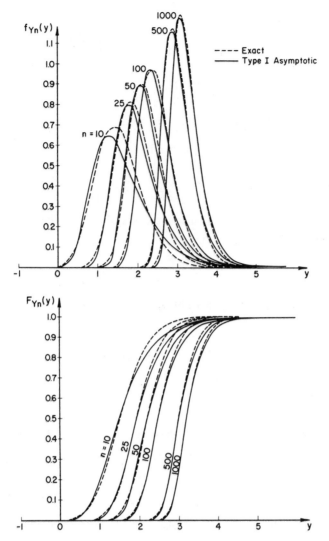

Figure E4.8 Exact and asymptotic distributions of the largest value from a Gaussian initial distribution.

Then,

$$F_{\xi_1}(\xi) = P(\xi_1 \leq \xi)$$

$$= P[nF_X(Y_1) \leq \xi]$$

$$= P\left[Y_1 \leq F_X^{-1}\left(\frac{\xi}{n}\right)\right]$$

$$= F_{Y_1}\left[F_X^{-1}\left(\frac{\xi}{n}\right)\right]$$

According to Eq. 4.4,

$$F_{\xi_1}(\xi) = 1 - \left\{1 - F_X\left[F_X^{-1}\left(\frac{\xi}{n}\right)\right]\right\}^n$$

$$= 1 - \left(1 - \frac{\xi}{n}\right)^n$$

and as $n \to \infty$ (see Eq. 4.7), we obtain

$$F_{\xi_1}(\xi) = 1 - e^{-\xi} \tag{4.13}$$

and

$$f_{\xi_1}(\xi) = e^{-\xi} \tag{4.14}$$

In this case, as ξ_1 increases, Y_1 also increases; therefore, the CDF of Y_1 follows from that of ξ_1 as follows:

$$F_{Y_1}(y) = F_{\xi_1}[g(y)]$$
$$= 1 - \exp[-g(y)] \tag{4.15}$$

where $g(y)$ is the function on the right side of Eq. 4.12.

The corresponding PDF of Y_1 is

$$f_{Y_1}(y) = \frac{dg(y)}{dy} \exp[-g(y)] \tag{4.16}$$

EXAMPLE 4.9

Consider the initial variate X with the exponential distribution of Example 4.5; that is,

$$f_X(x) = \lambda e^{-\lambda x}; \qquad x \geq 0$$

and

$$F_X(x) = 1 - e^{-\lambda x}$$

However, this time let us determine the distribution of the smallest value from samples of size n from X.

In this case, the transformed variate of Eq. 4.12 is

$$\xi_1 = n(1 - e^{-\lambda Y_1})$$

But, for large n, $Y_1 \to 0$, and thus,

$$e^{-\lambda Y_1} \simeq 1 - \lambda Y_1$$

Therefore,

$$\xi_1 \simeq n\lambda Y_1$$

and Eq. 4.15 then yields the following CDF:

$$F_{Y_1}(y) = 1 - \exp(-n\lambda y); \qquad y \geq 0$$

with the corresponding PDF

$$f_{Y_1}(y) = n\lambda e^{-n\lambda y}; \qquad y \geq 0$$

Comparing this CDF with that of Example 4.5, we see that the distribution of the smallest value from an initial exponential variate is different from that of the largest value from the same initial variate; that is, the distribution of the largest value (Example 4.5) will converge to a double exponential form, whereas the distribution of the smallest value from the same exponential variate will converge to a (single) exponential form.

EXAMPLE 4.10

In Example 4.8, we derive the asymptotic distribution of the largest value from a normal distribution. Let us now derive the corresponding asymptotic distribution for the smallest value.
For the smallest value, Eq. 4.12 becomes

$$\xi_1 = \frac{n}{\sqrt{2\pi}} \int_{-\infty}^{Y_1} e^{-(1/2)z^2} dz$$

$$= -\frac{n}{\sqrt{2\pi}} \int_{Y_1}^{-\infty} e^{-(1/2)z^2} dz$$

Using the results of Example 4.8, we obtain

$$\xi_1 = -\frac{n}{\sqrt{2\pi}} \cdot \frac{1}{Y_1} e^{-(1/2)Y_1^2}[1 + o(1/Y_1^2)]$$

For large n, Cramer (1946) gives the asymptotic solution for Y_1 as

$$Y_1 = -\sqrt{2 \ln n} + \frac{\ln \ln n + \ln 4\pi}{2\sqrt{2 \ln n}} + \frac{\ln \xi_1}{\sqrt{2 \ln n}}$$

Let

$$u_1 = -\sqrt{2 \ln n} + \frac{\ln \ln n + \ln 4\pi}{2\sqrt{2 \ln n}}$$

and

$$\alpha_1 = \sqrt{2 \ln n}$$

Then,

$$Y_1 = u_1 + \frac{\ln \xi_1}{\alpha_1}$$

and

$$\xi_1 = e^{\alpha_1(Y_1 - u_1)}$$

Thus, the CDF of Y_1, according to Eq. 4.15, is

$$F_{Y_1}(y) = 1 - \exp[-e^{\alpha_1(y - u_1)}]$$

and the corresponding PDF is

$$f_{Y_1}(y) = \alpha_1 e^{\alpha_1(y - u_1)} \exp[-e^{\alpha_1(y - u_1)}]$$

Therefore, the asymptotic distribution of the smallest value from a normal initial variate is of the same form as that of the largest value; that is, the double exponential.

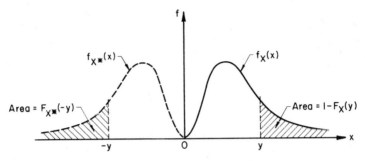

Figure 4.2   Symmetry principle.

4.2.3   The Symmetry Principle

The distributions of the largest and smallest values are related by certain conditions of symmetry, which Gumbel (1958) called the "symmetry principle."

Consider a random variable X with PDF $f_X(x)$. Introduce the "negative" variate X^* whose PDF $f_{X^*}(x)$ is the mirror image of $f_X(x)$, as shown in Fig. 4.2. As indicated in Fig. 4.2, at $X = y$ and $X^* = -y$, we have

$$1 - F_X(y) = F_{X^*}(-y)$$

Thus,

$$[1 - F_X(y)]^n = [F_{X^*}(-y)]^n$$

But, according to Eqs. 4.2 and 4.4, the two sides of the above equation are, respectively, the "survival" function of the smallest value from the variate X, and the distribution function of the largest value from the variate X^*; that is,

$$1 - F_{Y_1}(y) = F_{Y_n^*}(-y) \tag{4.17}$$

The corresponding PDFs, therefore, are also related by

$$f_{Y_1}(y) = f_{Y_n^*}(-y) \tag{4.18}$$

Through the symmetry principle, Eq. 4.17 or 4.18, the survival function of the smallest value of a variate X, therefore, may be obtained from the distribution of its negative largest value. The following examples should elucidate the meaning and potential usefulness of this principle.

EXAMPLE 4.11

Consider the exponential distribution of a random variable X, with PDF

$$f_X(x) = \lambda e^{-\lambda x}; \quad x \geq 0$$

and its CDF is

$$F_X(x) = 1 - e^{-\lambda x}$$

The "negative variate" X^*, whose PDF is the mirror image of $f_X(x)$, is (see Fig. E4.11)

$$f_{X^*}(x) = f_X(-x) = \lambda e^{-\lambda(-x)} = \lambda e^{\lambda x}; \quad x \leq 0$$

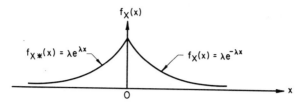

Figure E4.11

and the corresponding CDF is

$$F_{X*}(-x) = \int_{-\infty}^{-x} \lambda e^{\lambda x}\, dx = e^{-\lambda x}; \qquad x \geq 0$$

The survival function of the smallest value from samples of size n of X may be obtained through the symmetry principle as follows.

The negative largest value of X^*, according to Eq. 4.2, is

$$F_{Y_n^*}(-y) = [F_{X*}(-y)]^n = e^{-n\lambda y}$$

Therefore, according to Eq. 4.17, the survival function of the smallest value of X is

$$1 - F_{Y_1}(y) = e^{-n\lambda y}$$

This may be confirmed by observing that

$$[1 - F_X(x)]^n = e^{-n\lambda x}$$

thus demonstrating that the survival function of the smallest value is the same as the cumulative distribution of the negative largest value of X^*.

EXAMPLE 4.12

Next, consider a variate X with the Rayleigh distribution

$$f_X(x) = \frac{x}{\sigma^2} e^{-(1/2)(x/\sigma)^2}; \qquad x \geq 0$$

The corresponding PDF for X^* would be

$$f_{X*}(x) = \frac{-x}{\sigma^2} e^{-(1/2)(x/\sigma)^2}; \qquad x \leq 0$$

and

$$F_{X*}(-x) = \int_{-\infty}^{-x} \frac{-x}{\sigma^2} e^{-(1/2)(x/\sigma)^2}\, dx = e^{-(1/2)(x/\sigma)^2}; \qquad x \geq 0$$

On the basis of the symmetry principle, the survival function of the smallest value among samples of size n from X may be obtained as

$$1 - F_{Y_1}(y) = [1 - F_X(y)]^n = [F_{X*}(-y)]^n = e^{-(n/2)(y/\sigma)^2}$$

EXAMPLE 4.13

In Example 4.8, we found the asymptotic distribution of the largest value from the standard normal variate to be

$$F_{Y_n}(y) = \exp[-e^{-\alpha_n(y - u_n)}]$$

Applying the symmetry principle, we may obtain the asymptotic distribution of the smallest value from the same standard normal variate as follows:

$$1 - F_{Y_1}(y) = F_{Y_n^*}(-y)$$
$$= \exp[-e^{-\alpha_n(-y-u_n)}]$$
$$= \exp[-e^{\alpha_1(y-u_1)}]$$

where

$$\alpha_1 = \alpha_n$$
$$u_1 = -u_n$$

This is the same result as that derived in Example 4.10 using Cramer's method.

4.3 THE THREE ASYMPTOTIC FORMS

4.3.1 Gumbel's Classification

In Section 4.2.2, we observed that the asymptotic distributions of the extremes in the several examples illustrated tend to converge on certain limiting forms for large n; specifically to the double exponential form or to two different (single) exponential forms. Gumbel (1958) had classified these as the Type I, II, and III asymptotic forms:

Type I: The double exponential form, $\exp[-e^{-A(n)y}]$.

Type II: The exponential form, $\exp[-A(n)/y^k]$.

Type III: The exponential form with upper bound ω, $\exp[-A(n)(\omega - y)^k]$.

Similar asymptotic forms also apply to the smallest values.

Examples 4.5, 4.6, and 4.7 are, respectively, illustrations of the Type I, II, and III asymptotic distributions of the largest value; whereas Examples 4.10 and 4.9 are examples of the Type I and III asymptotic distributions of the smallest value. Example 4.8 also shows that the largest value from a normal distribution will converge asymptotically to the Type I limiting form.

Because the extreme values of a random variable are invariably associated with the tails of its probability density function, the convergence of the distribution function of its extreme (largest or smallest) value to a particular limiting form, therefore, will depend largely on the tail behavior of the initial distribution in the direction of the extreme. In particular, the extreme value from an initial distribution with an exponentially decaying tail (in the direction of the extreme) will converge asymptotically to the Type I limiting form; whereas, for an initial variate that decays with a polynomial tail, its extreme value will converge to the Type II asymptotic form. Finally, if the extreme value is limited; that is, the largest value has a finite upper bound, or the smallest value has a finite lower bound, the corresponding extremal distribution will converge (for large n) to the Type III asymptotic form.

It should be mentioned that the three asymptotic forms are not exhaustive (Gumbel, 1958); that is, there could be asymptotic distributions of the extreme values that do not belong to any of the three types described above. Moreover, the extremal distribution for the extreme value from a given initial variate may converge to one type of asymptotic form, whereas the distribution of the other extreme from the same initial variate could converge to another asymptotic form. This was illustrated earlier in Examples 4.5 and 4.9 for an exponential initial variate.

Notations For the sake of clarity and to distinguish the descriptions of the three types of asymptotic distributions, the following notations will be used to designate the respective extremal variates associated with these three asymptotic forms:

X_n, Y_n, Z_n: Extremal variates for the largest values with the Type I, II, and III asymptotic forms, respectively.

X_1, Y_1, Z_1: Extremal variates for the smallest values with the Type I, II, and III asymptotic forms, respectively.

For ready reference, the CDFs and statistics of these three asymptotic types of extremal distributions are summarized in Table A.3 of Appendix A.

4.3.2   The Type I Asymptotic Form

The cumulative distribution function (CDF) of the Type I asymptotic form for the distribution of the largest value, in the classification of Gumbel (1958), is as follows:

$$F_{X_n}(x) = \exp[-e^{-\alpha_n(x - u_n)}] \tag{4.19}$$

where u_n and α_n are, respectively, the location and scale parameters, defined as follows:

u_n: The *characteristic largest value* of the initial variate X.

α_n: An inverse measure of dispersion of X_n.

The corresponding probability density function (PDF) is

$$f_{X_n}(x) = \alpha_n e^{-\alpha_n(x - u_n)} \exp[-e^{-\alpha_n(x - u_n)}] \tag{4.20}$$

For the smallest value from an initial variate X with an "exponential tail" (see below), the corresponding Type I asymptotic form for the CDF is

$$F_{X_1}(x) = 1 - \exp[-e^{\alpha_1(x - u_1)}] \tag{4.21}$$

and the PDF is

$$f_{X_1}(x) = \alpha_1 e^{\alpha_1(x - u_1)} \exp[-e^{\alpha_1(x - u_1)}] \tag{4.22}$$

where the parameters are

u_1: The *characteristic smallest value* of the initial variate X.

α_1: An inverse measure of dispersion of X_1.

Exponential Tail For an initial distribution with an exponentially decaying tail (in the direction of the extreme), the distribution of its extreme value will converge to the Type I asymptotic form (Gumbel, 1958; Fisher and Tippett, 1927). Gumbel (1958) defined an initial distribution with an *exponential tail* to be unlimited in the direction of the extreme, and for large positive values of the initial variate X, the ordinates of $f_X(x)$ and $1 - F_X(x)$ are small, whereas $f'_X(x)$ is small and negative. Specifically, for large x

$$\frac{f(x)}{1 - F(x)} \approx -\frac{f'(x)}{f(x)} \tag{4.23}$$

Similarly, for large negative values of X, an exponential tail is defined as

$$\frac{f(x)}{F(x)} = \frac{f'(x)}{f(x)} \tag{4.24}$$

The Characteristic Extremes   The *characteristic largest value*, u_n, is a convenient measure of the central location of the possible largest values. In a sample of size n from an initial variate X, the expected number of sample values that are larger than x is $n[1 - F_X(x)]$. The characteristic largest value, u_n, is defined as the particular value of X such that in a sample of size n from the initial population X, the expected number of sample values larger than u_n is one; that is,

$$n[1 - F_X(u_n)] = 1.0$$

or

$$F_X(u_n) = 1 - \frac{1}{n} \tag{4.25}$$

In other words, u_n is the value of X with an exceedance probability of $1/n$; see Fig. 4.3.

With Eq. 4.25, Eq. 4.2 yields (with Y_n replaced by X_n)

$$F_{X_n}(u_n) = \left(1 - \frac{1}{n}\right)^n$$

Thus,

$$\ln F_{X_n}(u_n) = n \ln\left(1 - \frac{1}{n}\right) = \frac{\ln\left(1 - \dfrac{1}{n}\right)}{\dfrac{1}{n}}$$

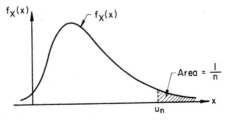

Figure 4.3   Definition of characteristic largest value, u_n.

As $n \to \infty$,

$$\lim_{n \to \infty} \frac{\ln \left(1 - \dfrac{1}{n}\right)}{\dfrac{1}{n}} = -1$$

Hence, for large n,

$$F_{X_n}(u_n) = e^{-1} \simeq 0.368$$

and

$$P(X_n > u_n) \simeq 0.632$$

This means, therefore, that among the population of possible largest values from samples of size n, about 37% are less than u_n, or 63% are greater than u_n.

Similarly, the characteristic smallest value, u_1, may be defined as

$$nF_X(u_1) = 1$$

or

$$F_X(u_1) = \frac{1}{n} \tag{4.26}$$

Eq. 4.4 then yields (with Y_1 replaced by X_1)

$$F_{X_1}(u_1) = 1 - \left(1 - \frac{1}{n}\right)^n$$

Therefore, for large n,

$$F_{X_1}(u_1) = 1 - e^{-1} \simeq 0.632$$

which means that among the population of possible smallest values of X, about 63% will be less than u_1.

The characteristic extremes, u_n and u_1, are also the modal values (i.e., most probable values) of the respective extremal variates X_n and X_1. That is, u_n is the most probable largest value from samples of X, and u_1 is the most probable smallest value from samples of X. This may be shown for the characteristic largest value u_n as follows.

Let $\tilde{x}$ be the modal value of X_n. Then, $\tilde{x}$ must be the solution of

$$f'_{X_n}(\tilde{x}) = 0 \tag{4.27}$$

Taking the derivative of Eq. 4.3, we obtain

$$f'_{X_n}(\tilde{x}) = n[F_X(\tilde{x})]^{n-1}f'_X(\tilde{x}) + n(n-1)[F_X(\tilde{x})]^{n-2}f_X^2(\tilde{x})$$

Eq. 4.27 then yields (dropping the subscript X on the initial distribution F_X and f_X),

$$\frac{f'(\tilde{x})}{f(\tilde{x})} + (n-1)\frac{f(\tilde{x})}{F(\tilde{x})} = 0$$

In light of Eq. 4.23, for initial distributions with exponential tails, the above equation becomes

$$-\frac{f(\tilde{x})}{1 - F(\tilde{x})} + (n - 1)\frac{f(\tilde{x})}{F(\tilde{x})} = 0$$

From this we obtain

$$F_X(\tilde{x}) = 1 - \frac{1}{n} = F_X(u_n)$$

Therefore, (denoting $\tilde{x}_n = \tilde{x}$ for the largest value)

$$u_n = \tilde{x}_n$$

Thus, we confirm that the characteristic value of the initial variate X is equal to the most probable largest value of X_n.

Using Eq. 4.5 in $f'_{X_1}(\tilde{x}) = 0$, we can similarly show that

$$u_1 = \tilde{x}_1$$

that is, the characteristic smallest value of samples of size n from the initial variate X is equal to the modal value of X_1.

The Hazard Function In reliability theory, the conditional probability of failure in time $(t, t + dt)$ given no failure in $(0, t)$ involves the *hazard function* $h(t) = f(t)/[1 - F(t)]$. As we shall see, such a function is also involved in the distribution of extreme values.

For the largest value, the hazard function of X may be defined as

$$h_n(x) = \frac{f_X(x)}{1 - F_X(x)} \tag{4.28}$$

whereas for the smallest value, the corresponding hazard function is

$$h_1(x) = \frac{f_X(x)}{F_X(x)} \tag{4.29}$$

At the characteristic extremes, u_n and u_1, the respective hazard functions of Eqs. 4.28 and 4.29 become

$$h_n(u_n) = nf_X(u_n) \tag{4.30}$$

and

$$h_1(u_1) = nf_X(u_1) \tag{4.31}$$

Consider the largest value from an initial distribution $f_X(x)$ with exponential tail, and assume that all the derivatives of $f_X(x)$ exist. Expand $F_X(x)$ in a Taylor series about the characteristic largest value u_n; that is,

$$F_X(x) = F(u_n) + (x - u_n)f(u_n) + \tfrac{1}{2}(x - u_n)^2 f'(u_n) + \frac{1}{3!}(x - u_n)^3 f''(u_n) + \cdots$$

From Eq. 4.30,

$$f(u_n) = \frac{1}{n}h(u_n)$$

whereas combining this with Eqs. 4.23 and 4.28, we obtain

$$f'(u_n) = -f(u_n)h(u_n) = -\frac{1}{n}h^2(u_n)$$

Taking the derivative of $f'(x)$ from Eq. 4.23 yields

$$f''(x) = f(x)h^2(x)$$

and at $x = u_n$

$$f''(u_n) = \frac{1}{n}h^3(u_n)$$

Higher derivatives may be similarly obtained. Thus, with Eq. 4.25,

$$F_X(x) = 1 - \frac{1}{n}\left[1 - (x - u_n)h(u_n) + \frac{1}{2}(x - u_n)^2 h^2(u_n) - \frac{1}{3!}(x - u_n)^3 h^3(u_n) + \cdots\right]$$

$$= 1 - \frac{1}{n}\exp[-h(u_n)(x - u_n)]$$

Therefore, the distribution of the largest value, Eq. 4.2, becomes

$$F_{X_n}(x) = \left[1 - \frac{1}{n}e^{-h(u_n)(x - u_n)}\right]^n$$

and

$$\ln F_{X_n}(x) = \frac{\ln\left[1 - \frac{1}{n}e^{-h(u_n)(x - u_n)}\right]}{\frac{1}{n}}$$

$$\lim_{n \to \infty} \frac{\ln\left[1 - \frac{a}{n}\right]}{\frac{1}{n}} = \left(e^{-a}\right)$$

As $n \to \infty$,

$$\lim_{n \to \infty} \ln F_{X_n}(x) = e^{-h(u_n)(x - u_n)}$$

Hence, for large n,

$$F_{X_n}(x) = \exp[-e^{-h(u_n)(x - u_n)}]$$

Comparing the above with Eq. 4.19, we observe that this is the Type I asymptotic distribution of the largest value with the second parameter

$$\alpha_n = h_n(u_n) = nf_X(u_n) \tag{4.32}$$

Similarly, it can be shown that for the Type I asymptotic distribution of the smallest value, the parameter α_1 is

$$\alpha_1 = h_1(u_1) = nf_X(u_1) \tag{4.33}$$

Standard Extremal Variates For the Type I largest value, we may introduce the *standardized extremal variate*

$$S = \alpha_n(X_n - u_n) \tag{4.34}$$

Its distribution (CDF) then is

$$F_S(s) = \exp(-e^{-s})$$ (4.35)

and its PDF is

$$f_S(s) = e^{-s} \exp(-e^{-s})$$ (4.36)

Equations 4.35 and 4.36 show that the distribution of S remains of the Type I asymptotic form, with parameters

$$u_n = 0 \quad \text{and} \quad \alpha_n = 1.0$$

The standard extremal variate serves the same purpose as the standard normal variate (Chapter 3, Vol. I). In particular, the probabilities of S, Eq. 4.35, may be tabulated (e.g., Table A.2); from which the cumulative probabilities of a general Type I extremal variate X_n may be evaluated from such a table of probabilities for S, using the relation of Eq. 4.34. That is, the CDF of X_n may be evaluated as

$$F_{X_n}(x) = F_S[\alpha_n(x - u_n)]$$ (4.37)

Similarly, for the Type I smallest value, the corresponding standardized extremal variate is

$$S = -\alpha_1(X_1 - u_1)$$ (4.38)

Then, cumulative probabilities associated with values of X_1 may be evaluated from those tabulated for S, Table A.2, using Eq. 4.38. That is,

$$F_{X_1}(x) = 1 - F_S[-\alpha_1(x - u_1)]$$ (4.39)

Moments and Moment-Generating Functions   The moment-generating function (see Chapter 3, Vol I) of the Type I standardized extremal variate S is

$$G_S(t) = E(e^{tS})$$

$$= \int_{-\infty}^{\infty} e^{ts} e^{-s} e^{-e^{-s}} \, ds$$

Let

$$r = e^{-s}$$

Then,

$$ds = -\frac{dr}{r}$$

and

$$G_S(t) = \int_0^{\infty} e^{ts} e^{-r} \, dr$$

$$= \int_0^{\infty} e^{-t \ln r} e^{-r} \, dr$$

$$= \int_0^{\infty} r^{-t} e^{-r} \, dr$$

This last integral is the *gamma function*† $\Gamma(1 - t)$; hence, the moment-generating function of S is

$$G_S(t) = \Gamma(1 - t) \tag{4.40}$$

The derivatives of $G_S(t)$, evaluated at $t = 0$, therefore yield the respective moments of S. The kth derivative of $\Gamma(1 - t)$ is (Cramer, 1946)

$$\Gamma^{(k)}(1 - t) = \int_0^\infty r^{-t} e^{-r} (\ln r)^k \, dr$$

At $t = 0$, the various derivatives of $\Gamma(1 - t)$ have been evaluated by Gumbel (1958); the first four of these are as follows:

$$\frac{d\Gamma(1)}{dt} = \gamma = 0.577216\ldots \text{(the Euler number)}$$

$$\frac{d^2\Gamma(1)}{dt^2} = \gamma^2 + \frac{\pi^2}{6}$$

$$\frac{d^3\Gamma(1)}{dt^3} = \gamma^3 + \frac{\pi^2}{2}\gamma + 2S_3$$

$$\frac{d^4\Gamma(1)}{dt^4} = \gamma^4 + \pi^2\gamma^2 + 8\gamma S_3 + \frac{3\pi^4}{20}$$

in which $S_3 = 1.20205690$. Hence, the mean and variance of S are

$$\mu_S = \gamma = 0.577216\ldots$$

$$\sigma_S^2 = \frac{\pi^2}{6}$$

The moments of a general Type I extremal variate may be evaluated from those of S using Eqs. 4.34 and 4.38. In particular, we obtain the mean and variance of X_n as follows.

From Eq. 4.34,

$$X_n = u_n + \frac{S}{\alpha_n}$$

Therefore, the mean value of X_n is

$$\mu_{X_n} = u_n + \frac{\gamma}{\alpha_n} \tag{4.41}$$

and the variance of X_n is

$$\sigma_{X_n}^2 = \frac{1}{\alpha_n^2}\sigma_S^2 = \frac{\pi^2}{6\alpha_n^2} \tag{4.42}$$

---

† Values tabulated in Appendix A.1.

Similarly, from Eq. 4.38, we have

$$X_1 = u_1 - \frac{S}{\alpha_1}$$

Therefore, the mean and variance of X_1 are

$$\mu_{X_1} = u_1 - \frac{\gamma}{\alpha_1} \tag{4.43}$$

and

$$\sigma_{X_1}^2 = \frac{\pi^2}{6\alpha_1^2} \tag{4.44}$$

The skewness coefficient of X_n can be shown to be equal to $\theta_S = 1.1414$, whereas the skewness coefficient of X_1 is equal to -1.1414.

The Extremal Parameters   The asymptotic form of the extremal distribution is largely independent of the exact form of the initial distribution; it depends only on the tail of the initial PDF in the direction of the extreme. However, the parameters u and α of the Type I asymptotic distribution, as well as those of the Type II and III asymptotic forms (described subsequently), are dependent on the initial distribution of X. The following examples should serve to emphasize this and also illustrate their determination for initial distributions with exponential tails.

EXAMPLE 4.14

A classic example of an initial distribution with an exponential tail is the exponential distribution, with PDF

$$f_X(x) = \frac{1}{\lambda} e^{-x/\lambda}; \quad x \geq 0$$

where $\lambda = \mu_X$, the mean value of the initial variate X.

The distribution of the largest value, therefore, will converge (for large n) to the Type I asymptotic form. The parameters, u_n and α_n, may be evaluated with Eqs. 4.25 and 4.32, respectively, as follows.

The CDF of the above exponential distribution is

$$F_X(x) = 1 - e^{-x/\lambda}$$

Then, according to Eq. 4.25,

$$1 - e^{-u_n/\lambda} = 1 - \frac{1}{n}$$

from which,

$$u_n = \lambda \ln n$$

Whereas, from Eq. 4.32,

$$\alpha_n = nf_X(u_n) = \frac{n}{\lambda} e^{-u_n/\lambda}$$

$$= \frac{n}{\lambda} e^{-\ln n}$$

Therefore,

$$\alpha_n = \frac{1}{\lambda}$$

The mean and standard deviation of the largest value X_n, Eq.

$$\mu_{X_n} = \lambda \ln n + 0.577215\lambda$$

$$\sigma_{X_n} = \frac{\pi}{\sqrt{6}} \lambda \qquad \mu_{X_n} = u_n + .577\frac{1}{\lambda}$$

EXAMPLE 4.15

An initial variate with the Rayleigh PDF is

$$f_X(x) = \frac{x}{\sigma^2} e^{-(1/2)(x/\sigma)^2}; \qquad x \geq 0$$

where σ = the modal value. This initial distribution has an exponential tail in the positive direction (see Example 4.20 for verification). The distribution of the largest value from this initial variate, therefore, will converge, for large n, to the Type I asymptotic form.

With Eqs. 4.25 and 4.32, the parameters, u_n and α_n, of this Type I asymptotic distribution are evaluated, respectively, as follows:

The CDF of X is

$$F_X(x) = 1 - e^{-(1/2)(x/\sigma)^2}$$

Therefore, according to Eq. 4.25,

$$1 - e^{-(1/2)(u_n/\sigma)^2} = 1 - \frac{1}{n}$$

from which we obtain

$$u_n = \sigma\sqrt{2 \ln n}$$

and according to Eq. 4.32,

$$\alpha_n = n f_X(u_n) = n \left(\frac{u_n}{\sigma^2}\right) e^{-(1/2)(u_n/\sigma)^2}$$

$$= \frac{n\sqrt{2 \ln n}}{\sigma} e^{-\ln n}$$

Thus,

$$\alpha_n = \frac{\sqrt{2 \ln n}}{\sigma}$$

The mean and standard deviation of the largest value X_n, therefore, are

$$\mu_{X_n} = \sigma\sqrt{2 \ln n} + 0.577215\left(\frac{\sigma}{\sqrt{2 \ln n}}\right)$$

$$\sigma_{X_n} = \frac{\pi}{\sqrt{6}} \frac{\sigma}{\sqrt{2 \ln n}} = \frac{\pi\sigma}{2\sqrt{3 \ln n}}$$

From the results of the last two examples (Examples 4.14 and 4.15), we see that the extremal parameters as well as the mean and standard deviation are functions of n; moreover, the specific form of these functions depends on the initial distribution. In particular, for the exponential initial distribution (Example 4.14) u_n and μ_{X_n} increase with $\ln n$, whereas α_n and σ_{X_n} are constants and therefore independent of n. However, in the case of the Rayleigh initial distribution (Example 4.15), the quantities u_n and μ_{X_n} increase with $\sqrt{\ln n}$, whereas the dispersion ($1/\alpha_n$ and σ_{X_n}) decreases with $\sqrt{\ln n}$.

For normal initial variates, see Example 4.8 for expressions of u_n and α_n.

4.3.3   The Type II Asymptotic Form

Extreme values from an initial distribution with a polynomial tail (in the direction of the appropriate extreme) will converge asymptotically, in distribution, to the Type II asymptotic form (Gumbel, 1958). For the largest value, the Type II asymptotic CDF is

$$F_{Y_n}(y) = \exp\left[-\left(\frac{v_n}{y}\right)^k\right]$$

(4.45)

and the corresponding PDF is

$$f_{Y_n}(y) = \frac{k}{v_n}\left(\frac{v_n}{y}\right)^{k+1}\exp\left[-\left(\frac{v_n}{y}\right)^k\right]$$

(4.46)

where

v_n: The characteristic largest value of the initial variate X.

k: The shape parameter; $1/k$ is a measure of dispersion.

v_n is defined in the same way as u_n; that is, for an initial variate X,

$$F_X(v_n) = 1 - \frac{1}{n}$$

(4.47)

Also, v_n is equal to the most probable value of Y_n, or

$$v_n = \tilde{y}_n$$

Polynomial Tail   An initial distribution with a polynomial tail, known also as the Cauchy/Pareto type, is also unlimited in the direction of the extreme; and for large positive values of x,

$$\lim_{x \to \infty} x^k[1 - F_X(x)] = a$$

(4.48)

where $k > 0$ and a is a positive constant.

For large n, using Eq. 4.47 in Eq. 4.48, we obtain

$$a = \frac{1}{n}v_n^k$$

Therefore, for large positive values of x,

$$F_X(x) = 1 - \frac{1}{n}\left(\frac{v_n}{x}\right)^k$$

(4.48a)

Hence, the distribution of the largest value, Eq. 4.2, becomes

$$F_{Y_n}(y) = \left[1 - \frac{1}{n}\left(\frac{v_n}{y}\right)^k \right]^n$$

Expanding this into a binomial series and observing that for large n, $y \to \infty$, we obtain

$$F_{Y_n}(y) = \exp\left[-\left(\frac{v_n}{y}\right)^k \right]$$

which is obviously the Type II asymptotic form of Eq. 4.45; this verifies that the distribution of the largest value from an initial distribution of the Cauchy/Pareto type, Eq. 4.48a, will converge, for large n, to the Type II asymptotic form.

The Type I and II asymptotic forms are related by the following *logarithmetic transformation.*

If Y_n has the Type II asymptotic distribution, Eq. 4.45, with parameters v_n and k, the distribution of $\ln Y_n$ *will have the Type I asymptotic form, Eq. 4.19, with parameters*

$$u_n = \ln v_n \quad \text{and} \quad \alpha_n = k$$

To show this, suppose that Y_n has a Type II asymptotic distribution with parameters v_n and k; that is, Eq. 4.45. The distribution of

$$X_n = \ln Y_n$$

may be derived as follows.

Observe that

$$Y_n = e^{X_n}$$

Then, the distribution of X_n is

$$F_{X_n}(x) = F_{Y_n}(e^x) = \exp\left[-\left(\frac{v_n}{e^x}\right)^k \right]$$

If we define

$$v_n = e^{u_n}$$

and

$$k = \alpha_n$$

the CDF of X_n becomes

$$F_{X_n}(x) = \exp[-e^{-\alpha_n(x - u_n)}]$$

thus verifying the above logarithmic transformation.

EXAMPLE 4.16

Suppose that the initial distribution of a random variable X is lognormal with parameters λ_X and ζ_X. It is well-known (see, e.g., Chapter 4, Vol. I) that the natural logarithm of the variate, that is,

$$X' = \ln X$$

is normal with parameters $\mu = \lambda_X$ and $\sigma = \zeta_X$. According to the results of Example 4.8, the largest value of X' will converge to the Type I asymptotic distribution with parameters

$$u'_n = \zeta_X \sqrt{2 \ln n} - \zeta_X \frac{\ln \ln n + \ln 4\pi}{2\sqrt{2 \ln n}} + \lambda_X$$

and

$$\alpha'_n = \sqrt{2 \ln n}/\zeta_X$$

Then, according to the above logarithmic transformation, the distribution of the largest value of X should converge to the Type II asymptotic form with parameters

$$v_n = e^{u'_n}$$

and

$$k = \alpha'_n$$

The Type II Asymptotic Form for Smallest Value   Applying the symmetry principle to Eqs. 4.45 and 4.46, we obtain the corresponding Type II asymptotic form for the smallest value as follows:

$$F_{Y_1}(y) = 1 - \exp\left[-\left(\frac{v_1}{y}\right)^k\right]; \qquad y \leq 0; v_1 > 0 \tag{4.49}$$

and

$$f_{Y_1}(y) = -\frac{k}{v_1}\left(\frac{v_1}{y}\right)^{k+1} \exp\left[-\left(\frac{v_1}{y}\right)^k\right] \tag{4.50}$$

where the parameter v_1 is the characteristic smallest value of the initial variate X, and k is the shape parameter; again, $1/k$ is a measure of dispersion.

As with Eq. 4.26, v_1 is defined as

$$F_X(v_1) = \frac{1}{n} \tag{4.51}$$

Moreover, v_1 is equal to the modal value of Y_1; that is,

$$v_1 = \tilde{y}_1$$

The logarithmic transformation between the Type I and II asymptotic distributions applies also to the distributions of the respective smallest positive values; that is, *if Y_1 has the Type II asymptotic distribution with positive parameters v_1 and k, then $\ln Y_1$ has the Type I asymptotic distribution with parameters $u_1 = \ln v_1$ and $\alpha_1 = k$.*

In light of the above logarithmic transformation, the standardized extremal variate of Eq. 4.34 becomes

$$\begin{aligned} S &= \alpha_n(X_n - u_n) \\ &= k(\ln Y_n - \ln v_n) \\ &= k \ln(Y_n/v_n) \end{aligned} \tag{4.52}$$

Equation 4.52 defines the standardized extremal variate for the Type II largest value. Accordingly, the cumulative probabilities of Y_n may be obtained from those tabulated for S, Table A.2, as follows:

$$F_{Y_n}(y) = F_S\left(k \ln \frac{y}{v_n}\right) \tag{4.53}$$

Similarly, from Eq. 4.38 we obtain the corresponding standardized extremal variate for the Type II smallest value as (for positive y and v_1)

$$S = -k \ln (Y_1/v_1) \tag{4.52a}$$

and the cumulative probabilities of Y_1 may be obtained from Table A.2 as follows:

$$F_{Y_1}(y) = 1 - F_S\left(-k \ln \frac{y}{v_1}\right) \tag{4.53a}$$

Moments and Moment-Generating Functions From Eq. 4.52, we obtain

$$Y_n = v_n e^{S/k}$$

Therefore, the tth moment of Y_n is

$$E(Y_n^t) = v_n^t E[e^{(t/k)S}]$$
$$= v_n^t G_S(t/k)$$

where $G_S(t/k)$ is the moment-generating function of S, given by Eq. 4.40. Hence, the tth moment of Y_n is

$$E(Y_n^t) = v_n^t \Gamma(1 - t/k) \tag{4.54}$$

Observe that only moments of order $t < k$ exist (Gumbel, 1958).

With Eq. 4.54, we obtain the mean value of Y_n as

$$\mu_{Y_n} = v_n \Gamma\left(1 - \frac{1}{k}\right) \tag{4.55}$$

and for $k \geq 2$,

$$E(Y_n^2) = v_n^2 \Gamma\left(1 - \frac{2}{k}\right)$$

Thus, the variance of Y_n is

$$\sigma_{Y_n}^2 = v_n^2 \left[\Gamma\left(1 - \frac{2}{k}\right) - \Gamma^2\left(1 - \frac{1}{k}\right)\right] \tag{4.56}$$

The coefficient of variation, δ_{Y_n}, therefore, is given by

$$1 + \delta_{Y_n}^2 = \frac{\Gamma\left(1 - \frac{2}{k}\right)}{\Gamma^2\left(1 - \frac{1}{k}\right)} \tag{4.57}$$

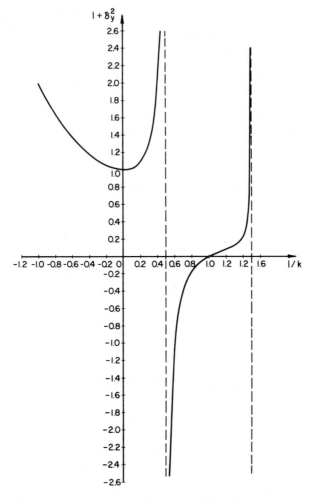

Figure 4.4   Relation between c.o.v. of Y_n and parameter k.

Equation 4.57 is shown graphically in Fig. 4.4, which may be used to evaluate $1/k$ for given δ_{Y_n}, or vice versa. Observe that at $1/k = \frac{1}{2}, \frac{3}{2}, \dots$, the value of $(1 + \delta_{Y_n}^2) \to \infty$.

Because of the similarity between Eqs. 4.52 and 4.52a, it follows that the moments of the smallest variate Y_1 may be deduced from those of Y_n. Thus,

$$E(Y_1^t) = v_1^t \Gamma(1 - t/k) \tag{4.54a}$$

From this the mean and variance of Y_1 are

$$\mu_{Y_1} = v_1 \Gamma\left(1 - \frac{1}{k}\right) \tag{4.55a}$$

$$\sigma_{Y_1}^2 = v_1^2 \left[\Gamma\left(1 - \frac{2}{k}\right) - \Gamma^2\left(1 - \frac{1}{k}\right)\right] \tag{4.56a}$$

whereas the coefficient of variation of Y_1 is the same as that of Y_n, or

$$\delta_{Y_1} = \delta_{Y_n}$$

EXAMPLE 4.17

An example of the Cauchy/Pareto type of initial distribution is the Pareto distribution with the following PDF:

$$f_X(x) = \frac{q}{x^{q+1}}; \qquad x \geq 1$$

Its CDF is

$$F_X(x) = 1 - \frac{1}{x^q}$$

The distribution of the largest value of X with the above initial distribution will converge, for large n, to the Type II asymptotic form. The associated parameters may be determined using its logarithmic relationship with the Type I asymptotic form, as follows.

Let

$$X' = \ln X$$

or

$$X = e^{X'}$$

Then, the initial distribution of X' is

$$F_{X'}(x) = F_X(e^x) = 1 - \frac{1}{e^{qx}}$$

Thus,

$$F_{X'}(x) = 1 - e^{-qx}; \qquad x \geq 0$$

which is of the exponential type; therefore, its largest value will converge in distribution to the Type I asymptotic form. Its parameters may be obtained from the results of Example 4.14, yielding

$$u_n = \frac{1}{q} \ln n$$

$$\alpha_n = q$$

Therefore, the parameters of the Type II asymptotic distribution of the largest value of X are as follows:

$$v_n = e^{u_n} = n^{1/q}$$

and

$$k = \alpha_n = q$$

The parameter v_n, of course, may be determined also through Eq. 4.47; that is,

$$F_X(v_n) = 1 - \frac{1}{v_n^q} = 1 - \frac{1}{n}$$

From this, the same expression as above for v_n is obtained.

For the particular Pareto distribution with $q = 3$; that is,

$$f_X(x) = \frac{3}{x^4}; \quad x \geq 1$$

the distribution of its largest value would be the Type II asymptotic form, with the parameters

$$v_n = n^{1/3}$$

and

$$k = 3$$

Thus,

$$F_{Y_n}(y) = \exp\left[-\left(\frac{n}{y^3}\right)\right]$$

The mean and standard deviation of Y_n are (using Table A.1):

$$\mu_{Y_n} = n^{1/3}\Gamma(1 - \tfrac{1}{3}) = n^{1/3}\Gamma(\tfrac{2}{3}) = 1.354n^{1/3}$$
$$\sigma_{Y_n} = n^{1/3}[\Gamma(1 - \tfrac{2}{3}) - \Gamma^2(1 - \tfrac{1}{3})]^{1/2}$$
$$= n^{1/3}[2.679 - (1.354)^2]^{1/2} = 0.846n^{1/3}$$

4.3.4   The Type III Asymptotic Form

Both the Type I and II asymptotic distributions are the limiting forms of the distribution of extreme values from initial distributions that are unlimited in the directions of the relevant extremes. In contrast, the Type III asymptotic form represents the limiting distribution of the extreme values from initial distributions that have a finite upper- or lowerbound value; that is, $F_X(\omega) = 1.0$, in the case of the largest value, whereas $F_X(\varepsilon) = 0$, in the case of the smallest value, in which ω = the upperbound value and ε = the lowerbound value of X.

For the largest value, the Type III asymptotic CDF is

$$F_{Z_n}(z) = \exp\left[-\left(\frac{\omega - z}{\omega - w_n}\right)^k\right]; \quad z \leq \omega \qquad (4.58)$$

and the corresponding PDF is

$$f_{Z_n}(z) = \frac{k}{\omega - w_n}\left(\frac{\omega - z}{\omega - w_n}\right)^{k-1}\exp\left[-\left(\frac{\omega - z}{\omega - w_n}\right)^k\right]; \quad z \leq \omega \qquad (4.59)$$

where the parameter w_n is the *characteristic largest value* of X, defined by

$$F_X(w_n) = 1 - \frac{1}{n} \qquad (4.60)$$

and k is the shape parameter. The characteristic extreme, w_n, is again equal to the modal value of Z_n; that is,

$$w_n = \tilde{z}_n$$

whereas $1/k$ is a measure of the dispersion of Z_n.

A limited initial distribution may be expressed as

$$F_X(x) = 1 - \frac{1}{n}\left(\frac{\omega - x}{\omega - w_n}\right)^k \qquad (4.61)$$

where ω is the upper limit value of X, such that $F_X(\omega) = 1.0$. The largest value from such an initial distribution will converge to the Type III asymptotic distribution, Eq. 4.58.

With Eq. 4.2, the CDF of the largest value of X is

$$F_{Z_n}(z) = \left[1 - \frac{1}{n}\left(\frac{\omega - z}{\omega - w_n}\right)^k\right]^n$$

Expanding the above into a binomial series, we observe that for large values of z, the series yields

$$F_{Z_n}(z) = \exp\left[-\left(\frac{\omega - z}{\omega - w_n}\right)^k\right]$$

which is the Type III asymptotic form of Eq. 4.58.

By the symmetry principle, Eq. 4.17, the Type III asymptotic distribution of the smallest value from an initial distribution with lower limit ε is

$$F_{Z_1}(z) = 1 - \exp\left[-\left(\frac{z - \varepsilon}{w_1 - \varepsilon}\right)^k\right]; \qquad z \geq \varepsilon \qquad (4.62)$$

and its corresponding PDF is

$$f_{Z_1}(z) = \frac{k}{w_1 - \varepsilon}\left(\frac{z - \varepsilon}{w_1 - \varepsilon}\right)^{k-1} \exp\left[-\left(\frac{z - \varepsilon}{w_1 - \varepsilon}\right)^k\right]; \qquad z \geq \varepsilon \qquad (4.63)$$

where the parameter w_1 is the *characteristic smallest value*, defined as

$$F_X(w_1) = \frac{1}{n} \qquad (4.64)$$

and k is the shape parameter. The characteristic value w_1 is also equal to the modal value of Z_1, or

$$w_1 = \tilde{z}_1$$

whereas $1/k$ is a measure of the dispersion of Z_1.

The Type III asymptotic distribution of the smallest value was developed also by Weibull (1951) in connection with the study of fatigue and fracture of materials, and is known also as the *Weibull distribution*.

The Type III asymptotic form is also related to the Type I asymptotic form in the following manner.

Suppose

$$X_n = a - \ln\left(\frac{\omega - Z_n}{\omega - w_n}\right)$$

from which

$$Z_n = \omega - (\omega - w_n)e^{-(X_n - a)}$$

Then, if the distribution of Z_n is of the Type III asymptotic form for the largest value, Eq. 4.58, with parameters ω, w_n, and k, the distribution of X_n would be

$$F_{X_n}(x) = \exp\left\{-\left[\frac{\omega - \omega + (\omega - w_n)e^{-(x-a)}}{\omega - w_n}\right]^k\right\}$$

$$= \exp[-e^{-k(x-a)}]$$

which means that X_n has the Type I asymptotic distribution with parameters

$$u_n = a \quad \text{and} \quad \alpha_n = k$$

Therefore, if Z_n has the Type III asymptotic distribution of the largest value with an upper bound ω and parameters w_n and k, then

$$X_n = a - \ln\left[(\omega - Z_n)/(\omega - w_n)\right]$$

has the Type I asymptotic distribution with parameters $u_n = a$ and $\alpha_n = k$.

Similarly, if the extremal variate Z_1 has the Type III asymptotic distribution of the smallest value with parameters ε, w_1, and k, the variate

$$X_1 = a - \ln\left(\frac{Z_1 - \varepsilon}{w_1 - \varepsilon}\right)$$

will have the Type I asymptotic distribution of the smallest value with parameters $u_1 = a$ and $\alpha_1 = k$.

In light of the above relationships between X_n and Z_n, and between X_1 and Z_1, we may define the corresponding standardized Type III extremal variate as follows.

For the Type III largest value,

$$S = \alpha_n(X_n - u_n)$$

$$= k\left[a - \ln\left(\frac{\omega - Z_n}{\omega - w_n}\right) - a\right]$$

$$= -k \ln\left(\frac{\omega - Z_n}{\omega - w_n}\right) \tag{4.65}$$

Therefore, the cumulative probability of Z_n may be obtained from Table A.2 as

$$F_{Z_n}(z) = F_S\left[-k \ln\left(\frac{\omega - z}{\omega - w_n}\right)\right] \tag{4.66}$$

By the same token, the standardized Type III extremal variate for the smallest value is

$$S = -k \ln\left(\frac{Z_1 - \varepsilon}{w_1 - \varepsilon}\right) \tag{4.67}$$

Hence, the corresponding probabilities of Z_1 may be obtained from Table A.2 as

$$F_{Z_1}(z) = 1 - F_S\left[-k \ln\left(\frac{z - \varepsilon}{w_1 - \varepsilon}\right)\right] \tag{4.68}$$

Moments and Moment-Generating Functions   From the standardized Type III extremal variate, Eq. 4.65, we have

$$\frac{\omega - Z_n}{\omega - w_n} = e^{-S/k}$$

and

$$(\omega - Z_n) = (\omega - w_n)e^{-S/k}$$

Taking the expectations of both sides, we obtain

$$\begin{aligned}
E(\omega - Z_n)^t &= (\omega - w_n)^t E(e^{-tS/k}) \\
&= (\omega - w_n)^t G_S(-t/k) \\
&= (\omega - w_n)^t \Gamma(1 + t/k)
\end{aligned} \qquad (4.69)$$

Equation 4.69, therefore, gives the tth moment of $(\omega - Z_n)$ from which the corresponding moment of Z_n may be obtained. On this basis, we obtain the mean value of Z_n

$$\mu_{Z_n} = \omega - (\omega - w_n)\Gamma(1 + 1/k) \qquad (4.70)$$

whereas, with Eq. 4.69, the variance of Z_n is obtained as follows:

$$E(\omega - Z_n)^2 = (\omega - w_n)^2 \Gamma(1 + 2/k)$$

Hence,

$$\sigma_{Z_n}^2 = \text{Var}(\omega - Z_n) = (\omega - w_n)^2[\Gamma(1 + 2/k) - \Gamma^2(1 + 1/k)] \qquad (4.71)$$

With Eqs. 4.70 and 4.71, we obtain the following ratio:

$$1 + \left(\frac{\sigma_{Z_n}}{\omega - \mu_{Z_n}}\right)^2 = \frac{\Gamma(1 + 2/k)}{\Gamma^2(1 + 1/k)} \qquad (4.72)$$

Such a relation is shown graphically in Fig. 4.5 as a function of $1/k$, thus confirming that $1/k$ is a measure of dispersion.

For the Type III asymptotic distribution of the smallest value, we obtain with Eq. 4.67

$$\begin{aligned}
E(Z_1 - \varepsilon)^t &= (w_1 - \varepsilon)^t E(e^{-tS/k}) \\
&= (w_1 - \varepsilon)^t G_S(-t/k) \\
&= (w_1 - \varepsilon)^t \Gamma(1 + t/k)
\end{aligned} \qquad (4.73)$$

From which we obtain the mean value and variance of Z_1 as follows:

$$\mu_{Z_1} = \varepsilon + (w_1 - \varepsilon)\Gamma(1 + 1/k) \qquad (4.74)$$

and

$$\sigma_{Z_1}^2 = \text{Var}(Z_1 - \varepsilon) = (w_1 - \varepsilon)^2\left[\Gamma\left(1 + \frac{2}{k}\right) - \Gamma^2\left(1 + \frac{1}{k}\right)\right] \qquad (4.75)$$

Thus,

$$1 + \left(\frac{\sigma_{Z_1}}{\mu_{Z_1} - \varepsilon}\right)^2 = \frac{\Gamma(1 + 2/k)}{\Gamma^2(1 + 1/k)} \qquad (4.76)$$

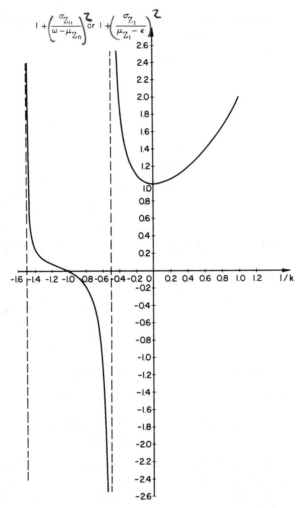

Figure 4.5 Relation between the mean and variance of Z_n and the parameter k.

which is also portrayed in Fig. 4.5. In this case, observe that if $\varepsilon = 0$, σ_{Z_1}/μ_{Z_1} is the coefficient of variation, δ_{Z_1}. Hence, for $\varepsilon = 0$, Fig. 4.5 gives the relationship between the coefficient of variation, δ_{Z_1}, and $1/k$.

EXAMPLE 4.18

Suppose the initial distribution between 0 and a is

$$F_X(x) = 1 - \left(\frac{a - x}{a}\right)^2; \qquad 0 \leq x \leq a$$

Its corresponding PDF is

$$f_X(x) = \frac{2}{a^2}(a - x)$$

which is a triangular density function as shown in Fig. E4.18.

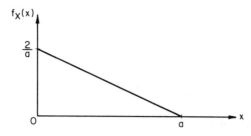

Figure E4.18

Observe that the above CDF is of the limited form of Eq. 4.61, with upper limit $\omega = a$.

The distribution of the largest value of X, therefore, will converge to the Type III asymptotic form of Eq. 4.58 with parameters

$$k = 2; \qquad \omega = a$$

and w_n determined with Eq. 4.60 as follows:

$$1 - \left(\frac{a - w_n}{a}\right)^2 = 1 - \frac{1}{n}$$

Then,

$$w_n = a\left(1 - \frac{1}{\sqrt{n}}\right)$$

Hence, the Type III asymptotic distribution, Eq. 4.58, becomes

$$F_{Z_n}(z) = \exp\left[-n\left(\frac{a - z}{a}\right)^2\right]$$

The smallest value of this same initial distribution is also limited, with lower limit value $\varepsilon = 0$. Therefore, its CDF is given by Eq. 4.62 with parameters

$$k = 2$$

and w_1 determined by Eq. 4.64 as follows:

$$1 - \left(\frac{a - w_1}{a}\right)^2 = \frac{1}{n}$$

Thus

$$w_1 = a\left(1 - \sqrt{\frac{n-1}{n}}\right)$$

Therefore, the asymptotic distribution, Eq. 4.62, of the smallest value becomes

$$F_{Z_1}(z) = 1 - \exp\left[\frac{-(z/a)^2}{\left(1 - \sqrt{\frac{n-1}{n}}\right)^2}\right]$$

The mean and standard deviation of the largest variate, Z_n, are as follows. From Eq. 4.70,

$$\mu_{Z_n} = a\left[1 - \frac{1}{\sqrt{n}}\Gamma\left(\frac{3}{2}\right)\right]$$

but

$$\Gamma\left(\frac{3}{2}\right) = \frac{1}{2}\Gamma\left(\frac{1}{2}\right) = \frac{1}{2}\sqrt{\pi}$$

Therefore,

$$\mu_{Z_n} = a\left(1 - \frac{1}{2}\sqrt{\frac{\pi}{n}}\right)$$

whereas with Eq. 4.71, we obtain

$$\sigma_{Z_n}^2 = \frac{a^2}{n}\left[\Gamma(2) - \Gamma^2\left(\frac{3}{2}\right)\right]$$

$$= \frac{a^2}{n}\left(1 - \frac{\pi}{4}\right)$$

Hence,

$$\sigma_{Z_n} = a\sqrt{\frac{1 - \pi/4}{n}}$$

The corresponding mean and standard deviation of the smallest variate Z_1 are similarly obtained with Eqs. 4.74 and 4.75, yielding

$$\mu_{Z_1} = \frac{\sqrt{\pi}}{2}a\left(1 - \sqrt{\frac{n-1}{n}}\right)$$

and

$$\sigma_{Z_1} = a\left(1 - \sqrt{\frac{n-1}{n}}\right)\left(1 - \frac{\pi}{4}\right)$$

4.3.5 Convergence Criteria

Depending on the tail behavior of the initial distribution (in the direction of the extreme), the extreme value may converge, as $n \to \infty$, to one or the other asymptotic forms of extremal distributions. In some cases, the tail behavior of the initial distribution may be obvious, and there would be no question about which asymptotic form is appropriate. For example, for initial variates with the exponential or normal PDF, the tails are clearly exponentially decaying in the direction of large values of the variate; hence, the distribution of the largest value from either of these initial variates will converge to the Type I asymptotic form. In the case of a normal initial variate, it is also unlimited in the negative direction; hence, the distribution of its smallest value will also converge to the Type I asymptotic form.

There are cases, however, in which the tail behavior is less obvious and consequently questions of the limiting form of the distributions of the extreme values may arise. Mathematical criteria are available to provide guidance for this purpose.

Summarized here are von Mises' criteria for determining the asymptotic form of distribution of the extreme value from an initial variate X with CDF $F_X(x)$.

Convergence to Type I Asymptotic Form   The extreme value of X will converge in distribution to the Type I asymptotic form for large n, if

$$\lim_{x \to \infty} \frac{d}{dx}\left[\frac{1}{h_n(x)}\right] = 0 \tag{4.77}$$

where $h_n(x)$ is the hazard function of Eq. 4.28. Equation 4.77 applies to the largest value of X; the corresponding criterion for the smallest value would be

$$\lim_{x \to -\infty} \frac{d}{dx}\left[\frac{1}{h_1(x)}\right] = 0 \tag{4.78}$$

where $h_1(x)$ is given by Eq. 4.29.

EXAMPLE 4.19

Consider the largest value from an initial variate X with the exponential distribution

$$f_X(x) = \lambda e^{-\lambda x}; \qquad x \geq 0$$

and

$$F_X(x) = 1 - e^{-\lambda x}$$

where λ = constant.
    For the largest value,

$$h_n(x) = \frac{f(x)}{1 - F(x)} = \frac{\lambda e^{-\lambda x}}{e^{-\lambda x}} = \lambda$$

Therefore, Eq. 4.77 yields

$$\lim_{x \to \infty} \frac{d}{dx}\left[\frac{1}{\lambda}\right] = 0$$

This confirms the fact that the largest value from an exponential distribution will converge to the Type I asymptotic form; that is, Eq. 4.19.
    To illustrate the criterion for the smallest value, consider X with the following exponential PDF:

$$f_X(x) = \lambda e^{\lambda x}; \qquad x \leq 0$$

Then,

$$1 - F_X(x) = \int_x^0 \lambda e^{\lambda x}\, dx$$

$$= 1 - e^{\lambda x}$$

Thus,

$$F_X(x) = e^{\lambda x}; \qquad x \leq 0$$

The hazard function, Eq. 4.29, therefore is

$$h_1(x) = \lambda$$

and thus Eq. 4.78 yields

$$\frac{d}{dx}\left(\frac{1}{\lambda}\right) = 0$$

    Therefore, the smallest value from the above "negative" initial variate will also converge to the Type I asymptotic form for the smallest value; namely, Eq. 4.21.

EXAMPLE 4.20

Consider an initial variate with the Rayleigh distribution

$$f_X(x) = \frac{x}{\sigma^2} e^{-(1/2)(x/\sigma)^2}; \qquad x \geq 0$$

and

$$F_X(x) = 1 - e^{-(1/2)(x/\sigma)^2}$$

In this case, it is not obvious that the tail of the PDF in the positive direction is exponential. Applying Eq. 4.77, we obtain

$$\frac{1}{h_n(x)} = \frac{1 - [1 - e^{-(1/2)(x/\sigma)^2}]}{\dfrac{x}{\sigma^2} e^{-(1/2)(x/\sigma)^2}} = \frac{\sigma^2}{x}$$

and thus,

$$\lim_{x \to \infty} \frac{d}{dx}\left(\frac{\sigma^2}{x}\right) = -\lim_{x \to \infty} \frac{\sigma^2}{x^2} = 0$$

Hence, the distribution of the largest value from a Rayleigh-distributed initial variate will converge to the Type I asymptotic form.

For the Rayleigh distribution, we observe that the lower limit of X is finite; namely, $x = 0$. Therefore, the distribution of the smallest value from this initial variate will not approach the same asymptotic form as that of the largest value; see Example 4.23 below.

Convergence to Type II Asymptotic Form   The extreme value from an initial variate X will converge (in distribution) to the Type II asymptotic form for the largest value if

$$\lim_{x \to \infty} x h_n(x) = k; \qquad k > 0, \text{ a constant} \tag{4.79}$$

whereas convergence to the Type II asymptotic form for the smallest value will result if

$$\lim_{x \to -\infty} -x h_1(x) = k; \qquad k > 0 \tag{4.80}$$

EXAMPLE 4.21

Suppose an initial variate X with the following Pareto distribution:

$$f_X(x) = \frac{1}{x^2}; \qquad x \geq 1$$

$$= 0; \qquad x < 1$$

Then,

$$F_X(x) = 1 - \frac{1}{x}$$

Eq. 4.79, therefore, yields

$$\lim_{x \to \infty} x h_n(x) = x \cdot \frac{\dfrac{1}{x^2}}{1 - \left(1 - \dfrac{1}{x}\right)} = 1$$

Hence, the distribution of the largest value from X will converge asymptotically to the Type II asymptotic form.

Next, consider the smallest value from X with the following Pareto PDF:

$$f_X(x) = \frac{1}{x^2}; \qquad x \le -1$$

$$= 0; \qquad \text{elsewhere}$$

In this case,

$$F_X(x) = -\frac{1}{x}$$

and the hazard function of Eq. 4.29 gives

$$h_1(x) = -1/x$$

Hence, Eq. 4.80 becomes

$$\lim_{x \to -\infty} \left(-x \cdot \frac{-1}{x} \right) = 1$$

Therefore, the smallest value from the above initial variate will also converge to the Type II asymptotic form.

Convergence to Type III Asymptotic Form   The largest value from an initial variate X will converge (in distribution) to the Type III asymptotic form if

$$F_X(\omega) = 1.0; \qquad \text{for upper bound } \omega$$

and

$$\lim_{x \to \omega} (\omega - x)h_n(x) = k; \qquad k > 0, \text{ a constant} \tag{4.81}$$

For the smallest value from an initial variate X with lower bound ε, the corresponding criterion is

$$F_X(\varepsilon) = 0$$

and

$$\lim_{x \to \varepsilon} (x - \varepsilon)h_1(x) = k; \qquad k > 0 \tag{4.82}$$

EXAMPLE 4.22

Consider an initial variate X with a uniform distribution between a and b, that is, with PDF,

$$f_X(x) = \frac{1}{b - a}; \qquad a \le x \le b$$

and

$$F_X(x) = \frac{x - a}{b - a}$$

In this case,

$$\omega = b; \qquad \varepsilon = a$$

Therefore, for the largest value, Eq. 4.81 yields

$$\lim_{x \to b} (b - x) \frac{1/(b - a)}{1 - (x - a)/(b - a)} = \frac{(b - x)}{(b - x)} = 1$$

whereas, for the smallest value, Eq. 4.82 becomes

$$\lim_{x \to a} (x - a) \frac{1/(b - a)}{(x - a)/(b - a)} = \left(\frac{x - a}{x - a}\right) = 1$$

Therefore, both the largest and smallest values from an initial variate with a uniform distribution will converge asymptotically to the respective Type III asymptotic forms.

EXAMPLE 4.23

In the case of the Rayleigh distribution, we saw in Example 4.20 that the distribution of its largest value will converge, for large n, to the Type I asymptotic form.

Since the lower limit of the Rayleigh distribution is finite; namely, at $x = 0$, the distribution of its smallest value will obviously not approach the Type I asymptotic form. It may converge on the Type III asymptotic form; we confirm this through Eq. 4.82 as follows:

For the smallest value of the Rayleigh distribution, with the PDF and CDF of Example 4.20, we have

$$xh_1(x) = \frac{\left(\dfrac{x^2}{\sigma^2}\right)e^{-(1/2)(x/\sigma)^2}}{1 - e^{-(1/2)(x/\sigma)^2}} = \frac{(x/\sigma)^2}{e^{(1/2)(x/\sigma)^2} - 1}$$

$$\lim_{x \to 0} \left[\frac{(x/\sigma)^2}{e^{(1/2)(x/\sigma)^2} - 1}\right] = \lim_{x \to 0} \left[\frac{2}{e^{(1/2)(x/\sigma)^2}}\right] = 2$$

Therefore, the distribution of the smallest value from an initial Rayleigh distribution will converge on the Type III asymptotic form.

The Stability Postulate   The existence of the asymptotic forms of the distribution of extremes is based on the premise that the form of a probability distribution is invariant under a linear transformation. Thus, the distribution of the largest value of an initial variate X may be expressed as a linear function of X; that is,

$$[F_X(x)]^n = F_X(a_n x + b_n) \tag{4.83}$$

where a_n and b_n are functions of n. Equation 4.83 is called the *stability postulate* (Gumbel, 1958); its solutions yield all the possible limiting forms of $F_X(x)$ as $n \to \infty$ (Fisher and Tippett, 1928).

Depending on the functions a_n and b_n that will satisfy Eq. 4.83, different limiting forms of the function $F_X(x)$ will result; three such forms were derived by Fisher and Tippett (1928) that Gumbel subsequently called the three types of asymptotic forms. However, there are forms of the initial distributions whose extreme values will have distributions that naturally belong to the three asymptotic forms irrespective of the sample size n, for example, n can be as small as 2. In particular, if the initial distribution is one of the three types of asymptotic forms, the distribution of its extreme value will also be of the same asymptotic form.

For example, if X has initial distribution

$$F_X(x) = \exp[-e^{-\alpha(x-u)}]$$

then the CDF of its largest value among samples of any size n is, according to Eq. 4.2,

$$F_{X_n}(x) = \exp[-ne^{-\alpha(x-u)}]$$
$$= \exp\{-e^{-\alpha[x-(u+\ln n/\alpha)]}\}$$

This means that the distribution of the largest value X_n is of the same shape as that of the initial variate X, except that the distribution is shifted to the right by $(\ln n)/\alpha$ as shown in Fig. 4.6 for the PDFs.

The parameters of X_n, therefore, are related to those of X as follows:

$$\alpha_n = \alpha$$

and

$$u_n = u + \frac{\ln n}{\alpha}$$

Then, the corresponding mean and standard deviation of X_n are

$$\sigma_{X_n} = \sigma_X$$

and

$$\mu_{X_n} = u + \frac{\ln n}{\alpha} + \frac{\gamma}{\alpha_n}$$

$$= \mu_X + \frac{\ln n}{\alpha}$$

$$= \mu_X\left(1 + \frac{\sqrt{6}}{\pi}\delta_X \ln n\right)$$

Similarly, if the initial distribution is the Type II asymptotic distribution,

$$F_X(x) = \exp\left[-\left(\frac{v}{x}\right)^k\right]$$

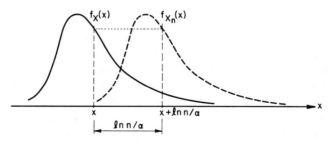

Figure 4.6 PDF of the largest value from the Type I asymptotic distribution.

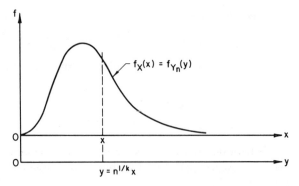

Figure 4.7 PDF of the largest value from the Type II asymptotic distribution.

the CDF of the largest value among samples of any size n from X would be, according to Eq. 4.2,

$$F_{Y_n}(y) = \exp\left[-n\left(\frac{v}{y}\right)^k\right]$$

$$= \exp\left[-\left(\frac{v}{y/n^{1/k}}\right)^k\right]$$

This simply means that the distribution of Y_n is of the same Type II asymptotic form as the initial distribution, with the values of the initial variate $X = x$ replaced by $y = n^{1/k}x$, as illustrated in Fig. 4.7 for the PDF.

Finally, if the initial variate has a distribution of the Type III asymptotic form, with upper limit ω,

$$F_X(x) = \exp\left[-\left(\frac{\omega - x}{\omega - w}\right)^k\right]$$

the corresponding CDF for the largest value of X is

$$F_{Z_n}(z) = \exp\left[-n\left(\frac{\omega - z}{\omega - w}\right)^k\right]$$

$$= \exp\left\{-\left[\frac{(\omega - z)n^{1/k}}{(\omega - w)}\right]^k\right\}$$

which is also the same Type III asymptotic distribution as that of the initial distribution, with values of $(\omega - x)$ replaced by $(\omega - z) = (\omega - x)n^{-1/k}$.

By virtue of the *stability postulate*, therefore, the extreme value from an initial variate whose distribution is one of the asymptotic forms (i.e., Type I, II, or III) will have a CDF that is of the same corresponding asymptotic form. In short, *the distribution of the extreme from an asymptotic form will remain of the same asymptotic form.*

It should be emphasized that the three types of asymptotic forms discussed thus far are not exhaustive; that is, the three types do not include all possible asymptotic distributions of extremes, or it is possible that the asymptotic distributions of the extremes from an initial variate may not belong to any of Gumbel's three types of asymptotic forms. Nevertheless, the three types of asymptotic extremal distributions presented here should cover most (if not all) of the possible distributions of interest in engineering applications. Furthermore, it is not unusual to have one extreme (e.g., the largest) of a given initial variate belonging to one asymptotic form, whereas the other extreme (e.g., the smallest) belonging to another asymptotic form. We saw this in Examples 4.20 and 4.23.

4.4 EXTREMAL PROBABILITY PAPERS

4.4.1 The Gumbel Extremal Probability Paper

With the procedure described in Chapter 6 of Volume I, a probability paper can be constructed for the Type I asymptotic distribution. In this case the standard variates for the largest and smallest values were given, respectively, in Eqs. 4.34 and 4.38 as

$$s = \alpha_n(x_n - u_n)$$

and

$$s = -\alpha_1(x_1 - u_1)$$

An extremal probability paper may then be constructed with values of s scaled on one axis (in arithmetic scale), and the associated cumulative probabilities $F_S(s) = \exp(-e^{-s})$ given on the same (or parallel) axis. The other (perpendicular) axis represents values of the extremal variate x_n or x_1 in arithmetic scale. The resulting probability paper is shown in Fig. 4.8. It is due to Gumbel and is accordingly known as the *Gumbel Probability Paper*.

A straight line with a positive slope on the Gumbel probability paper, therefore, represents a particular Type I asymptotic distribution for the largest value, Eq. 4.19; whereas a straight line with a negative slope would represent a Type I asymptotic distribution for the smallest value, that is, Eq. 4.21.

Accordingly, if N largest values from a population of the exponential type are plotted on this extremal probability paper, such that the mth value (in increasing order) is plotted at the probability $m/(N + 1)$, the results should show a linear trend with a positive slope. Similarly, if N smallest values are plotted on this paper with the mth value (also in increasing order) plotted at the position $[1 - m/(N + 1)]$, the results should also indicate a linear trend with a negative slope.

The statistical estimation of the parameters u_n and α_n, or u_1 and α_1, from a set of extreme value data is the subject of Section 4.6. However, these parameters may also be estimated approximately by graphically drawing a straight line through the plotted data points on the Gumbel probability paper (if a linear trend is observed). The parameter u_n would then be the value at a cumulative probability of 0.368 or $s = 0$, and $1/\alpha_n$ is the slope of the straight line; that is, $1/\alpha_n = (x_n - u_n)/s$ and $\alpha_n = s/(x_n - u_n)$.

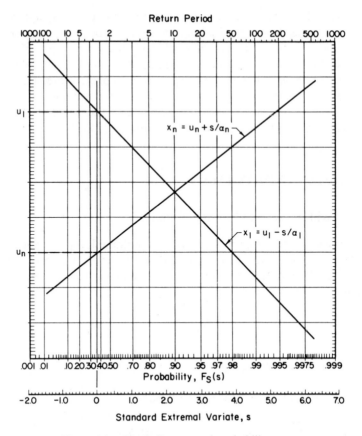

Return Period

Figure 4.8 Gumbel's extremal probability paper.

EXAMPLE 4.24 (*Type I Largest*)

Tabulated in Table E4.24 are the 31 annual maximum earthquake magnitudes (rearranged in increasing order) observed in California between 1932 to 1962 (data cited and discussed in Epstein and Lomnitz, 1966).

The values of x_n and the corresponding probabilities $m/(N + 1)$ are plotted on the Gumbel probability paper shown in Fig. E4.24. The values of the mode u_n and slope $1/\alpha_n$ may be determined approximately by passing a straight line through the data points. In this case, the line shown in Fig. E4.24 was determined through a regression analysis (Epstein and Lomnitz, 1966), from which we obtain

$$\hat{u}_n = 5.70$$

and the slope may be obtained as

$$\frac{1}{\hat{\alpha}_n} = \frac{8.00 - 5.70}{4.60} = 0.50$$

Therefore,

$$\hat{\alpha}_n = 2.00$$

Table E4.24 Annual Maximum Earthquake Magnitudes

m	Richter's Earthquake Magnitude, x_n	$\dfrac{m}{N+1}$ Plotting Position
1	4.9	0.03125
2	5.3	0.06250
3	5.3	0.09375
4	5.5	0.12500
5	5.5	0.15625
6	5.5	0.18750
7	5.5	0.21875
8	5.6	0.25000
9	5.6	0.28125
10	5.6	0.31250
11	5.8	0.34375
12	5.8	0.37500
13	5.8	0.40625
14	5.9	0.43750
15	6.0	0.46875
16	6.0	0.50000
17	6.0	0.53125
18	6.0	0.56250
19	6.0	0.59375
20	6.0	0.62500
21	6.2	0.65625
22	6.2	0.68750
23	6.3	0.71875
24	6.3	0.75000
25	6.4	0.78125
26	6.4	0.81250
27	6.5	0.84375
28	6.5	0.87500
29	7.1	0.90625
30	7.1	0.93750
31	7.7	0.96875

From the straight line graph, the probability associated with an earthquake of given magnitude may be read off directly. For example, the annual probability of earthquakes exceeding magnitude 7 is

$$P(X_n > 7) = 1 - F_{X_n}(7) = 1 - 0.93 = 0.07$$

EXAMPLE 4.25

Shown in Table E4.25 is the data series for the annual flood (maximum discharges) of the Wabash River at Mount Carmel, Illinois, measured between the years 1896 and 1950. In chronological order, the discharges, in 1000 cfs, are listed in the table below.

The data may be plotted on the Gumbel probability paper as shown in Fig. E4.25. This plot is obtained by plotting the discharges with the corresponding plotting positions (probabilities) given in the fourth and eighth columns of Table E4.25. The order m shown in the third and seventh columns is listed in increasing order.

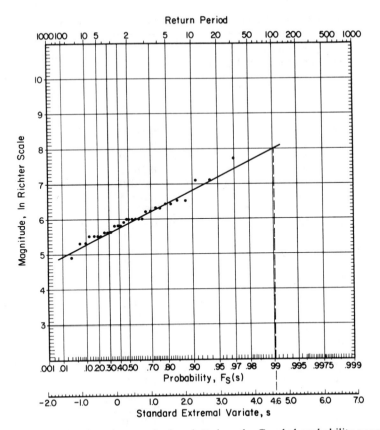

Figure E4.24 Earthquake magnitudes plotted on the Gumbel probability paper.

A straight line may be drawn (by eye) through the data points of Fig. E4.25. On this basis, the parameters of the corresponding Type I asymptotic distribution are obtained approximately as follows.

At $s = 0$, or probability 0.368, we obtain

$$\hat{u}_n = 120 \times 10^3 \text{cfs}$$

According to Eq. 4.34, the inverse slope of the straight line is α_n; thus,

$$\hat{\alpha}_n = \frac{4}{390 - 120} = 0.015 \times 10^{-3}$$

4.4.2 The Logarithmic Extremal Probability Paper

In Section 4.3.3, we saw that the Types I and II asymptotic largest variates, X_n and Y_n, are related by the logarithmic transformation

$$X_n = \ln Y_n$$

Furthermore, the Type II standardized extremal variate is given in Eq. 4.52 as

$$S = k \ln (Y_n/v_n)$$

Table E4.25 Annual Flood Flows of the Wabash River at Mount Carmel, Illinois (Data from Mitchell, 1954)

Year	Discharge in 1000 cfs	Order m	Plotting position $\dfrac{m}{N+1}$	Year	Discharge in 1000 cfs	Order m	Plotting position $\dfrac{m}{N+1}$
1896	57	5	0.0893	1924	116	20	0.3571
1897	252	47	0.8393	1925	101	16	0.2857
1898	272	49	0.8750	1926	139	29	0.5178
1899	90	11	0.1964	1927	192	41	0.7321
1900	86	9	0.1607	1928	128	26	0.4643
1901	75	7	0.1250	1929	155	32	0.5714
1902	57	4	0.0714	1930	279	50	0.8928
1903	124	25	0.4494	1931	30	1	0.0178
1904	280	51	0.9107	1932	157	34	0.6071
1905	67	6	0.1071	1933	232	44	0.7857
1906	154	31	0.5536	1934	43	3	0.0536
1907	182	40	0.7143	1935	122	22	0.3928
1908	195	42	0.7500	1936	123	24	0.4286
1909	92	13	0.2321	1937	285	52	0.9286
1910	120	21	0.3750	1938	162	37	0.6607
1911	92	12	0.2143	1939	197	43	0.7679
1912	146	30	0.5357	1940	100	15	0.2678
1913	428	55	0.9821	1941	39	2	0.0357
1914	88	10	0.1786	1942	108	18	0.3214
1915	84	8	0.1428	1943	305	54	0.9643
1916	261	48	0.8571	1944	167	39	0.6964
1917	131	28	0.5000	1945	162	36	0.6428
1918	106	17	0.3036	1946	92	14	0.2500
1919	164	38	0.6786	1947	110	19	0.3393
1920	156	33	0.5893	1948	162	35	0.6250
1921	122	23	0.4107	1949	236	46	0.8214
1922	232	45	0.8036	1950	288	53	0.9464
1923	128	27	0.4821				

$N = 55.$

Similarly, the Type III largest variate Z_n is related to the Type I largest variate X_n by

$$X_n - u_n = \ln \left(\frac{\omega - Z_n}{\omega - w_n} \right)$$

whereas the corresponding standardized extremal variate is

$$S = -k \ln \left(\frac{\omega - Z_n}{\omega - w_n} \right)$$

Therefore, by simply changing the x_n scale in the Gumbel probability paper (Fig. 4.8) into the logarithmic scale, we obtain the *logarithmic extremal probability paper* (or log-extremal probability paper) for the Type II and Type III asymptotic distributions. For a Type II largest variate, values of y_n are represented on the logarithmic scale, whereas the same probability scale $F_S(s) = e^{-e^{-s}}$ applies for

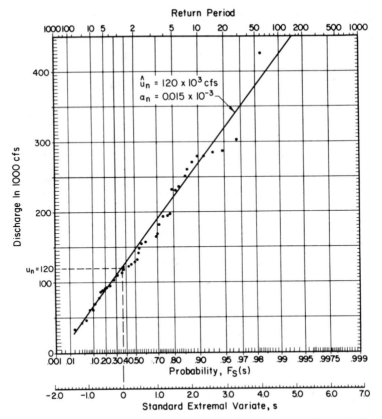

Figure E4.25 Annual flood flows of the Wabash River at Mount Carmel, Illinois.

the standard variate $S = k \ln (y_n/v_n)$. In the case of a Type III largest variate, the logarithmic scale represents values of $(\omega - z_n)$ and the same probability scale applies to the standard variate $S = -k \ln [(\omega - z_n)/(\omega - w_n)]$.

Accordingly, the Type II and Type III asymptotic distributions of the respective largest values, Eqs. 4.45 and 4.58, are represented by straight lines on the log-extremal probability paper as shown in Figs. 4.9a and 4.9b. The parameters v_n and $(\omega - w_n)$ are the values on the respective straight lines corresponding to $s = 0$ or $F_S(s) = e^{-1} = 0.368$; whereas the parameters k would be the slopes of these straight lines which may be obtained, respectively, as follows (according to Eqs. 4.52 and 4.65):

$$k = \frac{s}{\ln \left(\dfrac{y_n}{v_n}\right)} = \frac{s}{\ln y_n - \ln v_n}; \qquad \text{for the Type II largest} \qquad (4.84)$$

and

$$k = \frac{-s}{\ln \left(\dfrac{\omega - z_n}{\omega - w_n}\right)} = \frac{-s}{\ln (\omega - z_n) - \ln (\omega - w_n)}; \qquad \text{for the Type III largest} \qquad (4.85)$$

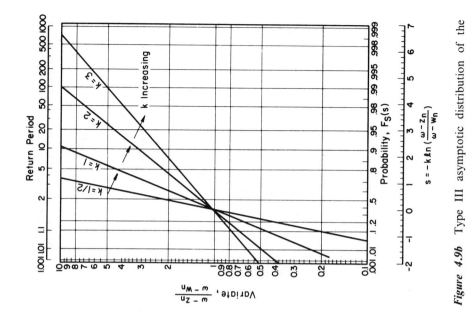

Figure 4.9b Type III asymptotic distribution of the largest value on log-extremal probability paper.

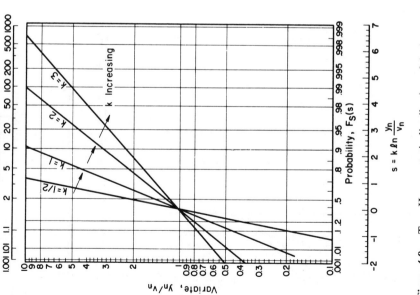

Figure 4.9a Type II asymptotic distribution of the largest value on log-extremal probability paper.

241

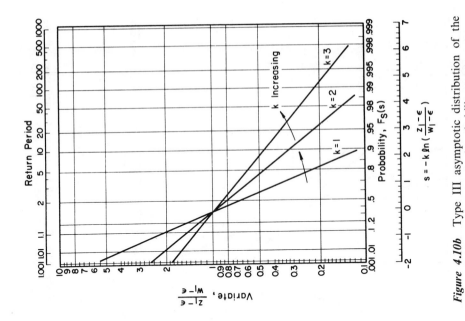

Figure 4.10b Type III asymptotic distribution of the smallest value on log-extremal probability paper.

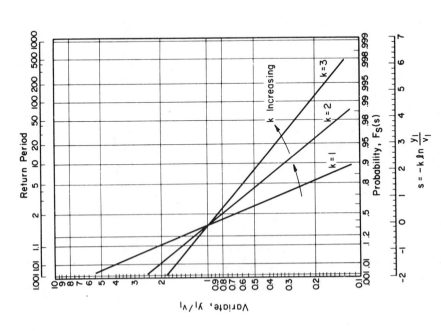

Figure 4.10a Type II asymptotic distribution of the smallest value on log-extremal probability paper.

242

Table E4.26 Annual Fastest Mile Wind Speed: Birmingham, Alabama

m	$V(63')$mph	$\dfrac{m}{N+1}$ Plotting Position
1	43	0.0454
2	43	0.0909
3	45	0.1364
4	47	0.1818
5	47	0.2273
6	47	0.2727
7	47	0.3182
8	48	0.3636
9	48	0.4091
10	49	0.4545
11	49	0.5000
12	52	0.5454
13	52	0.5909
14	54	0.6364
15	54	0.6818
16	56	0.7273
17	56	0.7727
18	59	0.8182
19	60	0.8636
20	65	0.9091
21	65	0.9545

The above log-extremal probability paper is also applicable for the Type II and Type III smallest variates. In these cases, the logarithmic ordinates represent the values of y_1 or $(z_1 - \varepsilon)$, whereas the standard variates are $s = -k \ln (y_1/v_1)$ for the Type II smallest and $s = -k \ln [(z_1 - \varepsilon)/(w_1 - \varepsilon)]$ for the Type III smallest value, as shown in Figs. 4.10a and 4.10b, respectively. Straight lines on this probability paper then represent the Type II and Type III asymptotic distributions of the smallest values. The values on the respective straight lines at $s = 0$, or $F_S(s) = 0.368$, are the values of the parameters v_1 and $(w_1 - \varepsilon)$. Also, the corresponding slopes of these lines are the respective parameters k that, according to Eqs. 4.52a and 4.67, may be obtained as

$$k = \frac{-s}{\ln \left(\dfrac{y_1}{v_1}\right)} = \frac{-s}{\ln y_1 - \ln v_1} ; \qquad \text{for the Type II smallest} \qquad (4.84a)$$

$$k = \frac{-s}{\ln \left(\dfrac{z_1 - \varepsilon}{w_1 - \varepsilon}\right)} = \frac{-s}{\ln (z_1 - \varepsilon) - \ln (w_1 - \varepsilon)} ; \qquad \text{for the Type III smallest}$$

$$(4.85a)$$

EXAMPLE 4.26 (*Type II Largest*)

Thom† reported a set of annual fastest mile wind speeds for the Birmingham, Alabama Airport from 1944 through 1964, as shown in Table E4.26. The data were measured at a height of 63 ft

† H. C. S. Thom. "Asymptotic Extreme Value Distributions Applied to Wind and Waves," unpublished manuscript.

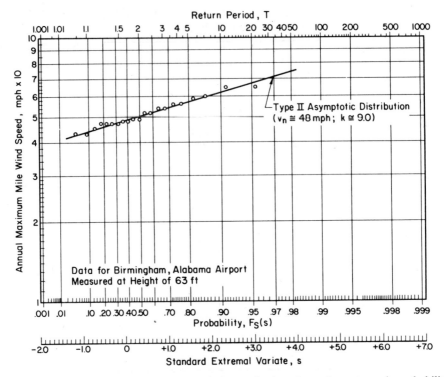

Figure E4.26 Annual maximum mile wind speed plotted on log-extremal probability paper.

(with the exception of three measurements at 62 ft). The data, rearranged in increasing order, are shown in Table E4.26 with the corresponding plotting positions.

Columns 2 and 3 of the above tabulated data are plotted on the log-extremal probability paper as shown in Fig. E4.26. The straight line shown in the figure is drawn visually through the data points; on this basis, we obtain graphically the characteristic largest wind speed as

$$\hat{v}_n \simeq 48 \text{ mph}$$

Similarly, the slope of the straight line according to Eq. 4.84 is

$$\hat{k} = \frac{s}{\ln y_n - \ln v_n}$$

$$= \frac{4}{\ln 75 - \ln 48} = \frac{4}{0.446} = 9.0$$

Therefore, in accordance with the available data set, the annual maximum mile wind speed in Birmingham, Alabama may be described with the Type II asymptotic distribution of the largest value (since the data points clearly show a linear trend) with parameters $v_n = 48$ mph and $k = 9.0$.

Probabilities associated with given wind speeds may then be obtained from the straight line constructed in Fig. E4.26. For example, the probability of the annual wind speed exceeding 70 mph is $1 - 0.965 = 0.035$. Conversely, the 20-year wind speed (corresponding to the cumulative probability of 0.95) is 67 mph.

EXAMPLE 4.27   (*Type III Smallest*)

To illustrate the use of the logarithmic extremal probability paper for a Type III smallest variate, consider the following set of fatigue life test data for metals (ASTM, 1963):

m	Cycles to Failure, z_1 (in 10^5 cycles)	$1 - \dfrac{m}{N+1}$
1	4.0	0.889
2	5.0	0.777
3	6.0	0.667
4	7.3	0.556
5	8.0	0.444
6	9.0	0.333
7	10.6	0.222
8	13.0	0.111

If we assume a minimum life $\varepsilon = 1 \times 10^5$ cycles (see Example 4.31), the results for $(z_1 - \varepsilon)$ are plotted on the log-extremal probability paper shown in Fig. E4.27. From the straight line graph of Fig. E4.27, we obtain visually (corresponding to $s = 0$)

$$(w_1 - \varepsilon) = 8.3 \times 10^5 \text{ cycles}$$

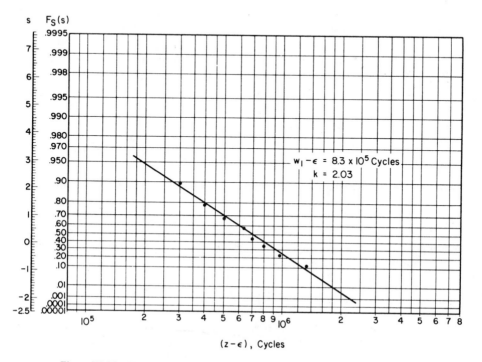

Figure E4.27   Fatigue life data plotted on log-extremal probability paper.

Therefore, $w_1 = 9.30 \times 10^5$ cycles. The slope of the straight line drawn through the data points may be obtained, according to Eq. 4.85a, as

$$k = \frac{-3}{\ln(1.90 \times 10^5) - \ln(8.30 \times 10^5)} = 2.03$$

Based on the linearity of the graph for the above set of fatigue data, it is reasonable to assume that the fatigue life of the particular material has a Weibull distribution (Type III smallest) with parameters $w_1 = 9.3 \times 10^5$ cycles and $k = 2.03$.

4.5 EXCEEDANCE PROBABILITY

4.5.1 Distribution-Free Probability

Of some importance in connection with extreme-value data is the problem of forecasting future exceedances of previously observed extremes; realistically, such forecasts have to be stated in terms of probability. Such problems could have implications in several areas of engineering applications, including the forecasting of extreme natural hazards. As we shall observe, the probability of such exceedances may be independent of the probability distribution functions; that is, distribution-free.

Suppose that there are n observations $(x_1, x_2, \ldots, x_n)$ from past records. In forecasting, we may wish to know the probability that the largest among these n previous observations will be exceeded in N future observations. For instance, the set $(x_1, x_2, \ldots, x_n)$ could be a sequence of the largest annual wind speeds observed in the last n years; in planning the design of an important structure in which wind damage is an important consideration, we might require a forecast of the maximum wind speed in the next N years. Such a forecast may be expressed in terms of the probability that the observed maximum wind in the previous n years will be exceeded in N future years. This probability is obtained as follows.

If y_n is the largest among the n prior observations, that is, $y_n = \max(x_1, x_2, \ldots, x_n)$, then there are $(n - 1)$ values that are less than y_n; therefore, with the assumption of statistical independence, the probability that y_n is the largest among n observed values from the same population X is

$$P[\max(X_1, X_2, \ldots, X_n) = y_n] = \binom{n}{n-1}[F_X(y_n)]^{n-1} \, dF_X(y_n)$$

Now denote the largest among N subsequent observations as X_N; then, for a given value y

$$P(X_N \le y) = [F_X(y)]^N$$

For any y_n that is the largest among n previous observations, we obtain by combining the above two equations

$$P(X_N \le y_n) = n \int_0^{1.0} [F_X(y_n)]^{N+n-1} \, dF_X(y_n)$$

The above integral is the *beta function* $B(N + n, 1) = 1/(N + n)$. Hence,

$$P(X_N \le y_n) = \frac{n}{N + n} \tag{4.86}$$

Equation 4.86 is the probability that the largest value among n previous observations will not be exceeded in N future observations. Conversely, the probability that this previously observed largest value will be exceeded in N subsequent observations, therefore, is

$$P(X_N > y_n) = \frac{N}{N + n} \tag{4.86a}$$

The above problem can be extended to the more general problem of determining the probability that the kth value y_k (counted from the bottom) will be exceeded r times in N future observations.

Assume that all the observations are independent samples from the same population X with the distribution function $F_X(x)$. In any future observation, the probability that the observed value will be less than y_k is $F_X(y_k)$ and the probability of it being larger than y_k is $1 - F_X(y_k)$. If y_k is the kth value among n, there must be $(k - 1)$ values that are smaller and $(n - k)$ values that are equal or larger than y_k. Therefore, the probability that y_k is the kth value among the n observations is

$$\frac{n!}{(n - k)!(k - 1)!} F^{k-1}(1 - F)^{n-k}\, dF = k\binom{n}{k} F^{k-1}(1 - F)^{n-k}\, dF$$

where $F = F_X(y_k)$. The probability that exactly r among N future observations will equal or exceed y_k is given by the binomial distribution or

$$p(r \mid N) = \binom{N}{r} F^{N-r}(1 - F)^r$$

Therefore, considering all possible values of y_k, we obtain the required probability as

$$p(r \mid n, k, N) = k\binom{n}{k}\binom{N}{r} \int_0^{1.0} F^{N-r+k-1}(1 - F)^{n-k+r}\, dF$$

We observe that the above integral is also the beta function; that is,

$$\int_0^1 F^{N-r+k-1}(1 - F)^{n-k+r}\, dF = B(w, v) = \frac{\Gamma(w)\Gamma(v)}{\Gamma(w + v)}$$

where

$$w = N + k - r$$
$$v = n + r - k + 1$$
$$w + v = N + n + 1$$

Since all the quantities are integers, it can be shown that

$$B(w, v) = \frac{1}{(N + k - r)\dbinom{N + n}{N + k - r}}$$

Hence, the probability that the kth value among n previous observations will be equaled or exceeded r times in N future observations is

$$p(r \mid n, k, N) = \frac{k}{(N + k - r)} \frac{\binom{n}{k}\binom{N}{r}}{\binom{N + n}{N + k - r}} \qquad (4.87)$$

The probability that a previously observed largest value will not be exceeded in N future observations, therefore, may also be obtained with Eq. 4.87 using $r = 0$ and $k = n$; thus,

$$p(0 \mid n, n, N) = \frac{n}{N + n}$$

which is the same as the result of Eq. 4.86. We observe that Eqs. 4.86 and 4.87 are both independent of the initial distribution $F_X(x)$.

In Eq. 4.86a, as $N \to \infty$, $N/(N + n) = 1.0$, which shows that in the absence of physical limitations, the probability of exceeding a previously observed largest value approaches certainty if the observational period is sufficiently long.

Equation 4.87 is also the probability function that r among N future observations are smaller than the kth smallest among n previous observations; that is, k is counted from the top. In this case, Eq. 4.86a is also the probability that in N future observations there will be at least one value that is smaller than the previously observed smallest value.

Equation 4.87 has been used as the basis for modeling hydrologic extremes in the design of water-control projects (Chow and Takase, 1977).

EXAMPLE 4.28

Suppose that a flood-control dike has been designed for the largest flood observed over the last 50 years. What is the probability that in the next 20 years the dike will be topped by flood water? This probability is given by Eq. 4.86a; that is,

$$p = \frac{20}{50 + 20} = \frac{20}{70} = 0.29$$

How long a period of time (in years) can we be assured (with a 90% probability) that the dike will not be topped? In this case, with Eq. 4.86,

$$0.90 = \frac{50}{50 + N}$$

Thus, the period is

$$N = \frac{5}{0.9} = 5.56 \text{ years.}$$

Therefore, over the next $5\frac{1}{2}$ years there is a 90% probability that the largest flood recorded in the last 50 years will not be exceeded.

4.5.2 Significance of Probability Distribution

In Section 4.5.1, the probability of exceeding a previously observed maximum (or minimum) in a future series of observations was determined without reference to the values of the underlying variate or its distribution. However, in many engineering situations, the distribution function of the prior observations as well as the distribution of the future observations may be available; for example, the distributions of the annual flood and annual maximum wind may be of the extremal types. In such cases, more accurate estimates of the required exceedance probabilities may be determined on the basis of the distribution information.

If X is the initial variate (or population) with distribution function $F_X(x)$, the CDF of the maximum in a sample of size n from X will, of course, have the CDF

$$F_{X_n}(x) = [F_X(x)]^n$$

Now, if the initial distribution of X is a Type I asymptotic distribution

$$F_X(x) = \exp[-e^{-\alpha(x-u)}]$$

the distribution of the largest among samples of size n from X would be

$$F_{X_n}(x) = \exp[-ne^{-\alpha(x-u)}]$$
$$= \exp[-e^{-\alpha(x-u-\ln n/\alpha)}]$$

However, if y_n is the observed largest value from a series of n previous observations $(x_1, x_2, \ldots, x_n)$, that is,

$$y_n = \max(x_1, x_2, \ldots, x_n)$$

then from this single sample of size n, we have

$$F_X(y_n) = 1 - \frac{1}{n}$$

which means that the best estimate for u_n is y_n.

In N future observations, the distribution of its largest value would be (using $u_n = y_n$) $\longleftarrow$ $(u_n = y_n)$

$$F_{X_N}(x) = [F_X(x)]^N$$
$$= \{[F_X(x)]^n\}^{N/n}$$
$$= \exp\left[-\frac{N}{n}e^{-\alpha(x-y_n)}\right]$$
$$= \exp[-e^{-\alpha(x-y_n-\ln(N/n)/\alpha)}] \qquad (4.88)$$

which means that $F_{X_N}(x)$ is also a Type I asymptotic distribution with parameters

$$\alpha_N = \alpha$$

and

$$u_N = y_n + \frac{\ln(N/n)}{\alpha}$$

Therefore, if the largest value among n previous observations from a Type I asymptotic variate is y_n, the distribution of the largest among N future observations is given by Eq. 4.88. On this basis, the probability that the largest in N future observations will exceed the previously observed maximum among n sample values, y_n, is

$$P(X_N > y_n) = 1 - \exp[-e^{-\alpha(y_n - y_n - (1/\alpha)\ln N/n)}]$$
$$= 1 - e^{-N/n}$$

(4.89)

This result may be compared with the distribution-free probability of Eq. 4.86a. The difference in the two results may be seen from the following:

	$P(X_N > y_n)$	
N/n	$1 - e^{-N/n}$	$\dfrac{N}{N + n}$
0.5	0.39	0.33
1	0.63	0.50
1.5	0.78	0.60
2	0.86	0.67
2.5	0.92	0.71
5	0.99	0.83
10	~ 1.00	0.91

These show that Eq. 4.86a consistently gives lower exceedance probabilities.

Conversely, if the probability of exceeding a previously observed large value in N future observations is specified, the value of the variate, x_N, may be determined as follows.

Suppose $p = $ the specified probability of exceedance; then,

$$\exp\{-e^{-\alpha[x_N - y_n - \ln(N/n)/\alpha]}\} = 1 - p$$

from which we obtain the "design value"

$$x_N(p) = y_n + \frac{1}{\alpha}\left[\ln (N/n) - \ln \ln \frac{1}{(1 - p)}\right]$$

(4.90)

For small p,

$$\ln \frac{1}{1 - p} \simeq p$$

Then,

$$x_N(p) = y_n + \frac{1}{\alpha}\left[\ln \left(\frac{N}{np}\right)\right]$$

(4.90a)

EXAMPLE 4.29

In Example 4.25, the largest flood flow observed over a period of 55 years (between 1896 and 1950) was 428,000 cfs.

The probability that this same maximum flood will be exceeded in the ensuing 50 years (i.e., following 1950) is, according to Eq. 4.89,

$$P(X_N > y_n) = 1 - e^{-50/55} = 0.597$$

However, for design purposes, the design flood may be higher or lower than the maximum flood observed between 1896 and 1950. If an allowable exceedance probability of $p = 10\%$ is specified for a period of 50 years and with the assumption that annual floods are distributed according to the Type I asymptotic distribution of Example 4.25, the "design flood" would be, according to Eq. 4.90a ($y_n = 428,000$ cfs and $\alpha = 0.015 \times 10^{-3}$ from Example 4.25),

$$x_N(0.10) = 428,000 + \frac{1}{0.015 \times 10^{-3}} \ln \left[\frac{50}{55 \times 0.10} \right]$$

$$= 575,000 \text{ cfs}$$

Alternatively, the design flood may be obtained on the basis of the distribution of annual maximum floods. In Example 4.25, the CDF of the annual maximum flood was shown to be a Type I asymptotic distribution, with parameters

$$u = 120 \times 10^3 \text{ cfs}$$

and

$$\alpha = 0.015 \times 10^{-3}$$

On this basis, it can be shown that the distribution of the maximum flood in N future years is

$$F_{X_N}(x) = \exp[-e^{-\alpha(x - u - \ln N/\alpha)}]$$

Conversely, the alternative design value (for small p) is

$$\boxed{x_N(p) \simeq u + \frac{1}{\alpha} \ln \frac{N}{p}}$$

Therefore, with this method we obtain the "design flood" for the 50 years following 1950 as

$$x_N(0.10) = 120,000 + \frac{1}{0.015 \times 10^{-3}} \ln \left(\frac{50}{0.1} \right)$$

$$= 534,300 \text{ cfs}$$

Clearly, the two results for the design flood obtained above are different; judgment is required to determine the more suitable value. In this regard, it may be emphasized that in one case, we used y_n as the most probable maximum in n years, whereas in the other case the parameters u and α were estimated statistically from the observed data series of annual maximas.

Exceedance of Other Asymptotic Extremes   If the initial variate has a Type II asymptotic distribution; that is,

$$F_X(x) = \exp\left[-\left(\frac{v}{x}\right)^k \right]$$

then the distribution of the largest value from a sample of size n is

$$F_{Y_n}(y) = [F_X(y)]^n$$

$$= \exp\left[-n\left(\frac{v}{y}\right)^k \right]$$

Therefore, Y_n remains a Type II asymptotic variate with parameters

$$k_n = k$$

and

$$v_n = n^{1/k}v$$

However, if y_n is the observed maximum value in the sample size n, then $v_n = y_n$.

In N future observations from the same initial variate X, we have (again invoking the stability postulate)

$$F_{Y_N}(y) = [F_X(y)]^N$$
$$= \{[F_X(y)]^n\}^{N/n}$$
$$= \exp\left[-\frac{N}{n}\left(\frac{v_n}{y}\right)^k\right] \qquad (4.91)$$

Again, with $v_n = y_n$, the probability of Y_N exceeding the previously observed largest value, y_n, is therefore

$$P(Y_N > y_n) = 1 - e^{-N/n}$$

which is the same as Eq. 4.89.

Conversely, the "design value" corresponding to a specified probability p that y_n will be exceeded in N future observations is determined as follows:

$$\exp\left[-\frac{N}{n}\left(\frac{y_n}{y_N}\right)^k\right] = 1 - p$$

From this we obtain the "design value"

$$y_N(p) = y_n\left[\frac{N/n}{-\ln(1-p)}\right]^{1/k} \qquad (4.92)$$

and for small p, $-\ln(1-p) \simeq p$; thus,

$$y_N(p) = y_n\left(\frac{N}{np}\right)^{1/k} \qquad (4.92a)$$

EXAMPLE 4.30

In Example 4.26 the 21-year series of data for the annual maximum wind speed recorded in Birmingham, Alabama were presented. It was also shown that the corresponding CDF of the annual maximum wind speed is reasonably represented by the Type II asymptotic distribution of the largest value.

The largest value recorded in the sample of 21 observations was 65 mph (i.e., $y_n = 65$), and the parameters were estimated graphically in Example 4.26 to be

$$v = 48 \text{ mph}$$

and

$$k = 9.0$$

In the next 25 years, the probability that this previously observed annual maximum wind of 65 mph will be exceeded is, according to Eq. 4.89,

$$P(y_N > y_n) = 1 - \exp\left[\left(\frac{25}{21}\right)\right]$$

$$= 0.696$$

If the above exceedance probability is to be limited to 10% over the next 25 years, the "design wind speed" is given by Eq. 4.92a as

$$y_N(0.10) = 65\left(\frac{25}{21 \times 0.10}\right)^{1/9}$$

$$= 85.6 \text{ mph}$$

In problems of obtaining less than a previously observed smallest value, consider an initial variate with the Type III asymptotic distribution of the smallest value with lower limit ε; that is,

$$F_X(x) = 1 - \exp\left[-\left(\frac{x - \varepsilon}{w - \varepsilon}\right)^k\right]$$

The smallest value among n observations taken from X, therefore, will have a distribution

$$F_{Z_{1n}}(z) = 1 - [1 - F_X(z)]^n$$

$$= 1 - \exp\left[-n\left(\frac{z - \varepsilon}{w - \varepsilon}\right)^k\right]$$

which remains a Type III asymptotic distribution with parameters

$$k_1 = k$$

and

$$(w_1 - \varepsilon) = (w - \varepsilon)n^{-1/k}$$

In N future observations, the smallest value would have

$$F_{Z_{1N}}(z) = 1 - [1 - F_X(z)]^N$$

$$= 1 - \{[1 - F_X(z)]^n\}^{N/n}$$

$$= 1 - \exp\left[-\frac{N}{n}\left(\frac{z - \varepsilon}{w_1 - \varepsilon}\right)^k\right] \qquad (4.93)$$

If $z_1 = \min(x_1, x_2, \ldots, x_n)$, that is, z_1 is the smallest value among n observed sample values, then on the basis of this single sample of size n

$$F_X(z_1) = \frac{1}{n}$$

which means $w_1 = z_1$.

Then, the probability that the smallest value among N future observations will be less than the smallest of n earlier values is

$$P(Z_{1N} < z_1) = 1 - e^{-N/n}$$

Again, this is the same as Eq. 4.89.

However, if the probability of being less than a given "design value" in N future observations is specified to be p, we obtain the corresponding "design smallest value" as follows:

$$F_{Z_{1N}}(z_{1N}) = p$$

or

$$1 - \exp\left[-\frac{N}{n}\left(\frac{z_{1N} - \varepsilon}{z_1 - \varepsilon}\right)^k\right] = p$$

From this we obtain the "design value"

$$z_{1N} = \varepsilon + (z_1 - \varepsilon)\left[\frac{-n}{N}\ln(1 - p)\right]^{1/k} \tag{4.94}$$

whereas, for small p,

$$z_{1N} = \varepsilon + (z_1 - \varepsilon)\left(\frac{np}{N}\right)^{1/k} \tag{4.94a}$$

EXAMPLE 4.31

In Example 4.35, the annual drought (lowest 7-day average flow) of the Iroquois River was modeled with the Type III asymptotic distribution of the smallest value. The parameters were also estimated in Example 4.35 as follows:

$$\varepsilon = 0$$
$$w = 61.88 \text{ cfs}$$
$$k = 1.83$$

The annual smallest flow of the river observed during the 32-year period was 14.1 cfs; that is, $z_1 = 14.1$.

In planning a water supply system, we would need to know the reliable minimum level of flow from the river. This minimum "design flow," z_{1N}, may be determined on the basis of Eq. 4.94. Suppose that the specified allowable probability is $p = 0.20$ over the next 20 years; then with Eq. 4.94a, we obtain

$$z_{1N} = 14.1\left(-\frac{32}{20}\ln 0.8\right)^{1/1.83}$$

$$= 8.03 \text{ cfs}$$

Alternatively, the "design flow" may be obtained using the parameters estimated in Example 4.35, in particular $w = 61.88$. In this case, we have

$$F_{Z_{1N}}(z) = 1 - \exp\left[-N\left(\frac{z - \varepsilon}{w - \varepsilon}\right)^k\right]$$

Then, specifying

$$F_{Z_{1N}}(z_{1N}) = p$$

we obtain

$$z_{1N} = \varepsilon + (w - \varepsilon)\left[\frac{-\ln (1 - p)}{N}\right]^{1/k}$$

$$\simeq \varepsilon + (w - \varepsilon)\left(\frac{p}{N}\right)^{1/k}$$

Therefore, for the Iroquois River, the "design flow" corresponding to $p = 0.20$ in 20 years is

$$z_{1N} = 61.88\left(\frac{-\ln 0.8}{20}\right)^{1/1.83}$$

$$= 5.30 \text{ cfs}$$

4.6 ESTIMATION OF EXTREMAL PARAMETERS

4.6.1 Remarks on Extreme Value Estimation

Extreme value data are, in most cases, time-consuming and/or costly to acquire. For example, with regard to the annual maximum flood of a river, it obviously requires one year to increase the sample size by one; whereas in the case of strong-motion earthquakes (say of Richter magnitude $\geq$ 5.0), it may take several years before another large earthquake will occur in a given region.

The sample size of available extreme value data, therefore, is usually small. Consequently, in estimating the parameters of an extreme value distribution, such as those for a given asymptotic form, the selection or use of the "best" method of estimation is more crucial than when data sampling or acquisition is less restrictive. That is, when the available data are limited, the need to optimize the accuracy of the estimation procedure becomes more important.

As with other situations requiring statistical sampling and estimation (Chapter 5, Vol. I), one is faced with the problem of making inferences about an underlying population (of extreme values in the present case) on the basis of finite samples of the population. The objective, of course, is to determine the values of the unknown parameters of the population probability distribution. For this purpose, there may be different ways (i.e., different estimator functions) of estimating a given parameter, with their respective qualities. The quality of an estimator is generally determined by its unbiasness (lack of bias) and efficiency (as measured by its variance). Ideally, an estimator should be unbiased and has as small a variance as possible.

The usual methods of parameter estimation are the *method of maximum likelihood* and the *method of moments*. For the estimation of the extremal parameters, maximum likelihood estimators have been developed, for example, by Kimball (1946); however, such estimators are not simple to evaluate and may be biased for small samples, even though they may be efficient. The method of moments has been used in a number of applications involving extremes; for example, Gumbel (1941) and Press (1950). In this latter method, the biases and efficiencies of the resulting estimators are complicated and difficult to evaluate, requiring multiple numerical integrations (Lieblein, 1954). An order-statistics method had been developed by Lieblein (1954) for estimating the parameters of the Type I (and Type II) asymptotic distributions. The method yields a minimum variance unbiased estimator.

Let us concentrate first on the estimation of the parameters of the Type I asymptotic distribution for the largest value; namely, the parameters u_n and α_n of Eq. 4.19, or

$$F_{X_n}(x) = \exp[-e^{-\alpha_n(x-u_n)}]$$

4.6.2   The Method of Moments

The estimation of the parameters by the method of moments involves the estimation of the appropriate sample moments from the sampled data—usually the sample mean, $\bar{x}$, and the sample variance, ∂_x^2, and on occasion the third moment or skewness—and then evaluating the parameters using the pertinent relationships between the parameters and the sample moments (see Chapter 5, Vol. I). In the case of the parameters for the Type I asymptotic distribution of the largest value, these relationships are as follows:

From Eqs. 4.41 and 4.42

$$\mu_{X_n} = u_n + \gamma/\alpha_n$$

and

$$\sigma_{X_n} = \frac{\pi}{\sqrt{6}\,\alpha_n}$$

If we replace the above population moments, μ_{X_n} and σ_{X_n}, by the corresponding sample moments, $\bar{x}_n$ and ∂_{x_n}, the moment estimators of the extremal parameters are, therefore,

$$\hat{u}_n = \bar{x}_n - \frac{\sqrt{6}}{\pi}\gamma \partial_{X_n} \tag{4.95}$$

and

$$\hat{\alpha}_n = \frac{\pi}{\sqrt{6}\,\partial_{X_n}} \tag{4.96}$$

See Chapter 5, Vol. I for the estimation of the sample mean, $\bar{x}$, and the sample standard deviation ∂_X from a set of sample values.

4.6.3   The Method of Order Statistics

A method, based on the theory of order statistics, was developed by Lieblein (1954) expressly for estimating the extremal parameters u_n and α_n of the Type I asymptotic distribution, Eq. 4.19. These *order statistics estimators* are unbiased and possess minimum variance; moreover, they are simple to apply. Their efficiencies can be accurately evaluated (Lieblein, 1954) and have been shown to be much higher than those of other estimators, including the moment estimators.

Lieblein's order statistics method is based on a linear function of a set of ordered sample values $(x_1, x_2, \ldots, x_n)$; that is,

$$L = \sum_{i=1}^{n} w_i x_i \tag{4.97}$$

where $x_1 \le x_2 \le \cdots \le x_n$, and w_i are weights that may be decomposed into

$$w_i = a_i + b_i s_p$$

where s_p is the value of the standardized variate S at an exceedance probability p; that is,

$$\exp(-e^{-s_p}) = 1 - p \quad \text{or} \quad s_p = -\ln\left[-\ln\left(1 - p\right)\right]$$

Then, the estimator of Eq. 4.97 becomes

$$L = \sum_{i=1}^{n} [a_i x_i + (b_i x_i) s_P] \tag{4.97a}$$

where the weights a_i and b_i are functions of n and p. In order to insure an unbiased and minimum variance estimator L (for given n and p), the following conditions may be imposed:

$$E(L) = u_n + \frac{1}{\alpha_n} s_p \tag{4.98}$$

and

$$\text{Var}(L) = \text{minimum} \tag{4.99}$$

Equation 4.97a with the values of a_i and b_i obtained on the basis of Eqs. 4.98 and 4.99, therefore, yields the desired estimator function L. Accordingly, the corresponding unbiased-minimum variance estimators for u_n and α_n would be

$$\hat{u}_n = \sum_{i=1}^{n} a_i x_i \tag{4.100}$$

and

$$\frac{1}{\hat{\alpha}_n} = \sum_{i=1}^{n} b_i x_i \tag{4.101}$$

The relevant set of weights, a_i and b_i, that satisfy Eqs. 4.98 and 4.99 have been obtained by Lieblein (1954) for $n = 2$ through 6, with $p \ge 0.90$; these values are reproduced in Table A.4 in Appendix A.

The efficiency of the resulting estimator L depends on n and the exceedance probability p, as shown in Fig. 4.11. For the purpose of extremal parameter estimation, we are interested only in the range $p \ge 0.90$. As may be seen from Fig. 4.11, the relative efficiency[†] of the estimator in this range of p is generally greater than 80% for a sample size of $n = 6$ or 5. Apparently, there is not much

---

[†] Efficiency is defined (Lieblein, 1954) relative to a lowerbound variance.

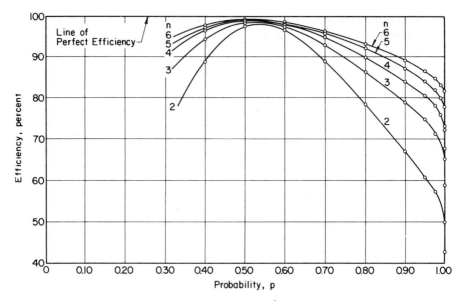

Figure 4.11 Efficiencies of order-statistics estimator $\hat{L}$ as functions of p and n (reproduced from Lieblein, 1954).

improvement in efficiency for $n > 6$; consequently, the values of a_i and b_i are limited to those for $n = 2$ through $n = 6$ in Table A.4.

For a sample size n that is larger than 6, Lieblein (1954) recommended that the entire set of sample values should be divided into subgroups of six, or (if necessary) into subgroups of five. However, if the available sample size is not divisible into subgroups of six or five, it should be divided into subgroups of either size plus a remainder subgroup. If possible, in order to maintain high efficiency, the subgroup size should be selected (between sizes 6 and 5) so that the size of the remainder subgroup is as close to six as possible. For example, if the total sample size is $n = 16$, it should be subdivided into two subgroups of size $n_s = 5$ plus a remainder subgroup of size $n_r = 6$; whereas if $n = 17$, it should be subdivided into two subgroups of $n_s = 6$ plus a remainder subgroup of $n_r = 5$.

From Fig. 4.11 we can see that efficiency would be sacrificed if the subgroup size is < 5, in the range $p > 0.90$; therefore, in order to maintain high efficiency, all subgroup sizes should be as close to six as possible.

Implementation of the Order Statistics Method   Lieblein's method of order statistics is based on the assumption that the set of sample (extreme) values constitutes a statistically independent series of observations. Therefore, in applying Lieblein's method, this assumption ought to be observed as closely as possible. Usually, the original sequence of the data series would satisfy this requirement, and therefore should be retained or preserved. If the original sequence is not known, the available data series may be randomized (e.g., rearranged through a random number generator; see Chapter 5).

In its original sequence (or randomized if necessary), the sample is then divided into consecutive subgroups of size $n_s = 6$ or 5, plus a remainder subgroup of size n_r. Within each subgroup, the sample values are then rearranged in increasing order; that is, $x_1 \leq x_2 \leq \cdots \leq x_{n_s}$ (or x_{n_r}).

For each subgroup, with the sample values rearranged in increasing order, the estimates for u_n and α_n are obtained with Eqs. 4.100 and 4.101, respectively, using the values of a_i and b_i from Table A.4 appropriate for the subgroup size n_s. The same calculations are performed for the remainder subgroup, using the a_i and b_i from Table A.4 corresponding to the sample size n_r. The final estimates for u_n and α_n then are the weighted averages of those obtained above for the subgroups (including the remainder subgroup), weighted by the respective subgroup sizes.

It may be observed that if there are k subgroups of the same size n_s, the estimates $\hat{u}_n$ and $1/\hat{\alpha}_n$ would be

$$\hat{u}_n = \frac{1}{k} \sum_{j=1}^{k} \left(\sum_{i=1}^{n_s} a_i x_i \right)_j = \frac{1}{k} \sum_{i=1}^{n_s} a_i \left(\sum_{j=1}^{k} x_{ij} \right)$$

and

$$\frac{1}{\hat{\alpha}_n} = \frac{1}{k} \sum_{j=1}^{k} \left(\sum_{i=1}^{n_s} b_i x_i \right)_j = \frac{1}{k} \sum_{i=1}^{n_s} b_i \left(\sum_{j=1}^{k} x_{ij} \right)$$

Denote

$$S_i = \sum_{j=1}^{k} x_{ij}$$

which is the sum of the x_i's for the k subgroups. Then, we may write

$$\hat{u}_n = \frac{1}{k} \sum_{i=1}^{n_s} a_i S_i$$

and

$$\frac{1}{\hat{\alpha}_n} = \frac{1}{k} \sum_{i=1}^{n_s} b_i S_i$$

However, if there is a remainder subgroup of size n_r, then the weighted averages of the estimates would be

$$\hat{u}_n = \frac{k n_s}{n} \left(\frac{1}{k} \sum_{i=1}^{n_s} a_i S_i \right) + \frac{n_r}{n} \sum_{i=1}^{n_r} a_i x_i$$

$$= \frac{n_s}{n} \sum_{i=1}^{n_s} a_i S_i + \frac{n_r}{n} \sum_{i=1}^{n_r} a_i x_i \qquad (4.100a)$$

and

$$\frac{1}{\hat{\alpha}_n} = \frac{n_s}{n} \sum_{i=1}^{n_s} b_i S_i + \frac{n_r}{n} \sum_{i=1}^{n_r} b_i x_i \qquad (4.101a)$$

EXAMPLE 4.32

In Example 4.25, we observed that the annual flood flows of the Wabash River may be modeled with the Type I asymptotic distribution of largest values. Let us now estimate the corresponding parameters for this distribution using the order statistics method.

In this case, the original sequence of the data series is available, and the sample size is $n = 55$. Therefore, we may divide the entire data series consecutively into 11 subgroups of $n_s = 5$. Within each subgroup, rearrange the sample values in increasing order, thus obtaining the 11 subgroups as follows (discharges in 1000 cfs):

Subgroup	x_1	x_2	x_3	x_4	x_5
1	57	86	90	252	272
2	57	67	75	124	280
3	92	120	154	182	195
4	84	88	92	146	428
5	106	131	156	164	261
6	101	116	122	128	232
7	128	139	155	192	279
8	30	43	122	157	232
9	100	123	162	197	285
10	39	108	162	167	305
11	92	110	162	236	288
S_i:	886	1131	1452	1945	3057

The corresponding sums S_i are as indicated above. From Table A.4, for $n = 5$, the values of a_i and b_i are as follows:

a_i: 0.41893   0.24628   0.16761   0.10882   0.05835

b_i: −0.50313   0.00653   0.13045   0.18166   0.18448

Then, the sums of products are

$$\sum_{i=1}^{5} a_i S_i = 1283.12 \times 10^3$$

$$\sum_{i=1}^{5} b_i S_i = 668.31 \times 10^3$$

In this case there are no remainder subgroups; therefore, the parameters are, according to Eqs. 4.100a and 4.101a (with $n_r = 0$):

$$\hat{u}_n = \frac{1}{11}(1283.12 \times 10^3) = 116.65 \times 10^3 \text{ cfs}$$

$$\frac{1}{\hat{\alpha}_n} = \frac{1}{11}(668.31 \times 10^3) = 60.76 \times 10^3$$

from which

$$\hat{\alpha}_n = \frac{1}{60.76 \times 10^3} = 0.01646 \times 10^{-3}$$

By the method of moments, Eqs. 4.95 and 4.96 would yield the following:

$$\bar{x}_n = 154.02 \times 10^3 \text{ cfs}$$

$$\delta_{x_n} = 80.32 \times 10^3 \text{ cfs}$$

Therefore, the corresponding estimates by the method of moments are

$$\hat{u}_n = 154.02 - 0.577\left(\frac{\sqrt{6}}{\pi}\right) \times 80.32 = 117.89 \times 10^3 \text{ cfs}$$

and

$$\hat{\alpha}_n = \frac{\pi}{\sqrt{6} \times 80.32 \times 10^3} = 0.01597 \times 10^{-3}$$

Finally, the parameters estimated above may be compared with the values obtained graphically in Example 4.25 using probability paper (see Section 4.6.5) which were

$$\hat{u}_n = 120 \times 10^3 \text{ cfs} \quad \text{and} \quad \hat{\alpha}_n = 0.015 \times 10^{-3}$$

4.6.4   Parameters of Other Asymptotic Distributions

The parameters of the other asymptotic extremal distributions may be estimated statistically using one of the methods described above for estimating u_n and α_n.

Estimating u_1 *and* α_1 Lieblein's method of order statistics, which was described in Section 4.6.3 for estimating the parameters of the asymptotic distribution of the largest value, is equally applicable for estimating the parameters u_1 and α_1 of the Type I asymptotic distribution of the smallest value. For this purpose, the sample values in each subgroup (including the remainder subgroup) must be rearranged in *decreasing order*; that is, $x_1 \geq x_2 \geq \cdots \geq x_{n_s}$ (or x_{n_r}). Otherwise, the procedure described in Section 4.6.3 for obtaining $\hat{u}_n$ and $\hat{\alpha}_n$ remains valid for obtaining the estimates of u_1 and α_1; in particular, Eqs. 4.100a and 4.101a are equally applicable for $\hat{u}_1$ and $\hat{\alpha}_1$.

Estimating Parameters of the Type II Asymptotic Distribution   In Section 4.3.3, we observed that if an extremal variate Y_n has the Type II asymptotic distribution with parameters v_n and k_n, then the distribution of $\ln Y_n$ will have the Type I asymptotic distribution with parameters

$$u_n = \ln v_n$$

and

$$\alpha_n = k_n$$

From this the pertinent extremal parameters for the Type II asymptotic distribution become

$$v_n = e^{u_n}; \quad k_n = \alpha_n$$

Therefore, the parameters of the Type II asymptotic distribution of the largest value may be obtained by first estimating $\hat{u}_n$ and $\hat{\alpha}_n$ for the logarithms of the sampled data, that is, $\ln y_1, \ln y_2, \ldots, \ln y_n$, using for example the order statistics method, therefore obtaining $\hat{u}_n$ and $\hat{\alpha}_n$. The estimates of v_n and k_n may then be obtained as

$$\hat{v}_n = e^{\hat{u}_n}; \qquad \hat{k}_n = \hat{\alpha}_n$$

Similarly, for the parameters of the Type II asymptotic distribution of the smallest value, we obtain $\hat{u}_1$ and $\hat{\alpha}_1$ as the sample parameters using the sample values $\ln y_1, \ln y_2, \ldots, \ln y_n$; and then the relevant parameters for the Type II asymptotic distribution of the smallest value are

$$\hat{v}_1 = e^{\hat{u}_1}; \qquad \hat{k}_1 = \hat{\alpha}_1$$

EXAMPLE 4.33

From Example 4.26, the distribution of the wind speeds for Birmingham, Alabama appears to be of the Type II asymptotic form. Let us now estimate the corresponding parameters v_n and k_n by the method of order statistics.

The entire set of data as reported by Thom (see Example 4.26) had already been rearranged in increasing order. For estimating the parameters by the method of order statistics, the ordered data series given in Table E4.26 is randomized, yielding the following:

	Velocity y at 63 ft (mph)	$\ln y$
Subgroup 1	49	3.8918
	47	3.8501
	47	3.8501
	56	4.0254
	59	4.0775
Subgroup 2	65	4.1744
	45	3.8067
	52	3.9512
	43	3.7612
	60	4.0943
Subgroup 3	56	4.0254
	47	3.8501
	47	3.8501
	65	4.1744
	52	3.9512
Remainder Subgroup	48	3.8712
	54	3.9890
	43	3.7612
	49	3.8918
	54	3.9890
	48	3.8712

The total sample size is $n = 21$; therefore, it may be divided into three subgroups of size $n_s = 5$ plus a remainder subgroup of $n_r = 6$, as indicated above. Rearranging the sample values y_i (and their logarithms $\ln y_i$) in increasing order within each subgroup, we have the values shown in the first four columns below:

	Subgroup 1	Subgroup 2	Subgroup 3	S_i	a_i	b_i
$\ln y_1$	3.8501	3.7612	3.8501	11.4614	0.41893	-0.50313
$\ln y_2$	3.8501	3.8067	3.8501	11.5069	0.24628	0.00653
$\ln y_3$	3.8918	3.9512	3.9512	11.7942	0.16761	0.13045
$\ln y_4$	4.0254	4.0943	4.0254	12.1451	0.10882	0.18166
$\ln y_5$	4.0775	4.1744	4.1744	12.4263	0.05835	0.18448

Column 5 in the above table lists the sums $S_i = \sum_{j=1}^{3} \ln y_{ij}$; whereas columns 6 and 7 contain the values of a_i and b_i from Table A.4 corresponding to $n = 5$.

For the remainder subgroup of size $n_r = 6$, we have (also rearranged in increasing order)

Remainder Subgroup	a_i	b_i
$\ln y_1 = 3.7612$	0.35545	-0.45928
$\ln y_2 = 3.8712$	0.22549	-0.03599
$\ln y_3 = 3.8712$	0.16562	0.07319
$\ln y_4 = 3.8918$	0.12105	0.12673
$\ln y_5 = 3.9890$	0.08352	0.14953
$\ln y_6 = 3.9890$	0.04887	0.14581

The values of a_i and b_i corresponding to $n = 6$ from Table A.4 are also shown above. Therefore, for the regular subgroups, we obtain from the first table above

$$\sum_{i=1}^{5} a_i S_i = 11.6590$$

and

$$\sum_{i=1}^{5} b_i S_i = 0.3458$$

whereas, for the remainder subgroup, we obtain from the second table above

$$\sum_{i=1}^{6} a_i \ln y_i = 3.8502$$

$$\sum_{i=1}^{6} b_i \ln y_i = 0.0879$$

Therefore, the estimates $\hat{u}_n$ and $1/\hat{\alpha}_n$ are, according to Eqs. 4.100a and 4.101a,

$$\hat{u}_n = \frac{5}{21} \times 11.6590 + \frac{6}{21} \times 3.8502$$

$$= 3.876$$

$$\frac{1}{\hat{\alpha}_n} = \frac{5}{21} \times 0.3458 + \frac{6}{21} \times 0.0879$$

$$= 0.10745$$

Thus, $\hat{\alpha}_n = 1/0.10745 = 9.307$

Hence, the appropriate parameters of the Type II asymptotic distribution for the annual fastest mile wind speed in Birmingham, Alabama are

$$\hat{v}_n = e^{3.876} = 48.23 \text{ mph}$$

$$\hat{k}_n = \hat{\alpha}_n = 9.307$$

These may be compared with the values obtained graphically in Example 4.26, which were $\hat{v}_n = 48$ mph and $k = 9.0$.

Estimating Parameters of the Type III Asymptotic Distribution   In the case of the Type III asymptotic distribution, there are three parameters: ω, w_n, k_n or ε, w_1, k_1. The Type III asymptotic distribution of the smallest value is of special significance in engineering applications, particularly in materials engineering where this distribution is known as the Weibull distribution (Weibull, 1951). For the Weibull distribution, the parameters ε, w_1, and k_1 may be estimated by the method of moments.

If all three parameters must be estimated from the sampled data, the sample skewness (third moment) will also be needed, in addition to the sample mean and sample variance. If the first three sample moments have been evaluated from a set of sample (smallest) values, namely, the sample mean $\bar{z}$, the sample standard deviation ∂_Z, and the sample skewness $\hat{\theta}_Z$, the three extremal parameters of the Weibull distribution may be estimated as follows.

First, we observe that with Eqs. 4.73 and 4.75, the skewness coefficient θ_Z can be shown to be

$$\theta_Z = \frac{E(Z_1 - \mu_Z)^3}{\sigma_Z^3}$$

$$= \frac{\Gamma(1 + 3/k) + 2\Gamma^3(1 + 1/k) - 3\Gamma(1 + 2/k)\Gamma(1 + 1/k)}{[\Gamma(1 + 2/k) - \Gamma^2(1 + 1/k)]^{3/2}} \qquad (4.102)$$

Similarly, it can be shown that

$$A(k) \equiv \frac{w_1 - \mu_Z}{\sigma_Z} = \frac{[1 + \Gamma(1 + 1/k)]}{[\Gamma(1 + 2/k) - \Gamma^2(1 + 1/k)]^{1/2}} \qquad (4.103)$$

and, of course, from Eq. 4.75

$$B(k) \equiv \frac{w_1 - \varepsilon}{\sigma_Z} = [\Gamma(1 + 2/k) - \Gamma^2(1 + 1/k)]^{-1/2} \qquad (4.104)$$

The values of θ_Z, as well as $A(k)$ and $B(k)$, may be evaluated as functions of $1/k$ according to the above equations. The results may be tabulated as shown in Table A.5 of Appendix A.

To estimate the parameter k, we simply find the value of $1/k$ in Table A.5 that corresponds to the sample skewness coefficient $\hat{\theta}_Z$, thus obtaining $\hat{k}$.

At this value of $1/\hat{k}$, we obtain the value of $A(\hat{k})$ from Table A.5 (third column); then, with the sample mean $\bar{z}$ and the sample standard deviation ∂_Z, the estimate for w_1 is obtained from

$$\frac{\hat{w}_1 - \bar{z}}{\partial_Z} = A(\hat{k})$$

Thus, the estimate of w_1 is

$$\hat{w}_1 = A(\hat{k})\sigma_Z + \bar{z}$$

Then, at the same value of $1/\hat{k}$, we also obtain the value $B(\hat{k})$ from Table A.5 (fourth column). With the above estimated $\hat{w}_1$, we obtain the estimate for ε from

$$\frac{\hat{w}_1 - \hat{\varepsilon}}{\sigma_Z} = B(\hat{k})$$

or

$$\hat{\varepsilon} = \hat{w}_1 - B(\hat{k})\sigma_Z$$

However, if the lower limit ε is known (e.g., $\varepsilon = 0$), the other two parameters of the Weibull distribution may be estimated alternatively as follows.

First, we observe from Eq. 4.76 that

$$1 + \left(\frac{\sigma_Z}{\mu_Z - \varepsilon}\right)^2 = \frac{\Gamma(1 + 2/k)}{\Gamma^2(1 + 1/k)}$$

Therefore, using the sample mean $\bar{z}$ and sample variance σ_Z^2 in place of μ_Z and σ_Z^2 in the above equation, the resulting solution yields the estimate $\hat{k}$. Figure 4.5 may be used to facilitate the determination of this solution.

With the above estimated $\hat{k}$, the remaining parameter may be estimated on the basis of Eq. 4.74; that is,

$$\hat{w}_1 = \varepsilon + \frac{\bar{z} - \varepsilon}{\Gamma(1 + 1/\hat{k})}$$

EXAMPLE 4.34

Consider the set of fatigue life data given earlier in Example 4.27, which may be represented by a Weibull distribution. Let us now estimate the parameters of the distribution in light of the sample data.

The first three sample moments can be shown to be as follows:

$$\bar{z} = 7.86 \times 10^5 \text{ cycles}$$

$$\sigma_Z = 2.99 \times 10^5 \text{ cycles}$$

$$\hat{\theta}_Z = \frac{10.05}{26.73} = 0.376$$

With $\hat{\theta}_Z = 0.376$, we obtain from Table A.5

$$1/\hat{k} \simeq 0.407$$

Thus,

$$\hat{k} = 1/0.407 = 2.46$$

and also from Table A.5

$$A(\hat{k}) \simeq 0.293$$

$$B(\hat{k}) \simeq 2.595$$

From which we obtain the estimates

$$\hat{w}_1 = 0.293 \times 2.99 \times 10^5 + 7.86 \times 10^5$$
$$= 8.74 \times 10^5 \text{ cyc}$$

and

$$\hat{\varepsilon} = 8.74 \times 10^5 - 2.595 \times 2.99 \times 10^5$$
$$= 0.98 \times 10^5 \text{ cyc}$$

The above results may be compared with those obtained graphically in Example 4.27, which were based on the assumption that $\varepsilon = 1 \times 10^5$ cycles; namely,

$$\hat{w}_1 = 9.3 \times 10^5 \text{ cycles}$$

and

$$\hat{k} = 2.03$$

4.6.5   Graphical Method

The parameters of the asymptotic distributions of extremes may also be obtained graphically by plotting the sample extreme values on an appropriate extremal probability paper. For this purpose, however, the plotted data points should exhibit the required linear trend; otherwise, the particular asymptotic distribution is not appropriate to model or describe the underlying population of extremes.

If a linear trend is observed for the data points on a given extremal probability paper, such that a straight line graph can be established, the parameters of the appropriate asymptotic distribution may then be obtained directly from the straight line. The graphical procedures for this purpose were described and illustrated earlier in Section 4.4.

EXAMPLE   4.35

The drought of a river is often expressed in terms of its 7-*day average flow*. The annual drought, therefore, may be defined as the annual lowest 7-day average flow. Such annual droughts for the Iroquois River, recorded near Chebanse, Illinois, between April 1924 and March 1956 are shown in Table E4.35; the data have been rearranged in increasing order.

The set of data is plotted on the log-extremal probability paper as shown in Fig. E4.35 with the assumption of $\varepsilon = 0$. A reasonably linear trend may be observed, suggesting that the distribution of the annual drought may be modeled by the Type III asymptotic distribution of the smallest value. Its parameters may be estimated by the method of moments as follows.

The first three sample moments are evaluated as

$$\bar{z} = 54.34 \text{ cfs}$$

$$\partial_z = 34.10 \text{ cfs}$$

$$\hat{\theta}_Z = 0.754$$

Then, from Table A.5, the value of $1/k$ corresponding to the above skewness of $\hat{\theta}_Z = 0.754$ is

$$1/k = 0.546$$

Thus,

$$\hat{k} = 1/0.546 = 1.83$$

Table E4.35 Recorded Annual Drought of the Iroquois River near Chebanse, Illinois

Year of Observation	Seven-Day Flow (cfs)	Order m	Plotting Position $1 - m/(N + 1)$
1934	14.1	1	0.96970
1941	15.2	2	0.93939
1925	18.0	3	0.90909
1936	19.1	4	0.87879
1944	20.7	5	0.84848
1932	22.6	6	0.81818
1940	23.9	7	0.78788
1930	27.3	8	0.75758
1954	33.3	9	0.72727
1933	33.4	10	0.69697
1939	33.6	11	0.66667
1929	34.3	12	0.63636
1935	37.1	13	0.60606
1953	41.1	14	0.57576
1927	42.7	15	0.54545
1948	45.1	16	0.51515
1946	45.3	17	0.48485
1949	47.4	18	0.45455
1942	48.0	19	0.42424
1955	51.0	20	0.39394
1952	55.1	21	0.36364
1935	63.3	22	0.33333
1931	70.3	23	0.30303
1924	74.0	24	0.27273
1947	77.7	25	0.24242
1943	77.9	26	0.21212
1928	83.0	27	0.18182
1945	93.4	28	0.15152
1950	106	29	0.12121
1938	109	30	0.09091
1926	130	31	0.06061
1951	146	32	0.03030

From Table A.5, we also obtain the following:

$$A(\hat{k}) = 0.221$$
$$B(\hat{k}) = 1.99$$

Therefore,

$$\hat{w}_1 = 0.221 \times 34.1 + 54.34$$
$$= 61.88 \text{ cfs}$$

and

$$\hat{\varepsilon} = 61.88 - 1.99 \times 34.10 = -5.98$$

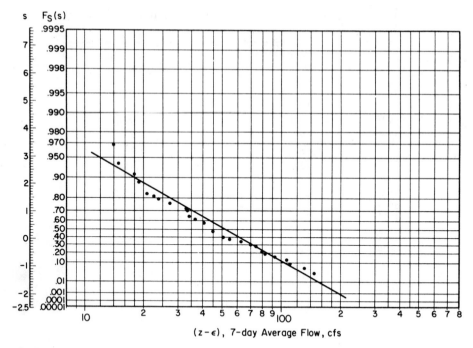

Figure E4.35   Annual lowest 7-day average flow of the Iroquois River, plotted on log-extremal probability paper.

But, in the case of droughts, $\varepsilon \geq 0$; therefore,

$$\hat{\varepsilon} = 0$$

Graphically, from the straight line constructed in Fig. E4.35, we obtain the most probable flow (at $s = 0$) of

$$\hat{w}_1 = 64 \text{ cfs}$$

and the slope of the line representing k (with Eq. 4.85a) as

$$\hat{k} = \frac{-3}{\ln 12 - \ln 64} = 1.79$$

PROBLEMS

4.1   The wave height X of the individual ocean waves may be modeled by the Rayleigh distribution with the PDF

$$f_X(x) = \frac{x}{a^2} e^{-(1/2)(x/a)^2}; \qquad x \geq 0$$

where a is the parameter. For a linear response, the wave-induced mid-ship bending moment of a ship may also be assumed to follow the same distribution.

(a)   Derive the exact distribution of the maximum wave height among n waves. Compare the form of this exact distribution with the corresponding asymptotic distribution form.

Type I Compare $n = 100$

$n = 10,000$

(b) For an ocean-going ship, the number of waves encountered will obviously increase with its exposure time; according to Ochi (1973), the number of waves is related to the operation (or exposure) time T as follows:

$$n = \frac{(60)^2}{2\pi} T \frac{\sqrt{m_2}}{a}$$

$$f_M(m) = \frac{m}{M^2} \exp\left[-\frac{1}{2}\left(\frac{m}{M}\right)^2\right]$$

where

T: Exposure time in hours.

m_2: The second moment of the response power spectral density (see Ochi, 1973).

Suppose a ship with length $L = 300$ ft and width $B = 40$ ft will be subjected to 75 hours of exposure (or operation) in storms over its life. If the standard deviation of the wave-induced bending moment is

$$\sigma_M = 3.44 \times 10^{-2} \rho g B L^2 \quad \text{Let } \rho g \, BL^2 = A$$

and

$$m_2 = 3.91 \times 10^{-2} (\rho g B L^2)^2 \quad \sigma_M = (2 - \frac{\pi}{2}) M$$

where ρ is the mass density and g is the gravitational acceleration, determine the mean lifetime maximum mid-ship bending moment and the corresponding c.o.v.

4.2 The sustained floor live loads surveyed for office buildings may be modeled by a lognormal distribution with a mean EUDL (equivalent uniformly distributed load) of 12 psf and an average c.o.v. of 30%. Over the life of a building, the live floor load intensity may fluctuate because of changes in occupancy and use of the floor space. Assuming an average rate of occupancy change once every two years, formulate a procedure for determining the exact distribution of the lifetime maximum sustained floor EUDL for office buildings over a life of 50 years.

Deduce the corresponding asymptotic form of distribution for the lifetime maximum EUDL, and the associated parameters for a 50-year life.

4.3 In the case of multistory factory buildings, the floor live load (in terms of EUDL) for a given floor area is likely to remain constant throughout its life. Suppose a survey of 50 room areas was conducted, yielding a mean and c.o.v. of 20 psf and 0.42, respectively. On the basis of the survey data and assuming appropriately a Rayleigh distribution for the floor live load intensity, determine the distribution and associated parameters for the maximum EUDL for a building of 1200 rooms.

4.4 Suppose the daily maximum wind speed at a given location has a Gaussian distribution with a mean speed of 15 mph and a c.o.v. of 67%.
  (a) Determine the asymptotic distribution of the annual maximum wind speed for the location, that is, specify the asymptotic form of the distribution and the associated parameters.
  (b) Using the results of (a), derive the corresponding distribution for the 50-year maximum wind speed and determine the 50-year most probable maximum wind speed.

4.5 Do Problem 4.4 if the daily maximum wind speed is lognormally distributed.

4.6 The PDF of an initial variate with a gamma distribution is

$$f_X(x) = \frac{v(vx)^{k-1} e^{-vx}}{\Gamma(k)}; \quad x \geq 0, k > 1$$

where v and k are the parameters, and $\Gamma(k)$ is the gamma function.
  (a) Determine the CDF and PDF of the largest value from a sample of size n.
  (b) To which of the three asymptotic forms will the distribution of the largest value converge?
  (c) Derive the expressions for the appropriate extremal parameters for $k = 2$, and assume small vu_n.

4.7   The significant wave height (in meters) at an offshore platform site for a 3-hour period during the winter months (October–April) is observed to follow a Type I largest value distribution with parameters $u = 2.2$ and $\alpha = 0.82$; the wave heights over the summer months (May–September) have a negligible contribution toward the annual maximum wave. Assume that the significant wave height is constant over each 3-hour period, and the significant wave heights between nonoverlapping 3-hour periods are statistically independent. Determine the distribution of the annual maximum significant height.

4.8   (a)   Suppose the joints on rock surfaces occur as a Poisson process with a mean density of 0.01 per square meter, and the length of each joint follows an exponential distribution with a mean of 4 meters. Determine the distribution of the maximum joint length over a rock surface of 10,000 square meters.

   (b)   Determine the distribution of the maximum joint length among a set of 100 joints. Plot the CDF and compare it with the CDF of part (a).

4.9   Truck axle loads higher than 18 tons may be described with a shifted exponential distribution as follows:

$$f_X(x) = \frac{1}{3.2} e^{-(x-18)/3.2}; \qquad x \geq 18$$

Suppose a culvert is subject to 1355 such axle loads (> 18 tons) in a year. Assume all axle loads are statistically independent and the traffic volume will remain constant over the years.

   (a)   Determine the mean and c.o.v. of the maximum axle load over periods of 1, 5, 10, 20, and 50 years.

   (b)   For an expected culvert life of 20 years, determine the probability that it will be subjected to an axle load of over 80 tons.

   (c)   Determine the "design axle load" corresponding to an allowable exceedance probability of $p = 0.10$ over the life of 20 years.

4.10   Individual wave heights during a short storm period, say 3 hours, can be modeled by a Rayleigh distribution with the following CDF:

$$F_H(h) = 1 - \exp\left(-\frac{2h^2}{H_S^2}\right); \qquad h \geq 0$$

where H_S is the average of the highest one-third of all the waves, defined as the *significant wave height* of the storm period.

   (a)   Consider a 3-hour storm period with $H_S = 10$m and an average wave period of 12 seconds. Estimate the probability that the maximum wave height during the 3-hour period will exceed 20m.

   (b)   Suppose the actual storm duration is 15 hours, which can be divided into five 3-hour periods with significant wave heights (in meters) of 6, 8, 10, 8, and 6, respectively. Estimate the probability that the maximum wave height during the 15-hour storm will exceed 20m. Assume the average wave periods are 9 and 8 secs for waves with $H_S = 8$m and 6m, respectively.

4.11   Suppose the performance of a pile foundation system is governed by the pile with the smallest capacity. Moreover the weakest pile associated with a foundation project has a mean value of 15 tons and a c.o.v. of 20%. Determine the probability that it will be less than the design load capacity of 10 ton if

   (a)   The weakest pile follows a type II smallest value distribution.

   (b)   The weakest pile follows a type III smallest value extremal distribution with a minimum threshold value of $\varepsilon = 2$ tons.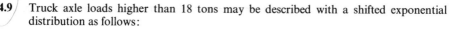

4.12   The daily level of dissolved oxygen (DO) concentration for a stream is assumed to be lognormally distributed with a mean of 2.52mg/l and a standard deviation of 2.05mg/l. Assume the DO concentrations between days are statistically independent.

(a) Determine the asymptotic distribution for the minimum daily DO in a month and in a year. Evaluate the probability that the DO concentration will be below 0.5mg/l in a day, in a month, in a year.

(b) Determine the mean and c.o.v. of the minimum daily DO concentration in a month, in a year.

4.13 Example 4.22 shows that the largest value from the following initial uniformly distributed random variable with

$$f_X(x) = \frac{1}{b-a}; \quad a \le x \le b$$

will converge asymptotically to the Type III asymptotic form. Derive the corresponding CDF for the asymptotic extremal distribution and its parameters.

4.14 Verify that for an initial variate X with the following triangular distribution

$$f_X(x) = \frac{2}{a^2}(a - x); \quad 0 \le x \le a$$

both the largest and smallest values will converge asymptotically to the respective Type III asymptotic forms.

4.15 (a) A structure consists of 100 column footings. Suppose the settlement of each footing is exponentially distributed with a mean of 2.0 inches and the settlements between footings are statistically independent. What is the probability that the maximum settlement S_{max} among the 100 footings will not exceed 5.0 inches? Compare this exact probability with the answer obtained assuming an appropriate asymptotic distribution for S_{max}.

(b) What is the probability that the minimum settlement S_{min} is at least 0.5 inch? Compare this with that determined through the asymptotic distribution of S_{min}.

(c) Determine the mean and c.o.v. of the maximum differential settlement Δ_{max} defined as $(S_{max} - S_{min})$.

4.16 The occurrence of significant earthquakes (of magnitude $\ge m_o$) in a region is often modeled with the Poisson process; that is, the probability of n earthquakes in duration t is

$$P(N_t = n) = \frac{(v_o t)^n}{n!} e^{-v_o t}$$

where v_o = the mean occurrence rate of $m \ge m_o$ in the region.

Also, the magnitude of an earthquake may be described by the shifted exponential CDF

$$F_M(m) = 1 - e^{-\beta(m - m_o)}; \quad m \ge m_o$$
$$= 0; \quad m < m_o$$

where β is a constant parameter. Assuming that M and N_t are statistically independent, derive the CDF of the annual maximum earthquake magnitude in the pertinent region. What is the asymptotic form?

4.17 The maximum ground motion intensity from an earthquake attenuates with the focal distance r from the source. If we assume that the total energy from an earthquake is radiated from a point source, as shown in Fig. P4.17, the attenuation of the maximum intensity may be represented by

$$y = b_1 e^{b_2 m} r^{-b_3}$$

where b_1, b_2, b_3 are constants. Then, if earthquakes at the source, 0, occur in accordance with Problem P4.16, derive the CDF of the annual maximum ground motion intensity at a site of focal distance r from the source. Which of the asymptotic forms does the result belong to?

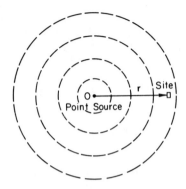

Figure P4.17   Spatial radiation of energy from point source.

4.18 In the development of advanced surface-effect ships, experimental tow tank tests were conducted of small-scale models of the ship. Continuous data of the mid-ship bending moments were recorded; the maximum moments for successive 10-second intervals (in appropriate units) were obtained as follows:

2163	1677	2526	2455	2940	1976
1957	2546	1597	2672	1909	2765
2611	2105	1985	1114	1154	1769
1710	1506	1816	1210	1518	1006
2819	2410	2197	3100	1224	2725
1848	1086	1370	2445	2970	750
1252	3481	1231	889	1248	2239
1794	2253	1485	1297	1174	2491
2056	1946	1651	2345	1884	3347
945	2387	1639	2233	2773	2158

    (a)   Plot the 10-second maximum values on a Type I extremal probability paper.

    (b)   Identify the maximum values in successive 40-second intervals and also plot these maximas on a Type I extremal probability paper.

    On the basis of the plots in (a) and (b), what can you infer about the distribution (or its convergence) of the maximum moments for increasing intervals?

4.19 The annual wind velocity has been modeled by the Type I asymptotic distribution of the largest value. According to the stability postulate, the lifetime (e.g., 50-year) maximum wind velocity will also have a Type I asymptotic distribution.

    Determine the mean and standard deviation of the 50-year maximum wind velocity in terms of the annual distribution parameters.

4.20 The annual maximum "mile wind" speed may be modeled by the Type I asymptotic distribution with parameters u, which depends on the location of a site, and $1/\alpha = 9$.

    (a)   On the basis of the above, if the maximum wind speed observed in the past n years is V_n, derive the relation for determining the most-probable maximum wind in the next N years.

    (b)   Specialize the relation in (a) for predicting the N-year wind speed using $n = 5$ and 10 years.

    (c)   Develop the corresponding relation for determining the most probable N-year maximum wind speed if the distribution of the annual maximum wind is Type II asymptotic. On this basis, specialize the relation for $n = 5$ and 10 years, and compare the results with those of (b) for $N = 25$, 50, and 100 years. Assume $k = 9.0$, and $V_5 = 60$ mph and $V_{10} = 70$ mph.

4.21 Evaluate the parameters for the distribution of the annual maximum earthquake magnitude of Example 4.24, using the order-statistics methods of Lieblein.

Determine the "design earthquake" corresponding to an allowable exceedance probability of $p = 0.10$ over 30 years.

4.22 From U.S. Geological Survey Water Resources Data, the annual maximum discharges for the gaging station of the Salt Fork River near St. Joseph, Illinois between 1959 and 1975 are listed below:

1959	6030 (cfs) 16
1960	1800 4
1961	2300 6
1962	3370 14
1963	2340 7
1964	5380 15
1965	1230 2
1966	950 1
1967	2640 11
1968	6860 17
1969	2630 10
1970	2600 9
1971	2350 8
1972	1350 3
1973	2990 13
1974	2750 12
1975	1920 5

$N = 17 + 1 = 18$

Suppose a Type I largest value extremal distribution is adopted for the annual maximum discharge at this station. Determine the parameters for the distribution using:
(a) Graphical method.
(b) Method of moments.
(c) Method of order statistics.

4.23 The annual fastest mile wind (in mph) at Midway Airport, Chicago has been observed between 1943 and 1977; the record is summarized below (Simiu et al., 1979):

1943	44 (mph)		1961	44 (mph)
1944	46		1962	42
1945	41		1963	54
1946	37		1964	51
1947	46		1965	51
1948	49		1966	43
1949	53		1967	56
1950	53		1968	46
1951	49		1969	42
1952	59		1970	43
1953	49		1971	47
1954	50		1972	44
1955	53		1973	46
1956	45		1974	45
1957	42		1975	47
1958	46		1976	44
1959	48		1977	51
1960	40			

Obtain the extremal distribution that best fits this set of data and estimate the corresponding parameters. In particular, examine the appropriateness of the Type I and Type II asymptotic distributions.

Determine the "design wind" corresponding to an exceedance probability of 0.05 for the next 20 years.

5. *Monte Carlo Simulation*

5.1 INTRODUCTION

Simulation is the process of replicating the real world based on a set of assumptions and conceived models of reality. It may be performed theoretically or experimentally. In practice, theoretical simulation is usually performed numerically; this has become a much more practical tool since the advent of computers. As with experimental methods, numerical simulation may be used to obtain (simulated) data, either in lieu of or in addition to actual real-world data. In effect, theoretical simulation is a method of "numerical or computer experimentation."

For engineering purposes, simulation may be applied to predict or study the performance and/or response of a system. With a prescribed set of values for the system parameters (or design variables), the simulation process yields a specific measure of performance or response. Through repeated simulations, the sensitivity of the system performance to variation in the system parameters may be examined or assessed. By this procedure, simulation may be used to appraise alternative designs or determine optimal designs.

For problems involving random variables with known (or assumed) probability distributions, *Monte Carlo simulation* is required. This involves repeating a simulation process, using in each simulation a particular set of values of the random variables generated in accordance with the corresponding probability distributions. By repeating the process, a sample of solutions, each corresponding to a different set of values of the random variables, is obtained. A sample from a Monte Carlo simulation is similar to a sample of experimental observations. Therefore, the results of Monte Carlo simulations may be treated statistically; such results may also be presented in the form of histograms, and methods of statistical estimation and inference are applicable. For these reasons, Monte Carlo is also a sampling technique, and as such shares the same problems of sampling theory; namely, the results are also subject to sampling errors. Generally, therefore, Monte Carlo solutions from finite samples are not "exact" (unless the sample size is infinitely large).

One of the main tasks in a Monte Carlo simulation is the generation of random numbers from prescribed probability distributions; for a given set of generated random numbers, the simulation process is deterministic.

In theory, simulation methods can be applied to large and complex systems; often the rigid idealizations and/or simplifications necessary of analytical models can be relaxed, resulting in more realistic simulation models. However, in practice, Monte Carlo simulations may be limited by constraints of economy and computer capability. Moreover, solutions obtained from simulations (particularly Monte

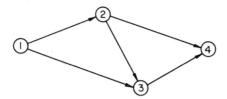

Figure 5.1   A construction network.

Carlo simulation) may not be amenable to generalization or extrapolation. Therefore, as a general rule, Monte Carlo methods should be used only as a last resort; that is, when and if analytical solution methods are not available or are ineffective (e.g., because of the need for gross idealizations). Monte Carlo solutions, however, are often the only means of verifying or validating approximate analytical solution methods.

5.1.1   An Example of Monte Carlo Simulation

The essence of Monte Carlo simulation may be described with a specific example. For this purpose, consider the activity network for a construction project shown in Fig. 5.1, in which the arrows represent activities, whereas the nodes represent the completion of one or more activities and the initiation of subsequent activities. The durations for the various activities are randomly distributed and statistically independent. The possible durations and associated probabilities are given in Table 5.1; these represent the respective probability distributions of the activity durations. Suppose that six days have been scheduled for the completion of the project. Determine the probability that the project will be completed on time.

Conceptually, the completion of the project within six days requires the completion of all the possible (three) activity paths between nodes 1 and 4, namely,

Table 5.1   Activity Durations

Activity	Duration (days)	Probability
1–2	2	$\frac{1}{3}$
	3	$\frac{1}{3}$
	4	$\frac{1}{3}$
1–3	4	$\frac{1}{2}$
	5	$\frac{1}{2}$
2–3	1	$\frac{2}{3}$
	2	$\frac{1}{3}$
2–4	2	$\frac{1}{2}$
	3	$\frac{1}{2}$
3–4	2	1

paths 1–2–4, 1–2–3–4, and 1–3–4. Denoting the duration of activity $i - j$ as T_{ij}, the probability of completing the project within six days is given by

$$P(T \leq 6) = P[T_{12} + T_{24} \leq 6) \cap (T_{12} + T_{23} + T_{34} \leq 6) \cap (T_{13} + T_{34} \leq 6)]$$

(5.1)

If activity 2–3 did not exist, only two activity paths, 1–2–4 and 1–3–4, need to be considered. Since the durations of these two paths are statistically independent, the probability of completing the project on time may be evaluated by first determining the probability mass functions of the durations of the two paths and then applying the method similar to Example 4.9 in Vol. I. The presence of activity 2–3 creates not only an additional path but also a correlation between the various activity paths. Hence, the calculation of the probability of Eq. 5.1 is complicated considerably. As an alternative, simulation may be used; for this purpose, we first generate possible sets of activity durations. For example, since the duration of activity 1–3 is equally likely to be 4 or 5 days, a coin may be tossed; depending on the outcome (i.e., a head or a tail) 4 or 5 days would be assigned accordingly as the duration of activity 1–3. Similarly, for activity 1–2, two of the six faces of a dice may be arbitrarily assigned for each of the durations: 2, 3, or 4 days. Depending on the outcome of a roll of the dice, the specific durations of activity 1–2 would be generated. Suppose the durations of activities 1–2, 1–3, 2–3, and 2–4 have been generated as 2, 5, 1, and 3 days, respectively, constituting sample #1 as shown in Table 5.2. The activity path 1–2–4 will, therefore, require five days; whereas paths 1–2–3–4 and 1–3–4 will require 5 and 7 days, respectively. Thus, the project will not be completed within 6 days for this set of activity durations. Another set of activity durations is similarly generated as shown for sample #2 in Table 5.2; in this case, the project will require only 6 days. Table 5.2 summarizes the results of 15 simulations, constituting a sample size of 15. Based on the observation that in

Table 5.2   Simulation of Project Duration

Sample #	1–2	1–3	2–3	2–4	3–4	1–2–4	1–3–4	1–2–3–4	Is Project Duration >6 Days?
1	2	5	1	3	2	5	7	5	Yes
2	2	4	1	3	2	5	6	5	No
3	4	4	2	2	2	6	6	8	Yes
4	4	5	1	2	2	6	7	7	Yes
5	3	5	1	3	2	6	7	6	Yes
6	2	5	2	2	2	4	7	6	Yes
7	3	5	1	3	2	6	7	6	Yes
8	3	5	2	2	2	5	7	7	Yes
9	3	4	1	2	2	5	6	6	No
10	3	5	1	2	2	5	7	6	Yes
11	4	5	2	2	2	6	7	8	Yes
12	4	5	1	3	2	7	7	7	Yes
13	3	5	2	2	2	5	7	7	Yes
14	3	5	1	3	2	6	7	6	Yes
15	2	5	1	2	2	4	7	5	Yes

only two cases out of 15 is the project completed within 6 days, the probability of project completion within 6 days is estimated to be $\frac{2}{15}$ or 0.13. If another 15 simulations were repeated, the resulting probability would most likely be different. In fact, 18 repetitions were performed independently for sample sizes 15 and 30; the probability of project completion based on these two sample sizes are summarized in Table 5.3. Although the estimated probability of project completion varies from one simulation solution to another, it may be observed that the range and standard derivation of the estimated probability decrease as the sample size increases. The confidence interval on the probability value estimated on the basis of sample size n may be established using Eq. 5.47 of Vol. I. For example, based on a sample size of 30 and an estimated probability of 0.23, the 95% confidence interval of the actual probability of project completion is given by

$$\langle p \rangle_{0.95} = 0.23 \pm 1.96 \sqrt{\frac{0.23 \times 0.77}{30}} = \{0.08; 0.38\}$$

On the other hand, if all 18 repetitions of sample size 30 were combined (giving an equivalent sample of size 540), the 95% confidence interval on the probability of project completion would be

$$\langle p \rangle_{0.95} = 0.27 \pm 1.96 \sqrt{\frac{0.27 \times 0.73}{540}} = \{0.23; 0.31\}$$

Table 5.3   Estimated Completion Probabilities

Repetition No.	Estimated Probability of Completion	
	Sample Size 15	Sample Size 30
1	0.13	0.23
2	0.33	0.30
3	0.40	0.30
4	0.27	0.30
5	0.20	0.27
6	0.33	0.33
7	0.33	0.30
8	0.27	0.34
9	0.13	0.20
10	0.44	0.35
11	0.33	0.30
12	0.27	0.17
13	0.20	0.27
14	0.07	0.10
15	0.20	0.33
16	0.20	0.23
17	0.33	0.37
18	0.27	0.23
Mean	0.26	0.27
Standard Deviation	0.10	0.07
Range	0.07–0.44	0.10–0.37

5.2   GENERATION OF RANDOM NUMBERS

A key task in the application of Monte Carlo simulation is the generation of the appropriate values of the random variables (i.e., random numbers) in accordance with the respective prescribed probability distributions. Special devices may be used for simple random variables, for example, tossing a fair coin for random variables with two equally likely values or rolling a 6-faced dice for a random variable with six equally likely possible values. Uniformly distributed random numbers within a given range may be obtained by spinning a wheel with the range subdivided equally on the circumference of the wheel. Uniformly distributed random numbers (to five significant digits) have also been tabulated (e.g., Rand Corp., 1955).

As alluded to earlier, Monte Carlo simulation is most effective and practical if used with digital computers; for this purpose, the automatic generation of the requisite random numbers with specified distributions will be necessary. This can be accomplished systematically for each variable by first generating a uniformly distributed random number between 0 and 1.0, and through appropriate transformations obtaining the corresponding random number with the specified probability distribution. The basis for this is as follows.

Suppose a random variable X with CDF $F_X(x)$. Then, at a given cumulative probability $F_X(x) = u$, the value of X is

$$x = F_X^{-1}(u) \tag{5.1}$$

Now suppose that u is a value of the *standard uniform variate*, U, with a uniform PDF between 0 and 1.0; then, as shown in Fig. 5.2b,

$$F_U(u) = u \tag{5.2}$$

that is, the cumulative probability of $U \le u$ is equal to u.

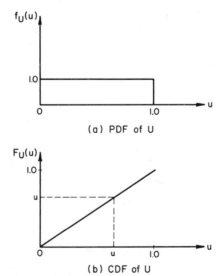

(a) PDF of U

(b) CDF of U

Figure 5.2   PDF and CDF of standard uniform variate U.

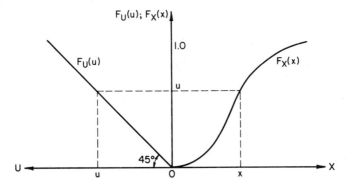

Figure 5.3 Relation between u and x.

Therefore, if u is a value of U, the corresponding value of the variate X obtained through Eq. 5.1 will have a cumulative probability,

$$
\begin{aligned}
P(X \le x) &= P[F_X^{-1}(U) \le x] \\
&= P[U \le F_X(x)] \\
&= F_U[F_X(x)] = F_X(x)
\end{aligned}
$$

which means that if $(u_1, u_2, \ldots, u_n)$ is a set of values from U, the corresponding set of values obtained through Eq. 5.1, that is,

$$
x_i = F_X^{-1}(u_i); \qquad i = 1, 2, \ldots, n \tag{5.3}
$$

will have the desired CDF $F_X(x)$. The relationship between u and x may be seen graphically in Fig. 5.3.

5.2.1   Random Numbers with Standard Uniform Distribution

From the above discussions, Eqs. 5.1 through 5.3, we observe that the generation of uniformly distributed random numbers between 0 and 1, that is, values of the *standard uniform variate*, is basic in the generation of random numbers with a general probability distribution. The methods described subsequently in Sections 5.2.2 and 5.2.3 for generating random numbers with other distributions are based on the assumption that uniformly distributed random numbers are already available.

Methods for generating uniformly distributed random numbers are generally based on recursive calculations of the residues of modulus m from a linear transformation. An example of a recursive relation for this purpose is

$$
x_{i+1} = (ax_i + c)(\bmod m) \tag{5.4}
$$

where a, c, and m are nonnegative integers. If k_i is the integer part of the ratio $(ax_i + c)/m$, that is,

$$
k_i = \mathrm{Int}\left(\frac{ax_i + c}{m}\right)
$$

then the corresponding residue of modulus m is

$$x_{i+1} = ax_i + c - mk_i \tag{5.5}$$

Normalizing the values obtained from Eq. 5.5 by the modulus m, we obtain

$$u_{i+1} = \frac{x_{i+1}}{m} \tag{5.6}$$

which constitute a set of random numbers between 0 and 1 with the standard uniform probability distribution.

In reality, random numbers generated by a systematic procedure, such as that described above, can be duplicated exactly and therefore constitute a deterministic set. Strictly speaking, therefore, such generated random numbers are really not random; for this reason, they may be called "pseudo random" numbers.

To demonstrate the procedure of generating a set of random numbers, suppose $a = 3$, $c = 1$, $m = 5$, and starting with an initial value $x_0 = 1.0$, Eqs. 5.5 and 5.6 yield the following pseudo random numbers:

$$k_0 = \text{Int}\left(\frac{3 \times 1 + 1}{5}\right) = \text{Int}(0.8) = 0$$

$$x_1 = 3 \times 1 + 1 - 5 \times 0 = 4$$

$$u_1 = \tfrac{4}{5} = 0.8$$

$$k_1 = \text{Int}\left(\frac{3 \times 4 + 1}{5}\right) = \text{Int}(2.6) = 2$$

$$x_2 = 3 \times 4 + 1 - 5 \times 2 = 3$$

$$u_2 = \tfrac{3}{5} = 0.6$$

The above procedure may be repeated; the results would be as follows:

i	u_i
1	0.8
2	0.6
3	0
4	0.2
5	0.8
6	0.6
7	0
8	0.2
$\vdots$	$\vdots$

Observe that the generated pseudo random numbers are cyclic; that is, they are repeated with a given period (four in the above case). The period of the cycle is less than m (Knuth, 1969); therefore, to insure randomness, the period should be as long as possible, and thus in practical applications a large value of m should be assigned in the generation of u_i.

Although the above procedure for generating random numbers is basically deterministic, it can be shown (Knuth, 1969) that the numbers generated with large m appear to be uniformly distributed and statistically independent. Furthermore, Greenberger (1961) has shown that the correlation coefficient between x_i and x_{i+1} is bounded as follows:

$$\rho = \frac{1}{a} - \left(\frac{6c}{a\,m}\right)\left(1 - \frac{c}{m}\right) \pm \frac{a}{m} \tag{5.7}$$

Clearly, $\rho \to 0$ for large values of a and m.

Satisfactory results have been observed (Rubinstein, 1981) with $m = 2^{35}$, $a = 2^7 + 1$, and $c = 1$ for binary computers; and $m = 2^\beta$, $a = 101$, and $c = 1$ for a decimal computer with a word length of β.

Another common recursive relation for generating random numbers is the multiplicative congruential generator

$$x_{i+1} = ax_i(\mathrm{mod}\ m) \tag{5.8}$$

and

$$u_i = \frac{x_i}{m} \tag{5.9}$$

Eqs. 5.8 and 5.9 are used in the IBM System 360 as a uniform random number generator with $a = 16807$ and $m = 2^{31} - 1$. Observe that Eq. 5.8 is a special case of Eq. 5.1 with $c = 0$.

Random numbers generated by the procedures described above may be tested for statistical independence and for uniform distribution. The *serial test* (Knuth, 1969) may be used to check the randomness between successive numbers in a sequence, whereas goodness-of-fit tests, such as the chi-square test, Kolmogorov-Smirnov test (see Chapter 6, Vol. I), and the Cramer-von Mises test, may be used to verify that the pseudo random numbers are uniformly distributed. The procedures using the parameter values suggested above have been tested statistically and shown to give satisfactory results (Rubinstein, 1981).

5.2.2 Continuous Random Variables

As indicated in Section 5.2, random numbers with a prescribed distribution may be generated through Eq. 5.3 once the standard uniformly distributed random numbers have been obtained. The generation of random numbers through Eq. 5.3 is known as the *inverse transform method*. The application of the inverse transform method is most effective if the inverse of the CDF of the random variable X can be expressed analytically; that is, if the inverse function $F_X^{-1}(u)$ is available.

EXAMPLE 5.1

Consider the exponential distribution with the CDF

$$F_X(x) = 1 - e^{-\lambda x}; \qquad x \geq 0$$

The inverse function is

$$x = F_X^{-1}(u) = -\frac{1}{\lambda}\ln(1 - u)$$

Therefore, in this case, once the standard uniformly distributed random numbers $u_i, i = 1, 2, \ldots$ are generated, we obtain the corresponding exponentially distributed random numbers, according to Eq. 5.3, as

$$x_i = -\frac{1}{\lambda} \ln (1 - u_i)$$

Since $(1 - u_i)$ is also uniformly distributed, the required random numbers may also be generated by

$$x_i = -\frac{1}{\lambda} \ln u_i; \qquad i = 1, 2, \ldots$$

EXAMPLE 5.2

Suppose $Y_n = \max(X_1, X_2, \ldots, X_n)$, where the X_i's are independent and identically distributed random variables with CDF $F_X(x)$. According to Eq. 4.2, the CDF of Y_n is

$$F_{Y_n}(y) = [F_X(y)]^n$$

Then,

$$[F_X(y)]^n = u$$

or

$$F_X(y) = u^{1/n}$$

Thus,

$$y = F_X^{-1}(u^{1/n})$$

Therefore, to obtain a value y of the random variable Y_n, we first generate a value u for the standard uniform variate U and then compute $F_X^{-1}(u^{1/n})$. In this case, the random values of Y_n can be generated directly based on $F_X(x)$, without having to determine the CDF $F_Y(y)$.

In particular, if the X_i's are standard uniform random variables, then

$$y = F_U^{-1}(u^{1/n}) = u^{1/n}$$

EXAMPLE 5.3

The CDF of the Type I asymptotic distribution of largest value is

$$F_X(x) = \exp[-e^{-\alpha(x - \beta)}]$$

where β is the most probable value of X, and α is the shape parameter.

At a given probability value $F_X(x) = u$, we have

$$x = \beta - \frac{1}{\alpha} \ln \left(\ln \frac{1}{u} \right)$$

Therefore, random numbers with the Type I asymptotic distribution can be generated from the corresponding uniformly distributed random numbers using the above relation.

The inverse transform method is effective if the inverse of the CDF can be expressed analytically. However, there are many probability distributions (e.g., normal and lognormal) in which the CDF cannot be inverted analytically. For such cases, other methods may be more efficient or effective; these include the *composition method* and the method of *functions of random variables*.

The Composition Method   The PDF of a random variable X may be expressed as a weighted sum of a set of other density functions through the theorem of total probability as follows:

$$f_X(x) = \sum_i f_X(x|E_i)P(E_i)$$

$$= \sum_i f_{X_i}(x)p_i \qquad (5.10)$$

where $f_{X_i}(x)$; $i = 1, \ldots, n$ is a set of component density functions, and p_i is the probability or relative weight associated with $f_{X_i}(x)$. The generation of a value for the random variable X may be performed in two steps: (i) a random number is first generated for the probability p_i and the corresponding component density function $f_{X_i}(x)$ is selected; (ii) another random number is then generated according to the PDF $f_{X_i}(x)$ selected in step (i).

    The advantage of this method is that a complex PDF $f_X(x)$ may be decomposed into a combination of simpler PDFs $f_{X_i}(x)$, whose corresponding CDFs $F_{X_i}(x)$ can be inverted analytically.

    A continuous version of Eq. 5.10 may be written as

$$f_X(x) = \int g(x|y)dF_Y(y) \qquad (5.11)$$

where $g(x|y)$ is a family of one-parameter density functions, in which y is the parameter. In this case, the first random number generated will be from the CDF $F_Y(y)$ to yield a value of y; the subsequent random number will be from $g(x|y)$ to obtain x.

EXAMPLE 5.4

A random variable X with a PDF

$$f_X(x) = \tfrac{2}{3} + x^2; \qquad 0 \le x \le 1$$

is shown in Fig. E5.4. This particular PDF may be decomposed as follows:

$$f_X(x) = \tfrac{2}{3}f_I(x) + \tfrac{1}{3}f_{II}(x); \qquad 0 \le x \le 1$$

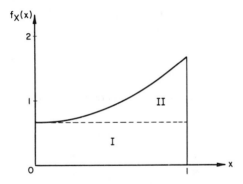

Figure E5.4   PDF of X.

where

$$f_I(x) = 1; \qquad 0 \le x \le 1$$
$$f_{II}(x) = 3x^2; \qquad 0 \le x \le 1$$

The relative weights p_I and p_{II} are $\frac{2}{3}$ and $\frac{1}{3}$, respectively, which correspond to the areas I and II of Fig. E5.4.

Two values u_1 and u_2 may be generated for the standard uniform variate. If $u_1 < \frac{2}{3}$, u_2 is then used to determine a value x from $f_I(x)$. Since $F_I(x) = x$, $x = F_I^{-1}(u_2) = u_2$. On the other hand, if $u_1 \ge \frac{2}{3}$, u_2 is used to determine a value for x from $f_{II}(x)$. In this case,

$$F_{II}(x) = x^3 = u_2$$

Thus

$$x = u_2^{1/3}$$

In short,

$$x = \begin{cases} u_2; & \text{if } u_1 < \frac{2}{3} \\ u_2^{1/3}; & \text{if } u_1 \ge \frac{2}{3} \end{cases}$$

Function of Random Variables Method   Suppose a random variable X can be expressed as a function of other random variables $Y_1, Y_2, \ldots, Y_m$; that is,

$$X = g(Y_1, Y_2, \ldots, Y_m)$$

and methods for generating values of Y_1 through Y_m are available. Then, a value of X may be determined as

$$x = g(y_1, y_2, \ldots, y_m) \tag{5.12}$$

where $(y_1, y_2, \ldots, y_m)$ are random values that have been generated for $Y_1, Y_2, \ldots, Y_m$. The following examples will illustrate the application of this method of random number generation.

EXAMPLE 5.5   (Normal Random Numbers)

Box and Muller (1958) have shown that if U_1 and U_2 are two independent standard uniform variates, then the following functions

$$S_1 = (-2 \ln U_1)^{1/2} \cos 2\pi U_2$$

and

$$S_2 = (-2 \ln U_1)^{1/2} \sin 2\pi U_2$$

constitute a pair of statistically independent standard normal variates. This can be verified as follows.

Let $U = -\ln U_1$ and $V = U_2$. As U_1 and U_2 are statistically independent, the joint PDF of U and V is

$$f_{U,V}(u, v) = f_U(u) f_V(v)$$

$$= f_{U_1}(e^{-u}) \left| \frac{de^{-u}}{du} \right| f_{U_2}(v)$$

$$= e^{-u}$$

Also, in terms of U and V,

$$S_1 = (2U)^{1/2} \cos 2\pi V$$

$$S_2 = (2U)^{1/2} \sin 2\pi V$$

Inversion then yields

$$U = \tfrac{1}{2}(S_1^2 + S_2^2)$$

and

$$V = \frac{1}{2\pi} \tan^{-1}\left(\frac{S_2}{S_1}\right)$$

The Jacobian of the transformation is

$$\mathbf{J} = \begin{vmatrix} \dfrac{\partial u}{\partial s_1} & \dfrac{\partial u}{\partial s_2} \\[2ex] \dfrac{\partial v}{\partial s_1} & \dfrac{\partial v}{\partial s_2} \end{vmatrix} = \begin{vmatrix} s_1 & s_2 \\[2ex] \dfrac{-s_2}{2\pi(s_1^2 + s_2^2)} & \dfrac{s_1}{2\pi(s_1^2 + s_2^2)} \end{vmatrix} = \frac{1}{2\pi}$$

Therefore, the joint PDF of S_1 and S_2 is (e.g. see Papoulis, 1965)

$$f_{S_1, S_2}(s_1, s_2) = f_{U, V}\left[\frac{1}{2}(s_1^2 + s_2^2), \frac{1}{2\pi}\tan^{-1}\left(\frac{s_2}{s_1}\right)\right]\frac{1}{2\pi}$$

$$= \frac{1}{2\pi}\exp\left(-\frac{s_1^2 + s_2^2}{2}\right); \qquad -\infty < s_1, s_2 < \infty$$

which is the joint PDF of two independent standard normal variates. Therefore, if u_1 and u_2 are a pair of independent uniformly distributed random numbers, a pair of independent random numbers from a normal distribution $N(\mu, \sigma)$ may be generated by

$$x_1 = \mu + \sigma\sqrt{-2\ln u_1}\,\cos 2\pi u_2$$

$$x_2 = \mu + \sigma\sqrt{-2\ln u_1}\,\sin 2\pi u_2$$

Lognormal Random Numbers For a lognormal variate X with parameters λ and ζ, it has been shown (Chapter 4, Vol. I) that $\ln X$ is a normal random variable with mean λ and standard deviation ζ. Therefore, if x' is a value from a normal distribution $N(\lambda, \zeta)$, then

$$x = e^{x'}$$

will be a random number from the lognormal distribution with parameters λ and ζ.

EXAMPLE 5.6 (Beta-Distributed Random Numbers)

The PDF of the standard Beta distribution with shape parameters q and r is given by

$$f_X(x) = \frac{1}{B(q, r)}x^{q-1}(1 - x)^{r-1}; \qquad 0 \le x \le 1$$

where $B(q, r)$ is the Beta function. A value from this distribution may be generated with

$$x = \frac{u_1^{1/q}}{u_1^{1/q} + u_2^{1/r}}$$

where u_1 and u_2 are two random values from the standard uniform distribution, and provided that $u_1^{1/q} + u_2^{1/r} \leq 1$. The basis for the above procedure is as follows (Jöhnk, 1964). Let

$$Y_1 = U_1^{1/q}$$

$$Y_2 = U_2^{1/r}$$

where U_1 and U_2 are two independent standard uniform variates. Then,

$$f_{Y_1}(y_1) = f_{U_1}(y_1^q) \left| \frac{d(y_1^q)}{dy_1} \right|$$

$$= qy_1^{q-1}; \qquad 0 \leq y_1 \leq 1$$

$$f_{Y_2}(y_2) = ry_2^{r-1}; \qquad 0 \leq y_2 \leq 1$$

and

$$f_{Y_1, Y_2}(y_1, y_2) = qry_1^{q-1}y_2^{r-1}$$

Moreover, define $X = Y_1/(Y_1 + Y_2)$ and $W = Y_1 + Y_2$, from which $Y_1 = WX$ and $Y_2 = W(1 - X)$. The Jacobian is

$$\mathbf{J} = \begin{vmatrix} \dfrac{\partial y_1}{\partial x} & \dfrac{\partial y_1}{\partial w} \\[2ex] \dfrac{\partial y_2}{\partial x} & \dfrac{\partial y_2}{\partial w} \end{vmatrix} = \begin{vmatrix} w & x \\ -w & 1-x \end{vmatrix} = w$$

Hence, the joint PDF of X and W becomes

$$f_{X, W}(x, w) = f_{Y_1, Y_2}[wx, w(1 - x)](w)$$

$$= qrx^{q-1}(1 - x)^{r-1}w^{q+r-1}; \qquad 0 \leq x \leq 1, 0 \leq w \leq 2$$

Observe that

$$f_{X, W}(x, 0 \leq W \leq 1) = \int_0^1 f_{X, W}(x, w) \, dw$$

$$= qrx^{q-1}(1 - x)^{r-1} \frac{1}{q + r}$$

and

$$P(0 \leq W \leq 1) = \int_0^1 f_{X, W}(x, 0 \leq W \leq 1) \, dx$$

$$= \frac{qr}{(q + r)} \frac{\Gamma(q)\Gamma(r)}{\Gamma(q + r)}$$

Hence, the marginal PDF of X conditional on $0 \leq W \leq 1$ is

$$f_X(x \mid 0 \leq W \leq 1) = \frac{f_{X, W}(x, 0 \leq W \leq 1)}{P(0 \leq W \leq 1)}$$

$$= \frac{\Gamma(q + r)}{\Gamma(q)\Gamma(r)} x^{q-1}(1 - x)^{r-1}; \qquad 0 \leq x \leq 1$$

which is the PDF of the standard Beta distribution.

Random numbers x' from a general Beta distribution, that is, with lower and upper bounds a and b, may then be obtained through

$$x' = \frac{x - a}{b - a}$$

where x is the corresponding random number from the standard Beta distribution.

EXAMPLE 5.7   (Gamma-Distributed Random Numbers)

The PDF of the gamma distribution is

$$f_X(x) = \frac{v(vx)^{k-1}e^{-vx}}{\Gamma(k)} ; \qquad x \geq 0$$

with parameters v and k.

For the case of integer k, it can be shown (Hoel et al, 1971) that X is a gamma-distributed random variable if it is the sum of k independent exponential variates; that is,

$$X = \sum_{i=1}^{k} Y_i$$

where Y_i is an exponential random variable with parameter v. With the results of Example 5.1, a value from the gamma distribution may be generated as

$$x = -\frac{1}{v} \sum_{i=1}^{k} \ln u_i$$

where $u_i, i = 1, 2, \ldots, k$ are k uniformly distributed random numbers.

For $k < 1$, Jöhnk (1964) has shown that if V is an exponential variate with a mean value of one and W is a standard Beta variate with parameters $q = k$ and $r = 1 - k$, then

$$X = \frac{1}{v} VW$$

is a gamma distributed random variable with parameters v and k. Hence, applying the results of Examples 5.1 and 5.6 yields

$$x = \frac{1}{v}(-\ln u_3) \frac{u_1^{1/k}}{u_1^{1/k} + u_2^{1/(1-k)}}$$

where u_1, u_2, and u_3 are values generated from the standard uniform distribution, and provided that $u_1^{1/k} + u_2^{1/(1-k)} \leq 1$.

Since the sum of two independent gamma distributed random variables with parameters v, k_1, and v, k_2 is also a gamma distributed variate with parameters v and $k_1 + k_2$, a value from a general gamma distribution (that is, with parameters v and general k) may be generated through a combination of the above results as follows:

$$x = x_1 + x_2$$
$$= -\frac{1}{v} \sum_{i=4}^{k'+3} \ln u_i + \frac{1}{v}(-\ln u_3) \frac{u_1^{1/k}}{u_1^{1/k} + u_2^{1/(1-k)}}$$

where $k' \geq 1$ is the integer part of k. With the above procedure, a total of $(k' + 3)$ uniformly distributed random numbers must be generated to obtain one gamma distributed random number.

5.2.3 Discrete Random Variables

In the case of a discrete random variable, the CDF is the sum of the probability mass function (PMF). Suppose a random variable X with PMF $P(X = x_i) = p_X(x_i)$; $i = 0, 1, \ldots, n$.

The generation of random numbers from a discrete distribution is also derived from uniformly distributed random numbers. A value u is first generated from the standard uniform distribution. The corresponding discrete random number is then determined as follows. Form

$$p_X(x_0) + p_X(x_1) + \cdots + p_X(x_{j-1}) < u \le p_X(x_0) + p_X(x_1) + \cdots + p_X(x_j)$$

$$(5.13a)$$

or in terms of the CDF

$$F_X(x_{j-1}) < u \le F_X(x_j) \tag{5.13b}$$

then x_j is the corresponding value of the discrete random variable. Generation of discrete random numbers, by the method described above, requires the calculation of the CDF for most if not all possible values of the random variable; this is followed by a search for x_j each time a number u is generated.

The method presented above is basically the inverse-transform procedure described earlier in Section 5.2.2, except that numerical search is necessary to determine the inverse of the CDF.

An alternative procedure for generating random numbers from a discrete distribution was first suggested by Marsaglia (1962) and further developed by Abramowitz (1965). The method is specifically developed for computer generation of discrete random numbers.

EXAMPLE 5.8 (*Binomial Distribution*)

The general method outlined above in Eq. 5.13 may be used to generate values from a binomial distribution. Consider a case with $n = 4$ and $p = 0.3$; the probability mass function is

$$p_X(i) = \binom{4}{i}(0.3)^i(0.7)^{4-i}; \qquad i = 0, 1, \ldots, 4$$

The corresponding CDF is, therefore,

$$F_X(j) = \sum_{i=0}^{j} \binom{4}{i}(0.3)^i(0.7)^{4-i}; \qquad j = 0, 1, \ldots, 4$$

which are tabulated in Table E5.8 and also shown graphically in Fig. E5.8. Suppose the value of u generated is 0.7000. Since

$$F_X(1) < 0.7000 < F_X(2)$$

the corresponding value of X is $x = 2$, as indicated in Fig. E5.8.

For large n, the basic procedure described above may become time-consuming; in such cases, the normal distribution may be used as an approximation to generate random numbers with a binomial distribution. For $np > 10$ and $p \ge \frac{1}{2}$, or $n(1 - p) > 10$ and $p < \frac{1}{2}$, it can be shown (e.g., Rubinstein, 1981) that a binomial random variable X approaches a normal variate with mean $(np - \frac{1}{2})$ and standard deviation $np(1 - p)$. Therefore, to generate a value x from a

Table E5.8 PMF and CDF of a Binomial Distribution

i	$p_X(i)$	$F_X(i)$
0	0.2401	0.2401
1	0.4116	0.6517
2	0.2646	0.9163
3	0.0756	0.9919
4	0.0081	1.0000

binomial distribution, a value x' may first be generated from the normal distribution $N[np - \frac{1}{2}, np(1 - p)]$, and then examine the following:

(i)   If $x' \leq 0$, set $x = 0$.
(ii) If $x' \geq n$, set $x = n$.
(iii) If $0 < x' < n$, round off x' to the nearest integer and set x equal to that integer.

EXAMPLE 5.9   (Poisson Distribution)

For a Poisson-distributed random variable X with parameter $\lambda = vt$, Eq. 5.13 becomes

$$e^{-vt} \sum_{i=0}^{j-1} \frac{(vt)^i}{i!} < u \leq e^{-vt} \sum_{i=0}^{j} \frac{(vt)^i}{i!}$$

in which j is the value of the random variable.

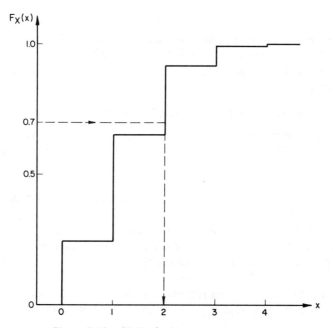

Figure E5.8   CDF of a binomial distribution.

Aside from the general method outlined in Section 5.2.3 for generating discrete random numbers, the following methods may also be used to generate values for a Poisson-distributed random variable.

Observe that the interarrival time (or distance between occurrences) of a Poisson process is exponentially distributed (see Chapter 3, Vol. I). Over a given time period (or distance) t, a sequence of interarrival times y_i may be generated from an exponential distribution with PDF

$$f_Y(y) = ve^{-vy}$$

according to the procedure of Example 5.1. Let these values be $y_i; i = 1, 2, \ldots$. Suppose

$$y_1 + y_2 + \cdots + y_i < t < y_1 + y_2 + \cdots + y_{i+1}$$

Then the appropriate value of the Poisson-distributed random variable is i, since i occurrences are observed within the period t, whereas the $(i + 1)$th occurrence will occur beyond the period t.

For large λ, say $\lambda > 10$, the Poisson distribution may be approximated also by the normal distribution (Rubinstein, 1981), with mean $(\lambda - \frac{1}{2})$ and standard deviation $\sqrt{\lambda}$. Therefore, in such a case, to generate a value x for a Poisson distributed random variable, a value x' may first be generated from the normal distribution $N[(\lambda - \frac{1}{2}), \sqrt{\lambda}]$ and

(i)   If $x' \leq 0$, set $x = 0$.
(ii) If $x' > 0$, round off x' to the nearest integer and set x equal to that integer.

5.2.4   Generation of Jointly Distributed Random Numbers

Let $X_1, X_2, \ldots, X_n$ be a set of n random variables. If the variables are statistically independent, the joint probability density function is

$$f_{X_1, X_2, \ldots, X_n}(x_1, \ldots, x_n) = \prod_{i=1}^{n} f_{X_i}(x_i) \tag{5.14}$$

where $f_{X_i}(x_i)$ is the marginal PDF of X_i. It is obvious, in this case, that the random numbers for each variate can be generated separately and independently of one another using the methods of Sections 5.2.2 and 5.2.3.

If the set of random variables $X_1, X_2, \ldots, X_n$ are dependent, the joint PDF may be expressed as

$$f_{X_1, \ldots, X_n}(x_1, \ldots, x_n) = f_{X_1}(x_1)f_{X_2}(x_2|x_1) \cdots f_{X_n}(x_n|x_1, \ldots, x_{n-1}) \tag{5.15}$$

where $f_{X_1}(x_1)$ is the marginal PDF of X_1, and $f_{X_k}(x_k|x_1, \ldots, x_{k-1})$ is the conditional PDF of X_k given $X_1 = x_1, \ldots, X_{k-1} = x_{k-1}$. The corresponding joint CDF is

$$F_{X_1, \ldots, X_n}(x_1, \ldots, x_n) = F_{X_1}(x_1)F_{X_2}(x_2|x_1) \cdots F_{X_n}(x_n|x_1, \ldots, x_{n-1}) \tag{5.16}$$

where $F_{X_1}(x_1)$ and $F_{X_k}(x_k|x_1, \ldots, x_{k-1})$ are the marginal and conditional CDF of X_1 and X_k, respectively.

In the case of dependent random variables, the required random numbers cannot be generated independently for each variable. However, with Eq. 5.16, the following provides a basis for generating the required set of dependent random numbers.

Suppose a set of uniformly distributed random numbers $(u_1, u_2, \ldots, u_n)$ has been generated. Then, a value x_1 may be determined independently as

$$x_1 = F_{X_1}^{-1}(u_1) \tag{5.17a}$$

With this value of x_1, the conditional CDF $F_{X_2}(x_2|x_1)$ is a function only of x_2, and hence a value x_2 may be determined from

$$x_2 = F_{X_2}^{-1}(u_2|x_1) \tag{5.17b}$$

Similarly, using the values $x_1, \ldots, x_{n-1}$ already obtained, we determine the value x_n as

$$x_n = F_{X_n}^{-1}(u_n|x_1, \ldots, x_{n-1}) \tag{5.17c}$$

Therefore, recursively the required set of dependent random numbers $(x_1, x_2, \ldots, x_n)$ can also be determined, one at a time. As implied in Eq. 5.17, the method will be efficient if the marginal and conditional CDFs can be inverted analytically.

EXAMPLE 5.10 (Bivariate Normal Random Numbers)

The joint PDF of a bivariate normal distribution for random variables X and Y can be written as

$$f_{X,Y}(x, y) = f_{Y|X}(y|x) f_X(x)$$

where the conditional PDF is (see Chapter 3, Vol. I)

$$f_{Y|X}(y|x) = \frac{1}{\sqrt{2\pi}\sigma_Y\sqrt{1-\rho^2}} \exp\left[-\frac{1}{2}\left(\frac{y - \mu_Y - \rho(\sigma_Y/\sigma_X)(x - \mu_X)}{\sigma_Y\sqrt{1-\rho^2}}\right)^2\right]$$

and the marginal PDF of X is

$$f_X(x) = \frac{1}{\sqrt{2\pi}\sigma_X} \exp\left[-\frac{1}{2}\left(\frac{x - \mu_X}{\sigma_X}\right)^2\right]$$

both of which are Gaussian, and μ_X, σ_X and μ_Y, σ_Y, are the means and standard deviations of X and Y, respectively, and ρ is the correlation coefficient. A number x is first generated from the normal random variable X with mean μ_X and standard deviation σ_X according to the procedure described in Example 5.5. Given this value of x, the conditional mean of Y is

$$E(Y|x) = \mu_Y + \rho\frac{\sigma_Y}{\sigma_X}(x - \mu_X)$$

whereas the conditional standard deviation of Y is

$$\sigma_{Y|x} = \sigma_Y\sqrt{1 - \rho^2}$$

A value y is then generated from the normal distribution with the above conditional mean and standard deviation. The pair of values (x, y) is therefore obtained from the bi-variate normal distribution.

5.2.5 Error Associated with Sample Size

Monte Carlo simulation is often used to estimate the probability of unsatisfactory performance of a system. It is desirable to know the error underlying an estimated probability; or more importantly, how many simulations (i.e., sample size) are required to obtain a certain accuracy. By approximating the binomial distribution with a normal distribution, Shooman (1968) has developed the following expression for the percent error

$$\% \text{ error} = 200 \sqrt{\frac{1 - p_F}{np_F}} \qquad (5.18)$$

where p_F is the estimated probability of unsatisfactory performance and n is the sample size. There is a 95% chance that the percent error in the estimated probability will be less than that given by Eq. 5.18. As an example, if 10,000 simulations were performed, obtaining an estimated probability of 0.01, Eq. 5.18 would yield a 20 percent error. Thus, it is 95% likely that the actual probability will be within 0.01 ± 0.002. On the other hand, if an accuracy of 0.01 ± 0.001 is desired, Eq. 5.18 can also be used to obtain the required number of simulations, which would be $n = 39,600$.

5.3   VARIANCE REDUCTION TECHNIQUES

Since Monte Carlo simulation is basically a sampling process, the results are subject to a sampling error that decreases with the sample size. In the case of Monte Carlo simulation, however, the error (or variance) may be reduced without increasing the sample size; procedures for this purpose are known as *variance reduction techniques*.

5.3.1   Antithetic Variates

Suppose Z' and Z'' are two unbiased estimators of Z; for example, from two separate samples. The two estimators may be combined to form another estimator

$$Z_A = \tfrac{1}{2}(Z' + Z'')$$

The expected value of Z_A, of course, is

$$E(Z_A) = \tfrac{1}{2}[E(Z') + E(Z'')] = \tfrac{1}{2}(Z + Z) = Z \qquad (5.19)$$

which means that Z_A is an unbiased estimator.

The corresponding variance is

$$\text{Var}(Z_A) = \tfrac{1}{4}[\text{Var}(Z') + \text{Var}(Z'') + 2\,\text{Cov}(Z', Z'')] \qquad (5.20)$$

If the estimators Z' and Z'' are statistically independent, for example, based on two separate and independent sets of random numbers,

$$\text{Var}(Z_A) = \tfrac{1}{4}[\text{Var}(Z') + \text{Var}(Z'')] \qquad (5.21)$$

However, if Z' and Z'' are negatively correlated; that is, $\text{Cov}(Z', Z'') < 0$, we would have

$$\text{Var}(Z_A) < \tfrac{1}{4}[\text{Var}(Z') + \text{Var}(Z'')] \qquad (5.22)$$

Therefore, the accuracy of the estimator Z_A can be improved, over that of the independent case, if Z' and Z'' are negatively correlated estimators. A numerical sampling procedure to insure negative correlation between Z' and Z'' is the following *antithetic variates* method (Hammersley and Morton, 1956).

Suppose that a set of n uniformly distributed random numbers $u_1, u_2, \ldots, u_n$ was generated to obtain the estimator Z'. The related set of random numbers $1 - u_1$, $1 - u_2, \ldots, 1 - u_n$ is subsequently generated to obtain Z''. The resulting Z' and Z'' will be negatively correlated.

The improvement in the accuracy of the above antithetic variates method will depend on the problem, as illustrated in the following examples.

EXAMPLE 5.11

The amount of pollutants remaining in an effluent after two successive waste treatment processes, as shown schematically in Fig. E5.11, may be modeled as

$$Y = X_1 X_2 Y_o$$

where Y_o is the amount of pollutants in the untreated influent; $(1 - X_1)$ is the fraction of the pollutant removed by treatment process I; and $(1 - X_2)$ is the fraction of the remaining pollutant removed by treatment process II. A quality indicator for the two processes may be defined as

$$Z = X_1 X_2$$

Suppose that X_1 and X_2 are statistically independent standard uniform variates; that is, $X_1 = U_1$ and $X_2 = U_2$. Estimating the mean value of Z using Monte Carlo simulation is desired.

Two sets of n uniformly distributed random numbers $u_{11}, u_{12}, \ldots, u_{1n}$ and $u_{21}, u_{22}, \ldots, u_{2n}$ are first generated. Then,

$$x'_{1i} = u_{1i}$$

$$x'_{2i} = u_{2i}$$

and

$$z'_i = u_{1i} \cdot u_{2i}; \qquad i = 1, 2, \ldots, n$$

Subsequently, calculate the corresponding antithetic sets of random numbers $1 - u_{11}$, $1 - u_{12}, \ldots, 1 - u_{1n}$ and $1 - u_{21}, 1 - u_{22}, \ldots, 1 - u_{2n}$, obtaining

$$z''_i = (1 - u_{1i})(1 - u_{2i}); \qquad i = 1, 2, \ldots, n$$

The combined estimator of Z, therefore, is

$$Z_A = \tfrac{1}{2}(Z' + Z'')$$

$$= \frac{1}{2n} \sum_{i=1}^{n} [U_{1i} U_{2i} + (1 - U_{1i})(1 - U_{2i})]$$

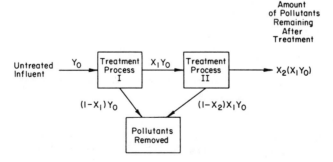

Figure E5.11   Two-stage waste treatment process.

whose variance is

$$\text{Var}(Z_A) = \frac{1}{4n^2} \text{Var}\left[\sum_i \{U_{1i}U_{2i} + (1 - U_{1i})(1 - U_{2i})\}\right]$$

$$= \frac{1}{4n^2} n \text{Var}[1 - U_{1i} - U_{2i} + 2U_{1i}U_{2i}]$$

$$= \frac{1}{4n} \{E(W^2) - E^2(W)\}$$

where $W = 1 - U_1 - U_2 + 2U_1U_2$. Observing that

$$E(W^2) = \int_0^1 \int_0^1 (1 + u_1^2 + u_2^2 + 4u_1^2 u_2^2 - 2u_1 - 2u_2 + 4u_1u_2 + 2u_1u_2$$

$$- 4u_1^2 u_2 - 4u_1 u_2^2)\, du_1\, du_2 = 0.278$$

and

$$E(W) = \int_0^1 \int_0^1 (1 - u_1 - u_2 + 2u_1u_2)\, du_1\, du_2 = 0.5$$

we have

$$\text{Var}(Z_A) = \frac{1}{4n} \{0.278 - 0.5^2\} = \frac{0.007}{n}$$

In the present case, if $2n$ independent pairs of random numbers were generated, the corresponding estimator would be

$$Z_B = \frac{1}{2n} \sum_{j=1}^{2n} X_{1j} X_{2j}$$

whose variance is given by

$$\text{Var}(Z_B) = \frac{1}{4n^2} \text{Var}\left[\sum_j X_{1j}X_{2j}\right] = \frac{1}{4n^2} \cdot n \text{Var}(X_{1j}X_{2j})$$

$$= \frac{1}{4n} \text{Var}(U_1U_2) = \frac{1}{4n} [E(U_1^2 U_2^2) - E^2(U_1U_2)]$$

$$= \frac{1}{4n} \left[\frac{1}{9} - \left(\frac{1}{4}\right)^2\right] = \frac{0.024}{n}$$

If we compare the above variances of Z_A and Z_B, the accuracy of the estimator Z_A is clearly better, demonstrating therefore that the antithetic variate method is superior to the method of random sampling.

EXAMPLE 5.12

Consider a transportation network of two connecting branches with respective travel times T_1 and T_2, such that the total travel time is

$$T = T_1 + T_2$$

Suppose T_1 follows a standard uniform distribution, whereas T_2 is exponentially distributed with a mean travel time of 1.0; T_1 and T_2 are statistically independent. It is desired to estimate the mean travel time $E(T)$ by Monte Carlo simulation.

Two sets of n random numbers $u_{11}, \ldots, u_{1n}$ and $u_{21}, \ldots, u_{2n}$ are first generated from the standard uniform distribution; then, with the results of Example 5.1, the values of T_1 and T_2 are

$$t'_{1i} = u_{1i}$$

$$t'_{2i} = -\ln u_{2i}$$

and

$$t'_i = u_{1i} - \ln u_{2i}; \qquad i = 1, 2, \ldots, n$$

With the corresponding sets of antithetic uniformly distributed random numbers, we obtain

$$t''_{1i} = 1 - u_{1i}$$

$$t''_{2i} = -\ln(1 - u_{2i})$$

and

$$t''_i = (1 - u_{1i}) - \ln(1 - u_{2i}); \qquad i = 1, 2, \ldots, n$$

The combined estimator is

$$Z_A = \tfrac{1}{2}(T' + T'')$$

$$= \frac{1}{2n} \sum_i [(T'_{1i} + T'_{2i}) + (T''_{1i} + T''_{2i})]$$

whose variance is

$$\mathrm{Var}(Z_A) = \frac{1}{4n^2} \cdot n \, \mathrm{Var}[(T_{1i} + T''_{1i}) + (T'_{2i} + T''_{2i})]$$

But

$$T'_{1i} + T''_{1i} = U_{1i} + (1 - U_{1i}) = 1$$

Therefore,

$$\mathrm{Var}(Z_A) = \frac{1}{4n} \, \mathrm{Var}(T'_{2i} + T''_{2i})$$

$$= \frac{1}{4n} [\mathrm{Var}(T'_{2i}) + \mathrm{Var}(T''_{2i}) + 2 \, \mathrm{Cov}(T'_{2i}, T''_{2i})]$$

Observe that

$$\mathrm{Var}(T'_{2i}) = \mathrm{Var}[-\ln U]$$

$$= \int_0^1 (\ln u)^2 \, du - \left[\int_0^1 -\ln u \, du \right]^2$$

$$= 2 - 1 = 1$$

$$\mathrm{Var}(T''_{2i}) = \mathrm{Var}[-\ln(1 - U)]$$

$$= \int_0^1 [\ln(1 - u)]^2 \, du - \left[\int_0^1 -\ln(1 - u) \, du \right]^2$$

$$= 2 - 1 = 1$$

and

$$\begin{aligned}
\text{Cov}(T'_{2i}, T''_{2i}) &= E[\{T'_{2i} - E(T'_{2i})\}\{T''_{2i} - E(T''_{2i})\}] \\
&= E[(T'_{2i} - 1)(T''_{2i} - 1)] \\
&= E[\{\ln U + 1\}\{\ln(1 - U) + 1\}] \\
&= 1 + \int_0^1 \ln(1 - u) \ln u \, du + \int_0^1 \ln(1 - u) \, du + \int_0^1 \ln u \, du \\
&= 1 + \left(2 - \frac{\pi^2}{6}\right) - 1 - 1 \\
&= 1 - \frac{\pi^2}{6}
\end{aligned}$$

Hence

$$\text{Var}(Z_A) = \frac{1}{4n}\left[1 + 1 + 2 - \frac{2\pi^2}{6}\right] = \frac{0.178}{n}$$

Alternatively, by random sampling, a set of $2n$ independent pairs of random numbers (u_{1j}, u_{2j}) are generated for T_1 and T_2. The estimator then is

$$Z_B = \frac{1}{2n}\sum_{j=1}^{2n}(T_{1j} + T_{2j})$$

whose variance would be

$$\begin{aligned}
\text{Var}(Z_B) &= \frac{1}{4n^2}(2n)\,\text{Var}(T_{1j} + T_{2j}) \\
&= \frac{1}{2n}[\text{Var}(T_{1j}) + \text{Var}(T_{2j})] \\
&= \frac{1}{2n}[\text{Var}(U) + \text{Var}(-\ln U)] \\
&= \frac{1}{2n}\left(\frac{1}{12} + 1\right) = \frac{0.542}{n}
\end{aligned}$$

which is more than three times the variance of Z_A. Therefore, for the same sample size, the application of the antithetic variate technique significantly improves the accuracy of the estimated expected travel time $E(T)$.

5.3.2 Correlated Sampling

Monte Carlo simulation may be used to study the difference in the performances between specific system designs. Suppose the performance of a design A is

$$Z_A = g(\mathbf{A}, \mathbf{X}) \tag{5.23}$$

where

$\mathbf{A} = (a_1, a_2, \ldots, a_m)$ is a set of design values for design A.

$\mathbf{X} = (X_1, X_2, \ldots, X_n)$ is a set of random variables.

Similarly, the performance of another design B is

$$Z_B = g(\mathbf{B}, \mathbf{X}) \qquad (5.24)$$

where

$\mathbf{B} = (b_1, b_2, \ldots, b_m)$ is a set of design values for design B.

The difference in performance between the two designs, therefore, is

$$Z = Z_A - Z_B \qquad (5.25)$$

Because Z_A and Z_B could be highly correlated, the method of *correlated sampling* may be used effectively to estimate the statistics of Z; for example, its mean-value, $\bar{Z}$. The variance of $\bar{Z}$ is

$$\text{Var}(\bar{Z}) = \text{Var}(\bar{Z}_A) + \text{Var}(\bar{Z}_B) - 2\,\text{Cov}(\bar{Z}_A, \bar{Z}_B)$$

Therefore, if Z_A and Z_B are positively correlated, that is, $\text{Cov}(Z_A, Z_B) > 0$,

$$\text{Var}(\bar{Z}) < \text{Var}(\bar{Z}_A) + \text{Var}(\bar{Z}_B) \qquad (5.26)$$

The random numbers z_{A_j} and z_{B_j} will be positively correlated if they are generated as follows:

$$z_{A_j} = g[\mathbf{A}, F_{X_1}^{-1}(u_1), F_{X_2}^{-1}(u_2), \ldots, F_{X_n}^{-1}(u_n)]$$
$$z_{B_j} = g[\mathbf{B}, F_{X_1}^{-1}(u_1), F_{X_2}^{-1}(u_2), \ldots, F_{X_n}^{-1}(u_n)]$$

where $(u_1, u_2, \ldots, u_n)$ is an independent set of uniformly distributed random numbers.

The method of *correlated sampling* is particularly effective for estimating the mean difference in system performance between proposed system designs.

EXAMPLE 5.13

Suppose the performance function of a system associated with design A is

$$Z_A = 3X_1 + 2X_2$$

and the corresponding performance function associated with design B is

$$Z_B = 2X_1 + 4X_2$$

that is, $\mathbf{A} = \{3, 2\}$ and $\mathbf{B} = \{2, 4\}$. Moreover, assume that X_1 and X_2 are uniformly distributed variates. Estimate the mean difference in performance between design A and design B.

Two sets of n uniformly distributed random numbers $u_{11}, u_{12}, \ldots, u_{1n}$ and $u_{21}, u_{22}, \ldots, u_{2n}$ are first generated for X_1 and X_2, respectively. Then, according to the method of correlated sampling,

$$z_{A_j} = 3u_{1j} + 2u_{2j}$$

and

$$z_{B_j} = 2u_{1j} + 4u_{2j}$$

The estimator of the mean difference is

$$\bar{Z} = \bar{Z}_A - \bar{Z}_B$$

$$= \frac{1}{n}\left[\sum_j (3U_{1j} + 2U_{2j}) - \sum_j (2U_{1j} + 4U_{2j})\right]$$

whose variance is

$$\mathrm{Var}(\bar{Z}) = \frac{1}{n^2}\left\{\mathrm{Var}\left[\sum_j (3U_{1j} + 2U_{2j})\right] + \mathrm{Var}\left[\sum_j (2U_{1j} + 4U_{2j})\right]\right.$$

$$\left. - 2\,\mathrm{Cov}\left[\sum_j (3U_{1j} + 2U_{2j}), \sum_j (2U_{1j} + 4U_{2j})\right]\right\}$$

$$= \frac{1}{n}\{[9\,\mathrm{Var}(U_1) + 4\,\mathrm{Var}(U_2)] + [4\,\mathrm{Var}(U_1) + 16\,\mathrm{Var}(U_2)]$$

$$- 2E[(3U_1 + 2U_2)(2U_1 + 4U_2)] + 2E(3U_1 + 2U_2)E(2U_1 + 4U_2)\}$$

$$= \frac{0.417}{n}$$

On the other hand, by random sampling, independent sets of uniformly distributed random numbers are generated for determining Z_A and Z_B, such that

$$\bar{Z} = \bar{Z}_A - \bar{Z}_B$$

$$= \frac{1}{n}\left[\sum_j (3U_{1j} + 2U_{2j}) - \sum_j (2U_{3j} + 4U_{4j})\right]$$

The corresponding variance may be shown to be

$$\mathrm{Var}(\bar{Z}) = \frac{2.750}{n}$$

Clearly, this is substantially higher than the variance of correlated sampling.

5.3.3   Control Variates

Suppose $\bar{Z}$ is an estimator. Sometimes, estimation accuracy may be gained through an indirect estimator $\bar{Y}$. For example,

$$\bar{Y} = \bar{Z} - \beta(C - \mu_C) \tag{5.27}$$

where C is a random variable with known mean μ_C and is correlated with $\bar{Z}$; β is a coefficient. C is called a *controlled variate* for $\bar{Z}$. Physically, C may represent the performance function of a crude model that has been simplified considerably to allow the determination of μ_C by analytical methods. Observe that

$$E(\bar{Y}) = E(\bar{Z}) - \beta[E(C) - \mu_C] = E(\bar{Z}) \tag{5.28}$$

Hence, if $\bar{Z}$ is an unbiased estimator, then $\bar{Y}$ is also an unbiased estimator. Moreover, the variance of $\bar{Y}$ is given by

$$\mathrm{Var}(\bar{Y}) = \mathrm{Var}(\bar{Z}) + \beta^2\,\mathrm{Var}(C) - 2\beta\,\mathrm{Cov}(\bar{Z}, C) \tag{5.29}$$

Therefore, if

$$2\beta \ \text{Cov}(\bar{Z}, C) > \beta^2 \ \text{Var}(C)$$

then

$$\text{Var}(\bar{Y}) < \text{Var}(\bar{Z})$$

implying that the indirect estimator $\bar{Y}$ is more accurate than the direct estimator $\bar{Z}$. In fact, the maximum reduction of variance may be achieved by selecting a value of β such that $\text{Var}(\bar{Y})$ is minimized; this is obtained as follows:

$$\frac{\partial \ \text{Var}(\bar{Y})}{\partial \beta} = 2\beta \ \text{Var}(C) - 2 \ \text{Cov}(\bar{Z}, C) = 0$$

From this the optimal value of β is

$$\beta^* = \frac{\text{Cov}(\bar{Z}, C)}{\text{Var}(C)} \tag{5.30}$$

and the corresponding minimum $\text{Var}(\bar{Y})$ is

$$\text{Var}(\bar{Y}) = \text{Var}(\bar{Z}) + \frac{\text{Cov}^2(\bar{Z}, C)}{\text{Var}^2(C)} \text{Var}(C) - \frac{2 \ \text{Cov}^2(\bar{Z}, C)}{\text{Var}(C)}$$

$$= \text{Var}(\bar{Z})[1 - \rho_{\bar{Z},c}^2] \tag{5.31}$$

where $\rho_{\bar{Z},c}$ is the correlation coefficient between $\bar{Z}$ and C. Observe that the reduction in variance increases as $\rho_{\bar{Z},c}$ increases. Therefore, if C can be selected so that it is heavily dependent on $\bar{Z}$, for example through a good approximate model, considerable improvement in the accuracy of estimation may be achieved. In order to apply the method of control variates, the value of β^* in Eq. 5.30 may have to be estimated from earlier simulations.

The control variate method may be generalized for a controlled vector $\mathbf{C} = (C_1, \ldots, C_q)$, consisting of q control variates. Equation 5.27 then becomes

$$\bar{Y} = \bar{Z} - \boldsymbol{\beta}^t(\mathbf{C} - \boldsymbol{\mu}_C) \tag{5.32}$$

where $\boldsymbol{\mu}_C$ is the vector of mean values of $C_j, j = 1, 2, \ldots, q$; and $\boldsymbol{\beta}$ is a vector of coefficients. The optimal vector $\boldsymbol{\beta}^*$ may be shown to be

$$\boldsymbol{\beta}^* = \boldsymbol{\sigma}_C^{-1} \begin{Bmatrix} \text{Cov}(\bar{Z}, C_1) \\ \vdots \\ \text{Cov}(\bar{Z}, C_q) \end{Bmatrix} \tag{5.33}$$

where $\boldsymbol{\sigma}_C$ is the covariance matrix of $\mathbf{C}$.

EXAMPLE 5.14 (excerpted from Gaver and Thompson, 1973)

Queueing models have been discussed in Chapter 3. Consider a single server queueing system. Suppose the interarrival time of the calling units is A with CDF $F_A(a)$, whereas the service time for each is S with corresponding CDF $F_S(s)$. Let W_i be the waiting time of the ith calling unit until it is served. For a system starting without any backlog, the waiting times may be determined with the following recursive relations:

$$W_0 = 0$$

$$W_i = \max(W_{i-1} - A_i + S_{i-1}, 0)$$
$$= (W_{i-1} - A_i + S_{i-1})^+; \quad i = 1, 2, \ldots$$

where A_i is the difference between the arrival time of the ith unit and that of the $(i - 1)$th unit; S_{i-1} is the service time for the $(i - 1)$th unit; $(x)^+$ denotes the function $\max(x, 0)$. It is desired to estimate the mean waiting time.

Random Sampling

Generate independently two sets of uniformly distributed random numbers $u_i^{(1)}$, $u_i^{(2)}$. The interarrival and service times are determined as

$$a_i = F_A^{-1}(u_i^{(1)})$$

$$s_i = F_S^{-1}(u_i^{(2)})$$

from which the waiting time of the ith unit is determined recursively as

$$W_0 = 0$$

$$W_i = [W_{i-1} - F_A^{-1}(u_i^{(1)}) + F_S^{-1}(u_{i-1}^{(2)})]^+$$

The above procedure is repeated n times such that n sample values of the waiting time for the ith unit are obtained. The estimator of the mean waiting time of the ith unit is then

$$\overline{W}_i = \frac{1}{n} \sum_{k=1}^{n} W_{ik}$$

where W_{ik} is the waiting time of the ith unit for the kth simulation. The variance of the estimator $\overline{W}_i$ is

$$\text{Var}(\overline{W}_i) = \frac{\text{Var}(W_i)}{n}$$

Antithetic Variate Technique

The first step associated with this technique is the same as that of random sampling, yielding the same value of W_i. However, the same sets of random numbers $u_i^{(1)}$ and $u_i^{(2)}$ are used again to obtain the corresponding antithetic sample values

$$a_i' = F_A^{-1}(1 - u_i^{(1)})$$

$$s_i' = F_S^{-1}(1 - u_i^{(2)})$$

and

$$w_i' = [w_{i-1}' - F_A^{-1}(1 - u_i^{(1)}) + F_S^{-1}(1 - u_i^{(2)})]^+$$

$$w_o' = 0$$

The entire procedure is then repeated n times. The estimator in this case is

$$\overline{W}_i = \frac{1}{n} \sum_{k=1}^{n} \frac{W_{ik} + W_{ik}'}{2}$$

where W_{ik} and W_{ik}' are the two waiting times of the ith unit for the kth simulation. The variance of this estimator can be shown to be

$$\text{Var}(\overline{W}_i) = \frac{1}{4n^2} \cdot n[\text{Var}(W_{ik}) + \text{Var}(W_{ik}') + 2\,\text{Cov}(W_{ik}, W_{ik}')]$$

$$= \frac{1}{2n} [\text{Var}(W_i) + \text{Cov}(W_i, W_i')]$$

Since W_i and W_i' are negatively correlated, $\text{Var}(\overline{W}_i)$ will be less than that of random sampling, as will be shown subsequently.

Control Variate Technique

For this problem, Gaver and Thompson (1973) suggested

$$C_0 = 0$$

$$C_i = C_{i-1} - A_i + S_{i-1}$$

be used as the control variates for the waiting time W_i. Observe that

$$C_1 = -A_1$$

$$C_2 = -A_1 - A_2 + S_1 = S_1 - (A_1 + A_2)$$

$$C_3 = (S_1 + S_2) - (A_1 + A_2 + A_3)$$

$$\vdots$$

$$C_i = (S_1 + S_2 + \cdots + S_{i-1}) - (A_1 + A_2 + \cdots + A_i)$$

Hence, C_i is simply an algebraic sum of the service times and interarrival times, whose probability distributions may be analytically determined. Furthermore, the mean value of C_i, namely μ_{C_i}, may be calculated as

$$\mu_{C_i} = \sum_{j=1}^{i-1} \mu_{S_j} - \sum_{j=1}^{i} \mu_{A_j}$$

$$= (i - 1)\mu_S - i\mu_A$$

According to Eq. 5.27, the indirect estimator is given by

$$\overline{Y}_i = \overline{W}_i - \beta(C_i - \mu_{C_i})$$

whose variance, Eq. 5.29, becomes

$$\text{Var}(\overline{Y}_i) = \text{Var}(\overline{W}_i) + \beta^2 \, \text{Var}(C_i) - 2\beta \, \text{Cov}(\overline{W}_i, C_i)$$

Numerical Results

In the numerical simulation, the service time S is assumed to be exponentially distributed with a mean of 1.11; whereas the interarrival time is assumed to be constant and equal to 1. Simulation is performed using the three methods described above. For the 200th calling unit, based on samples of 25 simulations (i.e., $n = 25$), the variance for the mean waiting time, associated with the different methods are as shown below:

Method	Variance of Estimator
Random sampling	0.427
Antithetic variate	0.071
Control variate ($\beta = 1$)	0.057

The above results indicate that substantial reduction in the variance of the estimator may be achieved with the antithetic variate or control variate technique.

5.4 APPLICATION EXAMPLES

The principal utility of Monte Carlo simulation is in the analysis of complex probability problems; that is, its major usefulness and advantage over analytical methods would be most apparent in problems where analytical models are mathematically intractable or (conversely) must be oversimplified. A number of such applications to major engineering systems are described below to illustrate specific simulation models and demonstrate the effectiveness of the Monte Carlo method.

The result of a Monte Carlo simulation is similar to that of a laboratory or field experiment, except that it is a synthetic or computer-generated experiment. In all cases, therefore, it may be reemphasized that irrespective of the complexity of the simulation model, it remains an *idealization of reality*, and as such the results are only as good (or credible) as the reliability of the assumptions underlying the model and the quality of the input information.

EXAMPLE 5.15

In Example 3.27, the corrosion of a pipeline was evaluated with the availability model. Analytic solutions were derived for exponential distributions. For general distributions, Monte Carlo simulation may be applied. Suppose the length of a corroded section is lognormally distributed with a mean of 100 ft and a c.o.v. of 50% (i.e., $\lambda = 4.72$; $\zeta = 0.47$); whereas the length of each noncorroded section is exponentially distributed with a mean of 400 ft.

Suppose that inspection at location A of the pipeline, as shown in Fig. E5.15a, reveals that no corrosion had occurred. It is desired to determine p, the probability that location B, 400 ft

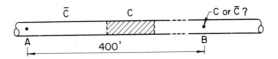

Figure E5.15a   A pipeline with corroded sections.

away from A, will be corroded. Furthermore, the fraction of the corroded sections of the pipeline between locations A and B, f_{AB} is also of interest.

Since the length of each corroded section is not exponentially distributed, analytical solutions become cumbersome. Monte Carlo simulation will be used to obtain estimates of p and f_{AB}.

The simulation procedure is proceeded as follows. Define c and $\bar{c}$ as the cumulative lengths of corroded and noncorroded sections, respectively. Initially, $c = \bar{c} = 0$. A value for the length of a noncorroded section to the right of location A is first generated from the exponential distribution with $\mu_c = 400$ ft, which is denoted as $\bar{c}_i$. The cumulative length of noncorroded sections is incremented by $\bar{c}_i$. The total length T of the pipeline from A is the sum of the current values of c and $\bar{c}$; that is, $T = c + \bar{c}$. At this stage, if $T > 400$ ft, it implies that location B is noncorroded and the fraction f_{AB} may be calculated as $c/400$. On the other hand, if $T < 400$ ft, a value for the length of the corroded section, c_i, is then generated from the lognormal distribution with $\lambda = 4.72$ and $\zeta = 0.47$. The cumulative length of the corroded sections is incremented by c_i. If the total pipe length is $T > 400$ ft, location B is corroded and the cumulative length of the corroded sections within the 400-foot section is simply 400 minus the current value of $\bar{c}$, which is then used to calculate f_{AB}. On the other hand, if $T < 400$ ft, another noncorroded section is generated and the process goes through another cycle until the total length, T, exceeds 400 ft. This process yields one simulated sample of a 400-foot long pipeline, with the following information:

(i)   Whether location B is corroded or not.
(ii) Fraction of corroded sections in the 400-foot pipeline, f_{AB}.

The above simulation process is performed for 100 400-foot pipelines, constituting, therefore, a sample of size 100, on the basis of which the probability of corrosion at B, namely p, is estimated. Moreover, the mean fraction of corroded sections between A and B, that is, $\bar{f}_{AB}$, may also be estimated with the sample mean of the 100 values of f_{AB}. The entire procedure is repeated 50 times, each with samples of 100, yielding histograms of p and $\bar{f}_{AB}$ as shown in the first of Fig. E5.15b.

To demonstrate the effect of the sample size in the estimation of p and $\bar{f}_{AB}$, the above simulation procedure is repeated using several other sample sizes; that is, 200, 500, and 1000. The respective histograms of p and $\bar{f}_{AB}$ with these other sample sizes are also shown in Fig. E5.15b. Clearly, decreasing scatters in p and $\bar{f}_{AB}$ are exhibited as the sample size increases. The same conclusion is also indicated in Fig. E5.15c, in which the sample standard deviations of p and $\bar{f}_{AB}$ are plotted as functions of the sample size (solid curves).

Monte Carlo results for p and $\bar{f}_{AB}$ were also obtained with the antithetic variate method, using a sample size of $n = 200$; these are shown as histograms in Fig. E5.15d. Comparing these

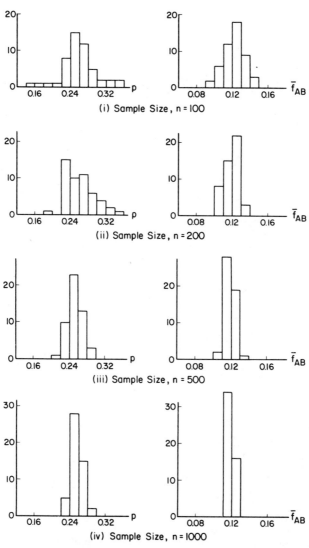

Figure E5.15b Histograms of p and f_{AB} as a function of sample size n.

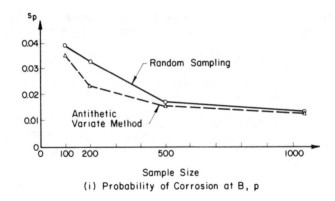

(i) Probability of Corrosion at B, p

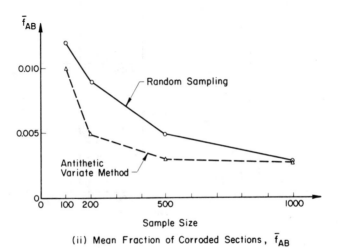

(ii) Mean Fraction of Corroded Sections, $\bar{f}_{AB}$

Figure E5.15c Sample standard deviation of p and f_{AB} as a function of sample size.

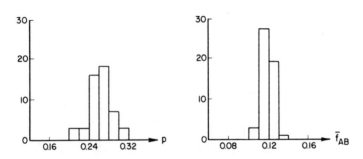

Figure E5.15d Histograms of p and $\bar{f}_{AB}$ based on the antithetic variate method (sample size, $n = 200$).

with the corresponding histograms of Fig. E5.15b, the effectiveness of this variance-reduction technique may be observed. The sample standard deviations of p and $\bar{f}_{AB}$ associated with the antithetic variate method are also plotted in Fig. E5.15c; these are clearly less than those of random sampling.

EXAMPLE 5.16

The shipping between two lakes, A and B, at different water elevations requires the use of a lock system, as shown in Fig. E5.16a. As a ship arrives at either end of the lock, she is processed immediately; however, if the lock system is occupied, she will have to wait for her turn. Suppose ships requiring the use of the lock arrive according to a Poisson process with a mean arrival rate of two per hour. Also, depending on the size of a ship and the skill of its operator, the time required to process it through the lock varies according to a gamma distribution with a mean of 20 minutes and a c.o.v. of 50%. Ships in either direction are processed through the lock system on a first-come-first-served basis and the lock system can process one ship at a time. Define T as the time a ship spends in the system consisting of the queueing time and the processing time through the lock system. It may be of interest to determine the following:

(i)   The mean time a ship spends in the system.
(ii) The probability that the total system time, T, will exceed some specified duration (e.g. one hour).

The operation of the above lock system is an example of the queueing model with one server (see Section 3.2.2). Since the process time is not exponentially distributed, it is difficult to analytically determine the mean value and distribution of T. Monte Carlo simulation may be used instead. The simulation procedure may be described as follows.

A set of exponentially distributed interarrival times for ships arriving at either end of the lock are generated, with a mean of 30 min. Based on these interarrival times, a list of the exact time of arrivals of the sequence of ships over a given operational period, say 100 hours, is obtained. Each ship on this list is then routed through the lock system according to the operation summarized in the flowchart shown in Fig. E5.16b. The lock is assumed to be available initially. Therefore, the first ship on the list is immediately processed through the lock. The process time is generated according to the gamma distribution with a mean of 20 min and a c.o.v. of 50%; that is, with parameters $v = 5$ and $k = 4$ (see Example 5.7). After recording the time this ship has spent in the system, the routing procedure returns to process the next ship on the list. During the 100 hours of operation, some of the ships will arrive when the lock is occupied. In this case, the ship will have to wait until the lock system becomes available. This

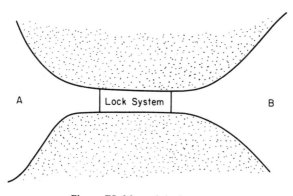

Figure E5.16a   A lock system.

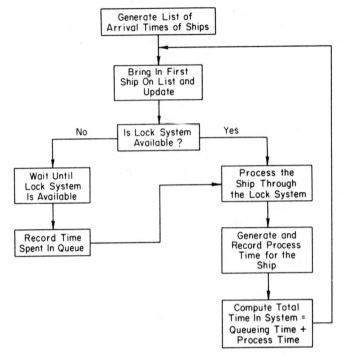

Figure E16b   Flowchart for simulation of lock system operation.

waiting time will be added to the process time to obtain the total time the ship spends in the system.

A conceivable scenario of the operation of the lock system is depicted in Fig. E5.16c for the first four hours of operation. A set of interarrival times (in minutes) of the ships was generated yielding 45, 18, 7, 53, 40, 21, 35, and 30. The corresponding arrival times (in minutes) at either end of the lock, relative to $t = 0$, are 45, 63, 70, 123, 163, 184, 219, 249... as shown in Fig. E5.16c. With the lock system idle at the beginning, the first ship is immediately processed

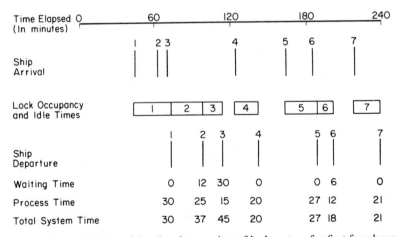

Figure E5.16c   Scenario of simulated operation of lock system for first four hours.

through the lock system. A value for the process time is then generated, which is 30 min. Therefore, the total system time for ship #1 is $T = 0 + 30 = 30$ min. Ship #2 arrives at $t = 63$ min. Since the lock system is occupied by ship #1 at the time the second ship arrives, she cannot be processed until $t = 75$. The waiting time for ship #2 is, therefore, $75 - 63 = 12$ min. A process time of 25 min is generated for ship #2; the total time ship #2 has spent in the system is thus $T = 12 + 25 = 37$ min.

A total of 168 ships arrived during the 100 hours of simulated operation. The histogram of the time spent in the system by these ships is shown in Fig. E5.16d, from which the probability that $(T > 60)$ may be estimated as $36/168 = 0.214$. Also, the sample mean and standard deviation of T are calculated to be 40.1 min and 29.1 min, respectively.

The above simulated operation is repeated 50 times. Based on these 50 repetitions of a 100-hour operation, histograms of the estimated exceedance probability p and of the mean system time T are plotted in Figs. E5.16e and E5.16f, respectively. The sample mean and standard deviation of p are calculated to be $\bar{p} = 0.225$ and $\sigma_p = 0.088$; whereas, the corresponding values for $\bar{T}$ are 43.13 min and 8.08 min, respectively.

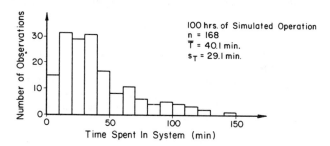

Figure 5.16d Histogram of total system time for a ship.

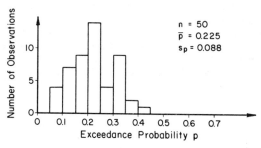

Figure 5.16e Histogram of exceedance probability (100 hrs of simulated operations).

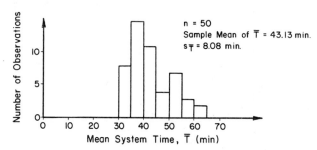

Figure 5.16f Histogram of mean system time (100 hrs of simulated operation).

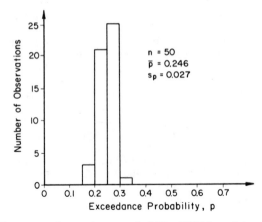

Figure 5.16g   Histogram of exceedance probability (1000 hrs of simulated operation).

To demonstrate the improvement in accuracy as the simulation time increases, the simulated operation was increased to 1000 hrs; and the simulations were also repeated 50 times. The corresponding histograms of the mean system time and exceedance probability are shown in Figs. E5.16g and E5.16h. It can be seen that the histograms are indeed narrower than those obtained earlier in Figs. E5.16e and E5.16f, indicating that the estimated mean time spent in the system and the exceedance probability are more accurate. However, if a very accurate estimate of p is required, for example, to within ± 0.01, a much longer simulation time will be necessary.

In order to reduce the bottleneck at the lock system, another parallel lock is proposed to be constructed as shown in Fig. E5.16i. Suppose each lock can process an incoming ship independently of the other, with an identical process time distribution, namely the gamma distribution with a mean of 20 min and c.o.v. of 50%. The proposed system then becomes a queueing model with two servers. Analytical solution of the time spent in the system is even more difficult to obtain than the case with one server. Again, Monte Carlo simulation is used. To illustrate the operation of the new parallel lock system, the same values of the interarrival times and process times generated for the single lock system illustrated earlier in Fig. E5.16c will be used.

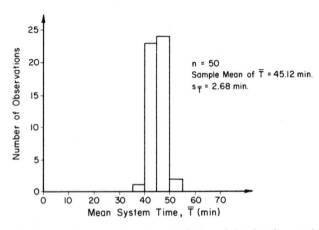

Figure 5.16h   Histogram of mean system time (1000 hrs of simulated operation).

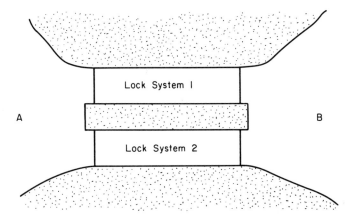

Figure E5.16i Two parallel lock systems.

The scenario of ship arrivals and occupancies of the parallel lock system for the first four hours of simulated operation is shown in Fig. E5.16j. The waiting times incurred previously by ships #2, 3, and 6 are now practically eliminated. Only ship #3 arrives with both locks occupied and has to wait for five minutes before lock system 1 becomes available.

The effectiveness of the proposed parallel lock system may be measured in terms of the reduction in the time spent by a ship in the system; that is,

$$Z = T_1 - T_2$$

where T_1 is the time spent by a ship in the single lock system, and T_2 is the corresponding time in the parallel lock system. In other words, Z is the saving in total system time of a ship.

The method of correlated variates, described in Section 5.3.2, may be used effectively to obtain estimates of the saving in mean system time, $\bar{Z}$. Basically, one set of random inter-arrival times and process times is generated in each 100 hours of simulated operation. The same

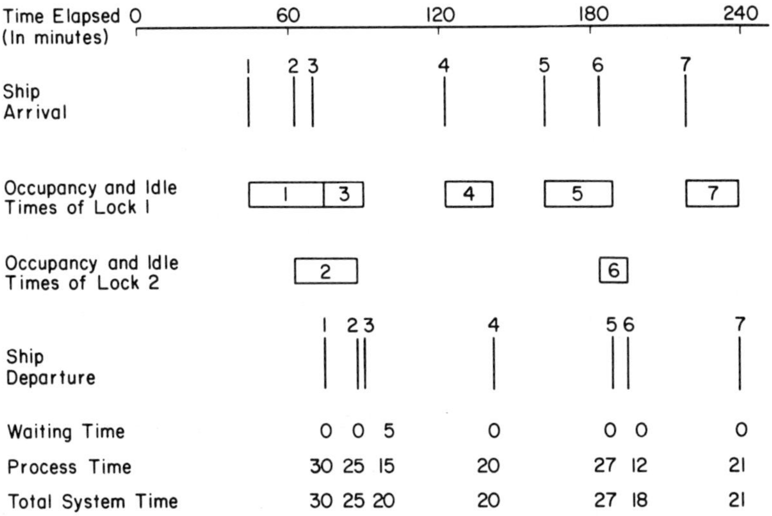

Time Elapsed 0 (In minutes)		60	120	180	240
Ship Arrival	1	2 3	4	5 6	7
Occupancy and Idle Times of Lock 1		1 3	4	5	7
Occupancy and Idle Times of Lock 2		2		6	
Ship Departure	1	2 3	4	5 6	7
Waiting Time		0 0 5	0	0 0	0
Process Time		30 25 15	20	27 12	21
Total System Time		30 25 20	20	27 18	21

Figure E5.16j Scenario of simulated operation of parallel lock system for first four hours.

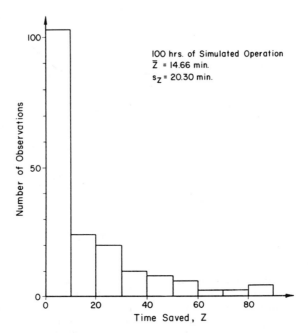

Figure E5.16k   Histogram of time saved by the parallel lock system.

set of random interarrival and process times is routed through the single lock system as well as the parallel lock system. The difference in the system time for the *i*th ship is $z_i = t_{1i} - t_{2i}$. A histogram of the time saved by a ship during a 100-hour simulated operation is shown in Fig. E5.16k. Again, 50 repetitions of the 100-hour operation were performed; the results of these 50 repetitions yield the histogram of $\bar{Z}$ as shown in Fig. E5.16l. According to these simulation results, the proposed parallel lock system could reduce the mean system time of the original single lock system by about 50%.

EXAMPLE 5.17

A simple ductile frame structure, shown in Fig. E5.17a, is subjected to a total gravity load W and an equivalent static earthquake load KW, where K is a seismic load coefficient. Assume that

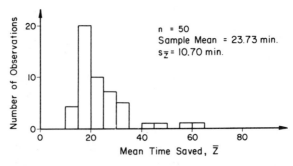

Figure E5.16l   Histogram of mean time saved by the parallel lock system (100 hrs of simulated operation).

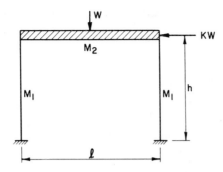

Figure E5.17a A simple frame structure.

the frame is built of ductile material with elastic-perfectly plastic behavior. Assume further that the failure of the frame through the formation of fully plastic hinge mechanisms, as shown in Fig. E5.17*b*, is of concern.

The fully plastic bending capacities of the columns, M_1, are identical and perfectly correlated, whereas the corresponding capacity of the beam, M_2, is partially correlated with M_1 with correlation coefficient $\rho_{1,2} = 0.8$. Each member is prismatic and the section capacities along the member are perfectly correlated.

All the variables are assumed to be normally distributed with the following statistics:

Variable	Mean	Standard Deviation
W	100 kips	0
K	0.3	0.1
M_1	300 ft-kip	45 ft-kip
M_2	450 ft-kip	45 ft-kip

with $\rho_{M_1, M_2} = 0.8$.

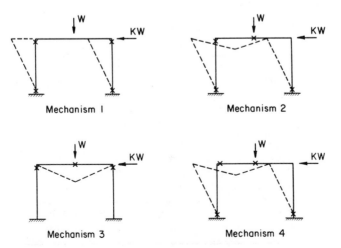

Figure E5.17b Major hinge mechanisms.

The major potential hinge mechanisms of the frame are indicated in Fig. E5.17b. By the *principle of virtual work*, the performance functions of the respective mechanisms can be shown to be as follows:

$$\text{Mechanism 1:} \quad g_1(X) = 4M_1 - KWh$$

$$\text{Mechanism 2:} \quad g_2(X) = 4M_1 + 2M_2 - KWh - W(l/2)$$

$$\text{Mechanism 3:} \quad g_3(X) = 2M_1 + 2M_2 - W(l/2)$$

$$\text{Mechanism 4:} \quad g_4(X) = 2M_1 + 4M_2 - KWh - W(l/2)$$

A number of other mechanisms are possible; however, the contributions of these other mechanisms to the total failure probability of the structure are less than those shown in Fig. E5.17b. The results presented subsequently, nevertheless, include the effects of the other mechanisms.

Simulation Procedure

The Monte Carlo calculations to evaluate the probability of failure are as follows:

(i) Generate random numbers for K and M_1 according to their respective normal distributions; based on the value for M_1, generate the corresponding correlated normal random number for M_2 (see Example 5.10).

(ii) For the particular set of values of K, M_1, and M_2 generated in (i), and with $W = 100$ kip, determine if any of the potential hinge mechanisms is formed; that is, $g_i(X) < 0$, for any i.

Steps (i) and (ii) are repeated n times, therefore constituting a sample of size n. If among this sample of size n, failure is observed n_f times, the probability of failure is the ratio n_f/n.

Results of Computer Calculations

The probability of failure for a structure with $h = 15$ ft and $l = 20$ ft, estimated with different sample sizes n, is partially summarized in Table E5.17.

Sample sizes ranging from 200 to 10,000 were used. For each sample size, the calculations were repeated 100 times; typical results for the first 10 repetitions are shown in Table E5.17. On the

Table E5.17 Estimated Failure Probability

Sample No.	Sample Size, n					
	200	500	1000	2000	5000	10,000
1	0.0300	0.0120	0.0150	0.0190	0.0160	0.0140
2	0.0100	0.0180	0.0120	0.0120	0.0152	0.0140
3	0.0050	0.0180	0.0170	0.0145	0.0134	0.0150
4	0.0150	0.0160	0.0150	0.0130	0.0138	0.0143
5	0.0250	0.0120	0.0150	0.0105	0.0138	0.0143
6	0.0200	0.0200	0.0150	0.0140	0.0152	0.0154
7	0.0000	0.0220	0.0100	0.0165	0.0168	0.0170
8	0.0005	0.0140	0.0220	0.0175	0.0128	0.0115
9	0.0050	0.0160	0.0180	0.0175	0.0170	0.0155
10	0.0250	0.0160	0.0130	0.0105	0.0150	0.0153
$\bar{p}_F{}^a$	0.0148	0.0137	0.0152	0.0147	0.0145	0.0151
$s_{p_F}{}^a$	0.0102	0.0054	0.0036	0.0026	0.0018	0.0011

[a] Based on 100 repetitions.

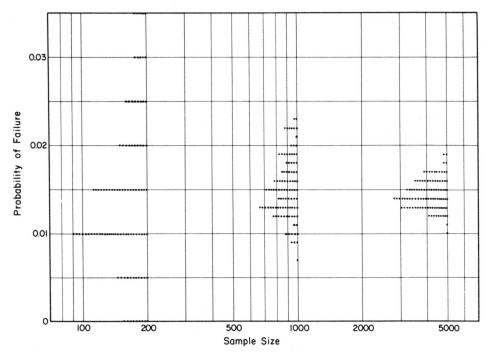

Figure E5.17c   Failure probability versus sample size.

basis of these 100 repetitions, the mean failure probability $\bar{p}_F$ and the associated standard deviation s_{p_F}, therefore, may be obtained; these are also shown in Table E5.17.

The results summarized in Table E5.17 are also shown graphically as histograms in Fig. E5.17c. The effect of the sample size is clearly demonstrated in Fig. E5.17c; in particular, the standard deviation of the failure probability decreases as the sample size increases.

EXAMPLE 5.18   Simulation of Air Traffic in a Two-Runway Airport (excerpted from Hall, 1974)

This example demonstrates the Monte Carlo simulation of airplane traffic through a busy two-runway airport, operating under Instrument Flight Rules. The general arrangement of the airport system is shown in Fig. E5.18a. An in-coming plane may be assigned to land on either

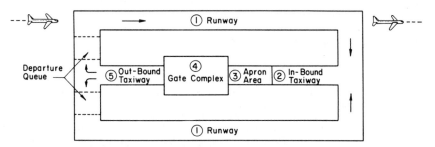

Figure E5.18a   General arrangement of airport system.

of the two runways ①, go through the in-bound taxiway ②, wait, if necessary, at the apron area ③, and proceed to the gate complex ④ for service. For departure, the plane advances to the out-bound taxiway ⑤, and waits for its turn to take off on one of the runways ①. The airport in this example is a large, modern jetport consisting of 100 loading gates, with room for 100 aircrafts in the apron queue, and an additional 100 in the departure queue.

The total time associated with an aircraft progressing through the system can be divided into the following components:

(1) Time between arrivals of aircrafts at the airport.
(2) Waiting time, if any, for the availability of a runway.
(3) Time between landing clearance and touchdown.
(4) Runway occupation time upon landing.
(5) Inbound taxiway service time.
(6) Waiting time, if any, at the apron queue for gate service.
(7) Loading gate service time.
(8) Time on out-bound taxiway.
(9) Waiting time, if any, at the departure queue for runway service.
(10) Runway occupation time during takeoff.
(11) Time between touchoff and the departing plane's clearoff from runway center line extension.

In addition to the waiting times (2, 6, and 9) that are functions of the traffic condition, all the other times are also random variables. Monte Carlo simulation may be used to generate simulated data on potential congestions associated with the airport operation and statistics on the delay time for a given aircraft. This information may then be used to determine the optimal scheduling of airport operations.

A flowchart outlining the simulation process of an aircraft arriving at and departing from the airport is shown in Fig. E5.18b.

Simulation Procedure

The simulation process may be described for an aircraft as it goes through the system.

Arrivals

An aircraft's arrival at the system is the time it is given a landing clearance. When an arrival occurs, the time interval to the next arrival is generated. This inter-arrival time between arriving aircrafts is assumed to follow the exponential distribution

$$f_T(t) = \lambda e^{-\lambda t}; \qquad t \geq 0$$

in which the parameter λ is governed by a diurnal profile of arrival rates shown in Fig. E5.18c. These data are estimates for the 1967–68 summer traffic at the JFK International Airport (Carlin and Park, 1969).

If a runway is available at the time of arrival, the arriving aircraft is given an immediate landing clearance and proceeds on final approach without delay. If neither runway is available, the aircraft joins whatever queue exists in the form of a "stack" and waits its turn. Aircrafts are served on a first-come first-served basis. The simulated time from arrival to landing clearance constitutes the aircraft's arrival delay.

Runway Service Intervals for Landings

An arriving aircraft occupies the runway for two consecutive time intervals; these intervals are generated when a landing clearance is issued.

The first runway service interval represents the time lapse from landing clearance until touchdown, denoted as L_1, whose CDF is given in Fig. E5.18d. During this interval the arriving aircraft has the exclusive use of the runway; in addition, it holds the alternate runway unavailable to subsequent arrivals. However, an arriving aircraft has no effect on departures from

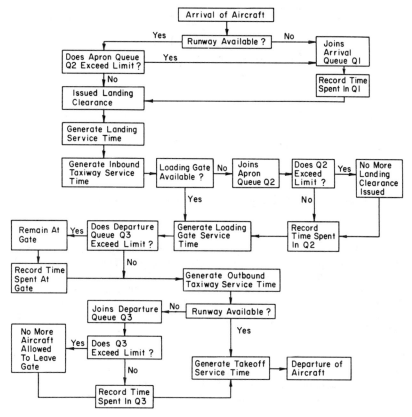

Figure E5.18b Flowchart of simulation process.

the alternate runway. Upon the completion of the first interval, the "temporary hold" on the alternate runway is released and the alternate runway becomes available to subsequent arrivals, unless it is occupied by an aircraft from an earlier arrival, or in the interim, it has become occupied by an aircraft in the early phase of departure. If the alternate runway becomes available for arrivals and such a function is in demand, a landing clearance will be issued and a set of runway service intervals generated for the next arriving aircraft.

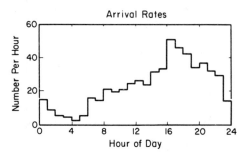

Figure E5.18c Diurnal profile of arrival rates.

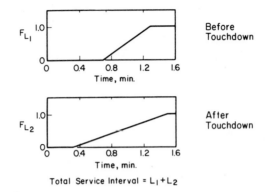

Total Service Interval = $L_1 + L_2$

Figure E5.18d Runway service interval distributions (Landing).

The second runway service interval represents the additional time required for an arriving aircraft to travel through the runway onto the inbound taxiway. This time is denoted by L_2, whose CDF is shown in Fig. 5.18d. During this interval, the runway just vacated becomes available to either arrivals or departures, except as these functions may be blocked by activities on the alternate runway. For example, an arrival in its first service interval on the alternate runway would prevent the runway just vacated from becoming available to subsequent arrivals at this time.

Inbound Taxiway

The taxiway service interval is generated at the time the arriving aircraft leaves the runway. It is uniformly distributed as shown in Fig. E5.18e. Any number of aircrafts may occupy the taxiway; although they are allowed to taxi at different speeds, passing is prohibited.

Apron Queue

Upon completion of the inbound taxiway service interval, the aircraft occupies a loading gate if one is available; if not, the aircraft joins the apron queue. This queue has a finite upper limit, namely 100 aircrafts in the present case. When the sum of this queue plus the current occupants of the inbound taxiway and the current inbound occupants of the runway is equal to the apron queue limit, no further landing clearances are issued until this congestion is reduced below the limit. During such times, the runways are available for departures only. Apron delay for an aircraft consists of the time spent in the apron queue plus any time lost on the inbound taxiway.

Loading Gate Service Interval

When an aircraft is assigned a loading gate, its gate service interval is generated. This time interval is assumed to have a shifted exponential distribution with a minimum of 20 min; the variation of the average time is as shown in Fig. E5.18f, depending on the hour of the day. This

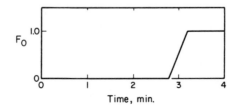

Figure E5.18e Inbound taxiway service interval distribution.

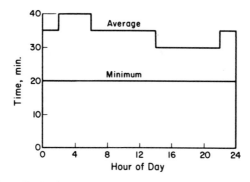

Figure E5.18f Diurnal profiles of gate occupation times.

diurnal profile is based on the assumption that an aircraft will tend to spend less time at the gate during busy periods then during slack times. When the gate service interval exceeds 40 min, the excess is counted as a delay at the gate.

Outbound Taxiway

An aircraft vacates its loading gate and enters the outbound taxiway upon completing its gate service interval. The outbound taxiway service interval is generated according to the uniform CDF shown in Fig. E5.18g. As with the inbound taxiway, any number of aircrafts may occupy the outbound taxiway and proceed at different speeds, but they may not pass each other.

Departure Queue

Upon expiration of the outbound taxiway service interval, an aircraft is given takeoff clearance and proceeds immediately if a runway is available for departures. If no runway is available, no clearance can be given and the aircraft joins the departure queue.

The departure queue has an upper limit, namely 100. When the sum of this queue and the outbound taxiway occupants equals this limit, no additional aircrafts are allowed to leave the gates until the congestion in the departure queue is relieved. The time spent in this manner at the loading gate plus the time in the departure queue and any time lost on the outbound taxiway constitute an aircraft's total departure delay.

Runway Service Interval for Takeoffs

When a departing aircraft is given a takeoff clearance, two consecutive runway service intervals are generated in the same manner as for an arrival following a landing clearance. These intervals are both uniformly distributed as shown in Fig. E5.18h.

The first runway service interval T_1 is the time lapse from takeoff clearance until the aircraft is airborne. During this time, a departing aircraft has exclusive use of the runway it is occupying; in addition, it holds the alternate runway unavailable to subsequent departures. A departing

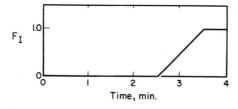

Figure E5.18g Outbound taxiway service interval distribution.

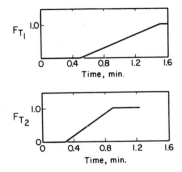

Figure E5.18h   Runway service interval distribution (takeoff).

aircraft, however, does not interfere with landing operations on the alternate runway. Upon completion of this interval, the alternate runway becomes available to subsequent departures, and the occupied runway becomes available to subsequent arrivals. To the extent that these functions are in demand, appropriate clearances will be issued at this time except for whatever blockage may have occurred as a result of additional operations not previously impeded. For example, an arrival in its first runway service interval on the alternate runway will preclude either clearance being issued at this time.

The second runway service interval, T_2, simulates an additional time period until the departing aircraft initiates it first course change, after vacating the runway center line extension. At this point, the departure is counted as complete and the departing aircraft is no longer within the control of the airport system. The runway so vacated now becomes available for subsequent departures unless it is otherwise restricted.

Computations and Results

With the simulation procedure outlined above, the landing and takeoffs of aircrafts are simulated in chronological order, making appropriate changes as each event occurs. The process may be used to simulate the activities of the airport for a day or number of days. The results will consist of profiles of aircraft activities, average system status and congestion at the end of each hour, and statistical distributions of the delays experienced during each hour.

Activities

Profiles for each phase of the airport activities are summarized in Fig. E5.18i. The levels shown are the averages from 100 repeated simulations for each hour.

System Status and Congestion

Hourly variations in the system's status and congestion are shown in Fig. E5.18j. Congestion develops whenever there is a difference in activities between two successive phases of operation. For example, the "stack" increases when there are more arrivals than landings.

Delays

The delays encountered by aircrafts, in terms of the median and the 90-percentile delays, are shown in Fig. E5.18k. The significance of the median curve is that half the aircrafts were delayed longer than the indicated time, whereas the 90-percentile curve means that only 10% of the aircraft population were delayed longer than the indicated delay time.

In this example, the arrival and departure congestions are slow to dissipate, and heavy delays are still being suffered several hours after the arrival rate has peaked and subsided. In this case, the relatively short ground dwell times assumed for all hours of the day result in considerable coincidence in the demand for both arrivals and departures. Conceivably, this

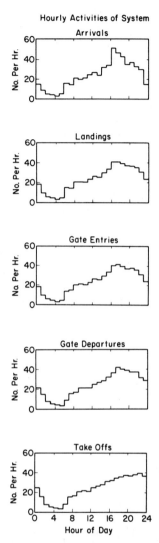

Figure E5.18i   Diurnal profiles of airport activities.

situation may be relieved by arranging the profile of ground dwell time so that departure demands will peak at a significantly different time from the peak arrival demand.

Importance of Repetitive Trials

Figure E5.18*l* presents the results for arrival delays for a particular hour of the day in which traffic is generally heavy. The distributions of delay for several trials are shown; it can be seen that the variation in delays between trials is significant, indicating that one trial is insufficient for determining the distribution of delay especially for a short time period. Also shown in Fig. E5.18*l* are the averages from five independent sets, each consisting of 100 trials. The scatter in the delay time distribution between the sets is dramatically reduced compared to the scatter between individual trials.

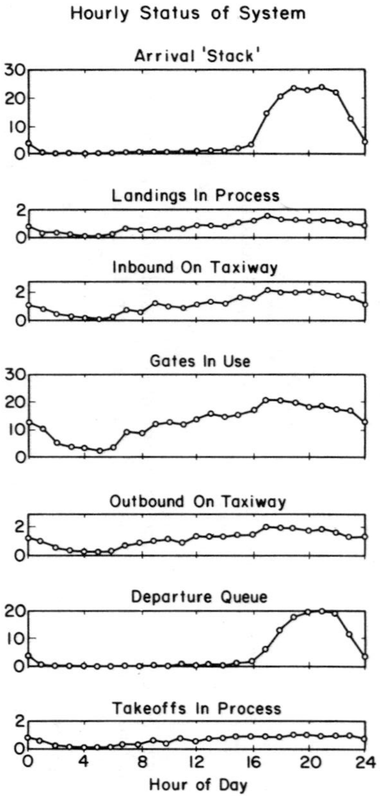

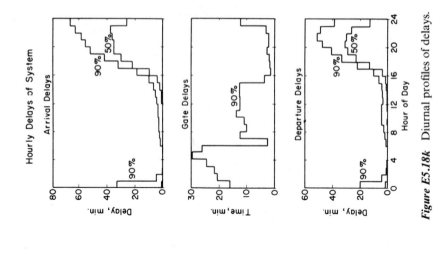

Figure E5.18k Diurnal profiles of delays.

Figure E5.18j Diurnal profiles of airport status.

Simulated Arrival Delays
During 2000–2100 Hour Interval

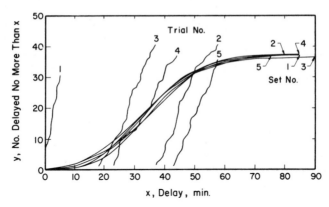

Figure E5.18l Comparison of single-trial results against averages of 100-trial sets.

EXAMPLE 5.19 Optimal Reservoir Design (excerpted from Fiering, 1961)

The design of a multipurpose reservoir (as shown in Fig. E5.19a) for irrigation, hydroelectric power generation, and flood control is considered for a given river basin. The performance of the reservoir may be the amount of available release each year. The available release depends heavily on the amount of inflow from the stream during the years. Moreover, the reservoir size and the operating policy in terms of the release (draft) level at various times and storage conditions will also affect the amount of annual available release. Unfortunately, the inflow from the stream exhibits considerable variability between the years; as a result, the amount of available release for a given reservoir size and the actual draft will also fluctuate between the years. Economic return from the reservoir will have to account for this variation.

Monte Carlo simulation may be used to obtain simulated data on the probability distribution of available release each year, which may be used to determine the optimal reservoir size. A critical step in the simulation study is the generation of a sequence of annual stream inflows and routing them through the reservoir system according to a set of operating rules, thus obtaining the annual draft probability distribution. For the sake of simplicity, seasonal draft patterns will not be considered.

In the following example, the unit of storage will be in millions of acre feet (maf), the draft in maf per year, and the cost and benefit in millions of dollars.

Inflow Probability Distribution

The annual inflow to the model reservoir system is assumed to follow a truncated normal distribution, with lower and upper limits of 0 and 4 units, respectively; the corresponding mean flow is 2 units, and the standard deviation is 1.4 units.

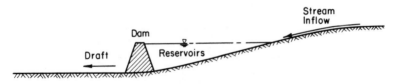

Figure E5.19a Proposed reservoir system.

Serial Correlation

If the level of ground water is drawn down because of drought conditions in a particular year, normal or excessive precipitation in the following year will tend first to replenish the ground water supply and not contribute significantly to stream flows. Conversely, if the level of ground water is elevated during a year, an unusually high proportion of subsequent precipitation will tend to run off into the streams. Thus, years of low flow will tend to be clustered together, as well as the years of high flow. A positive correlation, $\rho > 0$, therefore is used to describe the dependence of the inflows between two successive years.

Reservoir Size

Five discrete values of storage, namely 0, 1, 2, 3, and 4 units, are considered for the capacity of the reservoir, S_m. The case of $S_m = 0$ denotes no construction, whereas $S_m = 4$ denotes the largest feasible structure for the site of the model river basin. The active storage in the reservoir at the end of the ith year is denoted by S_i.

Operating Rules

Suppose I_i denotes the inflow in the ith year. The total amount of water available for distribution in the ith year is

$$W_i = I_i + S_{i-1}$$

Thus, the draft in the ith year, D_i, could not exceed W_i. Furthermore, because storage cannot exceed the reservoir capacity S_m, whenever $W_i > S_m$, an amount at least equal to $(I_i + S_{i-1} - S_m)$ has to be released in that year. Therefore, D_i has a minimum of zero or $(I_i + S_{i-1} - S_m)$, whichever is greater. The shaded portion of Fig. E5.19b delineates the zone of permissible drafts governed by these two natural bounds. It remains, however, to select an operating curve within this shaded region. Suppose an empirical rule is proposed as follows. Define a parameter D_L as the normal draft level. When the available amount of water in the ith year, W_i, is less than D_L, the reservoir is emptied. On the other hand, when W_i exceeds $D_L + S_m$, the reservoir is kept full. For intermediate values of W_i, the draft is set at D_L. In other words,

(i) For $S_{i-1} + I_i < D_L$; $D_i = S_{i-1} + I_i$
     (leave reservoir empty)

(ii) For $D_L < S_{i-1} + I_i < D_L + S_m$; $D_i = D_L$
     (constant draft)

(iii) For $I_i + S_{i-1} \geq D_L + S_m$; $D_i = S_{i-1} + I_i - S_m$
     (leave reservoir full)

The operating curve is shown (solid lines) in Fig. E5.19c.

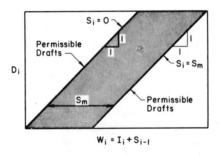

$$W_i = I_i + S_{i-1}$$

Figure E5.19b Zone of permissible drafts.

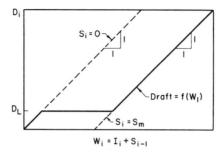

Figure E5.19c   Reservoir operating rules.

Simulation Procedure for Determining Draft Probability Distribution

The essence of the technique is the routing of a long hydrologic flow record through a proposed reservoir. Drafts are released in accordance with the prescribed operating rules, and a tally sheet is maintained to record the frequency of occurrence of particular draft levels. For simplicity, the draft is rounded off to the nearest million-acre-feet. Because long and complete stream gaging records are infrequently encountered as in this model river basin, synthetic hydrologic records may be generated by Monte Carlo simulation. The streamflow for the first year is simulated from the given inflow probability distribution. Given this generated value and the correlation coefficient ρ, the inflow for the subsequent year can be generated for jointly distributed normal random variables (see Example 5.10). The inflow value for the third year is similarly generated based on the inflow for the second year. If this process is repeated, a synthetic record of, say, 500 years is generated. For a given initial storage of 2 units, the annual draft and storage level of the reservoir at any subsequent years can be determined from the synthetic record of inflow I_i and routing through the operating rules specified earlier.

Table E5.19a summarizes the draft probabilities for a given synthetic record of 500 years, applied to the case $S_m = 4$, $D_L = 1$, and $\rho = 0.2$. For this assumed system, namely a reservoir size of 4 units and a normal draft of 1 unit, it is expected that in 47.8% of the years, the annual draft will be 1 unit; however, the annual drafts may also be 2, 3, or 4 units in 23%, 17.4%, and 11.8% of the time.

Five possible values of S_m (0, 1, 2, 3, 4) and three possible values of D_L (1, 2, 3) are considered as feasible design alternatives in this example. Each combination of S_m and D_L will yield a new table of draft probabilities associated with the given synthetic record of 500 years.

Benefit Functions

Recall that the model river basin provides three benefits: irrigation, power, and flood control. The gross benefit from irrigation can be computed as a function of the annual draft, such as that shown in Fig. E5.19d. The benefit function from irrigation may be expressed as

$$U_I = -(\alpha_{I2} - \alpha_{I1})D_s + \alpha_{I2}D; \qquad \text{for } D \le D_s$$
$$U_I = -(\alpha_{I3} - \alpha_{I1})D_s + \alpha_{I3}D; \qquad \text{for } D > D_s$$

where α_I's are the slopes of the lines shown in Fig. E5.19d. Observe that the utility increases generally with the draft; however, the rate of increase is smaller once D_s is reached. Furthermore, the benefit could be negative if the annual draft is very small. Conceptually, D_s may be

Table E5.19a   Draft Probabilities, $S_m = 4$, $D_L = 1$, $\rho = 0.2$

Draft (rounded)	0	1	2	3	4
Frequency	0	239	115	87	59
Probability	0.000	0.478	0.230	0.174	0.118

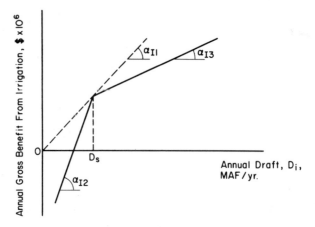

Figure E5.19d Typical benefit function from irrigation.

interpreted as the scale of development; D_s might be proportional to the total number of acres of land prepared for crops together with the cost of suitable structures for conveying and distributing the reservoir output. Once D_s is reached, a diminishing return will result from additional annual drafts.

Similarly, the benefit functions from hydroelectric power may be expressed as

$$U_p = -(\alpha_{p2} - \alpha_{p1})D_s + \alpha_{p2}D; \qquad \text{for } D \le D_s$$

$$U_p = -(\alpha_{p3} - \alpha_{p1})D_s + \alpha_{p3}D; \qquad \text{for } D > D_s$$

where α_p's are the slopes corresponding to the benefit functions for hydroelectric power. In this case, D_s might be associated with the generator capacity. Annual draft in excess of D_s will not yield significantly additional benefits.

In the present study, flood control benefits are assumed to be linearly proportional to the reduction in flood probability. For example, suppose a flow $D_i = 3.5$ is sufficiently large to cause flood damage in the model system. Let p_o be the probability that $D \ge 3.5$ prior to the construction of the reservoir. The probability that $D \ge 3.5$ after construction can be obtained from the draft probability distribution determined earlier for the assumed reservoir system. Hence, the benefit from flood control is given by

$$U_F = \alpha[p_o - P(D \ge 3.5)]$$

where α is the constant of proportionality.

In this example, values of α's used are $\alpha_{I1} = 1.0$, $\alpha_{I2} = 2.0$, $\alpha_{I3} = 0.2$, $\alpha_{p1} = 2.0$, $\alpha_{p2} = 3.0$, $\alpha_{p3} = 0.3$, and $\alpha = 55.2$.

Expected Annual Gross Benefits

For an assumed reservoir system, namely given S_m, D_L, and D_s, the annual gross benefit from irrigation, power supply, and flood control may be determined from the draft probability distribution and benefit function. For example, for the case $S_m = 4$, $D_L = 1$, and $D_s = 1$, with the results from Table E5.19a the expected annual gross benefit from irrigation is

$$E(U_I) = \sum_i U_I(D_i)P(D_i)$$

$$= U_I(0)P_{D_i}(0) + U_I(1)P_{D_i}(1) + \cdots + U_I(4)P_{D_i}(4)$$

$$= (-1.0)(0.0) + (1.0)(0.478) + (1.2)(0.230) + (1.4)(0.174) + (1.6)(0.118)$$

$$= 1.187$$

Similarly, the expected annual gross benefit from hydroelectric power is calculated to be 2.279. The expected benefit from flood control is

$$E(U_F) = 55.2(0.145 - 0.118) = 1.490$$

The expected annual gross benefit for all uses is assumed to be the sum of the three benefits considered independently. Social benefits, such as wildlife, fish and game preservation, and the recreational uses of the reservoir, are not considered in this model. Therefore, for the case $S_m = 4$, $D_L = 1$, and $D_s = 1$, the expected gross benefit is

$$E(U) = 1.187 + 2.279 + 1.490 = 4.956$$

But this sum must be converted to its present value in order to justify the present expenditure of funds that will generate the expected annual return over the economic life of the reservoir. Using 50 years as the system life and a discount rate of 2.5%, the present value of benefit is $4.956 \times 28.362 = 140.57$.

Costs

Costs are divided into two categories: reservoir cost and costs allocated to the specific uses. In this simple model, the reservoir cost is taken as a function of the reservoir capacity S_m. Likewise, it is assumed that the costs allocated to irrigation and hydroelectric power are functions only of the scale of development, D_s. The cost of flood control is not explicitly allocated in this model; it is implicitly included in the reservoir cost. All investments are assumed to be made in the beginning, whereas subsequent costs of operation, maintenance, and replacement are accounted for in the utility functions for the annual benefit described earlier. Table E5.19b summarizes the cost functions used in this example. For $S_m = 4$ and $D_s = 1$, the total cost is $51.05 + 5.67 + 8.51 = 65.23$.

Table E5.19b Reservoir, Irrigation, and Power Costs Versus Scale of Development, in $ Million

Size, S_m or D_s	0	1	2	3	4
Reservoir Cost	8.51	11.34	17.02	28.36	51.05
Irrigation Cost	0.00	5.67	11.34	17.02	22.69
Power Cost	0.00	8.51	17.02	25.53	34.03

Net Benefit and Optimal Design

The difference between the present value of the benefit and the total cost yields the present value of the expected net benefit. For the case $S_m = 4$, $D_s = 1$, and $D_L = 1$, the expected net benefit is $140.57 - 65.23 = 75.34$.

For each combination of S_m, D_s, and D_L, the expected net benefit is calculated. The single combination of design parameters that is expected to return the largest net benefit is then the optimal design, which is shown to be the following:

$$S_m = 3 \text{ maf}$$
$$D_L = 3 \text{ maf per yr}$$
$$D_s = 2 \text{ maf per yr}$$
$$\text{Total Cost} = \$56.74 \text{ million}$$
$$\text{Gross Annual Benefit} = \$329.97 \text{ million}$$
$$\text{Net Annual Benefit} = \$273.23 \text{ million}$$

PROBLEMS

5.1   During the critical period of a severe storm, the height of wave peaks, X, at a site may be assumed to follow a Rayleigh distribution whose PDF is

$$f_X(x) = \frac{x}{\alpha^2} e^{-1/2(x/\alpha)^2}; \qquad x \geq 0$$

where $\alpha = 10$m.

(a)   Devise a procedure for generating random wave height values based on the Inverse Transform method.

(b)   The square root of the sum-of-squares of two independent zero-mean normal random variables is also Rayleigh-distributed. Devise an alternative procedure for generating random wave heights based on this result.

5.2   (a)   Obtain a histogram of Z_i using Monte Carlo simulation with 1000 samples if

(i) $Z_1 = X - Y$, where X and Y are independent lognormal variates with medians 5 and 2, respectively, and a c.o.v. of 10% for both X and Y.

(ii) $Z_2 = X/Y$, where X and Y are independent normal variates with mean values 5 and 2, respectively, and a c.o.v. of 20% for both X and Y.

(b)   Repeat part (a) using a sample size of 10,000.

(c)   By repeating parts (a) and (b) 100 times, estimate the probability of $Z < 0$ for both functions Z_1 and Z_2.

5.3   Suppose X_1 and X_2 are two independent standard normal variates. If $Y_1 = X_1/X_2$,

(a)   Verify that Y_1 follows a Cauchy distribution with the following PDF

$$f_{Y_1}(y_1) = \frac{1}{\pi} \frac{1}{(1 + y_1^2)}; \qquad -\infty < y_1 < \infty$$

(*Hint*: Introduce $Y_2 = X_1 + X_2$ and apply the Jacobian transformation method.) On this basis, devise a procedure for generating random values for a Cauchy random variable.

(b)   Devise a procedure for generating random numbers from a Cauchy distribution based on the inverse transform method.

5.4.   (a)   Obtain a histogram of Y using Monte Carlo simulation, in which

$$Y = X_1 + X_2 + X_3 + X_4 + X_5 + X_6$$

where

X_1 follows an exponential distribution with mean 1.0.

X_2 follows a normal distribution with mean 1.0 and c.o.v. 0.2.

X_3 follows a lognormal distribution with mean 1.0 and c.o.v. 0.2.

X_4 follows a uniform distribution between 0 and 2.

X_5 follows a symmetrical triangular distribution between 0 and 2.0.

X_6 follows a Poisson distribution with mean 1.0.

Assume that the X_i's are statistically independent.

(b)   Suppose

$$Y = \sum_{i=1}^{10} (X_1 + X_2 + X_3 + X_4 + X_5 + X_6)_i$$

where the six random variables X_1 to X_6 follow the respective distributions as defined in part (a). Furthermore, the 10 sets of six random variables are statistically independent. Generate 50 values of Y. Verify that Y is approximately normally distributed (according to the Central Limit theorem) using the Kolmogorov–Smirnov test.

5.5   The occurrence of floods (assumed at most once a year) between years at a site may be modeled as a homogeneous Markov chain with the following transition probability matrix

$$\mathbf{P} = \begin{bmatrix} p_{11} & p_{12} \\ p_{21} & p_{22} \end{bmatrix}$$

where states 1 and 2 correspond to the events of no flood and flood, respectively, in a year. Because of insufficient records of previous floods, the probabilities p_{11} and p_{22} could not be accurately determined. Assume that p_{11} follows a triangular distribution between 0 and 1 with the mode at 0.9; and p_{22} is similarly distributed between 0 and 1 with the mode at 0.5. Suppose no flood occurred last year; formulate a Monte Carlo simulation to estimate the following:
(a)   Probability of having, at most, three floods in the next 20 years.
(b)   Mean number of years until the next flood occurs.

5.6   A simple structural system is shown in Fig. P5.6a in which a two-bar system is pulled by load W.

Figure P5.6a   Two-bar structural system.

(a)   Suppose R_1 and R_2 denote the capacities of the two bars.
For the PMFs shown in Fig. P5.6b and assuming brittle failure (i.e. rupture) of the

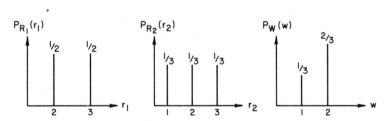

Figure P5.6b   PMF of R_1, R_2, and W.

bars, in which R_1, R_2, and W are statistically independent, estimate the probability of failure of the system using a Monte Carlo simulation and compare it with the corresponding analytical solution.
(b)   Suppose R_1 and R_2 are independent lognormal variates with mean values 2.5 and 2.0, and c.o.v. of 30% and 40%, respectively; W is also lognormally distributed with mean 2 and c.o.v. 25%. Calculate the probability of failure of the system using a Monte Carlo simulation.
(c)   Repeat part (b) if R_1 and R_2 are correlated with $\rho_{R_1, R_2} = 0.7$.

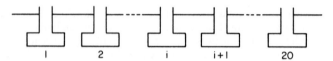

Figure P5.7   Sequence of footings.

5.7   A sequence of footings are shown in Fig. P5.7.

Let X_i be the settlement of the ith footing. Suppose the settlement of each footing follows a normal distribution with a mean of 3 in. and c.o.v. of 20%. Moreover, the correlation coefficient between X_i and X_{i+1} is 0.5, whereas there is negligible correlation between settlements of nonadjacent footings. Using Monte Carlo simulation, estimate:

(a)   The probability that the maximum settlement among the footings will exceed 5 in.

(b)   The probability that the maximum differential settlement between adjacent footings will exceed 2 in.

(c)   The probability that the maximum differential settlement among the 20 footings will exceed 2 in.

(d)   Repeat parts (a), (b), and (c) if the settlement between footings are statistically independent.

5.8   Suppose the random variables X and Y are jointly distributed with the following joint PDF:

$$f_{X,Y}(x, y) = \begin{cases} \frac{3}{4}y; & \text{if } x + y \le 2, x \ge 0; y \ge 0 \\ 0; & \text{otherwise} \end{cases}$$

(a)   Determine the marginal PDF of X, $f_X(x)$.

(b)   Determine the conditional PDF of Y given X, that is, $f_{Y|X}(y|x)$.

(c)   Derive the corresponding $F_X(x)$ and $F_{Y|X}(y|x)$.

(d)   Devise a Monte Carlo procedure for generating pairs of random values (x, y).

5.9   (a)   Estimate the probability of failure of the structural system shown in Fig. P5.9. The

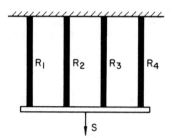

Figure P5.9   Parallel bar system.

system consists of four parallel bars. Suppose the load-carrying capacity of each bar is R_i, $i = 1$ to 4, and S is the applied load. Assume the load is always equally divided among the surviving (unfailed) bars. Assume the following statistics:

Variable	Distribution
R_1	$N(10, 2)$
R_2	$N(12, 3)$
R_3	$N(8, 2)$
R_4	$N(10, 2)$
S	$N(28, 4)$

Assume S and the R_i's are statistically independent.

(b) Repeat part (a) assuming R_1 and R_4 are perfectly correlated, whereas R_2 and R_3 are correlated with $\rho = 0.8$. Statistical independence is assumed between capacities of all other pairs of bars.

5.10 For the construction network shown in Fig. P5.10, the duration of each activity is a continuous random variable uniformly distributed over the range indicated on each branch.

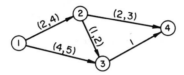

Figure P5.10   Activity network.

(a) Determine the distribution of completion time of the network using random sampling approach.
(b) Repeat part (a) using the antithetic variate approach.
(c) Estimate the probability that completion time will exceed six days using both random sampling and the antithetic variate approach and compare the sample variance of the estimate for a given sample size.

5.11 The event of overtopping of a dam because of extreme flood inflow may be analyzed with the following model:

(i) The geometrical dimensions of a reservoir and dam are shown in Figs. P5.11a, b, c. The dam is assumed to have a right triangular shape (Fig. P5.11b), and the reservoir is assumed to be a prism with triangular cross-section (Fig. P5.11c). The notations are:

H_D = Height of the top of the dam.

H_o = Initial reservoir elevation before the occurrence of a flood.

H_F = Maximum reservoir elevation during the occurrence of a flood.

H_P = Height of the spillway.

L_S = Length of the spillway.

H_S = Free board between spillway and top of dam = $H_D - H_P$.

Q_I = Flood peak discharge into the reservoir.

Q_S = Peak discharge at the spillway.

L = Length of the reservoir.

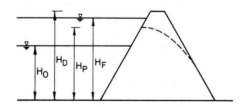

Figure P5.11a   Side view of reservoir and dam.

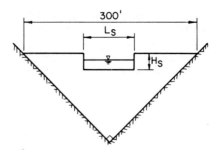

Figure P5.11b   End view of dam.

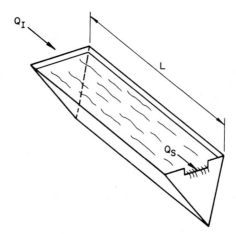

Figure P5.11c   Top view of reservoir.

(ii)   The inflow hydrograph into the reservoir is shown in Fig. 5.11*d*, which has a triangular shape with peak Q_I and duration T_D; T_P is the time at which peak flow Q_I occurs.

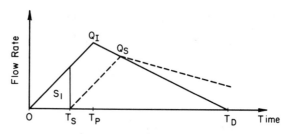

Figure P5.11d   Inflow and outflow hydrograph.

(iii)   The available reservoir storage S_1 for the flood inflow before flow at the spillway takes place can be calculated by the stage-storage curve which yields

$$S_1 = L(H_p^2 - H_0^2)$$

(iv)   Let T_S be the time until flow occurs at the spillway. If $T_S < T_P$, the outflow hydrograph at the spillway is shown in dotted lines in Fig. P5.11d. Line $T_S Q_S$ is assumed to be parallel to line OQ_I.
It can be shown that

$$Q_S = Q_I - \sqrt{\frac{S_1 Q_I}{T_D}} \qquad \text{for } T_S < T_P$$

On the other hand, if S_1 is large, T_S may exceed T_P, in which case the outflow hydrograph is shown in dotted lines in Fig. P5.11e. In this case,

$$Q_S = Q_I \sqrt{0.5 - \frac{S_1}{Q_I T_D}} \qquad \text{for } T_S \geq T_P$$

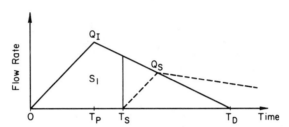

Figure P5.11e   Inflow and outflow hydrograph.

(v)   At peak outflow over the spillway, the relationship between the reservoir elevation H_F and the peak flow Q_S is governed by the spillway discharge formula as

$$Q_S = C_d L_S (H_F - H_P)^{3/2}$$

where C_d is the discharge coefficient.

(vi)   Overtopping occurs if the final reservoir elevation H_F exceeds the height of the dam H_D, that is, $H_F > H_D$.
Suppose the following data are given for the specific reservoir-dam system:

C_d: Uniformly distributed between 3.4 and 4.3.

H_P: Triangularly distributed between 140.25ft and 142.5ft with mode at 141.0ft.

Q_I: Log normally distributed with mean 1000 cfs and c.o.v. of 0.6.

T_D: Triangularly distributed between 0.1 and 3.5 days with mode at 0.5 day.

H_0: Uniformly distributed between 0 and 137ft.

$L = 1$ mile (deterministic).

$T_P = T_D/2.$

$H_D = 150$ft (deterministic).

L_S: Normally distributed with mean 20ft and c.o.v. of 5%.

(a)   Use Monte Carlo simulation to estimate the probability of overtopping during an extreme flood.
(b)   How would the estimated probability of overtopping in part (a) change if the mean spillway length is increased to 35ft; the c.o.v. of L_S remains at 5%.

5.12 Consider a soil stratum that is composed of silt and clay layers (e.g., varved clay formation). Two Markov chains could be used to generate the stratification of the soil stratum
(Ali et al, 1980). The first chain models the occurrences of silt (*S*) or clay (*C*) layers
along a vertical line. Suppose each stage of transition is taken to be 2.5mm, and the
one-stage transition probability matrix is

$$\mathbf{P}_1 = \begin{bmatrix} 0.3 & 0.7 \\ 0.1 & 0.9 \end{bmatrix}$$

where states 1 and 2 correspond to silt and clay, respectively.
(a) Simulate a vertical soil profile for a 30cm deep stratum and estimate the fraction of
silt.
(b) Estimate the mean and c.o.v. of the fraction of silt based on
(i) 10 simulated soil profiles.
(ii) 50 simulated soil profiles.
(c) A second Markov chain may be used to model the horizontal variations in the
thickness of each layer of silt. Suppose 25mm is taken as a stage in the horizontal
direction, the thickness of silt remains constant over each stage. The transitional
probability matrix between the thicknesses of two adjacent stages is given as
follows:

$$\mathbf{P}_2 = \begin{bmatrix} 1 & 0 & 0 & 0 & 0 & 0 \\ 0.1 & 0.7 & 0.2 & 0 & 0 & 0 \\ 0 & 0.15 & 0.7 & 0.1 & 0.05 & 0 \\ 0 & 0.05 & 0.15 & 0.6 & 0.15 & 0.05 \\ 0 & 0 & 0.1 & 0.2 & 0.6 & 0.1 \\ 0 & 0 & 0.05 & 0.15 & 0.3 & 0.5 \end{bmatrix}$$

where the six states of thickness at each stage correspond to 0, 2.5, 5.0, 7.5, 10.0,
and 12.5mm, respectively.

Besides the thickness of the silt layer at each stage, the elevation of the bottom
boundary is also required to define the horizontal profiles of each silt layer.
Suppose the change in the elevation of the bottom boundary of a silt layer at each
stage is also dependent on that of the previous stage according to the following
transitional probability matrix:

$$\mathbf{P}_3 = \begin{bmatrix} 0.3 & 0.6 & 0.1 \\ 0.1 & 0.8 & 0.1 \\ 0.1 & 0.6 & 0.3 \end{bmatrix}$$

where states 1, 2, and 3 correspond to the changes of the bottom elevation of
-2.5, 0, and 2.5mm, respectively, between two adjacent stages.

Combining the Markov chain model for initial thickness of each layer in part
(a) with the Markov chain models for the horizontal variation of layer thickness
and changes in the bottom elevation of the silt layer between stages, simulate a
soil profile that is 50cm wide and 30cm deep. Assume the bottom elevation of the
first stage is the same as that of the previous stage.

6. *Reliability and Reliability-Based Design*

6.1 RELIABILITY OF ENGINEERED SYSTEMS

One of the principal aims of engineering design is the assurance of system performance within the constraint of economy. Indeed, the assurance of performance, including safety, is primarily (if not solely) the responsibility of engineers. The achievement of this objective, however, is generally not a simple problem, particularly for large engineering systems. Engineering systems occasionally fail to perform their intended function, including rare instances of collapse of major structures. In this light, risk is generally implicit in all engineered systems; in fact, it has been stated (e.g., Pugsley, 1966) that "a profession that never has accidents is unlikely to be serving its country efficiently."

Most planning and design of engineering systems must be accomplished without the benefit of complete information; consequently, the assurance of performance can seldom be perfect. Moreover, many decisions that are required during the process of planning and design are invariably made under conditions of uncertainty. Therefore, there is invariably some chance of nonperformance or failure and of its associated adverse consequences; hence, risk is often unavoidable. Under such conditions, it is not feasible (practically or economically) to assure absolute safety or performance of engineered systems (Freudenthal, 1947). In the case of a structure, its safety is clearly a function of the maximum load (or combination of loads) that may be imposed over the useful life of the structure. Structural safety will also depend on the strength or load-carrying capacity of the structure or its components. As the lifetime maximum load and the actual capacity of a structure are difficult to predict exactly, and any prediction is subject to uncertainties, the absolute assurance of the safety of a structure is not possible. Realistically, safety (or serviceability) may be assured only in terms of the probability that the available strength (or structural capacity) will be adequate to withstand the lifetime maximum load.

In the case of hydrologic and water resources systems, the available supply of water (from all sources) relative to the maximum demand or usage of water is of concern in the planning and design of a system. The available supply from the different sources may be highly variable, whereas the actual usage may also fluctuate significantly, such that predictions in either case may be subject to significant uncertainties. Consequently, in the design of a water supply system, the question of adequacy of water is a matter of the probability of available supply relative to the demand.

Such reliability problems are also common to many other engineering systems; they involve a determination and assurance of the adequacy of a system over its useful life. The problem may be time-invariant, that is, the states of the system and of the requirements do not change with time; however, reliability problems may also be time-dependent, such as in the case of a structure whose capacity deteriorates with time and usage (e.g., due to fatigue).

In the sequel, we shall define *reliability* as the probabilistic measure of assurance of performance. That is, in light of uncertainty, the assurance of performance may be realistically stated only in terms of probability.

6.2   ANALYSIS AND ASSESSMENT OF RELIABILITY

6.2.1   Basic Problem

As alluded to in Section 6.1, the problems of reliability of engineering systems may be cast essentially as a problem of *supply* versus *demand*. In other words, problems of engineering reliability may be formulated as the determination of the (supply) capacity of an engineering system to meet certain (demand) requirements. In the consideration of the safety of a structure, we are concerned with insuring that the strength of the structure (supply) is sufficient to withstand the lifetime maximum applied load (demand). In the case of the reliability of a flood control system, our concern is the adequacy of the reservoir capacity (supply) to control the largest flood (demand) that may occur over the life of the system, whereas in the case of a water supply system the availability of water (supply) relative to the requirements for water (demand) is the pertinent problem.

Traditionally, the reliability of engineering systems is achieved through the use of factors or margins of safety and adopting conservative assumptions in the process of design; that is, by ascertaining that a "worst," or minimum, supply condition will remain adequate (by some margin) under a "worst," or maximum, demand requirement. What constitutes minimum supply and maximum demand conditions, however, is often defined on the basis of subjective judgments; also, the adequacy or inadequacy of the applied "margins" may be evaluated or calibrated only in terms of past experience with similar systems. The traditional approach is difficult to quantify and lacks the logical basis for addressing uncertainties; consequently, the level of safety or reliability cannot be assessed quantitatively. Moreover, for new systems in which there is no prior basis for calibration, the problem of assuring performance would obviously be difficult.

In reality, of course, the determination of the available supply as well as the determination of the maximum demand are not simple problems. Estimation and prediction are invariably necessary for these purposes; in these processes (which may include judgmental methods), uncertainties are unavoidable for the simple reason that engineering information is invariably incomplete. In light of such uncertainties, the available supply and actual demand cannot be determined precisely; they may be described as belonging to the respective ranges (or populations) of possible supply and demand. In order to explicitly represent or reflect the significance of uncertainty, the available supply and required demand may be modeled as *random variables*. In these terms, therefore, the reliability of a system may be more realistically measured in terms of probability. For this purpose, define the following random variables:

$$X = \text{The supply capacity.}$$

$$Y = \text{The demand requirement.}$$

Then, the objective of reliability analysis is to insure the event $(X > Y)$ throughout the useful life, or some specified life, of the engineering system. This assurance is possible only in terms of the probability $P(X > Y)$. This probability, therefore, represents a realistic measure of the reliability of the system; conversely, the probability of the complimentary event $(X < Y)$ is the corresponding measure of unreliability.

Assume, for the moment, that the necessary probability distributions of X and Y are available, that is, $F_X(x)$ or $f_X(x)$ and $F_Y(y)$ or $f_Y(y)$ are known. The required probabilities may then be formulated as follows:

$$p_F = P(X < Y) = \sum_{\text{all } y} P(X < Y | Y = y) P(Y = y) \tag{6.1}$$

If the supply and demand, X and Y, are statistically independent; that is,

$$P(X < Y | Y = y) = P(X < y)$$

Equation 6.1, for continuous X and Y, becomes (Freudenthal, Garrelt, and Shinozuka, 1966)

$$p_F = \int_0^\infty F_X(y) f_Y(y)\, dy \tag{6.2}$$

Equation 6.2 is the convolution with respect to y and may be explained also with reference to Fig. 6.1 as follows: If $Y = y$, the conditional probability of failure would be $F_X(y)$, but since $Y = y$ (or more precisely $y < Y \le y + dy$) is associated with probability $f_Y(y)\, dy$, integration over all values of Y yields Eq. 6.2. Alternatively, the reliability may be formulated also by convolution with respect to x, yielding

$$p_F = \int_0^\infty [1 - F_Y(x)] f_X(x)\, dx \tag{6.2a}$$

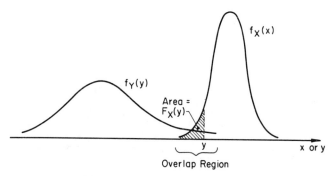

Figure 6.1   PDFs $f_X(x)$ and $f_Y(y)$.

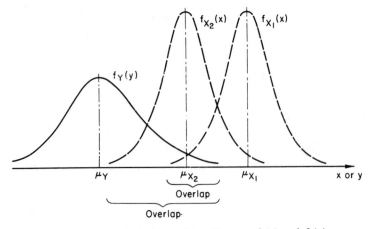

Figure 6.2a   Effect of relative position between $f_X(x)$ and $f_Y(y)$ on p_F.

The corresponding probability of nonfailure, therefore, is

$$p_S = 1 - p_F \tag{6.3}$$

As portrayed graphically in Fig. 6.1, the overlapping of the curves $f_X(x)$ and $f_Y(y)$ represents a qualitative measure of the failure probability p_F. In this regard, we observe the following:

(1)   The overlap region depends on the relative positions of $f_X(x)$ and $f_Y(y)$, as may be seen in Fig. 6.2a; that is, as the two curves become further apart, p_F decreases, whereas p_F increases as $f_X(x)$ and $f_Y(y)$ become closer. The relative position between $f_X(x)$ and $f_Y(y)$ may be measured by the ratio μ_X/μ_Y, which may be called the "central safety factor" or the difference $(\mu_X - \mu_Y)$ which is the mean "safety margin."

(2)   The overlap region also depends on the degree of dispersions in $f_X(x)$ and $f_Y(y)$, as shown in Fig. 6.2b—compare the overlap of the solid curves versus that of the dash curves. These dispersions may be expressed in terms of the c.o.v.'s δ_X and δ_Y.

In short,

$$p_F \sim g(\mu_x/\mu_Y; \delta_X, \delta_Y)$$

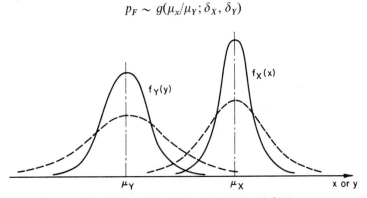

Figure 6.2b   Effect of dispersions in $f_X(x)$ and $f_Y(y)$ on p_F.

It follows, therefore, that any measure of safety or reliability properly ought to be a function of the *relative positions* of $f_X(x)$ and $f_Y(y)$ as well as of the *degree of dispersions.*

Theoretically, the failure probability p_F will also depend on the form of $f_X(x)$ and $f_Y(y)$. In practice, however, information is often limited; available information may be sufficient only to evaluate the main statistics (or first few moments) of X and Y, such as the mean values μ_X and μ_Y and the corresponding c.o.v.'s δ_X and δ_Y. The quantitative evaluation of the true p_F often poses major problems— for example, the determination of the correct forms of $f_X(x)$ and $f_Y(y)$ would be necessary which may not be a simple task.

In Eqs. 6.2 and 6.3, we assume that X and Y are statistically independent random variables. In general, however, these variables may be correlated; that is,

$$P(Y < X \mid X = x) \neq P(Y < x)$$

and

$$P(X < Y \mid Y = y) \neq P(X < y)$$

In such cases, the probability of failure may be expressed in terms of the joint PDF as follows:

$$p_F = \int_0^\infty \left[\int_0^y f_{X,Y}(x, y)\, dx \right] dy \qquad (6.4)$$

whereas the corresponding reliability is

$$p_S = \int_0^\infty \left[\int_0^x f_{X,Y}(x, y)\, dy \right] dx \qquad (6.5)$$

Margin of Safety   The above supply-demand problem may be formulated in terms of the *safety margin*, $M = X - Y$. As X and Y are random variables, M is also a random variable with corresponding PDF $f_M(m)$. In this case, failure is clearly the event $(M < 0)$, and thus the probability of failure is

$$p_F = \int_{-\infty}^0 f_M(m)\, dm = F_M(0) \qquad (6.6)$$

Graphically, this is represented by the area under $f_M(m)$ below 0, as shown in Fig. 6.3.

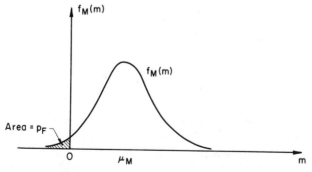

Figure 6.3   PDF of safety margin M.

EXAMPLE 6.1

Consider a structure whose strength (load-carrying capacity), R, is a normal random variable $N(\mu_R, \sigma_R)$. Similarly, the load is also normal $N(\mu_Q, \sigma_Q)$. The probability distribution of the safety margin $M = R - Q$ is also normal $N(\mu_M, \sigma_M)$, in which

$$\mu_M = \mu_R - \mu_Q$$

and, for statistically independent R and Q

$$\sigma_M^2 = \sigma_R^2 + \sigma_Q^2$$

Furthermore, $(M - \mu_M)/\sigma_M$ is $N(0, 1)$. Then, Eq. 6.6 yields

$$p_F = F_M(0) = \Phi\left(\frac{-\mu_M}{\sigma_M}\right) = 1 - \Phi\left(\frac{\mu_M}{\sigma_M}\right)$$

and

$$p_S = 1 - p_F = \Phi\left(\frac{\mu_M}{\sigma_M}\right)$$

It may be observed that the reliability is a function of the ratio μ_M/σ_M, which is the safety margin expressed in units of σ_M and may be called the *reliability index* or *safety index* (see Section 6.2.2) and denoted by β. In the present example,

$$\beta = \frac{\mu_M}{\sigma_M} = \frac{\mu_R - \mu_Q}{\sqrt{\sigma_R^2 + \sigma_Q^2}}$$

The probability of safety, therefore, becomes

$$p_S = \Phi(\beta)$$

and the corresponding probability of failure is

$$p_F = 1 - \Phi(\beta)$$

In the case of normal R and Q, the quantitative relation between the failure probability p_F and the reliability index β is as follows:

p_F	β
0.5	0
0.25	0.67
0.16	1.00
0.10	1.28
0.05	1.65
0.01	2.33
10^{-3}	3.10
10^{-4}	3.72
10^{-5}	4.25
10^{-6}	4.75

This illustrates the fact that the level of reliability is a function of both the relative position of $f_R(r)$ and $f_Q(q)$ as measured by the mean safety margin, $\mu_M = \mu_R - \mu_Q$, and the degree of dispersion as measured in terms of the standard deviation, $\sigma_M = \sqrt{\sigma_R^2 + \sigma_Q^2}$. Clearly, in this case a quantity that includes the effect of both these factors is the reliability index β, as indicated above.

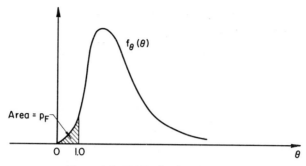

Figure 6.4   PDF of safety factor Θ.

The Factor of Safety   Another familiar term in engineering is the *factor of safety*, which is defined as

$$\Theta = X/Y$$

If the supply X and/or demand Y are random variables, the safety factor Θ will also be a random variable. Its distribution function, for example, PDF, may be derived from those of X and Y (see Chapter 4, Vol. I). In this case, failure would be the event $(\Theta < 1)$; the corresponding probability of failure, therefore, is

$$p_F = \int_0^1 f_\Theta(\theta)\, d\theta = F_\Theta(1.0) \tag{6.6a}$$

which is the area under $f_\Theta(\theta)$ between 0 and 1, as shown in Fig. 6.4.

EXAMPLE 6.2

For convenience of illustration, suppose that the structural strength and applied load, R and Q, in Example 6.1 are independent lognormal variates, with means and standard deviations μ_R, μ_Q and σ_R, σ_Q. The corresponding parameters of the lognormal distributions would be

$$\lambda_R = \ln \mu_R - \tfrac{1}{2}\zeta_R^2; \qquad \lambda_Q = \ln \mu_Q - \tfrac{1}{2}\zeta_Q^2$$

$$\zeta_R^2 = \ln(1 + \sigma_R^2/\mu_R^2); \qquad \zeta_Q^2 = \ln(1 + \sigma_Q^2/\mu_Q^2)$$

The safety factor $\Theta = R/Q$ is also a lognormal variate with parameters

$$\lambda_\Theta = \lambda_R - \lambda_Q$$

$$\zeta_\Theta^2 = \zeta_R^2 + \zeta_Q^2$$

Alternatively, in terms of the medians r_m and q_m,

$$\lambda_R = \ln r_m$$

$$\lambda_Q = \ln q_m$$

Then, $\lambda_\Theta = \ln r_m - \ln q_m = \ln r_m/q_m$. The ratio r_m/q_m may be called the "median safety factor."

Therefore, $\ln \Theta$ is normal with mean λ_Θ and standard deviation ζ_Θ; moreover, $(\ln \Theta - \lambda_\Theta)/\zeta_\Theta$ is $N(0, 1)$. Hence, according to Eq. 6.6a, the failure probability is

$$p_F = F_\Theta(1.0) = \Phi\left(\frac{\ln 1.0 - \lambda_\Theta}{\zeta_\Theta}\right) = \Phi\left(\frac{-\lambda_\Theta}{\zeta_\Theta}\right)$$

$$= 1 - \Phi\left(\frac{\ln r_m/q_m}{\sqrt{\zeta_R^2 + \zeta_Q^2}}\right)$$

In this case, the reliability is a function of the median safety factor $\theta_m = r_m/q_m$ and the standard deviation ζ_Θ. The ratio $\lambda_\Theta/\zeta_\Theta$ is also the reliability index; that is

$$\beta = \frac{\lambda_\Theta}{\zeta_\Theta}$$

6.2.2   Second-Moment Formulation

The calculation of the probability of safety, or probability of failure, requires the knowledge of the distributions $f_X(x)$ and $f_Y(y)$, or the joint distribution $f_{X,Y}(x, y)$. In practice, this information is often unavailable or difficult to obtain for reasons of insufficient data. Furthermore, even when the required distributions can be specified, the exact evaluation of the probabilities, generally requiring the numerical integration of Eqs. 6.1 through 6.5, may be impractical; as a practical alternative, equivalent normal distributions may be resorted to in approximation.

Not infrequently, the available information or data may be sufficient only to evaluate the first and second moments; namely, the mean values and variances of the respective random variables (and perhaps the covariances between pairs of variables). Practical measures of safety or reliability, therefore, must often be limited to functions of these first two moments. Under this condition, the implementation of reliability concepts must necessarily be limited to a formulation based on the first and second moments of the random variables—that is, restricted to the *second-moment formulation* (Cornell, 1969; Ang and Cornell, 1974). It may be emphasized that the second-moment approach is consistent also with the equivalent normal representation of nonnormal distributions (Section 6.2.4).

With the second-moment approach, the reliability may be measured entirely with a function of the first and second moments of the design variables; namely, the *reliability index*, β, when there is no information on the probability distributions; whereas, if the appropriate forms of the distributions are prescribed, the corresponding probability may be evaluated on the basis of equivalent normal distributions (see Section 6.2.4 and Appendix B).

Recall the safety margin $M = X - Y$. In this term, the "safe state" of a system may be defined as $(M > 0)$, whereas the "failure state" is $(M < 0)$. The boundary separating the safe and failure states is the "limit-state" defined by the equation $M = 0$.

Introduce the reduced variates

$$X' = \frac{X - \mu_X}{\sigma_X}$$

$$Y' = \frac{Y - \mu_Y}{\sigma_Y}$$

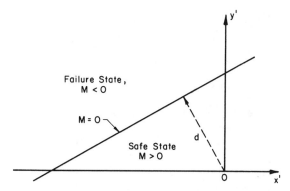

Figure 6.5 Space of reduced variates X' and Y'.

In the space of these reduced variates, the safe state and failure state may be represented as shown in Fig. 6.5. Also, in terms of the reduced variates, the limit-state equation, $M = 0$, becomes

$$\sigma_X X' - \sigma_Y Y' + \mu_X - \mu_Y = 0$$

which is a straight line as shown in Fig. 6.5. The distance from the (linear) failure line to the origin, 0, is in itself a measure of reliability; this distance, d, is given (from analytic geometry) as

$$d = \frac{\mu_X - \mu_Y}{\sqrt{\sigma_X^2 + \sigma_Y^2}}$$

Indeed, according to the results of Example 6.1, it can be observed that for normal X and Y, this distance d is also the safety index β; that is, $d = \beta$ and thus the reliability is $p_S = \Phi(d)$.

Generalization   The reliability of an engineering system may involve multiple variables. In particular, the supply and demand may, respectively, be functions of several other variables. For such cases, the supply-demand problem in Section 6.2.1 must be generalized. This generalization is often necessary in engineering, particularly when the problem must be formulated in terms of the basic design variables.

In a broader sense, the reliability of an engineering system may be defined as the probability of performing its intended function or mission. The level of performance of a system will obviously depend on the properties of the system. In this context and for the purpose of a generalized formulation, we define a *performance function*, or *state function*,

$$g(\mathbf{X}) = g(X_1, X_2, \ldots, X_n) \tag{6.7}$$

where $\mathbf{X} = (X_1, X_2, \ldots, X_n)$ is a vector of basic state (or design) variables of the system, and the function $g(\mathbf{X})$ determines the performance or state of the system. Accordingly, the limiting performance requirement may be defined as $g(\mathbf{X}) = 0$, which is the "limit-state" of the system.

It follows, therefore, that

$$[g(\mathbf{X}) > 0] = \text{the "safe state"}$$

and

$$[g(\mathbf{X}) < 0] = \text{the "failure state"}$$

Geometrically, the limit-state equation, $g(\mathbf{X}) = 0$, is an n-dimensional surface that may be called the "failure surface." One side of the failure surface is the safe state, $g(\mathbf{X}) > 0$, whereas the other side of the failure surface is the failure state, $g(\mathbf{X}) < 0$.

Hence, if the joint PDF of the design variables $X_1, X_2, \ldots, X_n$ is $f_{X_1, \ldots, X_n}(x_1, \ldots, x_n)$, the probability of the safe state is

$$p_S = \int \cdots \int_{\{g(\mathbf{x}) > 0\}} f_{X_1, \ldots, X_n}(x_1, \ldots, x_n)\, dx_1 \cdots dx_n$$

which may be written, for brevity, as

$$p_S = \int_{g(\mathbf{x}) > 0} f_{\mathbf{X}}(\mathbf{x})\, d\mathbf{x} \tag{6.8a}$$

Equation 6.8a is simply the volume integral of $f_{\mathbf{X}}(\mathbf{x})$ over the safe region $g(\mathbf{X}) > 0$. Conversely, the probability of the failure state, or failure probability, would be the corresponding volume integral over the failure region

$$p_F = \int_{g(\mathbf{x}) < 0} f_{\mathbf{X}}(\mathbf{x})\, d\mathbf{x} \tag{6.8b}$$

The evaluation of the probability p_S or p_F through Eq. 6.8a or 6.8b, however, is generally a formidable task.

For practical purposes, alternative methods of evaluating p_S or p_F (or its equivalent) are necessary. The remainder of this chapter, and part of Chapter 7, are devoted to the presentation and illustration of practical methods for calculating p_S or p_F.

Uncorrelated Variates   In general, the basic variables $(X_1, X_2, \ldots, X_n)$ in Eqs. 6.7 and 6.8 may be correlated. However, we consider first the case of uncorrelated variates; extension to correlated variates will follow subsequently.

Introduce the set of uncorrelated reduced variates (Freudenthal, 1956)

$$X_i' = \frac{X_i - \mu_{X_i}}{\sigma_{X_i}}; \qquad i = 1, 2, \ldots, n \tag{6.9}$$

Obviously, the safe state and failure state may also be portrayed in the space of the above reduced variates, separated by the appropriate limit-state equation. (In the two-variable case, this would be as shown in Fig. 6.6.) In terms of the reduced variates, X_i', the limit-state equation would be

$$g(\sigma_{X_1} X_1' + \mu_{X_1}, \ldots, \sigma_{X_n} X_n' + \mu_{X_n}) = 0 \tag{6.10}$$

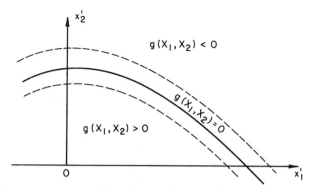

Figure 6.6 Safe and failure states in space of reduced variates.

Observe from Fig. 6.6 that as the limit-state surface (or failure surface), $g(\mathbf{X}) = 0$, moves further or closer to the origin, the safe region, $g(\mathbf{X}) > 0$, increases or decreases accordingly. Therefore, the position of the failure surface relative to the origin of the reduced variates should determine the safety or reliability of the system. The position of the failure surface may be represented by the minimum distance from the surface $g(\mathbf{X}) = 0$ to the origin of the reduced variates (Hasofer and Lind, 1974; Ditlevsen, 1979); indeed, Shinozuka (1983) has shown that the point on the failure surface with minimum distance to the origin is the *most probable failure point*.† Thus, in some approximate sense, this minimum distance may be used as a measure of reliability; the precise nature of this approximation is discussed in Section 6.2.4 for a general (nonlinear) performance function.

Following Shinozuka (1983), the required minimum distance may be determined as follows. The distance from a point $\mathbf{X}' = (X_1', X_2', \ldots, X_n')$ on the failure surface $g(\mathbf{X}) = 0$ to the origin of $\mathbf{X}'$ is, of course,

$$D = \sqrt{X_1'^2 + \cdots + X_n'^2} = (\mathbf{X}'^t\mathbf{X}')^{1/2}$$

The point on the failure surface, $(x_1'^*, x_2'^*, \ldots, x_n'^*)$, having the minimum distance to the origin may be determined by minimizing the function D, subject to the constraint $g(\mathbf{X}) = 0$; that is,

Minimize D

subject to $g(\mathbf{X}) = 0$.

For this purpose, the method of Lagrange's multiplier may be used. Let

$$L = D + \lambda g(\mathbf{X})$$

or

$$L = (\mathbf{X}'^t\mathbf{X}')^{1/2} + \lambda g(\mathbf{X})$$

---

† The notion of the "most probable failure point" was first introduced by Freudenthal (1956) for statistically independent normal variates.

In scalar notation,

$$L = \sqrt{X_1'^2 + X_2'^2 + \cdots + X_n'^2} + \lambda g(X_1, X_2, \ldots, X_n)$$

in which, $X_i = \sigma_{X_i} X_i' + \mu_{X_i}$.

Minimizing L, we obtain the following set of $n + 1$ equations with $n + 1$ unknowns

$$\frac{\partial L}{\partial X_i'} = \frac{X_i'}{\sqrt{X_1'^2 + X_2'^2 + \cdots + X_n'^2}} + \lambda \frac{\partial g}{\partial X_i'} = 0; \quad i = 1, 2, \ldots, n \quad (6.11)$$

and

$$\frac{\partial L}{\partial \lambda} = g(X_1, X_2, \ldots, X_n) = 0 \quad (6.12)$$

The solution of the above set of equations should yield the most probable failure point $(x_1'^*, x_2'^*, \ldots, x_n'^*)$.

Introduce the gradient vector

$$\mathbf{G} = \left(\frac{\partial g}{\partial X_1'}, \frac{\partial g}{\partial X_2'}, \ldots, \frac{\partial g}{\partial X_n'} \right)$$

in which

$$\frac{\partial g}{\partial X_i'} = \frac{\partial g}{\partial X_i} \frac{dX_i}{dX_i'} = \sigma_{X_i} \frac{\partial g}{\partial X_i}$$

The above set of equations, Eq. 6.11, can then be written in matrix notation as

$$\frac{\mathbf{X}'}{(\mathbf{X}'^t \mathbf{X}')^{1/2}} + \lambda \mathbf{G} = 0$$

from which

$$\mathbf{X}' = -\lambda D \mathbf{G} \quad (6.13)$$

Therefore,

$$D = [(\lambda D \mathbf{G}^t)(\lambda D \mathbf{G})]^{1/2} = \lambda D (\mathbf{G}^t \mathbf{G})^{1/2}$$

and thus,

$$\lambda = (\mathbf{G}^t \mathbf{G})^{-1/2}$$

Using this last result in Eq. 6.13 yields

$$\mathbf{X}' = \frac{-\mathbf{G}D}{(\mathbf{G}^t \mathbf{G})^{1/2}} \quad (6.14)$$

Conversely (premultiplying Eq. 6.14 by $\mathbf{G}^t$),

$$D = \frac{-\mathbf{G}^t \mathbf{X}'}{(\mathbf{G}^t \mathbf{G})^{1/2}} \quad (6.15)$$

Substituting Eq. 6.14 in Eq. 6.12 results in a single equation with the unknown D; solution of the resulting equation then yields the minimum distance $d_{min} = \beta$; thus,

$$\beta = \frac{-\mathbf{G}^{*t}\mathbf{X}'^*}{(\mathbf{G}^{*t}\mathbf{G}^*)^{1/2}} \qquad (6.15a)$$

in which $\mathbf{G}^*$ is the gradient vector at the most probable failure point $(x_1'^*, x_2'^*, \ldots, x_n'^*)$. In scalar form, Eq. 6.15a is

$$\beta = \frac{-\sum_i x_i'^*\left(\dfrac{\partial g}{\partial X_i'}\right)_*}{\sqrt{\sum_i \left(\dfrac{\partial g}{\partial X_i'}\right)_*^2}} \qquad (6.15b)$$

where the derivatives $(\partial g/\partial X_i')_*$ are evaluated at $(x_1'^*, x_2'^*, \ldots, x_n'^*)$. Using the above β in Eq. 6.14, the most probable point on the failure surface becomes

$$\mathbf{X}'^* = \frac{-\mathbf{G}^*\beta}{(\mathbf{G}^{*t}\mathbf{G}^*)^{1/2}} \qquad (6.14a)$$

In scalar form, the components of $\mathbf{X}'^*$, Eq. 6.14a, are

$$x_i'^* = -\alpha_i^*\beta; \qquad i = 1, 2, \ldots, n \qquad (6.16a)$$

in which

$$\alpha_i^* = \frac{\left(\dfrac{\partial g}{\partial X_i'}\right)_*}{\sqrt{\sum_i \left(\dfrac{\partial g}{\partial X_i'}\right)_*^2}} \qquad (6.16b)$$

are the direction cosines along the axes x_i'.

First-Order Interpretation The results derived above, Eqs. 6.14a and 6.15a, may be interpreted on the basis of first-order approximations for the function $g(\mathbf{X})$ as follows.

Expand the performance function $g(\mathbf{X})$ in a Taylor series at a point $\mathbf{x}^*$, which is on the failure surface $g(\mathbf{x}^*) = 0$; that is,

$$g(X_1, X_2, \ldots, X_n) = g(x_1^*, x_2^*, \ldots, x_n^*) + \sum_{i=1}^n (X_i - x_i^*)\left(\frac{\partial g}{\partial X_i}\right)_*$$

$$+ \sum_{j=1}^n \sum_{i=1}^n (X_i - x_i^*)(X_j - x_j^*)\Bigg/\left(\frac{\partial^2 g}{\partial X_i \partial X_j}\right)_* + \cdots$$

where the derivatives are evaluated at $(x_1^*, x_2^*, \ldots, x_n^*)$. But $g(x_1^*, x_2^*, \ldots, x_n^*) = 0$ on the failure surface; therefore,

$$g(X_1, X_2, \ldots, X_n) = \sum_{i=1}^n (X_i - x_i^*)\left(\frac{\partial g}{\partial X_i}\right)_*$$

$$+ \sum_{j=1}^n \sum_{i=1}^n (X_i - x_i^*)(X_j - x_j^*)\Bigg/\left(\frac{\partial^2 g}{\partial X_i \partial X_j}\right)_* + \cdots$$

Recall that

$$X_i - x_i^* = (\sigma_{X_i} X_i' + \mu_{X_i}) - (\sigma_{X_i} x_i'^* + \mu_{X_i}) = \sigma_{X_i}(X_i' - x_i'^*)$$

and

$$\frac{\partial g}{\partial X_i} = \frac{\partial g}{\partial X_i'} \left(\frac{dX_i'}{dX_i} \right) = \frac{1}{\sigma_{X_i}} \left(\frac{\partial g}{\partial X_i'} \right)$$

Then,

$$g(X_1, X_2, \ldots, X_n) = \sum_{i=1}^{n} (X_i' - x_i'^*) \left(\frac{\partial g}{\partial X_i'} \right)_* + \cdots$$

In first-order approximation, that is, truncating the above series at the first-order term, the mean value of the function $g(\mathbf{X})$, therefore, is

$$\mu_g \simeq - \sum_{i=1}^{n} x_i'^* \left(\frac{\partial g}{\partial X_i'} \right)_* \tag{6.17}$$

whereas the corresponding first-order approximate variance (for uncorrelated variates) is

$$\sigma_g^2 \simeq \sum_{i=1}^{n} \sigma_{X_i}^2 \left(\frac{\partial g}{\partial X_i'} \right)_*^2 = \sum_{i=1}^{n} \left(\frac{\partial g}{\partial X_i'} \right)_*^2 \tag{6.18}$$

From Eqs. 6.17 and 6.18, the ratio

$$\frac{\mu_g}{\sigma_g} = \frac{- \sum_{i=1}^{n} x_i'^* \left(\frac{\partial g}{\partial X_i'} \right)_*}{\sqrt{\sum_{i=1}^{n} \left(\frac{\partial g}{\partial X_i'} \right)_*^2}} \tag{6.19}$$

Comparing this with Eq. 6.15b, we see that the above ratio is the same as Eq. 6.15b, and thus μ_g/σ_g is also the distance from the tangent plane of the failure surface at $\mathbf{x}^*$ to the origin of the reduced variates. Therefore, the reliability index is also

$$\beta = \mu_g/\sigma_g \tag{6.20}$$

It may be emphasized that the first-order approximation of μ_g and σ_g derived above must be evaluated at a point on the failure surface $g(\mathbf{X}) = 0$. In earlier works (e.g., Cornell, 1969; Ang and Cornell, 1974) first-order approximations were evaluated at the mean values $(\mu_{X_1}, \mu_{X_2}, \ldots, \mu_{X_n})$, where significant errors could be incurred [for nonlinear $g(\mathbf{X})$] for the reason that the corresponding ratio μ_g/σ_g evaluated at the mean values may not be the distance to the nonlinear failure surface from the origin of the reduced variates.

Furthermore, first-order approximations evaluated at the mean values of the basic variates will give rise to the problem of invariance for equivalent limit states (Hasofer and Lind, 1974); that is, the result will depend on how a given limit-state event is defined. For example, for the equivalent limit-state events $(R - Q < 0)$ and $(R/Q < 1)$, the mean value first-order approximation will give different values of the safety index. Such an invariance problem is circumvented if the first-order approximations are evaluated at a point on the failure surface.

6.2.3 Linear Performance Functions

Consider a specialized class of performance functions, namely the linear performance function. Aside from its own usefulness, certain aspects of the linear case would be the basis for an approximation to nonlinear performance functions, as we shall see later in Section 6.2.4.

A linear performance function may be represented as

$$g(\mathbf{X}) = a_o + \sum_i a_i X_i$$

where a_o and a_i's are constants. The corresponding limit-state equation, therefore, is

$$a_o + \sum_i a_i X_i = 0 \tag{6.22}$$

In terms of the reduced variates, Eq. 6.9, the limit-state equation becomes

$$a_o + \sum_i a_i(\sigma_{X_i} X'_i + \mu_{X_i}) = 0 \tag{6.22a}$$

In three dimensions, Eq. 6.22a is

$$a_0 + a_1(\sigma_{X_1} X'_1 + \mu_{X_1}) + a_2(\sigma_{X_2} X'_2 + \mu_{X_2}) + a_3(\sigma_{X_3} X'_3 + \mu_{X_3}) = 0$$

which is a plane surface in the x'_1, x'_2, x'_3 space as shown in Fig. 6.7.

The distance of the failure plane, Eq. 6.22a, to the origin of the reduced variates $\mathbf{X}'$ is

$$\beta = \frac{a_o + \sum_i a_i \mu_{X_i}}{\sqrt{\sum_i (a_i \sigma_{X_i})^2}} \tag{6.23}$$

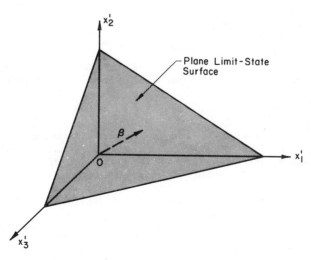

Figure 6.7 Linear limit-state surface in x'_1, x'_2, x'_3 space.

Equation 6.23 can also be obtained directly from Eq. 6.15b. In Section 6.2.2, we saw that for two uncorrelated normal variates, X and Y, the probability of the safe state, p_S, is directly a function of the distance of the failure line to the origin of the reduced variates; that is, the distance β. This result can be generalized— that is, if the random variables $X_1, X_2, \ldots, X_n$ are uncorrelated normal variates, the probability of the safe state is (in the present case)

$$
\begin{aligned}
p_S &= P\left(a_o + \sum_i a_i X_i > 0\right) \\
&= 1 - \Phi\left[\frac{-(a_o + \sum a_i \mu_{X_i})}{\sqrt{\sum (a_i \sigma_{X_i})^2}}\right] \\
&= \Phi\left[\frac{a_o + \sum a_i \mu_{X_i}}{\sqrt{\sum (a_i \sigma_{X_i})^2}}\right]
\end{aligned}
\tag{6.24}
$$

Comparing Eqs. 6.23 and 6.24, we see that the argument inside the brackets of Eq. 6.24 is the distance β. Therefore, the probability p_S is again a function of the distance from the failure plane $g(\mathbf{X}) = 0$ to the origin of the reduced variates. Therefore, in the general case of n uncorrelated normal variates, the probability of safety is

$$
p_S = \Phi(\beta)
\tag{6.24a}
$$

EXAMPLE 6.3   (Structural Elements)

The safety of a structural element may be evaluated on the basis of a linear performance function. For example,

$$
g(\mathbf{X}) = R - Q
$$

where:

R = The resistance of the element.
Q = The total load effect on the element; e.g. $Q = D + L$, in which
D = The dead load effect.
L = The lifetime maximum live load effect.

Indeed, the recommendation for the design of reinforced concrete structures of the American Concrete Institute, ACI-318(81) is implicitly based on such a linear performance function; namely, the requirement is

$$
0.9R_n \geq 1.4D_n + 1.7L_n
$$

where R_n, D_n, and L_n are the respective nominal values of resistance and loads.

Structural engineers often use "nominal" values of loads and resistances, which are generally different from the corresponding mean values. For example, according to Galambos et al. (1982), the ratios of the mean loads to the respective specified nominal loads for office buildings are as follows:

$$
\frac{\overline{D}}{D_n} = 1.05
$$

$$
\frac{\overline{L}}{L_n} = 1.15
$$

whereas in the case of the flexural capacity of reinforced concrete beams, the corresponding ratio is

$$\frac{\bar{R}}{R_n} = 1.05$$

The corresponding c.o.v.'s associated with these variables are (Galambos et al., 1982)

$$\Omega_R = 0.11$$
$$\Omega_D = 0.10$$
$$\Omega_L = 0.25$$

The nominal dead load effect D_n is due to the weight of the structure; this would include the partitions and permanent equipment. However, the nominal live load is usually specified by codes or standards for design. For buildings, the live load intensity, L_o, in the U.S. is specified by the American National Standards Institute, for example, ANSI A5.8 (1972). A load-reduction is also permitted yielding a nominal live load as follows:

$$L_n = L_o \left\{ 1 - \min\left[0.0008 A_T; 0.23\left(1 + \frac{D_n}{L_o}\right); 0.6 \right] \right\}$$

in which

A_T = The tributary floor area.

$\dfrac{L_o}{D_n}$ = The live load-dead load ratio.

For a live load-dead load ratio of $L_o/D_n = 1.0$ and a tributary area $A_T = 400$ ft^2, the above load-reduction factor would be

$$\frac{L_n}{L_o} = 1 - 0.32 = 0.68$$

In summary, therefore, the reliability underlying a reinforced concrete beam in an office building designed in accordance with the ACI-318(81) provision and the ANSI A5.8(72) load specification is as follows:

$$\frac{\bar{L}/1.15}{\bar{D}/1.05} = 0.68 \qquad \text{or} \qquad \bar{L} = 0.745\bar{D}$$

whereas the ACI 318(81) requirement becomes

$$0.9 \frac{\bar{R}}{1.05} = 1.4 \frac{\bar{D}}{1.05} + 1.7 \frac{\bar{L}}{1.15}$$

from which the required mean resistance is

$$\bar{R} = 2.831\bar{D}$$

Therefore, according to Eq. 6.23, the safety index of the beam is

$$\beta = \frac{\bar{R} - \bar{D} - \bar{L}}{\sqrt{\sigma_R^2 + \sigma_D^2 + \sigma_L^2}}$$

$$= \frac{2.831\bar{D} - \bar{D} - 0.745\bar{D}}{\sqrt{(0.11 \times 2.831\bar{D})^2 + (0.10\bar{D})^2 + (0.25 \times 0.745\bar{D})^2}}$$

$$= 2.885$$

If the variables can be assumed to be normal, the underlying probability of nonperformance or failure would be

$$p_F = 1 - \Phi(2.885) = 1.96 \times 10^{-3}$$

Equivalent Normal Distributions   If the probability distributions of the random variables $X_1, X_2, \ldots, X_n$ are not normal, the probability p_F or p_S may be evaluated through Eq. 6.8 (invariably, numerical integration would be necessary), where in the case of linear performance functions, $g(\mathbf{X}) = a_o + \sum a_i X_i$. However, p_S may be evaluated also using equivalent normal distributions (Paloheimo, 1974; Rackwitz, 1976). Theoretically, such equivalent normal distributions may be obtained through the Rosenblatt transformation (Rosenblatt, 1952); see Appendix B. With such equivalent normal distributions, the calculation of p_S follows the same procedure as that for normal variates; that is, through Eq. 6.24 for linear performance functions.

For an individual variate, the equivalent normal distribution for a nonnormal variate may be obtained such that the cumulative probability as well as the probability density ordinate of the equivalent normal distribution are equal to those of the corresponding nonnormal distribution at the appropriate point, x_i^*, on the failure surface.

Equating the cumulative probabilities as described above at the failure point x_i^*, we have

$$\Phi\left(\frac{x_i^* - \mu_{X_i}^N}{\sigma_{X_i}^N}\right) = F_{X_i}(x_i^*)$$

where

$\mu_{X_i}^N, \sigma_{X_i}^N =$ The mean value and standard deviation, respectively, of the equivalent normal distribution for X_i.

$F_{X_i}(x_i^*) =$ The original CDF of X_i evaluated at x_i^*.

$\Phi(-) =$ The CDF of the standard normal distribution.

The above equality then yields

$$\mu_{X_i}^N = x_i^* - \sigma_{X_i}^N \Phi^{-1}[F_{X_i}(x_i^*)] \tag{6.25}$$

whereas equating the corresponding probability density ordinates at x_i^* means

$$\frac{1}{\sigma_{X_i}^N} \phi\left(\frac{x_i^* - \mu_{X_i}^N}{\sigma_{X_i}^N}\right) = f_{X_i}(x_i^*)$$

where $\phi(-) =$ the PDF of the standard normal distribution; from this, we obtain

$$\sigma_{X_i}^N = \frac{\phi\{\Phi^{-1}[F_{X_i}(x_i^*)]\}}{f_{X_i}(x_i^*)} \tag{6.26}$$

In the case of a linear performance function, the appropriate point on the failure surface is given by Eq. 6.16, where the direction cosines, α_i, Eq. 6.16b, are

$$\alpha_i = \frac{a_i}{\sqrt{\sum_i a_i^2}}$$

and according to Eq. 6.23, the safety index is

$$\beta = \frac{a_o + \sum a_i \mu_{X_i}^N}{\sqrt{\sum\limits_i (a_i \sigma_{X_i}^N)^2}} \tag{6.27}$$

where the superscript N denotes the statistics for the equivalent normal distribution. Therefore, the failure point is

$$x_i^* = \sigma_{X_i}^N x_i'^* + \mu_{X_i}^N = -\alpha_i \beta \sigma_{X_i}^N + \mu_{X_i}^N \tag{6.28}$$

It may be emphasized that replacing the actual distribution with an equivalent normal distribution requires replacing the actual mean and standard deviation with those of the equivalent normal distribution, that is, Eqs. 6.25 and 6.26. Using these in Eq. 6.27, we obtain the safety index β, and the corresponding probability of safety, p_S, is given by Eq. 6.24a.

EXAMPLE 6.4

In Example 6.3, the variables R, D, and L were all assumed to be normally distributed. If some of the variables are nonnormal variates, equivalent normal distributions may be used to calculate the corresponding probability of failure or safety as illustrated below.

Suppose the distributions of R, D, and L are as follows:

R is lognormal.

D is normal.

L is Type I asymptotic.

The statistics (means and standard deviations) are the same as those given in Example 6.3; assume that the variables are statistically independent.

For the variable R, the parameters of the lognormal distribution are:

$$\zeta_R \simeq 0.11$$

$$\lambda_R = \ln \bar{R} - \tfrac{1}{2}\zeta_R^2 \simeq \ln \bar{R} = \ln 2.831\bar{D}$$

and

$$F_R(r) = \Phi\left(\frac{\ln r - \lambda_R}{\zeta_R}\right)$$

$$f_R(r) = \frac{1}{r\zeta_R} \phi\left(\frac{\ln r - \lambda_R}{\zeta_R}\right)$$

Then, Eqs. 6.26 and 6.25, respectively, yield

$$\sigma_R^N = \frac{1}{f_R(r^*)} \phi\left\{\Phi^{-1}\left[\Phi\left(\frac{\ln r^* - \lambda_R}{\zeta_R}\right)\right]\right\}$$

$$= \frac{1}{f_R(r^*)} \phi\left(\frac{\ln r^* - \lambda_R}{\zeta_R}\right)$$

$$= r^*\zeta_R$$

and

$$\mu_R^N = r^* - \sigma_R^N \Phi^{-1} \left[\Phi \left(\frac{\ln r^* - \lambda_R}{\zeta_R} \right) \right]$$

$$= r^* - r^* \zeta_R \left(\frac{\ln r^* - \lambda_R}{\zeta_R} \right)$$

$$= r^*(1 - \ln r^* + \lambda_R)$$

For the variable L, the extremal parameters are

$$\alpha = \frac{\pi}{\sqrt{6}} \cdot \frac{1}{\sigma_L} = \frac{\pi}{\sqrt{6}} \frac{1}{0.186 \bar{D}} = \frac{6.886}{\bar{D}}$$

$$u = \bar{L} - \frac{0.577}{\alpha} = 0.745 \bar{D} - \frac{0.577}{6.886} \bar{D} = 0.661 \bar{D}$$

and, from Eqs. 4.19 and 4.20,

$$F_L(l) = \exp[-e^{-\alpha(l-u)}]$$
$$f_L(l) = \alpha \exp[-\alpha(l-u) - e^{-\alpha(l-u)}]$$

In this case, that is, involving nonnormal distributions, even though the performance function is linear, the mean values and standard deviations required in Eq. 6.27 are unknown as these are now functions of the respective failure point values. An iterative solution, using Eqs. 6.25 and 6.26, is therefore necessary.

For the first iteration, assume (see Example 6.3)

$$r^* = \bar{R} = 2.831 \bar{D}$$

$$l^* = \bar{L} = 0.745 \bar{D}$$

Then,

$$\sigma_R^N = 2.831 \bar{D}(0.11) = 0.311 \bar{D}$$

and

$$\mu_R^N = 2.831 \bar{D}(1 - \ln 2.831 \bar{D} + \ln 2.831 \bar{D}) = 2.831 \bar{D}$$

For the Type I extremal distribution of L,

$$F_L(l^*) = 0.570$$

$$f_L(l^*) = 2.205/\bar{D}$$

and thus Eqs. 6.25 and 6.26 yield

$$\sigma_L^N = \frac{\frac{1}{\sqrt{2\pi}} \exp[-\frac{1}{2}\{\Phi^{-1}(0.570)\}^2]}{2.205/\bar{D}}$$

$$= \frac{\bar{D}}{2.205\sqrt{2\pi}} \exp[-\frac{1}{2}(0.178)^2] = 0.178 \bar{D}$$

and

$$\mu_L^N = 0.745 \bar{D} - 0.178 \bar{D} \times 0.178 = 0.713 \bar{D}$$

Since D is normal,

$$\sigma_D^N = \sigma_D = 0.10 \bar{D}; \qquad \mu_D^N = \bar{D}$$

Then, Eq. 6.27 becomes

$$\beta = \frac{\mu_R^N - \mu_D^N - \mu_L^N}{\sqrt{(\sigma_R^N)^2 + (\sigma_D)^2 + (\sigma_L^N)^2}}$$

$$= \frac{2.831\bar{D} - \bar{D} - 0.713\bar{D}}{\sqrt{(0.311\bar{D})^2 + (0.10\bar{D})^2 + (0.178\bar{D})^2}} = 3.00$$

The failure point is

$$r^* = \mu_R^N - \alpha_R^* \beta \sigma_R^N$$

where

$$\alpha_R^* = \frac{\sigma_R^N}{\sqrt{(\sigma_R^N)^2 + \sigma_D^2 + (\sigma_L^N)^2}} = \frac{0.311\bar{D}}{\sqrt{(0.311\bar{D})^2 + (0.10\bar{D})^2 + (0.178\bar{D})^2}}$$

$$= 0.836$$

Therefore,

$$r^* = 2.831\bar{D} - 0.836 \times 3.00 \times 0.311\bar{D} = 2.05\bar{D}$$

Similarly,

$$l^* = \mu_L^N - \alpha_L^* \beta \sigma_L^N$$

where

$$\alpha_L^* = \frac{-\sigma_L^N}{\sqrt{(\sigma_R^N)^2 + \sigma_D^2 + (\sigma_L^N)^2}} = \frac{-0.178\bar{D}}{\sqrt{(0.311\bar{D})^2 + (0.10\bar{D})^2 + (0.178\bar{D})^2}}$$

$$= -0.478$$

Hence,

$$l^* = 0.713\bar{D} + 0.478 \times 3.00 \times 0.178\bar{D} = 0.968\bar{D}$$

The second and subsequent iterations may be summarized as follows:

Iteration No.	Assumed Failure Point	σ^N	μ^N	β
2	$r^* = 2.05\bar{D}$ $l^* = 0.968\bar{D}$	$0.226\bar{D}$ $0.262\bar{D}$	$2.712\bar{D}$ $0.652\bar{D}$	2.947
3	$r^* = 2.295\bar{D}$ $l^* = 1.213\bar{D}$	$0.252\bar{D}$ $0.348\bar{D}$	$2.777\bar{D}$ $0.513\bar{D}$	2.865
4	$r^* = 2.363\bar{D}$ $l^* = 1.298\bar{D}$	$0.260\bar{D}$ $0.378\bar{D}$	$2.790\bar{D}$ $0.449\bar{D}$	2.856

Based on the results of the above 4th iteration,

$$p_F = 1 - \Phi(2.856) = 0.00214$$

Correlated Variates The procedure described above for evaluating the probability of safety or failure is tacitly based on the assumption that the random variables $X_1, X_2, \ldots, X_n$ are uncorrelated or statistically independent. For random variables that are correlated, the original variates may be transformed to a set of uncorrelated variables. The procedure described above, namely Eq. 6.15a, may then be applied to the uncorrelated set of transformed variables.† Indeed, this has been explicitly shown by Shinozuka (1983).

The required transformation is necessarily dependent on the covariances, or covariance matrix, of the original variates and may be obtained as follows.

Suppose the covariance matrix of the original variates $X_1, X_2, \ldots, X_n$ is

$$[C] = \begin{bmatrix} \sigma_{X_1}^2 & \text{Cov}(X_1, X_2) & \text{Cov}(X_1, X_3) & \cdots & \text{Cov}(X_1, X_n) \\ \text{Cov}(X_2, X_1) & \sigma_{X_2}^2 & \text{Cov}(X_2, X_3) & \cdots & \text{Cov}(X_2, X_n) \\ \vdots & \vdots & \vdots & \cdots & \vdots \\ \text{Cov}(X_n, X_1) & \text{Cov}(X_n, X_2) & \text{Cov}(X_n, X_3) & \cdots & \sigma_{X_n}^2 \end{bmatrix} \quad (6.29)$$

where the elements, $\text{Cov}(X_i, X_j)$, are the respective covariances between the pairs of variables X_i and X_j. The corresponding covariance between a pair of reduced variates, X_i' and X_j', is

$$\begin{aligned} \text{Cov}(X_i', X_j') &= E[X_i' - \mu_{X_i})(X_j' - \mu_{X_j})] \\ &= \frac{E[X_i - \mu_{X_i})(X_j - \mu_{X_j})]}{\sigma_{X_i}\sigma_{X_j}} \\ &= \frac{\text{Cov}(X_i, X_j)}{\sigma_{X_i}\sigma_{X_j}} \\ &= \rho_{X_i, X_j} \end{aligned}$$

which means that the covariance between a pair of reduced variates, X_i' and X_j', is equal to the correlation coefficient between the corresponding pair of original variates X_i and X_j. Therefore, the covariance matrix of the reduced variates $(X_1', X_2', \ldots, X_n')$ is the corresponding correlation matrix of the original variates $(X_1, X_2, \ldots, X_n)$; that is, the covariance matrix of $X_1', X_2', \ldots, X_n'$ is

$$[C'] = \begin{bmatrix} 1 & \rho_{12} & \rho_{13} & \cdots & \rho_{1n} \\ \rho_{21} & 1 & \rho_{23} & \cdots & \rho_{2n} \\ \cdot & \cdot & \cdot & \cdots & \cdot \\ \rho_{n1} & \rho_{n2} & \rho_{n3} & \cdots & 1 \end{bmatrix} \quad (6.30)$$

The required set of (uncorrelated) transformed variates can be obtained from $\mathbf{X}'$ through the following orthogonal transformation:

$$\mathbf{Y} = \mathbf{T}^t \mathbf{X}' \quad (6.31)$$

in which

$\mathbf{X}' = \{X_1', X_2', \ldots, X_n'\}$.
$\mathbf{Y} = \{Y_1, Y_2, \ldots, Y_n\}$ is the required set of uncorrelated transformed variates.
$\mathbf{T} = $ An orthogonal transformation matrix (superscript t represents the *transpose*).

---

† Strictly speaking, the transformation presented here is applicable only to normal variates; for correlated nonnormal variates, the *Rosenblatt transformation* is required (see Appendix B).

T will be an orthogonal matrix if it is composed of the eigenvectors corresponding to the eigenvalues of the correlation matrix $[C']$. Specifically, **T** is such that

$$\mathbf{T}^t[C']\mathbf{T} = [\lambda] \tag{6.32}$$

in which $[\lambda]$ is a diagonal matrix of the eigenvalues of $[C']$. It may be emphasized that the matrix $[C']$ is real and symmetric, as $\rho_{ij} = \rho_{ji}$; and thus the eigenvectors are mutually orthogonal.

With the orthogonal transformation of Eq. 6.31, it can be shown (Shinozuka, 1983) that the safety index of Eq. 6.15a becomes

$$\beta = \frac{-\mathbf{G}^{*t}\mathbf{X}'^*}{(\mathbf{G}^{*t}[C']\mathbf{G}^*)^{1/2}} \tag{6.15c}$$

The reduced variates **X**' and original variates **X** are related to **Y** as follows. Since **T** is orthogonal, $\mathbf{T}^{-1} = \mathbf{T}^t$; inversion of Eq. 6.31 yields

$$\mathbf{X}' = \mathbf{T}\mathbf{Y}$$

and

$$\begin{aligned}
\mathbf{X} &= [\sigma_X]\mathbf{X}' + \boldsymbol{\mu}_X \\
&= [\sigma_X]\mathbf{T}\mathbf{Y} + \boldsymbol{\mu}_X
\end{aligned}$$

in which

$$[\sigma_X] = \begin{bmatrix} \sigma_{X_1} & 0 & \cdots & 0 \\ 0 & \sigma_{X_2} & \cdots & 0 \\ \vdots & \vdots & \vdots & \vdots \\ 0 & 0 & \cdots & \sigma_{X_n} \end{bmatrix}$$

and

$$\boldsymbol{\mu}_X = \begin{Bmatrix} \mu_{X_1} \\ \mu_{X_2} \\ \vdots \\ \mu_{X_n} \end{Bmatrix}$$

Observe that the covariance matrix of **Y** is

$$\begin{aligned}
[C_Y] = E(\mathbf{Y}\mathbf{Y}^t) &= E(\mathbf{T}^t\mathbf{X}'\mathbf{X}'^t\mathbf{T}) \\
&= \mathbf{T}^t E(\mathbf{X}'\mathbf{X}'^t)\mathbf{T}
\end{aligned}$$

but

$$E(\mathbf{X}'\mathbf{X}'^t) = [C']$$

Thus, with Eq. 6.32,

$$[C_Y] = \mathbf{T}^t[C']\mathbf{T} = [\lambda] \tag{6.33}$$

Hence, the eigenvalues of $[C']$ are also the variances of the respective variates $Y_1, Y_2, \ldots, Y_n$.

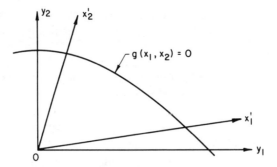

Figure 6.8 Rotation of coordinates **X′** to **Y**.

In the space of the transformed variates **Y**, the derivatives may be obtained through the chain rule of partial differentiation

$$\frac{\partial g}{\partial Y_i} = \sum_{j=1}^{n} \frac{\partial g}{\partial X'_j} \frac{\partial X'_j}{\partial Y_i} \tag{6.34}$$

Moreover,

$$\frac{\partial g}{\partial X'_j} = \frac{\partial g}{\partial X_j} \cdot \frac{dX_j}{dX'_j} = \sigma_{X_j} \left(\frac{\partial g}{\partial X_j} \right)$$

The transformation of Eq. 6.31 represents a rotation of the coordinates from **X′** to **Y**; for the two-variable case, this transformation is illustrated in Fig. 6.8. The origin of the **Y** axes remains the same as that of the **X′** axes.

The above transformation obviously applies also to linear performance functions. In this case, the partial derivatives of Eq. 6.34 are independent of the variables, and thus the failure point **y*** and **x*** can be determined directly; that is, one iteration of the numerical algorithm is sufficient. Alternatively, for linear performance functions of correlated (normal) variates, the safety index may also be determined directly from Eq. 6.15c or on the basis of Eq. 6.20, yielding

$$\beta = \frac{\mu_g}{\sigma_g} = \frac{a_o + \sum_{i=1}^{n} a_i \mu_{X_i}}{\sqrt{\sum_{i=1}^{n} \sum_{j=1}^{n} a_i a_j \rho_{ij} \sigma_{X_i} \sigma_{X_j}}} \tag{6.35}$$

in which ρ_{ij} is the correlation coefficient between X_i and X_j.

Again, if the distributions of the original random variables are nonnormal, the corresponding probability of safety or failure may be evaluated using equivalent normal distributions; in such a case, the mean values and standard deviations of the equivalent normal distributions, $\mu_{X_i}^N$ and $\sigma_{X_i}^N$, must be used in place of μ_{X_i} and σ_{X_i} in Eq. 6.35.

EXAMPLE 6.5

In Example 6.3, the variables R, D, and L were implicitly assumed to be statistically independent. In reality, the resistance R and the dead load D may be positively correlated (the weight of the structure will tend to increase with the resistance); similarly, D and L may also be partially correlated. Suppose these variables are correlated as follows:

$$\rho_{R,D} = 0.80$$

$$\rho_{R,L} = 0$$

$$\rho_{D,L} = 0.30$$

Then, according to Eq. 6.35, the safety index is

$$\beta = \frac{\bar{R} - \bar{D} - \bar{L}}{\sqrt{\sigma_R^2 + \sigma_D^2 + \sigma_L^2 - 2(0.80)\sigma_R\sigma_D + 2(0.30)\sigma_D\sigma_L}}$$

$$= \frac{2.831\bar{D} - \bar{D} - 0.745\bar{D}}{\sqrt{\begin{array}{c}(0.11 \times 2.831\bar{D})^2 + (0.10\bar{D})^2 + (0.25 \times 0.745\bar{D})^2 \\ - 2(0.80)(0.311\bar{D} \times 0.10\bar{D}) + 2(0.30)(0.10\bar{D} \times 0.186\bar{D})\end{array}}}$$

$$= \frac{1.086\bar{D}}{0.321\bar{D}}$$

$$= 3.383$$

If the variates are normal, the corresponding failure probability would be

$$p_F = 1 - \Phi(3.383) = 0.000363$$

For the purpose of demonstrating the validity of Eq. 6.31 and illustrating the general procedure for correlated variates, let us also evaluate the safety index β through the orthogonal transformation of Eq. 6.31. In the present case, the correlation matrix of the variates is

$$[C'] = \begin{bmatrix} 1.0 & 0.8 & 0 \\ 0.8 & 1.0 & 0.3 \\ 0 & 0.3 & 1.0 \end{bmatrix}$$

The corresponding determinantal equation is

$$\text{Det} \begin{bmatrix} (1 - \lambda) & 0.8 & 0 \\ 0.8 & (1 - \lambda) & 0.3 \\ 0 & 0.3 & (1 - \lambda) \end{bmatrix} = 0$$

which yields

$$(1 - \lambda)[(1 - \lambda)^2 - 0.73] = 0$$

The roots of this equation are the eigenvalues of $[C']$, which are

$$\lambda_1 = 1.8544$$

$$\lambda_2 = 0.1456$$

$$\lambda_3 = 1.0000$$

and the corresponding normalized eigenvectors are

$$\Phi_1 = \begin{Bmatrix} 0.6621 \\ 0.7071 \\ 0.2483 \end{Bmatrix}; \quad \Phi_2 = \begin{Bmatrix} 0.6621 \\ -0.7071 \\ 0.2483 \end{Bmatrix}; \quad \Phi_3 = \begin{Bmatrix} 0.3511 \\ 0 \\ -0.9363 \end{Bmatrix}$$

Therefore, the orthogonal transformation matrix is

$$\mathbf{T} = \begin{bmatrix} 0.6621 & 0.6621 & 0.3511 \\ 0.7071 & -0.7071 & 0 \\ 0.2483 & 0.2483 & -0.9363 \end{bmatrix}$$

The transformed orthogonal coordinates $\mathbf{Y}$ may then be obtained through Eq. 6.31, whose covariance matrix, according to Eq. 6.33, is

$$[C_Y] = \begin{bmatrix} 1.8544 & 0 & 0 \\ 0 & 0.1456 & 0 \\ 0 & 0 & 1.0000 \end{bmatrix}$$

The original variates $\mathbf{X} = \{R, D, L\}$ may be obtained through

$$\mathbf{X} = [\sigma_X]\mathbf{X}' + \boldsymbol{\mu}_X = [\sigma_X]\mathbf{T}\mathbf{Y} + \boldsymbol{\mu}_X$$

where

$$[\sigma_X] = \begin{bmatrix} 0.311\bar{D} & 0 & 0 \\ 0 & 0.100\bar{D} & 0 \\ 0 & 0 & 0.186\bar{D} \end{bmatrix}$$

and

$$\boldsymbol{\mu}_X = \begin{Bmatrix} 2.831\bar{D} \\ \bar{D} \\ 0.745\bar{D} \end{Bmatrix}$$

From which the limit-state equation becomes

$$R - D - L = 0.0890\bar{D}Y_1 + 0.2304\bar{D}Y_2 + 0.2834\bar{D}Y_3 + 1.086\bar{D} = 0$$

Since the variables Y_1, Y_2, and Y_3 are uncorrelated, the safety index is also

$$\beta = \frac{1.086\bar{D}}{\sqrt{(0.0890\bar{D})^2(1.8544) + (0.2304\bar{D})^2(0.1456) + (0.2834\bar{D})^2(1.0)}}$$

$$= \frac{1.086}{0.321} = 3.383$$

which is the same as the result obtained earlier with Eq. 6.35.

EXAMPLE 6.6   (*Travel Time*)

A limousine operates from an airport, travels to towns I and II, and then returns to the airport as shown in Fig. E6.6. Because of varied traffic conditions, the travel time required in each leg

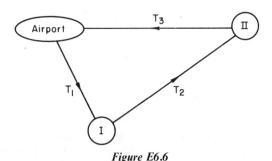

Figure E6.6

of the journey is random, with the following statistics:

Travel Time	Mean (min)	Standard Deviation (min)
T_1	40	10
T_2	15	5
T_3	50	10

Since the traffic conditions over the three legs of the journey can be expected to be correlated, a correlation coefficient of 0.5 is assumed between all pairs of T_j's, $j = 1, 2$, and 3; that is,

$$\rho_{T_1, T_2} = \rho_{T_2, T_3} = \rho_{T_1, T_3} = 0.5$$

The scheduled time for each round trip around the loop is two hours. Determine the probability that a round trip will not be completed on schedule.

The performance function is given by

$$Z = 120 - (T_1 + T_2 + T_3)$$

Equation 6.35 then yields the safety index

$$\beta = \frac{120 - (40 + 15 + 50)}{\sqrt{10^2 + 5^2 + 10^2 + 2 \times 0.5 \times 10 \times 5 + 2 \times 0.5 \times 10 \times 10 + 2 \times 0.5 \times 5 \times 10}}$$

$$= \frac{15}{20.62} = 0.727$$

Hence, if the travel times are normal variates, the required probability is

$$p_F = 1 - \Phi(0.727) = 0.233$$

6.2.4 Nonlinear Performance Functions

For performance functions, $g(\mathbf{X})$, that are nonlinear, the evaluation of the exact probability of safety or failure will generally be involved. The limit-state equation, $g(\mathbf{X}) = 0$, will also be nonlinear as shown in Fig. 6.6; unlike the linear case, there is no unique distance from the failure surface to the origin of the reduced variates. As indicated in Section 6.2.2, the evaluation of the exact probability of safety will involve the integration of the joint probability density function over the nonlinear region $g(\mathbf{X}) > 0$; generally, this will require multiple numerical quadrature.

For practical purposes, approximation to the exact probability will be necessary. According to the results of Section 6.2.2, the point $(x_1'^*, x_2'^*, \ldots, x_n'^*)$ on the failure surface with minimum distance to the origin of the reduced variates is the most probable failure point (Shinozuka, 1983). The tangent plane to the failure surface at $(x_1'^*, x_2'^*, \ldots, x_n'^*)$ may then be used to approximate the actual failure surface, and the required reliability index or probability of safety may be evaluated as in the linear case of Section 6.2.3. Depending on whether the exact nonlinear failure surface is convex or concave toward the origin, this approximation will be on the safe side or unsafe side, respectively, as may be seen in Fig. 6.9 for the two-variable case.

The pertinent tangent plane at $\mathbf{x}'^* = (x_1'^*, x_2'^*, \ldots, x_n'^*)$ is

$$\sum_{i=1}^{n} (X_i' - x_i'^*) \left(\frac{\partial g}{\partial X_i'} \right)_* = 0 \tag{6.36}$$

where the partial derivatives $(\partial g / \partial X_i')_*$ are evaluated at $(x_1'^*, x_2'^*, \ldots, x_n'^*)$.

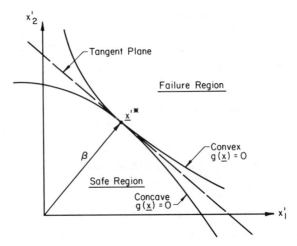

Figure 6.9 Tangent plane to $g(\mathbf{X}) = 0$ at $\mathbf{x}'^*$.

On the basis of the above approximation, the distance from the "minimum" tangent plane, Eq. 6.36, to the origin of the reduced variates is the appropriate reliability index, which may be used to represent the measure of reliability.

In the present case (in which the performance function is nonlinear), the pertinent point of tangency on the failure surface is not, a priori, known. Consequently, the determination of the required reliability index would not be as simple as in the linear case (Section 6.2.3), even though linear approximation is invoked. The "minimum" point of tangency on the failure surface may be determined through the Lagrange multiplier method as described in Section 6.2.2. The relevant results of Section 6.2.2 may be summarized as follows.

The most probable failure point of Eq. 6.16a is

$$x_i'^* = -\alpha_i^* \beta$$

in which α_i are the direction cosines of Eq. 6.16b

$$\alpha_i^* = \frac{\left(\dfrac{\partial g}{\partial X_i'}\right)_*}{\sqrt{\displaystyle\sum_i \left(\dfrac{\partial g}{\partial X_i'}\right)_*^2}}$$

where the derivatives are evaluated at $(x_1'^*, x_2'^*, \ldots, x_n'^*)$. Then,

$$x_i^* = \sigma_{X_i} x_i'^* + \mu_{X_i} = \mu_{X_i} - \alpha_i^* \sigma_{X_i} \beta \tag{6.37}$$

The solution of the limit-state equation

$$g(x_1^*, x_2^*, \ldots, x_n^*) = 0$$

then yields β.

Numerical Algorithm   The results summarized above would suggest the following simple algorithm (Rackwitz, 1976):

(1)   Assume initial values of x_i^*; $i = 1, 2, \ldots, n$ and obtain

$$x_i'^* = \frac{x_i^* - \mu_{x_i}}{\sigma_{x_i}}$$

(2)   Evaluate $(\partial g/\partial X_i')_*$ and α_i^* at x_i^*.
(3)   Form $x_i^* = \mu_{x_i} - \alpha_i^* \sigma_{X_i} \beta$.
(4)   Substitute above x_i^* in $g(x_1^*, x_2^*, \ldots, x_n^*) = 0$ and solve for β.
(5)   Using the β obtained in Step 4, reevaluate $x_i'^* = -\alpha_i \beta$.
(6)   Repeat Steps 2 through 5 until convergence is obtained.

EXAMPLE 6.7   (*Uncorrelated Normals*)

The fully plastic flexural capacity of a steel beam section may be given as YZ, where

$$Y = \text{The yield strength of steel.}$$

$$Z = \text{Section modulus of the section.}$$

Then, if the applied bending moment at the pertinent section is M, the performance function may be defined as

$$g(\mathbf{X}) = YZ - M$$

Assume that the variables are uncorrelated (see Example 6.9 for the effects of correlations).
   Suppose a beam with $\bar{Y} = 40$ ksi and $\bar{Z} = 50$ in.^{3} is subjected to $\bar{M} = 1000$ in-kip; and the corresponding c.o.v.'s are

$$\Omega_Y = 0.125; \qquad \Omega_Z = 0.05; \qquad \Omega_M = 0.20$$

Determine the reliability of the beam.
   The corresponding standard deviations are

$$\sigma_Y = 0.125 \times 40 = 5.0 \text{ ksi}$$

$$\sigma_Z = 0.05 \times 50 = 2.5 \text{ in.}^3$$

$$\sigma_M = 0.20 \times 1000 = 200 \text{ in-kip}$$

In this case, the partial derivatives are

$$\left(\frac{\partial g}{\partial Y'}\right) = \sigma_Y Z$$

$$\left(\frac{\partial g}{\partial Z'}\right) = \sigma_Z Y$$

$$\left(\frac{\partial g}{\partial M'}\right) = -\sigma_M$$

   For the first iteration, assume $y^* = \bar{Y} = 40$ ksi; $z^* = \bar{Z} = 50$ in.^{3}; and $m^* = \bar{M} = 1000$ in-kip. Then,

$$\left(\frac{\partial g}{\partial Y'}\right)_* = 5.0 \times 50 = 250$$

$$\left(\frac{\partial g}{\partial Z'}\right)_* = 2.5 \times 40 = 100$$

$$\left(\frac{\partial g}{\partial M'}\right)_* = -200$$

The direction cosines of Eq. 6.16b, therefore, are

$$\alpha_{Y'}^* = \frac{250}{\sqrt{(250)^2 + (100)^2 + (200)^2}} = \frac{250}{335.41} = 0.745$$

$$\alpha_{Z'}^* = \frac{100}{335.41} = 0.298$$

$$\alpha_{M'}^* = \frac{-200}{335.41} = -0.596$$

Hence, the components of the failure point, Eq. 6.37, are

$$y^* = 40 - 0.745 \times 5.0\beta = 40 - 3.725\beta$$
$$z^* = 50 - 0.298 \times 2.5\beta = 50 - 0.745\beta$$
$$m^* = 1000 + 0.596 \times 200\beta = 1000 + 119.20\beta$$

Substituting these into the limit-state equation, $y^* z^* - m^* = 0$, yields the following quadratic equation:

$$2.775\beta^2 - 335.25\beta + 1000 = 0$$

from which we obtain the solution

$$\beta = 3.06$$

The revised failure point then becomes

$$y^* = 40 - 3.725 \times 3.06 = 28.60$$
$$z^* = 50 - 0.745 \times 3.06 = 47.72$$
$$m^* = 1000 + 119.20 \times 3.06 = 1364.75$$

Repeating the procedure for subsequent iterations, the results are summarized in Table E6.7.

Table E6.7 Summary of Iterations for Example 6.7

Iteration No.	Variable	Assumed Failure Point	$\left(\dfrac{\partial g}{\partial X_i'}\right)_*$	$\alpha_{X_i}^*$	New x_i^*
1	Y	40	250	0.745	$40 - 3.725\beta$
	Z	50	100	0.298	$50 - 0.745\beta$
	M	1000	-200	-0.596	$1000 + 119.20\beta$
		$\beta = 3.06$			
2	Y	28.60	238.60	0.747	$40 - 3.735\beta$
	Z	47.72	71.50	0.224	$50 - 0.560\beta$
	M	1364.75	-200.00	-0.626	$1000 + 125.20\beta$
		$\beta = 3.05$			
3	Y	28.61	238.60	0.747	$40 - 3.735\beta$
	Z	48.29	71.50	0.224	$50 - 0.560\beta$
	M	1381.86	-200.00	-0.626	$1000 + 125.20\beta$
		$\beta = 3.05$			

Therefore, the underlying probability of failure is

$$p_F = 1 - \Phi(3.05) = 0.00114$$

EXAMPLE 6.8 (Uncorrelated Nonnormals)

Suppose the random variables in Example 6.7 are nonnormal, with the following distributions:

$$Y = \text{Lognormal}.$$

$$Z = \text{Lognormal}.$$

$$M = \text{Type I asymptotic extreme}.$$

Assume that the variates are uncorrelated (see Example 6.10 for the effects of correlations).

Using the same statistics of the design variables as in Example 6.7, namely, $\bar{Y} = 40$ ksi, $\Omega_Y = 0.125; \bar{Z} = 50$ in.$^3, \Omega_Z = 0.05; \bar{M} = 1000$ in-kips, $\Omega_M = 0.20$, we obtain the parameters of the respective distributions as follows.

For Y and Z, the parameters of the lognormal distributions are

$$\zeta_Y \simeq \Omega_Y = 0.125; \qquad \lambda_Y = \ln 40 - \tfrac{1}{2}(0.125)^2 = 3.681$$

$$\zeta_Z \simeq \Omega_Z = 0.05; \qquad \lambda_Z = \ln 50 - \tfrac{1}{2}(0.05)^2 = 3.911$$

The corresponding parameters of the Type I asymptotic distribution for M are

$$\alpha = \frac{\pi}{\sqrt{6}} \cdot \frac{1}{\sigma_M} = \frac{\pi}{\sqrt{6} \times 200} = 0.006413$$

$$u = \bar{M} - \frac{0.577}{\alpha} = 1000 - \frac{0.577}{0.006413} = 910.02$$

As in Example 6.7, the partial derivatives with respect to the reduced variates are

$$\left(\frac{\partial g}{\partial Y'}\right) = \sigma_Y Z; \qquad \left(\frac{\partial g}{\partial Z'}\right) = \sigma_Z Y; \qquad \left(\frac{\partial g}{\partial M'}\right) = -\sigma_M$$

Using the results derived in Example 6.4 for the lognormal distribution, we have

$$\sigma_Y^N = y^* \zeta_Y; \qquad \mu_Y^N = y^*(1 - \ln y^* + \lambda_Y)$$

$$\sigma_Z^N = z^* \zeta_Z; \qquad \mu_Z^N = z^*(1 - \ln z^* + \lambda_Z)$$

whereas for the Type I asymptotic distribution of M,

$$F_M(m) = \exp[-e^{-\alpha(m-u)}]$$

$$f_M(m) = \alpha \exp[-\alpha(m-u) - e^{-\alpha(m-u)}]$$

and

$$\mu_M^N = m^* - \sigma_M^N \Phi^{-1}[F_M(m^*)]; \qquad \sigma_M^N = \frac{\phi\{\Phi^{-1}[F_M(m^*)]\}}{f_M(m^*)}$$

Using these relationships and assuming for the first iteration,

$$y^* = \bar{Y} = 40; \qquad z^* = \bar{Z} = 50; \qquad m^* = \bar{M} = 1000$$

we obtain

$$\sigma_Y^N = 40 \times 0.125 = 5.0$$

$$\mu_Y^N = 40(1 - \ln 40 + 3.681) = 39.69$$

$$\sigma_Z^N = 50 \times 0.05 = 2.5$$

$$\mu_Z^N = 50(1 - \ln 50 + 3.911) = 49.95$$

$$F_M(m^*) = \exp[-e^{-0.006413(1000-910.02)}] = e^{-0.5616} = 0.5703$$

$$f_M(m^*) = 0.006413 \exp[-0.006413(1000 - 910.02) - 0.5616] = 0.002054$$

$$\sigma_M^N = \frac{\phi[\Phi^{-1}(0.5703)]}{0.002054} = 191.14$$

$$\mu_M^N = 1000 - 191.14\Phi^{-1}(0.5703) = 965.78$$

As in Example 6.7, the partial derivatives are

$$\left(\frac{\partial g}{\partial Y'}\right)_* = \sigma_Y^N z^* = 50 \times 5.0 = 250$$

$$\left(\frac{\partial g}{\partial Z'}\right)_* = \sigma_Z^N y^* = 40 \times 2.5 = 100$$

$$\left(\frac{\partial g}{\partial M'}\right)_* = -\sigma_M^N = -191.14$$

From which the direction cosines are

$$\alpha_Y^* = \frac{250}{\sqrt{(250)^2 + (100)^2 + (191.14)^2}} = \frac{250}{330.2} = 0.757$$

$$\alpha_Z^* = \frac{100}{330.2} = 0.303$$

$$\alpha_M^* = \frac{-191.14}{330.2} = -0.580$$

Therefore,

$$y^* = \mu_Y^N - \alpha_Y^* \beta\sigma_Y^N = 39.69 - 0.757 \times 5.0\beta = 39.69 - 3.786\beta$$

$$z^* = \mu_Z^N - \alpha_Z^* \beta\sigma_Z^N = 49.85 - 0.303 \times 2.5\beta = 49.85 - 0.757\beta$$

$$m^* = \mu_M^* - \alpha_M^* \beta\sigma_M^N = 965.78 + 0.580 \times 191.14\beta = 965.78 + 110.78\beta$$

The limit-state equation then becomes

$$2.866\beta^2 - 329.67\beta + 1011.86 = 0$$

The pertinent solution is

$$\beta = 3.156$$

Thus, the new failure point is

$$y^* = 27.69$$

$$z^* = 47.50$$

$$m^* = 1315.40$$

The results of the iterations may be summarized as shown in Table E6.8. Therefore, the probability of failure is

$$p_F = 1 - \Phi(2.745) = 0.00302$$

Table E6.8 Iterative Calculations for Example 6.8

Iteration No.	Variable	Assumed Failure Point	$\sigma_{X_i}^N$	$\mu_{X_i}^N$	$\left(\dfrac{\partial g}{\partial X_i'}\right)_*$	$\alpha_{X_i}^*$	New x_i^*
1	Y	40	5.0	39.69	250	0.757	$39.69 - 3.786\beta$
	Z	50	2.5	49.95	100	0.303	$49.95 - 0.757\beta$
	M	1000	191.14	965.78	-191.14	-0.580	$965.78 + 110.78\beta$

Failure equation: $2.866\beta^2 - 329.67\beta + 1011.86 = 0$

$$\beta = 3.156$$

	Y	27.69	3.461	37.63	164.4	0.462	$37.63 - 1.599\beta$
2	Z	47.50	2.375	49.84	65.76	0.185	$49.84 - 0.439\beta$
	M	1315.40	308.8	863.3	-308.8	-0.867	$863.3 + 267.73\beta$

Failure equation: $0.702\beta^2 - 363.94\beta + 1012.18 = 0$

$$\beta = 2.796$$

	Y	33.16	4.145	39.08	201.5	0.432	$39.08 - 1.791\beta$
3	Z	48.61	2.431	49.88	80.61	0.173	$49.88 - 0.421\beta$
	M	1611.9	412.3	667.8	-412.3	-0.885	$667.8 + 364.9\beta$

Failure equation: $0.754\beta^2 - 470.6\beta + 1281.5 = 0$

$$\beta = 2.735$$

	Y	34.18	4.273	39.29	208.22	0.432	$39.29 - 1.846\beta$
4	Z	48.73	2.437	49.93	83.30	0.173	$49.93 - 0.422\beta$
	M	1665.8	426.3	634.2	-426.3	-0.885	$634.2 + 377.3\beta$

Failure equation: $0.779\beta^2 - 486.05\beta + 1327.6 = 0$

$$\beta = 2.745$$

EXAMPLE 6.9 (Correlated Normals)

Consider the problem of Example 6.7 again. However, assume that the design variables Y and Z are now partially correlated with $\rho_{Y,Z} = 0.40$.

Again, the statistics of the design variables are:

$$\bar{Y} = 40 \text{ ksi}$$

$$\Omega_Y = 0.125; \qquad \sigma_Y = 5 \text{ ksi}$$

$$\bar{Z} = 50 \text{ in.}^3$$

$$\Omega_Z = 0.05; \qquad \sigma_Z = 2.5 \text{ in.}^3$$

$$\bar{M} = 1000 \text{ in-kips}$$

$$\Omega_M = 0.20; \qquad \sigma_M = 200 \text{ in-kips}$$

With the notation $X = \{Y, Z, M\}$, the pertinent correlation matrix, therefore, is,

$$[C'] = \begin{bmatrix} 1.0 & 0.40 & 0 \\ 0.40 & 1.0 & 0 \\ 0 & 0 & 1.0 \end{bmatrix}$$

The eigenvalues of $[C']$ are the solution of the following determinantal equation:

$$\det \begin{bmatrix} (1-\lambda) & 0.4 & 0 \\ 0.4 & (1-\lambda) & 0 \\ 0 & 0 & (1-\lambda) \end{bmatrix} = 0$$

or

$$(1-\lambda)^3 - (0.4)^2(1-\lambda) = 0$$

The solution (roots) of which are the three eigenvalues:

$$\lambda_1 = 1.4$$
$$\lambda_2 = 0.6$$
$$\lambda_3 = 1.0$$

The corresponding eigenvectors are obtained as solutions to

$$\begin{bmatrix} (1-\lambda_i) & 0.40 & 0 \\ 0.40 & (1-\lambda_i) & 0 \\ 0 & 0 & (1-\lambda_i) \end{bmatrix} \begin{Bmatrix} e_1 \\ e_2 \\ e_3 \end{Bmatrix} = 0$$

For example, for $\lambda_1 = 1.4$, the normalized eigenvector is $\{0.7071, 0.7071, 0\}$. The two other normalized eigenvectors are, respectively, $\{0.7071, -0.7071, 0\}$ and $\{0, 0, 1\}$.

The orthogonal transformation matrix, T, therefore is

$$T = \begin{bmatrix} 0.7071 & 0.7071 & 0 \\ 0.7071 & -0.7071 & 0 \\ 0 & 0 & 1 \end{bmatrix}$$

Thus, the transformed variates Y of Eq. 6.31 are

$$Y = T^t X'$$

and

$$X' = TY$$

Also,

$$X = [\sigma_X]X' + \mu_X$$

where

$$[\sigma_X] = \begin{bmatrix} 5 & 0 & 0 \\ 0 & 2.5 & 0 \\ 0 & 0 & 200 \end{bmatrix}$$

$$\mu_X = \begin{Bmatrix} 40 \\ 50 \\ 1000 \end{Bmatrix}$$

from which we obtain the components of **X**

$$Y = 3.536\,Y_1 + 3.536\,Y_2 + 40$$

$$Z = 1.768\,Y_1 - 1.768\,Y_2 + 50$$

$$M = 200\,Y_3 + 1000$$

The performance function then becomes

$$g(\mathbf{X}) = YZ - M$$
$$= 6.252\,Y_1^2 - 6.252\,Y_2^2 + 247.52\,Y_1 + 106.08\,Y_2 - 200\,Y_3 + 1000$$

and the partial derivatives are

$$\frac{\partial g}{\partial Y_1} = 12.504\,Y_1 + 247.52$$

$$\frac{\partial g}{\partial Y_2} = -12.504\,Y_2 + 106.08$$

$$\frac{\partial g}{\partial Y_3} = -200$$

For the first iteration, assume

$$x_1^* = \bar{Y} = 40; \qquad x_2^* = \bar{Z} = 50; \qquad x_3^* = \bar{M} = 1000$$

Since the corresponding mean values of **Y** are zero,

$$\left(\frac{\partial g}{\partial Y_1}\right)_* = 247.52$$

$$\left(\frac{\partial g}{\partial Y_2}\right)_* = 106.08$$

$$\left(\frac{\partial g}{\partial Y_3}\right)_* = -200$$

giving the respective direction cosines

$$\alpha_{Y_1}^* = 0.805; \qquad \alpha_{Y_2}^* = 0.226; \qquad \alpha_{Y_3}^* = -0.549$$

Hence,

$$y_1^* = -\alpha_1^* \beta \sigma_{Y_1} = -0.805\sqrt{1.4}\,\beta = -0.952\beta$$
$$y_2^* = -\alpha_2^* \beta \sigma_{Y_2} = -0.226\sqrt{0.6}\,\beta = -0.175\beta$$
$$y_3^* = -\alpha_3^* \beta \sigma_{Y_3} = 0.549\sqrt{1.0}\,\beta = 0.549\beta$$

Substituting these into the failure equation, $g(\mathbf{X}) = 0$, we obtain

$$5.475\beta^2 - 364.0\beta + 1000 = 0$$

From this, the safety index is

$$\beta = 2.871$$

The above calculations and those of the subsequent iterations may be summarized as shown in Table E6.9.

Table E6.9 Summary of Iterations for Example 6.9

Iteration No.	Variable	Assumed y_i^*	$\left(\dfrac{\partial g}{\partial Y_i}\right)_*$	$\alpha_{Y_i}^*$	New y_i^*
1	Y_1	0	247.52	0.805	-0.952β
	Y_2	0	106.08	0.226	-0.175β
	Y_3	0	-200	-0.549	0.549β

$$\text{Failure equation: } 5.475\beta^2 - 364.0\beta + 1000 = 0$$

$$\beta = 2.871$$

	Y_1	-2.733	213.35	0.757	-0.894β
2	Y_2	-0.502	112.36	0.261	-0.202β
	Y_3	1.576	-200	-0.599	0.599β

$$\text{Failure equation: } 4.742\beta^2 - 362.51\beta + 1000 = 0$$

$$\beta = 2.865$$

	Y_1	-2.561	215.50	0.760	-0.899β
3	Y_2	-0.579	113.32	0.261	-0.202β
	Y_3	1.716	-200	-0.596	0.596β

$$\text{Failure equation: } 4.798\beta^2 - 363.15\beta + 1000 = 0$$

$$\beta = 2.862$$

The most probable failure point in the **Y** space, therefore, is

$$y_1^* = -0.899 \times 2.862 = -2.573$$
$$y_2^* = -0.202 \times 2.862 = -0.578$$
$$y_3^* = 0.596 \times 2.862 = 1.706$$

whereas in the space of the original variates,

$$y^* = -3.536 \times 2.573 - 3.536 \times 0.578 + 40 = 28.86 \text{ ksi}$$
$$z^* = -1.768 \times 2.573 + 1.768 \times 0.578 + 50 = 46.47 \text{ in.}^3$$
$$m^* = 200 \times 1.706 + 1000 = 1341.20 \text{ in-kip}$$

The corresponding probability of failure is

$$p_F = 1 - \Phi(2.862) = 0.00212$$

This result may be compared with that of Example 6.7 in which the variates are uncorrelated.

EXAMPLE 6.10   (Correlated Nonnormals†)

Consider the same problem of Example 6.9, except that the variables are now distributed as follows:

$$Y = \text{Lognormal.}$$
$$Z = \text{Lognormal.}$$
$$M = \text{Type I asymptotic extreme.}$$

† In general, for random variables that are not normal or lognormal, the dependence between the variables may be expressed in terms of conditional probabilities. In these cases, the required independent set of equivalent (standard) normal variates can be obtained through the Rosenblatt transformation, as described in Appendix B. However, the *rotation of coordinates* represented by Eq. 6.31 may also be used in approximation; the results obtained here with Eq. 6.31 for the present example may be compared with those using the Rosenblatt transformation (see Example B.4 of Appendix B).

Otherwise, the same statistics apply; namely,

$$\bar{Y} = 40 \text{ ksi}; \qquad \Omega_Y = 0.125$$
$$\bar{Z} = 50 \text{ in.}^3; \qquad \Omega_Z = 0.05$$
$$\bar{M} = 1000 \text{ in-kip}; \qquad \Omega_M = 0.20$$

and the correlation matrix is also

$$[C'] = \begin{bmatrix} 1.0 & 0.40 & 0 \\ 0.40 & 1.0 & 0 \\ 0 & 0 & 1.0 \end{bmatrix}$$

The eigenvalues and respective eigenvectors, therefore, are also the same as those obtained earlier in Example 6.9; thus, the diagonal matrix of eigenvalues and orthogonal transformation matrix are

$$[\lambda] = \begin{bmatrix} 1.4 & 0 & 0 \\ 0 & 0.6 & 0 \\ 0 & 0 & 1.0 \end{bmatrix}$$

and

$$T = \begin{bmatrix} 0.707 & 0.707 & 0 \\ 0.707 & -0.707 & 0 \\ 0 & 0 & 1.0 \end{bmatrix}$$

Applying Eqs. 6.25 and 6.26, we obtain the means and standard deviations for the equivalent normal distributions of Y and Z as follows:

$$\sigma_Y^N = \zeta_Y y^*$$
$$\mu_Y^N = y^*(1 - \ln y^* + \lambda_Y)$$

and

$$\sigma_Z^N = \zeta_Z z^*$$
$$\mu_Z^N = z^*(1 - \ln z^* + \lambda_Z)$$

in which

$$\lambda_Y = 3.681; \qquad \zeta_Y = 0.125$$
$$\lambda_Z = 3.910; \qquad \zeta_Z = 0.05$$

whereas for M

$$f_M(m) = \alpha \exp[-\alpha(m - u) - e^{-\alpha(m-u)}]$$

and

$$F_M(m) = \exp[-e^{-\alpha(m-u)}]$$

in which

$$\alpha = 0.006413; \qquad u = 910$$

For the first iteration, assume

$$y^* = \bar{Y} = 40$$
$$z^* = \bar{Z} = 50$$
$$m^* = \bar{M} = 1000$$

thus obtaining

$$\sigma_X^N = 5.0; \qquad \mu_X^N = 39.64$$
$$\sigma_Z^N = 2.5; \qquad \mu_Z^N = 49.89$$
$$\sigma_M^N = 191.14; \qquad \mu_M^N = 965.8$$

Then, using the notation $\mathbf{X} = \{Y, Z, M\}$, yields

$$\mathbf{X} = [\sigma_X^N]\mathbf{TY} + \mu_X^N$$

$$= \begin{bmatrix} 5.0 & 0 & 0 \\ 0 & 2.5 & 0 \\ 0 & 0 & 191.14 \end{bmatrix} \begin{bmatrix} 0.707 & 0.707 & 0 \\ 0.707 & -0.707 & 0 \\ 0 & 0 & 1.0 \end{bmatrix} \begin{Bmatrix} Y_1 \\ Y_2 \\ Y_3 \end{Bmatrix} + \begin{Bmatrix} 39.64 \\ 49.89 \\ 965.8 \end{Bmatrix}$$

From which we obtain

$$Y = 3.536\,Y_1 + 3.536\,Y_2 + 39.64$$
$$Z = 1.768\,Y_1 - 1.768\,Y_2 + 49.89$$
$$M = 191.14\,Y_3 + 965.8$$

The performance function then becomes

$$g(\mathbf{X}) = YZ - M$$
$$= 6.252\,Y_1^2 - 6.252\,Y_2^2 + 246.5\,Y_1 + 106.3\,Y_2 - 191.14\,Y_3 + 1011.8$$

The partial derivatives are

$$\frac{\partial g}{\partial Y_1} = 12.504\,Y_1 + 246.5$$

$$\frac{\partial g}{\partial Y_2} = -12.504\,Y_2 + 106.3$$

$$\frac{\partial g}{\partial Y_3} = -191.14$$

The reduced variates at the failure point are

$$y'^* = \frac{40 - 39.64}{5.0} = 0.072$$

$$z'^* = \frac{50 - 49.89}{2.5} = 0.044$$

$$m'^* = \frac{1000 - 965.8}{191.14} = 0.179$$

From these, the transformed variates are

$$\begin{Bmatrix} y_1^* \\ y_2^* \\ y_3^* \end{Bmatrix} = \begin{bmatrix} 0.707 & 0.707 & 0 \\ 0.707 & -0.707 & 0 \\ 0 & 0 & 1.0 \end{bmatrix} \begin{Bmatrix} 0.072 \\ 0.044 \\ 0.179 \end{Bmatrix} = \begin{Bmatrix} 0.082 \\ 0.019 \\ 0.179 \end{Bmatrix}$$

Then,

$$\left(\frac{\partial g}{\partial Y_1}\right)_* = 12.504 \times 0.082 + 246.5 = 247.5$$

$$\left(\frac{\partial g}{\partial Y_2}\right)_* = -12.504 \times 0.019 + 106.3 = 106.06$$

$$\left(\frac{\partial g}{\partial Y_3}\right)_* = -191.14$$

and the direction cosines become

$$\alpha^*_{Y_1} = \frac{247.5\sqrt{1.4}}{\sqrt{(247.5)^2(1.4) + (106.06)^2(0.6) + (191.14)^2(1.0)}} = \frac{247.5 \times 1.183}{359.225} = 0.815$$

$$\alpha^*_{Y_2} = \frac{106.06\sqrt{0.6}}{359.225} = 0.229$$

$$\alpha^*_{Y_3} = \frac{-191.14}{359.225} = -0.532$$

Table E6.10 Summary of Iterative Calculations for Example 6.10

Iteration No.	X_i	Assumed x_i^*	$\sigma^N_{X_i}$	$\mu^N_{X_i}$	y_i^*	$\left(\dfrac{\partial g}{\partial Y_i}\right)_*$	$\alpha^*_{Y_i}$	New y_i^*
1	Y	40	5.0	39.64	0.082	247.50	0.815	-0.964β
	Z	50	2.5	49.89	0.019	106.06	0.229	-0.176β
	M	1000	191.14	965.80	0.179	-191.14	-0.532	0.532β

Failure equation $5.616\beta^2 - 358.02\beta + 1011.8 = 0$

$$\beta = 2.963$$

	Y	27.69	3.46	37.63	-3.235	157.0	0.532	-0.629β
2	Z	45.76	2.29	49.73	-0.826	67.30	0.149	-0.115β
	M	1267	291	888	1.303	-291	-0.834	0.834β

Failure equation: $1.515\beta^2 - 364.5\beta + 983.3 = 0$

$$\beta = 2.728$$

	Y	32.66	4.08	38.98	-1.802	191.71	0.495	-0.586β
3	Z	47.46	2.37	49.83	-0.388	82.11	0.139	-0.108β
	M	1550	392.3	712.50	2.135	-392.3	-0.857	0.857β

Failure equation: $1.604\beta^2 - 467.1\beta + 1230 = 0$

$$\beta = 2.657$$

	Y	33.66	4.208	39.20	-1.568	198.4	0.492	-0.582β
4	Z	47.70	2.385	49.85	-0.294	85.16	0.138	-0.107β
	M	1606	410.6	671.9	2.275	-410.6	-0.860	0.860β

Failure equation: $1.642\beta^2 - 486.7\beta + 1282 = 0$

$$\beta = 2.658$$

Therefore,

$$y_1^* = -\alpha_{Y_1}^* \beta \sigma_{Y_1} = -0.815 \times 1.183\beta = -0.964\beta$$

$$y_2^* = -\alpha_{Y_2}^* \beta \sigma_{Y_2} = -0.229 \times 0.774\beta = -0.176\beta$$

$$y_3^* = -\alpha_{Y_3}^* \beta \sigma_{Y_3} = 0.532 \times 1.0\beta = 0.532\beta$$

and the failure equation $g(\mathbf{X}) = 0$ becomes

$$5.616\beta^2 - 358.02\beta + 1011.8 = 0$$

yielding

$$\beta = 2.963$$

For the second iteration, the new failure point is calculated to be $y_1^* = -2.856$, $y_2^* = -0.5215$, $y_3^* = 1.576$; and in terms of the original variables,

$$y^* = -3.536 \times 2.856 - 3.536 \times 0.5215 + 39.64 = 27.69$$

$$z^* = -1.768 \times 2.856 + 1.768 \times 0.5215 + 49.89 = 45.76$$

$$m^* = 191.14 \times 1.576 + 965.8 = 1267.00$$

The above results and those of the subsequent iterations are summarized in Table E6.10. Thus, the probability of failure is

$$p_F = 1 - \Phi(2.658) = 0.00394$$

The above result may be compared with that of Example B.4 in Appendix B that gives $\beta = 2.663$.

EXAMPLE 6.11   (Settlement in Consolidated Clay)

The settlement of a point A in Fig. E6.11 caused by the construction of a structure can be shown to be primarily caused by the consolidation of the clay layer. Suppose the contribution of settlement due to secondary consolidation is negligible. According to Peck et al. (1974), for a normally loaded clay, the settlement S is given by

$$S = \frac{C_c}{1 + e_o} H \log \frac{p_o + \Delta p}{p_o}$$

where C_c is the compression index of the clay; e_o is the void ratio of the clay layer before loading; H is the thickness of the clay layer; p_o is the original effective pressure at point B (midheight

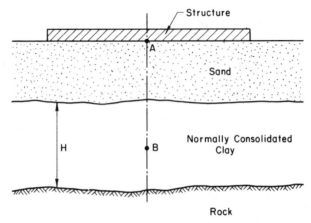

Figure E6.11   Soil profile.

of clay layer) before loading; and Δp is the increase in pressure at point B caused by the construction of the structure; "log" denotes logarithm to the base 10. Because of the nonuniform thickness and lack of homogeneity of the clay layer, the settlement predicted by the empirical formula could be subject to model error, which may be corrected by a factor N.

Suppose satisfactory performance requires that the settlement be less than 2.5 inches and the variables have the following statistics:

	Mean	c.o.v.	Standard Deviation
N	1.0	0.10	0.10
C_c	0.396	0.25	0.099
e_o	1.19	0.15	0.1785
H	168 in.	0.05	8.4
p_o	3.72 ksf	0.05	0.186
Δp	0.50 ksf	0.20	0.10

Since the probability distributions of the variables are unknown, they will be assumed to be normally distributed. Determine the probability of excessive settlement at point A.

The performance function is given by

$$g(\mathbf{X}) = 2.5 - N \frac{C_c}{1 + e_o} H \log \frac{p_o + \Delta p}{p_o}$$

The partial derivatives of the performance function with respect to each of the variables evaluated at the failure point are as follows:

$$\left(\frac{\partial g}{\partial N'}\right)_* = -\frac{c_c^*}{1 + e_o^*} h^* \log \frac{p_o^* + \Delta p^*}{p_o^*} \sigma_N$$

$$\left(\frac{\partial g}{\partial C_c'}\right)_* = -\frac{n^*}{1 + e_o^*} h^* \log \frac{p_o^* + \Delta p^*}{p_o^*} \sigma_{c_c}$$

$$\left(\frac{\partial g}{\partial e_o'}\right)_* = n^* \frac{c_c^*}{(1 + e_o^*)^2} h^* \log \frac{p_o^* + \Delta p^*}{p_o^*} \sigma_{e_o}$$

$$\left(\frac{\partial g}{\partial H'}\right)_* = -n^* \frac{c_c^*}{1 + e_o^*} \log \frac{p_o^* + \Delta p^*}{p_o^*} \sigma_H$$

$$\left(\frac{\partial g}{\partial p_o'}\right)_* = n^* \frac{c_c^*}{1 + e_o^*} h^* \frac{1}{\ln 10} \frac{\Delta p^*}{(p_o^* + \Delta p^*)p_o^*} \sigma_{p_o}$$

$$\left(\frac{\partial g}{\partial \Delta p'}\right)_* = -n^* \frac{c_c^*}{1 + e_o^*} h^* \frac{1}{\ln 10} \frac{1}{p_o^* + \Delta p^*} \sigma_{\Delta p}$$

For the first iteration, assume

$$n^* = \bar{N} = 1.0$$
$$c_c^* = \bar{C}_c = 0.396$$
$$e_o^* = \bar{e}_o = 1.19$$
$$h^* = \bar{H} = 168.0$$
$$p_o^* = \bar{p}_o = 3.72$$
$$\Delta p^* = \overline{\Delta p} = 0.50$$

The calculations for two iterations are summarized in Table E6.11.

Table E6.11   Summary of Iterative Calculations for Example 6.11

Iteration No.	Variable X_i	Assumed Failure Point x^*	$\left(\dfrac{\partial g}{\partial X_i'}\right)_*$	$\alpha^*_{X_i}$	New x_i^*
1	N	1.0	−0.166	−0.290	1.00 $+ 0.029\beta$
	C_c	0.396	−0.416	−0.724	0.396 $+ 0.0717\beta$
	e_o	1.19	0.136	0.236	1.19 $- 0.042\beta$
	H	168.00	−0.084	−0.145	168.00 $+ 1.218\beta$
	p_o	3.72	0.078	0.136	3.72 $- 0.0253\beta$
	Δp	0.50	−0.313	−0.544	0.5 $+ 0.0544\beta$

Failure equation:

$$2.5 - (1 + 0.029\beta)\,\frac{0.396 + 0.0717\beta}{1 + (1.19 - 0.042\beta)}\,(168 + 1.218\beta)$$

$$\times \log\left[\frac{(3.72 - 0.0253\beta) + (0.5 + 0.0544\beta)}{3.72 - 0.0253\beta}\right] = 0$$

By trial-and-error: $\beta = 1.27$

Iteration No.	Variable X_i	Assumed Failure Point x^*	$\left(\dfrac{\partial g}{\partial X_i'}\right)_*$	$\alpha^*_{X_i}$	New x_i^*
2	N	1.037	−0.241	−0.323	1.00 $+ 0.0323\beta$
	C_c	0.487	−0.508	−0.681	0.396 $+ 0.0674\beta$
	e_o	1.137	0.209	0.280	1.19 $- 0.05\beta$
	H	169.60	−0.126	−0.166	168.00 $+ 1.394\beta$
	p_o	3.688	0.117	0.157	3.72 $- 0.0292\beta$
	Δp	0.569	−0.409	−0.548	0.5 $+ 0.0548\beta$

Failure equation:

$$2.5 - (1 + 0.0323\beta)\,\frac{0.396 + 0.0674\beta}{1 + (1.19 - 0.05\beta)}\,(168 + 1.394\beta)$$

$$\times \log\left[\frac{(3.72 - 0.0292\beta) + (0.5 + 0.0548\beta)}{3.72 - 0.0292\beta}\right] = 0$$

By trial-and-error: $\beta = 1.27$

The failure point, therefore, is

$$n^* = 1.041,\ c_c^* = 0.482,\ e_o^* = 1.127,\ h^* = 169.8,\ p_o^* = 3.683,$$

$$\Delta p^* = 0.570$$

The probability of excessive settlement then is

$$p_F = 1 - \Phi(1.27) = 0.102$$

EXAMPLE 6.12   *(Spillway Capacity)*

Inadequate spillway capacity to carry the inflow water during an extreme flood is a major cause of dam failure. For a spillway with an uncontrolled overflow ogee crest, the discharge is given by (Bureau of Reclamation, 1977)

$$Q_c = NCLH^{3/2}$$

where C is the discharge coefficient that is influenced by a number of factors such as depth of approach, upstream face slope, etc.; L is the effective length of the crest; H is the total head on the crest, including the velocity of approach head; and N is the correction for imperfections in the empirical formula. The inflow rate at the spillway can be modeled by

$$Q_L = RQ_I$$

where Q_I is the peak flow upstream to the reservoir and R is the attenuation factor to account for the volume effect of the reservoir. Inadequate spillway capacity is defined here as the event $Q_C < Q_L$. Suppose the variables have the following statistics:

	Mean	c.o.v.	Distribution
N	1.0	0.20	Normal
C	3.85	0.07	Normal
L	93.4 ft	0.06	Normal
H	12.0 ft	0.06	Normal
R	0.7	0.14	Normal
Q_I	9146 cfs	0.35	Type I Largest

Determine the probability of unsatisfactory performance.
   The performance function is given by

$$g(\mathbf{X}) = Q_C - Q_L = NCLH^{3/2} - RQ_I$$

For Q_I, which has the Type I largest value distribution, the parameters of the distribution, u and α, are

$$\alpha = \frac{1}{\sqrt{6}} \frac{1}{\sigma_{Q_I}} = 0.0003895$$

$$u = \bar{Q}_I - \frac{0.577}{\alpha} = 7665$$

Then, with Eqs. 4.19 and 4.20 for the CDF of Q_I, the mean and standard deviation of the equivalent normal distribution for Q_I are obtained through Eqs. 6.27 and 6.28.
   The partial derivatives of the performance function with respect to each of the variables evaluated at the failure point are as follows:

$$\left(\frac{\partial g}{\partial N'}\right)_* = c^* l^* (h^*)^{3/2} \sigma_N$$

$$\left(\frac{\partial g}{\partial C'}\right)_* = n^* l^* (h^*)^{3/2} \sigma_C$$

$$\left(\frac{\partial g}{\partial L'}\right)_* = n^* c^* (h^*)^{3/2} \sigma_L$$

$$\left(\frac{\partial g}{\partial H'}\right)_* = 1.5 n^* c^* l^* (h^*)^{1/2} \sigma_H$$

$$\left(\frac{\partial g}{\partial R'}\right)_* = -q_I^* \sigma_R$$

$$\left(\frac{\partial g}{\partial Q_I'}\right)_* = -r^* \sigma_{Q_I}$$

Assuming for the first iteration,

$$n^* = \bar{N} = 1.0$$

$$c^* = \bar{C} = 3.85$$

$$l^* = \bar{L} = 93.4$$

$$h^* = \bar{H} = 12.0$$

$$r^* = \bar{R} = 0.7$$

$$q_I^* = \bar{Q}_I = 9146.0$$

Table E6.12   Summary of Iterations for Example 6.12

Variable X_i	Assumed Failure Point x_i^*	σ_{X_i}	u_{X_i}	$\left(\dfrac{\partial g}{\partial X_i'}\right)_*$	$\alpha_{X_i}^*$	New Failure Point x_i^*	
1st Iteration							
N	1.0	0.20	1.0	2989.6	0.699	1.0 −	0.1398β
C	3.85	0.27	3.85	1047	0.245	3.85 −	0.0662β
L	93.4	5.60	93.4	896	0.209	93.4 −	1.172β
H	12.0	0.72	12.0	1345	0.314	12.0 −	0.266β
R	0.7	0.098	0.7	−896.3	−0.209	0.7 +	0.0205β
Q_I	9146	3149	8589	−2204	−0.515	8589 +	1622β

Failure equation:

$$(1 - 0.1398\beta)(3.85 - 0.0662\beta)(93.4 - 1{,}172\beta)(12 - 0.226\beta)^{3/2} - (0.7 + 0.0205\beta)(8589 + 1622\beta) = 0$$

Trial-and-error solution: $\beta = 2.21$

2nd Iteration							
N	0.691	0.20	1.0	2623	0.588	1.0 −	0.1175β
C	3.704	0.27	3.85	661	0.148	3.85 −	0.04β
L	90.8	5.60	93.4	559	0.125	93.4 −	0.70β
H	11.5	0.72	12.0	851	0.191	12.0 −	0.137β
R	0.7453	0.098	0.7	−1193	−0.267	0.7 +	0.0262β
Q_I	12,174	4275	7899	−3185	−0.714	7899 +	3051β

Failure equation:

$$(1 - 0.1175\beta)(3.85 - 0.04\beta)(93.4 - 0.70\beta)(12 - 0.137\beta)^{3/2} - (0.7 + 0.0262\beta)(7899 + 3051\beta) = 0$$

Trial-and-error solution: $\beta = 2.025$

3rd Iteration							
N	0.762	0.20	1.0	2783	0.549	1.0 −	0.1098β
C	3.769	0.27	3.85	759	0.150	3.85 −	0.0405β
L	92.0	5.60	93.4	646	0.127	92.0 −	0.712β
H	11.72	0.72	12.0	977	0.193	12.0 −	0.139β
R	0.753	0.098	0.7	−1380	−0.272	0.7 +	0.0266β
Q_I	14,077	4987	7036	−3755	−0.740	7036 +	3690β

Failure equation:

$$(1 - 0.1098\beta)(3.85 - 0.0405\beta)(93.4 - 0.712\beta)(12 - 0.139\beta)^{3/2} - (0.7 + 0.0266\beta)(7036 + 3690\beta) = 0$$

Trial-and-error solution: $\beta = 2.00$

we obtain

$$F_{Q_I}(q_I^*) = \exp[-e^{-0.0003895(9146-7665)}] = 0.5703$$

$$f_{Q_I}(q_I^*) = 0.0003895 \exp[-0.0003895(9146 - 7665) - e^{-0.0003895(9146-7665)}]$$
$$= 0.0001247$$

and

$$\sigma_{Q_I}^N = \frac{\phi[\Phi^{-1}(0.5703)]}{0.0001247} = 3149.2$$

$$\mu_{Q_I}^N = 9146 - 3149.2 \times 0.177 = 8589$$

The values of the pertinent partial derivatives are

$$\left(\frac{\partial g}{\partial N'}\right)_* = 3.85 \times 93.4 \times (12)^{1.5} \times 0.2 = 2989.6$$

$$\left(\frac{\partial g}{\partial C'}\right)_* = 1 \times 93.4 \times (12)^{1.5} \times 0.27 = 1047$$

$$\left(\frac{\partial g}{\partial L'}\right)_* = 1 \times 3.85 \times (12)^{1.5} \times 5.60 = 896$$

$$\left(\frac{\partial g}{\partial H'}\right)_* = 1.5 \times 1 \times 3.85 \times 93.4 \times (12)^{0.5} \times 0.72 = 1345$$

$$\left(\frac{\partial g}{\partial R'}\right)_* = -9146 \times 0.098 = -896.3$$

$$\left(\frac{\partial g}{\partial Q_I'}\right)_* = -0.7 \times 3149 = -2204$$

from which the direction cosines are

$$\alpha_N^* = 0.699$$
$$\alpha_C^* = 0.245$$
$$\alpha_L^* = 0.209$$
$$\alpha_H^* = 0.314$$
$$\alpha_R^* = -0.209$$
$$\alpha_{Q_I}^* = -0.515$$

Therefore,

$$n^* = \mu_N - \alpha_N^* \beta \sigma_N = 1 - 0.699 \times 0.2\beta = 1 - 0.1398\beta$$

$$c^* = \mu_C - \alpha_C^* \beta \sigma_C = 3.85 - 0.245 \times 0.27\beta = 3.85 - 0.0662\beta$$

$$l^* = \mu_L - \alpha_L^* \beta \sigma_L = 93.4 - 0.209 \times 5.60\beta = 93.4 - 1.172\beta$$

$$h^* = \mu_H - \alpha_H^* \beta \sigma_H = 12.0 - 0.314 \times 0.72\beta = 12.0 - 0.226\beta$$

$$r^* = \mu_R - \alpha_R^* \beta \sigma_R = 0.7 + 0.209 \times 0.098\beta = 0.7 + 0.0205\beta$$

$$q^* = \mu_{Q_I}^N - \alpha_{Q_I}^* \beta \sigma_{Q_I}^N = 8589 + 0.515 \times 3149\beta = 8589 + 1622\beta$$

The limit-state equation then becomes

$$(1 - 0.1398\beta)(3.85 - 0.0662\beta)(93.4 - 1,172\beta)(12 - 0.226\beta)^{1.5}$$
$$-(0.7 + 0.0205\beta)(8589 + 1622\beta) = 0$$

By trial and error, we have $\beta = 2.21$. Thus, the new failure point has components

$$n^* = 0.691$$
$$c^* = 3.704$$
$$l^* = 90.8$$
$$h^* = 11.5$$
$$r^* = 0.7453$$
$$q^* = 12,174.0$$

Repeating the procedure, the results of the iterations may be summarized as shown in Table E6.12. Based on the results of the third iteration, the probability of unsatisfactory performance is

$$p_F = 1 - \Phi(2.00) = 0.023$$

and the most probable failure point is

$$n^* = 0.78$$
$$c^* = 3.77$$
$$l^* = 92.0$$
$$h^* = 11.72$$
$$r^* = 0.75$$
$$Q_i^* = 14,416$$

EXAMPLE 6.13   (Horizontal Displacement of Offshore Platform)

During a storm, the cyclic wave loading induces horizontal and rotational cyclic displacements in an offshore platform. A finite element model, with soil properties obtained from laboratory tests, has been used to calculate the maximum cyclic displacements for given maximum wave forces on a gravity platform (Anderson and Aas, 1979). Since the stiffness of the supporting soil generally decreases with increasing stress level, the induced displacement is expected to increase more rapidly at higher loads. The required relationship can be approximated by a quadratic equation of the form

$$D = AF + BF^2 + \varepsilon$$

where D is the maximum horizontal displacement, F is the maximum force, A and B are coefficients, and ε is the error of the quadratic model.

The soil properties assumed in the finite element analysis could be subject to considerable uncertainties; hence, the results of the equation as determined by the coefficients A and B could vary considerably. Nevertheless, for simplicity, the above quadratic model may be used, with the coefficients A and B assumed to be random variables. Suppose D is measured in cm and the random variables have the following statistics:

	Mean	c.o.v. (or σ)	Distribution
F	25	0.23	Lognormal
A	0.0113	0.30	Normal
B	0.0006	0.30	Normal
ε	0	$\sigma = 0.10$	Normal

Table E6.13 Summary of Iterations for Example 6.13

Variable X_i	Assumed Failure Point x_i^*	σ_{X_i}	μ_{X_i}	$\left(\dfrac{\partial g}{\partial X_i'}\right)_*$	$\alpha_{X_i}^*$	New Failure Point x_i^*
1st Iteration						
F	25.0	5.75	24.33	-0.237	-0.808	$24.33 + 4.646\beta$
A	0.0113	0.00339	0.0113	-0.0848	-0.289	$0.0113 + 0.000979\beta$
B	0.0006	0.00018	0.0006	-0.113	-0.384	$0.0006 + 0.000691\beta$
ε	0.0	0.1	0.0	-0.1	-0.341	0.0341β

Failure Equation:

$$1 - (0.0113 + 0.000979\beta)(24.33 + 4.646\beta)$$
$$-(0.0006 + 0.0000691\beta)(24.33 + 4.646\beta)^2 - 0.0341\beta = 0$$

By trial-and-error: $\beta = 1.134$

Variable X_i	x_i^*	σ_{X_i}	μ_{X_i}	$\left(\dfrac{\partial g}{\partial X_i'}\right)_*$	$\alpha_{X_i}^*$	x_i^*
2nd Iteration						
F	29.598	6.807	23.81	-0.357	-0.86	$23.81 + 5.854\beta$
A	0.0124	0.00339	0.0113	-0.100	-0.241	$0.0113 + 0.000817\beta$
B	0.000678	0.00018	0.0006	-0.158	-0.379	$0.0006 + 0.0000682\beta$
ε	0.0387	0.1	0.0	-0.1	-0.241	0.0241β

Failure Equation:

$$1 - (0.0113 + 0.000817\beta)(23.81 + 5.854\beta)$$
$$-(0.0006 + 0.0000682\beta)(23.81 + 5.854\beta) - 0.0241\beta = 0$$

By trial-and-error: $\beta = 1.07$

Variable X_i	x_i^*	σ_{X_i}	μ_{X_i}	$\left(\dfrac{\partial g}{\partial X_i'}\right)_*$	$\alpha_{X_i}^*$	x_i^*
3rd Iteration						
F	30.07	6.916	23.71	-0.364	-0.859	$23.71 + 5.94\beta$
A	0.01217	0.00339	0.0113	-0.102	-0.241	$0.0113 + 0.000817\beta$
B	0.000673	0.00018	0.0006	-0.163	-0.384	$0.0006 + 0.0000691\beta$
ε	0.0258	0.1	0.0	-0.1	-0.236	0.0236β

Failure Equation

$$1 - (0.0113 + 0.000817\beta)(23.71 + 5.94\beta)$$
$$-(0.0006 + 0.0000691\beta)(23.71 + 5.94\beta)^2 - 0.0236\beta = 0$$

By trial-and-error: $\beta = 1.07$

Assume that satisfactory performance requires that D should be less than 1 cm. Determine the probability of unsatisfactory performance.

The performance function, therefore, is given by

$$g(\mathbf{X}) = 1 - AF - BF^2 - \varepsilon$$

For the lognormal distribution of F, the parameters are

$$\zeta_F \simeq 0.23 \quad \text{and} \quad \lambda_F = \ln 25 - \tfrac{1}{2}(0.23)^2 = 3.192$$

The standard deviation and mean of the equivalent normal distribution for F are obtained as follows (see Example 6.4):

$$\sigma_F^N = f^* \zeta_F$$

and

$$\mu_F^N = f^*(1 - \ln f^* + \lambda_F)$$

The pertinent partial derivatives are

$$\left(\frac{\partial g}{\partial F'}\right)_* = (-a^* - 2b^*f^*)\sigma_F$$

$$\left(\frac{\partial g}{\partial A'}\right)_* = -f^*\sigma_A$$

$$\left(\frac{\partial g}{\partial B'}\right)_* = -(f^*)^2\sigma_B$$

$$\left(\frac{\partial g}{\partial \varepsilon'}\right)_* = -\sigma_\varepsilon$$

With the above relationships and assuming for the first iteration,

$$f^* = \bar{F} = 25$$

$$a^* = \bar{A} = 0.0113$$

$$b^* = \bar{B} = 0.0006$$

$$\varepsilon^* = \bar{\varepsilon} = 0.0$$

the calculations for three iterations are summarized in Table E6.13. Thus,

$$f^* = 30.06, \ a^* = 0.01217, \ b^* = 0.000674, \ \varepsilon^* = 0.0253 \quad \text{and} \quad p_F = 1 - \Phi(1.07) = 0.142$$

Hence, the probability of nonperformance is approximately 14%.

Accuracy of Linear Approximation   As alluded to earlier, the "linear" approximation of nonlinear performance functions is tantamount to replacing an n-dimensional failure surface (a hyper-surface) with a hyper-plane tangent to the failure surface at the "most probable failure point." In effect, this changes the boundary between the safe state, $g(\mathbf{X}) > 0$, and the failure state, $g(\mathbf{X}) < 0$, from a general curvilinear surface to a plane surface; the failure probability, p_F, is then the generalized volume integral of the joint PDF over the failure region $g(\mathbf{X}) < 0$. As observed earlier from Fig. 6.9, the reliability p_S estimated on the basis of this approximate planar failure surface will be on the conservative or unconservative side depending on whether the actual failure surface is convex or concave toward the origin of the reduced variates. The accuracy may be improved through quadratic or polynomial approximation (e.g. Fiessler et al., 1979) at the cost of mathematical and computational complications.

For a concave failure surface, the safe state, $g(\mathbf{X}) > 0$, is furthermore bounded between the half-space with the tangent plane (of distance β) and the hyper-sphere of radius β (as illustrated in Fig. 6.10 for two dimensions). The failure equation corresponding to the hyper-sphere is (Hasofer, 1974)

$$\sum_{i=1}^{n} X_i'^2 - \beta^2 = 0$$

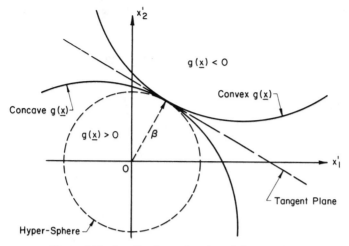

Figure 6.10 Implications of various failure surfaces.

If the variates X_i' are uncorrelated standard normal variates, the sum of squares $\sum_{i=1}^{n} X_i'^2$ has a chi-square distribution with n degrees-of-freedom (Vol. I, Chapter 5). Therefore, the probability of failure becomes

$$p_F = 1 - \chi_n^2(\beta^2)$$

where $\chi_n^2(-)$ is the CDF of the chi-square distribution with n degrees of freedom.

Accordingly, for concave failure surfaces the failure probability is bounded as follows:

$$\Phi(-\beta) < p_F < 1 - \chi_n^2(\beta^2)$$

In general, the accuracy of the second-moment linear approximation is difficult to assess; this will depend on the degree of nonlinearity of the function $g(\mathbf{X})$. Obviously, the method is mathematically exact if $g(\mathbf{X})$ is linear. For a general nonlinear $g(\mathbf{X})$, the accuracy may only be appraised numerically for specific forms of nonlinear performance functions. For this purpose, consider first the following nonlinear performance function for which we know the exact answer.

$$g(\mathbf{X}) = \frac{X^2}{Y} - 1$$

Assume that X and Y are statistically independent lognormal variates, with

$$\mu_X = 6 \text{ units}; \qquad \Omega_X = 0.08$$

$$\mu_Y = 12 \text{ units}; \qquad \Omega_Y = 0.30$$

The parameters of the respective lognormal distributions are then

$$\zeta_X = \sqrt{\ln(1 + 0.08^2)} = 0.08; \qquad \zeta_Y = \sqrt{\ln(1 + 0.30^2)} = 0.294$$

$$\lambda_X = \ln 6 - 1/2(0.08)^2 = 1.789; \qquad \lambda_Y = \ln 12 - 1/2(0.294)^2 = 2.442$$

In this case, the equivalent normal distribution for X will have mean and standard deviation as follows:

$$\mu_X^N = x^*[1 - (\ln x^* - \lambda_X)]$$
$$\sigma_X^N = \zeta_X x^*$$

Similarly, for Y,

$$\mu_Y^N = y^*[1 - (\ln y^* - \lambda_Y)]$$
$$\sigma_Y^N = \zeta_Y y^*$$

The pertinent partial derivatives are

$$\frac{\partial g}{\partial X'} = \frac{2X}{Y}\sigma_X^N; \qquad \frac{\partial g}{\partial Y'} = -\frac{X^2}{Y^2}\sigma_Y^N$$

Table 6.1 Summary of Iterations

Iteration No.	X_i	x_i^*	$\sigma_{X_i}^N$	$\mu_{X_i}^N$	$\left(\dfrac{\partial g}{\partial X_i'}\right)_*$	α_i^*	New x_i^*
1	X	6	0.480	5.983	0.480	0.478	$5.983 - 0.229\beta$
	Y	12	3.528	11.485	-0.882	-0.878	$11.485 + 3.098\beta$
				Failure equation: $\beta^2 - 111.41\beta + 463.95 = 0$			
				$\beta = 4.33$			
2	X	4.99	0.399	6.172	0.160	0.478	$6.172 - 0.191\beta$
	Y	24.90	7.321	5.656	-0.294	0.878	$5.656 + 6.428\beta$
				Failure equation: $\beta^2 - 240.71\beta + 888.71 = 0$			
				$\beta = 3.75$			
3	X	5.46	0.437	5.960	0.160	0.478	$5.960 - 0.209\beta$
	Y	29.76	8.749	1.453	-0.294	0.878	$1.453 + 7.664\beta$
				Failure equation $\beta^2 - 232.38\beta + 772.75 = 0$			
				$\beta = 3.375$			
4	X	5.25	0.420	5.937	0.161	0.477	$5.937 - 0.200\beta$
	Y	27.32	8.032	3.671	-0.297	0.879	$3.671 + 7.060\beta$
				Failure equation: $\beta^2 - 235.88\beta + 789.43 = 0$			
				$\beta = 3.395$			
5	X	5.26	0.420	5.938	0.160	0.478	$5.938 - 0.201\beta$
	Y	27.64	8.126	3.392	-0.294	0.878	$3.392 + 7.135\beta$
				Failure equation: $\beta^2 - 235.69\beta + 788.81 = 0$			
				$\beta = 3.395$			

With the above relationships, the results of the various iterations are summarized in Table 6.1. Therefore, with the result of Iteration No. 5, the calculated probability of failure is

$$p_F = 1 - \Phi(3.395) = 3.43 \times 10^{-4}$$

In the present case, the failure equation may also be expressed as

$$g(\mathbf{X}) = \ln \frac{X^2}{Y} = 2 \ln X - \ln Y = 0$$

Since X and Y are lognormal variates, $2 \ln X - \ln Y$ is normal and linear in the space of $\ln X$ and $\ln Y$. Hence, the exact probability of failure may be obtained as

$$p_F = P\left(\ln \frac{X^2}{Y} < 0\right) = \Phi\left[\frac{-(2\lambda_X - \lambda_Y)}{\sqrt{(2\zeta_X)^2 + \zeta_Y^2}}\right] = 1 - \Phi\left[\frac{2 \times 1.789 - 2.442}{\sqrt{(2 \times 0.08)^2 + (0.294)^2}}\right]$$

$$= 1 - \Phi(3.394) = 3.44 \times 10^{-4}$$

Thus, in this case, the approximation yields almost the exact answer.

For a general nonlinear performance function, the "correct" probability of failure may be evaluated through large-sample Monte Carlo calculations (see Chapter 5). In this regard, results of Monte Carlo calculations for Examples 6.7, 6.11, 6.12, and 6.13, all involving nonlinear $g(\mathbf{X})$, are summarized in Table 6.2. On this basis, the accuracy of the second-moment approximate method may be inferred. The results (albeit limited) provide evidence of the validity and accuracy of the second-moment approximation. Moreover, all the nonlinear performance functions examined are typical of those found in practical engineering problems.

Table 6.2 Comparison of Calculated Failure Probabilities

Example No.	$g(\mathbf{X})$	p_F by Second Moment	Monte Carlo Expected p_F	Monte Carlo 95% Confidence Interval	Monte Carlo Sample Size
6.7	$X_1 X_2 - X_3$	0.0031	0.0034	0.0029 – 0.0040	50,000
6.11	$2.5 - N\left[\frac{C_c}{1 + c_o} H \log\left(\frac{p_o + \Delta p}{p_o}\right)\right]$	0.102	0.096	0.087 – 0.105	4000
6.12	$2NCLH^{3/2} - RQ_I$	0.023	0.026	0.024 – 0.029	20,000
6.13	$1 - (AF + BF^2 + \varepsilon)$	0.142	0.128	0.116 – 0.140	3000

6.3 MODELING AND ANALYSIS OF UNCERTAINTY

It is easily observed from Section 6.2 that the evaluation of safety and reliability requires information on (random) uncertainty, as represented by the standard deviation or coefficient of variation; indeed, the determination of these uncertainties constitutes an essential task in the evaluation of engineering reliability.

This is, of course, not surprising as questions of safety or reliability arise principally because of the presence of uncertainty. An analysis and assessment of uncertainty must include the uncertainties in the design variables as well as in the prediction models.

6.3.1 Sources and Types of Uncertainty

Uncertainties in engineering may be associated with physical phenomena that are inherently random or with predictions and estimations of reality (i.e., state of nature) performed under conditions of incomplete or inadequate information. From this standpoint, uncertainty may be associated with the inherent variability of the physical process or with the imperfection in the modeling of the physical process; that is, randomness in a physical process contributes to uncertainty because it is inherently not possible to ascertain the realization or outcome of the process, whereas potential errors of an imperfect prediction model cannot be entirely corrected deterministically. Moreover, prediction or modeling error may contain two components: the *systematic component* and the *random component*. In measurement theory these are known as the "systematic error" and "random error," respectively. From a practical standpoint, inherent variability is essentially a state of nature and the resulting uncertainty may not be controlled or reduced; that is, the uncertainty associated with inherent variability is something we have to live with. The uncertainty associated with prediction or modeling error may be reduced through the use of more accurate models or the acquisition of additional data.

Uncertainty Due to Inherent Variability Uncertainty is naturally associated with random phenomena because the exact realization of a phenomenon cannot be determined with certainty. The conceivable or possible realizations may be described in terms of a range of possibilities, with their respective relative likelihoods of occurrence (e.g., with a probability density function). In other words, if the state of nature is basically random, it cannot be described with a deterministic model; its description must include a measure of its inherent variability and thus uncertainty. For practical purposes, the required description may have to be limited to the main descriptors of interest, which are the *central value* (e.g., the mean or median) and its *measure of dispersion* (e.g., standard deviation or coefficient variation). Available observational data are normally used to estimate the central value and the degree of dispersion of the possible realizations.

Consider, for example, the compressive strength of concrete, f_c, used in a major structure. For purposes of illustration, suppose that 15 control cylinders were sampled from the concrete mixes used in the construction and subsequently tested in compression with the following results (in ksi):

$$
\begin{array}{ccc}
6.5 & 6.1 & 4.7 \\
4.3 & 4.8 & 5.7 \\
5.2 & 5.5 & 5.2 \\
5.8 & 4.2 & 4.1 \\
5.0 & 5.1 & 6.3 \\
\end{array}
$$

On the basis of these observations, the sample mean and sample standard deviation (see Chapter 5, Vol. I) are obtained as follows:

Sample mean, $\bar{f}_c = 5.23$ ksi

Sample standard deviation, $\sigma_{f_c} = 0.75$ ksi

The corresponding coefficient variation, therefore, is

$$\delta_{f_c} = \frac{0.75}{5.23} = 0.14$$

The above sample mean, $\bar{f}_c = 5.23$ ksi, is of course an estimated value of the true mean strength of the concrete (which remains unknown). The fact that there is significant scatter in the observed strength of the fifteen different cylinders gives rise to uncertainty in the actual strength of any given section or part of the structure; this is the uncertainty associated with the inherent variability of the concrete strength, as measured by the c.o.v. of $\delta_{f_c} = 0.14$.

Uncertainty Associated with Prediction Error   Errors of prediction include estimation error (such as statistical sampling error) as well as the imperfection of the prediction model. As used here, a model is meant to be any technique or method for predicting or estimating the real-world condition. As indicated earlier, such prediction error may include a systematic component (i.e., bias) as well as a random component (random error).

In the above example, the mean strength of the concrete was estimated from the observed experimental data to be $\bar{f}_c = 5.23$ ksi. Conceivably, this estimate of the true mean value would contain error. If the experiment is repeated and other sets of data obtained, the sample mean estimated from other sets of data would most likely be different; the collection of all the sample means will also have a mean value, which may well be different from the individual mean values, and a corresponding standard deviation. Conceptually, the mean value of the collection of sample means may be assumed to be close to the true mean value (assuming that the estimator is unbiased). Then, the difference (or ratio) of the estimated sample mean (i.e., $\bar{f}_c = 5.23$ ksi) to the true mean is the systematic error, whereas the c.o.v. or standard deviation of the collection of sample means represents a measure of the random error.

In effect, random error is involved whenever there is a range of possible error. One source of random error is the error due to sampling (see Eq. 5.11, Vol. I) which is a function of the sample size. In the above example, the random sampling error (expressed in terms of c.o.v.) would be

$$\Delta_1 = 0.14/\sqrt{15} = 0.04$$

Systematic error or bias may arise from factors not accounted for in the prediction model that tend to consistently bias the estimate in one direction or the other. For instance, the concrete cylinders in the above example may be cured under laboratory conditions that would tend to raise the strength over that in the field; also, compaction and direction of casting may influence the strength of field concrete, whereas confinement of the material will tend to increase its strength.

It is, of course, difficult to evaluate or determine the individual effects of these additional factors on the strength of field concrete. However, there is evidence to indicate that the strength of laboratory concrete cylinders tends to be higher than the strength of field concrete; for example, Bloem (1968) reported that field concrete is 10 to 21 % lower than laboratory concrete. This information, therefore, would suggest that the mean strength of field concrete is invariably lower than laboratory concrete materials; specifically, it will range between 10 and 21 % less than that of corresponding laboratory concrete. With this information, a realistic prediction of field concrete strength, therefore, requires a correction of from 79 to 90 % of the corresponding laboratory concrete (of the same mix). If a uniform PDF between this range of correction factors is assumed (see Section 6.3.2), the systematic error in the estimated mean concrete strength of $\bar{f}_c = 5.23$ ksi will need to be corrected by a mean bias factor of

$$v = \tfrac{1}{2}(0.79 + 0.90) = 0.85$$

whereas the corresponding random error in the estimated mean value, expressed in c.o.v., is

$$\Delta_2 = \frac{1}{\sqrt{3}} \left(\frac{0.90 - 0.79}{0.90 + 0.79} \right) = 0.04$$

The total random error in the estimated mean value, therefore, is

$$\Delta = \sqrt{\Delta_1^2 + \Delta_2^2} = \sqrt{0.04^2 + 0.04^2} = 0.06$$

The concepts presented above are extensions or generalizations of the notion of estimation error that is well-established, for example, in measurement theory (Parratt, 1961). In measurement theory, the estimated mean value from a set of observations is usually used to represent the true measurement (a state of nature); the error of the estimated mean value consists of systematic and random components. The systematic component may be the result of certain well-identified factors whose effects can be determined (perhaps judgmentally) and thus corrected through a constant bias factor, whereas the random component, called *standard error* in measurement theory, may be represented as a range of possible errors, requiring statistical treatment. The systematic error, therefore, is a bias in the prediction or estimation, whereas the random error represents the degree of dispersiveness of the range of possible errors.

In general, therefore, the systematic error in prediction may be corrected by applying a constant bias factor; however, the random error requires a statistical treatment and may be represented by the standard deviation or coefficient of variation of the estimated mean value. In other words, the systematic bias is a definite fault in the model that will consistently underestimate or overestimate the state of nature, whereas the random error is the statistical imperfection in the estimated central value. An objective determination of the bias, as well as the random error, will require repeated data on the sample means (or medians) as illustrated previously; data for these purposes, however, may be difficult to obtain in practice and hence must often be augmented by judgments.

In practice, if the underlying phenomenon is random, prediction or estimation is usually limited to the determination of a central value (e.g., the mean or median)

Total uncertainty $\Omega_{f_c} = \sqrt{\Delta^2 + \delta_{f_c}^2}$

and its associated standard deviation or coefficient of variation, the latter representing the inherent variability of the random phenomenon. It may be emphasized that the uncertainty associated with the error in the prediction of the central value is of first-order importance; although there may also be error in the estimation of the degree of dispersion, the uncertainty associated with this latter error is of secondary importance. Therefore, in our discussion of uncertainty analysis, the error of prediction will be limited to that of the mean value.

The above discussions may be summarized as follows. Through methods of prediction, we obtain

$\bar{x}$ = Estimate of the mean value.

σ_x = Estimate of the standard deviation (representing inherent variability).

An assessment of the accuracy or inaccuracy of the above prediction for the mean value is performed, obtaining

v = Bias correction for systematic error in the predicted mean value $\bar{x}$.

Δ = Measure of random error in $\bar{x}$.

In other words, prediction (in the case of a random phenomenon) usually yields an estimate of the mean value, $\bar{x}$, and associated c.o.v. $\delta = \sigma_x/\bar{x}$; the systematic error in the estimated mean value may be corrected through a bias factor v, whereas its random error Δ may be expressed in terms of c.o.v. and treated statistically.

6.3.2 Quantification of Uncertainty Measures

The uncertainties associated with inherent variability as well as random error may be expressed in terms of the coefficient of variation. The systematic error, however, may be adjusted by a bias correction factor, v. Again, we emphasize that for practical purposes, errors of prediction will be limited to the errors in the estimation of the respective mean values; that is, the systematic and random errors shall refer to the bias and standard error, respectively, in the estimated mean value of a variable (or function of variables).

For obvious reasons, the credibility of the uncertainty measures is important; in particular, the validity of a calculated probability will depend on credible assessments of the individual uncertainty measures. Methods for evaluating uncertainty measures will depend on the form of the available data and information (or the lack thereof); some of these methods are described below. However, depending on the situation at hand, ad hoc methods may sometimes have to be devised when necessary.

Sample Data Available When a set of observational data is available, for example, a set of sample data $(x_1, x_2, \ldots, x_n)$, common statistical estimation techniques (see Chapter 5, Vol. I) may be used to estimate the mean value, obtaining

$$\bar{x} = \frac{1}{n} \sum_{i=1}^{n} x_i$$

and the corresponding variance

$$\sigma_X^2 = \frac{1}{n-1} \sum_{i=1}^{n} (x_i - \bar{x})^2$$

from which the uncertainty associated with inherent randomness is given by the c.o.v.

$$\delta_X = \sigma_X/\bar{x}$$

The above estimated mean value may not be totally accurate relative to the true mean (especially for small sample size n). The estimated mean value given above is unbiased as far as sampling is concerned; however, the random error of the above $\bar{x}$ is the standard error of $\bar{x}$, which is

$$\sigma_{\bar{x}} = \sigma_X/\sqrt{n}$$

Hence, the uncertainty associated with the random sampling error is

$$\Delta_X = \sigma_{\bar{x}}/\bar{x}$$

It should be emphasized that the above random error in $\bar{x}$ is limited only to the sampling error. In particular, there may be other biases and random errors in $\bar{x}$, such as the effects of factors not included in the observational program.

Range of Values Known   In engineering, the information may often have to be expressed in terms of the lower and upper limits of a variable. For example, when judgment is necessary, it is often convenient and perhaps more realistic to express judgmental information in the form of a range of possibilities.

Given the range of possible values of a random variable, the mean value of the variable and the underlying uncertainty may be evaluated by prescribing a suitable distribution within the range (Yucemen, Tang, and Ang, 1976). For example, for a variable X, if the lower and upper limits of its values are x_l and x_u, the mean and c.o.v. of X may be determined as follows.

If a uniform PDF is prescribed between x_l and x_u, as shown in Fig. 6.11a, the mean value is

$$\bar{x} = \tfrac{1}{2}(x_l + x_u)$$

whereas the c.o.v. is

$$\delta_X = \frac{1}{\sqrt{3}} \left(\frac{x_u - x_l}{x_u + x_l} \right)$$

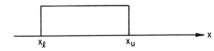

Figure 6.11a   Uniform PDF between x_l and x_u.

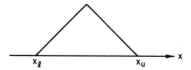

Figure 6.11b Symmetric triangular PDF between x_l and x_u.

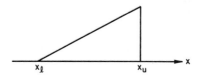

Figure 6.11c Upper triangular PDF between x_l and x_u.

Alternatively, if a symmetric triangular distribution is prescribed within the limits x_l and x_u, as in Fig. 6.11b, the corresponding c.o.v. would be

$$\delta_X = \frac{1}{\sqrt{6}}\left(\frac{x_u - x_l}{x_u + x_l}\right)$$

With either the uniform or the symmetric triangular distribution, it is implicitly assumed that there is no bias within the prescribed range of values of X. On the other hand, if there is bias, the following skewed distributions may be more appropriate. For example, if it is judged that there is a bias toward the higher values within the specified range, then the upper triangular distribution, as shown in Fig. 6.11c, would be appropriate; in such a case, the mean value is

$$\bar{x} = \tfrac{1}{3}(x_l + 2x_u)$$

and the corresponding c.o.v. becomes

$$\delta_X = \frac{1}{\sqrt{2}}\left(\frac{x_u - x_l}{2x_u + x_l}\right)$$

Conversely, if there is a bias toward the lower range of values, the appropriate distribution may be a lower triangular distribution, as shown in Fig. 6.11d, giving the mean value as

$$\bar{x} = \tfrac{1}{3}(2x_l + x_u)$$

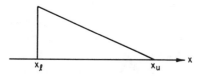

Figure 6.11d Lower triangular PDF between x_l and x_u.

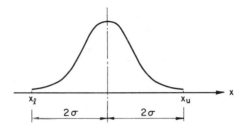

Figure 6.11e   Normal PDF between x_l and x_u.

and in such a case, the corresponding c.o.v. would be

$$\delta_X = \frac{1}{\sqrt{2}} \left(\frac{x_u - x_l}{x_u + 2x_l} \right)$$

Still other distributions may be assumed between the specified lower and upper limits x_l and x_u. For example, the given limits may be assumed to cover ± 2 standard deviations from the mean value of an underlying normal distribution (see Fig. 6.11e); in such a case, the mean value would be

$$\bar{x} = \tfrac{1}{2}(x_u + x_l)$$

and the corresponding c.o.v. is

$$\delta_X = \frac{1}{2} \left(\frac{x_u - x_l}{x_u + x_l} \right)$$

The means and c.o.v.'s corresponding to the various assumed distributions described above are summarized in Table 6.3.

6.3.3   Analysis of Uncertainty

The seemingly different types and sources of uncertainty described above may be analyzed in a unified manner with the following model.

Suppose that the true state of nature is X, whose actual realization is unknown. Prediction or estimation of X, therefore, is necessary; for this purpose, a model denoted as $\hat{X}$, may be used. As $\hat{X}$ is a model of the real world, imperfections in the model may be expected; the resulting predictions, therefore, will contain error and a correction N may be applied. Consequently, the true state of nature may be represented by (Ang, 1973)

$$X = N\hat{X} \qquad (6.38)$$

If the state of nature is random, the model $\hat{X}$ should naturally be a random variable. The estimated mean value $\bar{x}$ and variance $\sigma_{\hat{X}}^2$ (e.g., from a set of sample observations) are those of $\hat{X}$; from which the c.o.v. $\delta_X = \sigma_X/\bar{x}$ represents the inherent variability.

The necessary correction N may also be considered a random variable, whose mean value v represents the mean correction for systematic error in the predicted mean value $\bar{x}$, whereas the c.o.v. of N, Δ, represents the random error in the predicted mean value $\bar{x}$.

Table 6.3 Statistics of Prescribed Distributions

PDF	Mean Value, $\bar{x}$	c.o.v., δ_X or Δ_X
	$\frac{1}{2}(x_l + x_u)$	$\frac{1}{\sqrt{3}}\left(\dfrac{x_u - x_l}{x_u + x_l}\right)$
	$\frac{1}{2}(x_l + x_u)$	$\frac{1}{\sqrt{6}}\left(\dfrac{x_u - x_l}{x_u + x_l}\right)$
	$\frac{1}{3}(x_l + 2x_u)$	$\frac{1}{\sqrt{2}}\left(\dfrac{x_u - x_l}{2x_u + x_l}\right)$
	$\frac{1}{3}(2x_l + x_u)$	$\frac{1}{\sqrt{2}}\left(\dfrac{x_u - x_l}{x_u + 2x_l}\right)$
	$\frac{1}{2}(x_l + x_u)$	$\frac{1}{k}\left(\dfrac{x_u - x_l}{x_u + x_l}\right)$

It is reasonable to assume that N and $\hat{X}$ are statistically independent; on this basis, the mean value of X, following Eq. 6.38, is

$$\mu_X = \nu\bar{x} \tag{6.39}$$

The total uncertainty in the prediction of X then becomes

$$\Omega_X \simeq \sqrt{\delta_X^2 + \Delta_X^2} \tag{6.40}$$

The above pertains to a single variable. Often, the systematic and random errors of a function would also be of interest. For example, if Y is a function of several random variables $X_1, X_2, \ldots, X_n$, that is,

$$Y = g(X_1, X_2, \ldots, X_n)$$

the mean value and associated uncertainty of Y are of concern. In this case, an idealized (or model) function $\hat{g}$ would ordinarily be used, and a correction N_g may be necessary, such that

$$Y = N_g\hat{g}(X_1, X_2, \ldots, X_n) \tag{6.41}$$

in which N_g has mean value v_g and c.o.v. Δ_g. On the basis of first-order approximations (Chapter 3, Vol. I), the mean value of Y is

$$\mu_Y \simeq v_g \cdot \hat{g}(\mu_{X_1}, \mu_{X_2}, \ldots, \mu_{X_n}) \qquad (6.42)$$

where v_g is the bias in $\hat{g}(\ldots)$, and $\mu_{X_i} = v_i \bar{x}_i$. Also, the total c.o.v. of Y is

$$\Omega_Y^2 \simeq \Delta_g^2 + \frac{1}{\mu_{\hat{g}}^2} \sum_i \sum_j \rho_{ij} c_i c_j \sigma_{X_i} \sigma_{X_j} \qquad (6.43)$$

in which

$$c_i = \frac{\partial g}{\partial X_i}, \text{ evaluated at } (\mu_{X_1}, \mu_{X_2}, \ldots, \mu_{X_n}).$$

ρ_{ij} = Correlation coefficient between X_i and X_j.

To illustrate the general concepts presented above, let us examine the following examples.

EXAMPLE 6.14

Consider the estimation of the maximum deflection of a cantilever wood beam as shown in Fig. E6.14.

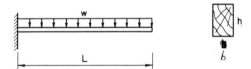

Figure E6.14   A cantilever wood beam.

where

w = Equivalent uniformly distributed load.

L = Span length.

b, h = Width and depth of a rectangular wood beam.

Assume that the beam is part of a roof that will be subjected to the lifetime maximum snow load besides the dead weight of the roof. Suppose further that the maximum deflection is calculated by the equation

$$d = \frac{w_n L^4}{8 E_n I_n} \qquad I = \frac{1}{12} b h^3 \qquad (i)$$

in which

E_n = The nominal modulus of elasticity.

I_n = The cross-sectional moment of inertia, based on nominal dimensions of the lumber. whereas the nominal load w_n is composed of

$$w_n = D_n + S_n$$

where

D_n = The nominal dead load.

S_n = The nominal (code-specified) snow load; for example, in the U.S. this may be the snow load specified in ANSI A5.8 (1972).

Determine the bias and uncertainty in the calculated deflection of the beam.
First, let us evaluate the bias and uncertainty in each of the variables.

On w_n

According to Ellingwood et al. (1980), there is a general tendency among designers to underestimate the dead load, and thus the ratio of $\bar{D}/D_n = 1.05$ may be assumed, whereas the c.o.v. of dead load is $\Omega_D = 0.10$.

The pertinent snow load would be the lifetime maximum snow load (e.g., over 50 years). However, most data on snow load are for ground snow, from which the roof snow load is determined as

$$S = C_s q \tag{ii}$$

where

q = Ground snow load.

C_s = A snow load coefficient, relating ground to roof snow load; dependent on the roof exposure, geometry, and thermal factors of the roof.

The annual maximum snow load, q_{an}, for various sites in the U.S. is reported by CRREL (1980); the lognormal distribution is used as the appropriate PDF and corresponding parameters are presented for a number of cities (e.g., for Green Bay, Wisconsin, $\lambda = 2.01$ and $\zeta = 0.70$).

Converting ground snow load to roof snow load with Eq. (ii) will entail additional uncertainty, namely that associated with C_s. Information from limited field surveys of flat roofs, augmented by professional judgments (Ellingwood et al., 1980), yielded

$$\bar{C}_s = 0.5$$

and

$$\Omega_{C_s} = 0.23$$

If C_s is also assumed to be lognormally distributed, its parameters would be

$$\zeta_{C_s} \simeq \Omega_{C_s} = 0.23$$

$$\lambda_{C_s} = \ln 0.5 - \tfrac{1}{2}(0.23)^2 = -0.667$$

Then, the distribution of the annual maximum roof snow load S_{an} will also be lognormal with the parameters (for Green Bay, WI)

$$\lambda_{S_{an}} = 2.01 - 0.667 = 1.34$$

$$\zeta_{S_{an}} = \sqrt{(0.70)^2 + (0.23)^2} = 0.74$$

On the bases of the above and according to Section 4.3.3, the lifetime maximum roof snow load will approach the Type II asymptotic extremal distribution. Thus, for $n = 50$ years, we obtain (see Example 4.16)

$$u'_n = 0.74\sqrt{2\ln 50} - 0.74\,\frac{\ln\ln 50 + \ln 4\pi}{2\sqrt{2\ln 50}} + 1.34 = 2.895$$

$$\alpha'_n = \frac{\sqrt{2\ln 50}}{0.74} = 3.78$$

The parameters of the appropriate Type II distribution for S_{50} are, therefore,

$$v_{50} = e^{2.895} = 18.08 \text{ psf}$$
$$k_{50} = 3.78$$

The mean and c.o.v. of S_{50} may then be obtained using Eqs. 4.55 and 4.57 as follows:

$$\bar{S}_{50} = 18.08\Gamma\left(1 - \frac{1}{3.78}\right)$$

$$= 18.08\Gamma(0.735) = 18.08\,\frac{\Gamma(1.735)}{0.735}$$

$$= 18.08\left(\frac{0.9157}{0.735}\right) = 22.52 \text{ psf}$$

$$\Omega_{S_{50}}^2 = \frac{\Gamma(1 - 2/3.78)}{\Gamma^2(1 - 1/3.78)} - 1$$

$$= \frac{\Gamma(0.471)}{\Gamma^2(0.735)} - 1 = \frac{(0.8856)(0.735)^2}{(0.471)(0.9157)^2} - 1$$

$$= 0.211$$

Thus,

$$\Omega_{S_{50}} = 0.46$$

According to ANSI A5.8(1972), the nominal ground snow load for Green Bay, WI is 28 psf and a value of $C_s = 0.8$ is recommended. Then, the ratio

$$\frac{S_n}{\bar{S}_{50}} = \frac{28 \times 0.80}{22.52} = 0.99$$

Consider a case in which the load ratio is $\bar{S}_{50}/\bar{D} = 1.0$; then, the total nominal load is

$$w_n = D_n + S_n$$

$$= \frac{\bar{D}}{1.05} + 0.99\bar{D} = 1.94\bar{D}$$

whereas the actual mean maximum load would be

$$\bar{w} = \bar{D} + \bar{S}_{50} = 1.99\bar{D}$$

Hence,

$$\frac{w_n}{\bar{w}} = \frac{1.94\bar{D}}{1.99\bar{D}} = 0.97$$

The total c.o.v. in w is obtained as

$$\sigma_w = \sqrt{\sigma_D^2 + \sigma_S^2} = \sqrt{(0.10\bar{D})^2 + (0.46\bar{D})^2} = 0.47\bar{D}$$

and

$$\Omega_w = \frac{0.47\bar{D}}{1.99\bar{D}} = 0.24$$

On E

The modulus of elasticity of wood varies widely; a c.o.v. of $\delta_E = 0.20$ for construction grade lumber is not unusual. However, the mean value, $\bar{E}$, will also depend on the moisture content and the quality of the lumber. Suppose (for the purpose of illustration) that conditions in the field could affect the laboratory-measured mean, $\bar{E}$, by as much as $\pm 15\%$; moreover, the field values of $\bar{E}$ will tend toward the lower values. Then, it may be appropriate to prescribe the lower triangular distribution for $\bar{E}$ as follows:

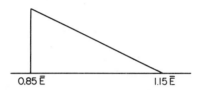

$$0.85\,\bar{E} \qquad\qquad\qquad 1.15\,\bar{E}$$

Thus, the best estimate for the actual mean is

$$\mu_E = \tfrac{1}{3}(2 \times 0.85\bar{E} + 1.15\bar{E}) = 0.95\bar{E}$$

meaning that the appropriate bias correction factor is $v_E = 0.95$.

The corresponding uncertainty (c.o.v.) in $\bar{E}$ would be

$$\Delta_E = \frac{1}{\sqrt{2}}\left(\frac{1.15 - 0.85}{1.15 + 2 \times 0.85}\right) = 0.07$$

Then the total c.o.v. in E becomes

$$\Omega_E = \sqrt{(0.20)^2 + (0.07)^2} = 0.21$$

On I

Suppose the cross-sectional dimensions, b and h, could vary by $\pm 10\%$ of the nominal. Assuming further that this variability corresponds to $\pm 2\sigma$ (two standard deviations), we have

$$\delta_b = \delta_h = 0.05$$

The moment of inertia of a rectangular section is

$$I = \tfrac{1}{12}bh^3$$

It is reasonable to assume that b and h are perfectly correlated; then, the c.o.v. in I would be

$$
\begin{aligned}
\delta_I &= \sqrt{\delta_b^2 + (3\delta_h)^2 + 2(\delta_b \times 3\delta_h)} \\
&= \sqrt{(0.05)^2 + (0.15)^2 + 2(0.05 \times 0.15)} \\
&= 0.20
\end{aligned}
$$

whereas

$$v_I = 1.00 \text{ (no bias)}$$

On ĝ

The deflection equation, $\hat{g} = wL^4/8EI$, is based on the assumption of linearly elastic material and a fully fixed support of the cantilever beam. Obviously, in a real beam, these conditions are seldom satisfied.

If tests of "fixed" cantilever beams were performed, the ratio of the test to calculated results would be a basis for determining the bias and uncertainty in the function $\hat{g}$. Suppose (hypothetically) that six such tests yielded the following ratios:

1.07	1.12
0.95	1.06
0.98	1.10

These would give a mean ratio of (which is the mean bias factor)

$$v_g = 1.05$$

and a c.o.v. of (representing the random model error in $\hat{g}$)

$$\Delta_g = 0.07$$

In summary, we have for each of the variables the following:

Nominal Variable	v_i	Ω_i
w_n	0.97	0.24
$\bar{E}$	0.95	0.21
I_n	1.00	0.20
$\hat{g}$	1.05	0.07

With the assumption that $E_n = \bar{E}$ and $I_n = \bar{I}$, the mean maximum deflection of the beam would be

$$\mu_d = 1.05 \left[\frac{(w_n/0.97)L}{8 \times 0.95 E_n \times I_n} \right]$$

$$= \frac{1.05}{0.97 \times 0.95} \left(\frac{w_n L}{8 E_n I_n} \right) = 1.14d$$

meaning, therefore, that the calculated deflection should be corrected with a bias factor of

$$v_d = 1.14$$

whereas the c.o.v. of the calculated deflection is

$$\Omega_d = \sqrt{\Omega_w^2 + \Omega_E^2 + \Omega_I^2 + \Delta_g^2}$$
$$= \sqrt{0.24^2 + 0.21^2 + 0.20^2 + 0.07^2}$$
$$= 0.38$$

EXAMPLE 6.15 (Storm Sewer; Excerpted from Yen and Tang, 1976)

A storm sewer is designed such that its flow capacity will be able to handle the maximum peak discharge over a given period T, in years. For a nearly full gravitational pipe flow, the Darcy-Weisbach formula gives the flow capacity as

$$Q_C = \frac{\pi}{4} \left(\frac{2gS}{f} \right)^{1/2} (4R)^{2.5} \tag{i}$$

where f = the Weisbach resistance coefficient; S = the friction slope; g = gravitational acceleration; and R = the hydraulic radius, assumed (in this example) to be $D/4$ in which D is the diameter of the pipe. However, Eq. (i) is only an approximation for unsteady nonuniform flow; a term N_C may be introduced to account for any modeling error in the flow capacity.

Furthermore, using the rational formula to predict the peak discharge, the maximum discharge over T years is given by

$$Q_L = N_L C i A \qquad \text{(ii)}$$

where i = the maximum rainfall intensity over T years; C = a runoff coefficient; A = the drainage area; and N_L corrects any error in the peak discharge formula. The parameters contributing to the uncertainty in the sewer capacity and maximum peak discharge are evaluated in the following, for a 5-foot diameter concrete pipe located in Urbana, Illinois for a period of 10 years.

Hydraulic Radius R

A major source of uncertainty may be associated with the manufacturing tolerance of the pipe. For large concrete pipes, say pipes with larger than 4-foot diameters, the manufacturer's tolerance is about ± 0.5 in. Assuming a uniform distribution over this range, that is, from 59.5 to 60.5 in. for a nominal diameter of $D = 5.0$ ft, we obtain

$$\bar{R} = \frac{\bar{D}}{4} = 1.25 \text{ ft}$$

and

$$\delta_R = \delta_D = \frac{1}{\sqrt{3}} \left(\frac{60.5 - 59.5}{60.5 + 59.5} \right) = 0.01$$

After a period of use, the size of the pipe could decrease as a consequence of deposition. This effect is traditionally accounted for through a change in the resistance coefficient f.

Resistance Coefficient f

The Weisbach resistance coefficient f is obtained by using the Moody diagram in Fig. E6.15a. Consider a fully developed turbulent flow with N_R greater than 10^6. In this region f is mainly a function of e/D. For new concrete pipes, e may range from 0.001 to 0.01 ft (see the recommended values in Fig. E6.15a). This implies that f would be between 0.0140 and 0.0235 for a 5-foot diameter pipe. Moreover f would be more likely to take on higher values within its range. Thus, it is reasonable to assume an upper triangular distribution over this range of f with the mode at 0.0235; from which, $\bar{f} = 0.0203$ and $\delta_f = 0.11$.

As the sewer ages, organic and mineral depositions will increase the value of f that fluctuates with time as certain storm runoffs tend to contribute deposition while others have a cleaning effect. According to data reported by Horowitz and Lee (1971), the increase in Manning's roughness factor is observed to have a mean of 9.7% and a standard deviation of 2.3%. For a given hydraulic radius, f is proportional to the square of Manning's roughness factor; hence, the corresponding increase in f has a mean of 19.4% and a standard deviation of 4.6%, respectively. Based on this information, we estimate

$$v_f = 1.194$$

and

$$\Delta_f = \frac{0.046}{1.194} = 0.039$$

Thus, the total uncertainty in f is

$$\Omega_f = \sqrt{0.11^2 + 0.039^2} = 0.117$$

Values of (DV) for Water at 60°F (Diameter in inches, Velocity in ft/sec)

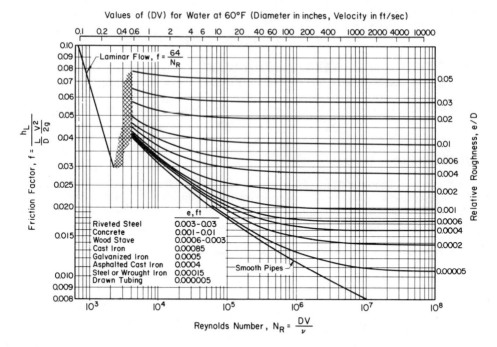

Figure E6.15a   Moody diagram for estimating f for pipes (after Linsley and Franzini, 1964).

Slope S

Uncertainties in S arise mainly from sewer misalignment and crookedness of the pipe and are worse for shallower slopes. A slope with a 6 in. (150 mm) drop in 500 ft (150 m) is not uncommon for the plain regions of central Illinois. If we assume a deviation of ± 1 in. for the 6 in. drop and a symmetric triangular distribution over this range, the variability in S is $\delta_S = 0.068$ for $\bar{S} = 0.001$.

Equation Error N_C

Urban storm flood flows are highly unsteady and nonuniform; thus, error in the Darcy–Weisbach formula can be expected. A statistical analysis of the storm sewer flows by Yen and Sevuk (1975) indicates that the required bias correction is $\nu_C = 1.1$. Assuming a symmetric triangular distribution of N_C from 0.8 to 1.4 with the mode at 1.1 yields $\Delta_C = 0.11$.

Runoff Coefficient C

Since an inlet basin usually consists of a variety of ground covers, this coefficient is generally computed from the weighted average of the runoff coefficients for the various ground covers as

$$C = \sum_i C_i \alpha_i$$

where α_i is the fraction of the drainage area with the same runoff coefficient C_i. Because of the variability of the surface roughness, and permeability of the drainage surfaces, C_i is a random variable. α_i may not be precisely determined either because of the approximate grouping of, and any future change in, the type of ground coverings of an area. The uncertainties of the variables

Table E6.15a Component Errors for Runoff Coefficient

	Types of Surface		
	Driveways and Sidewalks	Roofs	Streets
$\bar{\alpha}_i$	0.40	0.40	0.20
Δ_{α_i}	0.10	0.10	0.10
Range of C_i^*	0.75–0.85	0.75–0.95	0.70–0.95
$\bar{C}_i^\dagger$	0.800	0.850	0.825
$\delta_{C_i}^\dagger$	0.036	0.068	0.087
$\Delta_{C_i}^\ddagger$	0.012	0.023	0.029
$\Omega_{C_i} = \sqrt{\delta_{C_i}^2 + \Delta_{C_i}^2}$	0.038	0.072	0.092

* Obtained from standard references; for example, Chow (1964), p. 14.8.

† Assume uniform distribution over the range.

‡ Assume $\bar{C}_i$ varies uniformly within the middle third of the range.

contributing to Ω_C are summarized in Table E6.15a. The prediction error for α_i is subjectively and conservatively assumed to be 0.10. Based on the following first-order equations

$$\bar{C} = \sum_i \bar{C}_i \bar{\alpha}_i$$

$$\Omega_C = \frac{1}{\bar{C}} \left[\sum_i \bar{\alpha}_i^2 \, \bar{C}_i^2 (\Delta_{\alpha_i}^2 + \Omega_{C_i}^2) \right]^{1/2}$$

and the values of $\bar{\alpha}_i$, $\bar{C}_i$, Δ_{α_i} in Table E6.15a, we obtain $\bar{C} = 0.825$ and $\Omega_C = 0.071$.

Area A

The drainage area is usually determined from a map. To estimate the error for A, 34 engineers were asked to inspect a drainage basin and determine the area from the United States Geological Survey map. The average error was found to be $\Delta_A = 0.045$. The error in the map is usually small and is assumed to be 0.001. Thus, $\Omega_A = (0.045^2 + 0.001^2)^{1/2} = 0.045$ for an area $\bar{A} = 10$ acres.

Rainfall Intensity i

The uncertainty in the rainfall intensity, i, varies with the design period T and the assumed duration t_d of the severe part of the storm. In this example, a period of $T = 10$ yrs is considered and the duration t_d is assumed to be equal to the time of concentration of the drainage area, which is taken to be 30 min.

Based on data from Hershfield (1963) for the Urbana basin with $t_d = 30$ min, the annual maximum rainfall intensity i_a is observed to be lognormal with parameters $\lambda = 0.843$ and $\zeta = 0.278$ (see Fig. E6.15b). The maximum rainfall intensity i over a ten-year period may be assumed to approach a Type II (largest value) asymptotic distribution. The mean and c.o.v. of i are given by (see Chapter 4)

$$\bar{i} = v\Gamma\left(1 - \frac{1}{k}\right)$$

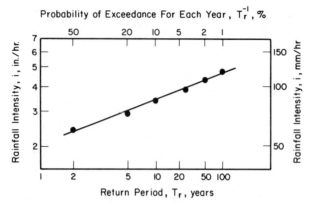

Figure E6.15b   Variation of rainfall intensity with return period for 30-min duration.

and

$$\delta_i = \sqrt{\frac{\Gamma\left(1 - \dfrac{2}{k}\right)}{\Gamma^2\left(1 - \dfrac{1}{k}\right)} - 1}$$

where

$$v = \exp\left[\zeta\sqrt{2 \ln 10} - \zeta \frac{\ln \ln 10 + \ln 4\pi}{2\sqrt{2 \ln 10}} + \lambda\right]$$

and

$$k = \frac{1}{\zeta}\sqrt{2 \ln 10}$$

Hence $\bar{i} = 3.71$ in./hr and $\delta_i = 0.183$. There are three major components of prediction error for the rainfall intensity i.

(a) Effect of Duration

Rainfall intensity i is related to the duration t_d as

$$i = \frac{a}{b + t_d}$$

where a and b are evaluated as 180 and 27, respectively, based on the Urbana data with $T = 10$ years [Hershfield, 1963]. With the prediction error of duration estimated to be 10%, the corresponding error on i can be shown to be

$$\Delta_{i_1} = 0.1 \frac{\bar{t}_d}{27 + \bar{t}_d} = 0.053$$

where $\bar{t}_d$ is 30 min.

(b) Effect of Limited Rainfall Record

Because of limited rainfall records available to establish the values in the Atlas (Hershfield, 1963), a prediction error will be incurred mainly from the estimation of the parameter λ.

Table E6.15*b* Summary of Uncertainty Analysis

Variable	Estimated Mean	v_i	Ω_i
R	1.25 ft	1.0	0.010
f	0.0203	1.194	0.117
S	0.001	1.0	0.068
N_C	—	1.1	0.110
C	0.825	1.0	0.071
i	3.71 in./hr	1.0	0.209
A	10 acres	1.0	0.045
N_L	—	1.0	0.100

For a 50-year record, the error can be shown to be

$$\Delta_{i_2} = \frac{\zeta}{\sqrt{n}} = \frac{0.278}{\sqrt{50}} = 0.03$$

(c)   Effect of Errors in Instrumentation and Data Manipulation

Available information for the example is inadequate for a detailed analysis of the individual effects. The gross error in this group is subjectively estimated to be $\Delta_{i_3} = 0.08$.

In summary, the total uncertainty in i is

$$\Omega_i = \sqrt{0.183^2 + 0.053^2 + 0.03^2 + 0.08^2} = 0.21$$

Equation Error

The rational formula may over- or underestimate the peak runoff rate depending on the conditions encountered. Thus, $v_{N_L} = 1.0$ and $\Delta_{N_L} = 0.1$ are assumed

The results of the above uncertainty analysis for a storm sewer in Urbana, Illinois, are summarized in Table E6.15*b*.

EXAMPLE 6.16   (*Stability of Earth Slope; Excerpted from Tang, Yucemen, and Ang, 1976*)

When a saturated soil stratum is excavated at a rate so that no significant dissipation of the excess pore pressure takes place, a short-term stability analysis of the resulting slope (see Fig. E6.16*a*) is required. The potential failure surface of the slope is usually assumed to be cylindrical. The resisting moment M is provided by the shear strength mobilized along the potential failure surface. For each unit section of the slope,

$$M = r \sum_{j=1}^{m} l_j S_j \tag{i}$$

where S_j is the spatial average in-situ undrained shear strength along the jth segment of the failure surface with length l_j, and r is the radius of the cylindrical failure surface. The main variable in the resisting moment is the undrained shear strength.

The average shear strength is generally estimated from laboratory-measured strengths on a limited set of soil specimen. Furthermore, the laboratory-measured strength is generally based on a standard triaxial shear test that does not necessarily reproduce the actual in-situ conditions; factors such as sampling disturbance, rate of shearing, progressive failure effect, etc. would tend to cause a discrepancy between the laboratory-measured and in-situ strengths. Therefore, the in-situ spatial average undrained strength S may be modeled as

$$S = N_0(N_1 N_2 \cdots N_k)\hat{S} \tag{ii}$$

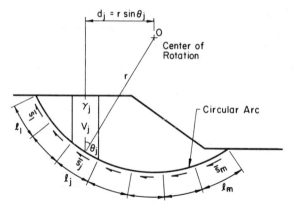

Figure E6.16a   Circular arc analysis of slope stability.

where $\hat{S}$ is the spatial average undrained shear strength uncorrected for discrepancies; N_o is the corrective factor, with mean 1.0 and c.o.v. Δ_o accounting for error resulting from insufficient sampling; and N_j's, $j = 1$ to k, are other corrective factors with respective mean $\bar{v}_j$ and c.o.v. Δ_j. The detailed statistical evaluation of the corrective factors is subsequently presented below. By first-order approximation, the mean in-situ average strength is

$$\mu_S \simeq \bar{v}_1 \bar{v}_2 \ldots \bar{v}_k \bar{s} = \bar{v}_s \bar{s} \tag{iii}$$

where $\bar{s}$ is the mean value of $\hat{S}$. The various factors may reasonably be assumed to be statistically independent; hence, the overall c.o.v. of S is

$$\Omega_S \simeq \sqrt{\Delta_o^2 + \sum_{j=1}^{k} \Delta_j^2 + \delta_s^2} \tag{iv}$$

where δ_s is the c.o.v. of $\hat{S}$.

The individual sources of uncertainty may be assessed as follows.

Inherent Spatial Variability

The average undrained shear strength $\bar{s}$ and corresponding c.o.v. δ_S characterizing the inherent variability of a specific soil layer can be estimated from strengths measured in a typical soil exploration program. The value of δ_S reported for various sites reveals that it can range from 0.11 to 0.41. This inherent variability contributes to Ω_S of Eq. (iv) in two ways:

(i) $\Delta_0 = \delta_S / \sqrt{n}$ where n is the number of specimens tested for the layer, if we assume that the specimens are independent.

(ii) $\delta_s = \delta_S / \sqrt{n_e}$, where n_e is the equivalent number of independent soil elements along a given segment of the failure surface. The value of n_e depends on the length of the segment as well as the correlation between the undrained shear strengths of adjacent soil elements. The concept underlying the factor $1/\sqrt{n_e}$ is that when the strengths are averaged over a domain, the variability in the average strength will be less than that at a given point. This factor is related to the variance function defined by Vanmarcke (1977).

Discrepancy Between Laboratory and In-situ Undrained Strengths

A number of factors contribute to the discrepancy between laboratory-measured and in-situ shear strengths. Experimental data giving the required correction factor are generally not available for a specific site. However, based on the results reported in the literature, the range of the required correction N_j may be established. With appropriate distributions (such as triangular or uniform) prescribed over the respective ranges, estimates of v_j and Δ_j may be determined. A detailed analysis of the published data is presented in Tang et al. (1976). Table E6.16a summarizes the statistics evaluated for each correction factor. The significance of each factor may be briefly described as follows:

(i) Disturbance during sampling—The disturbances during sampling can be of two types: (1) the inevitable changes in the stress system because of the removal of the sample from the ground; (2) the mechanical disturbances of sampling and preparation of the specimen. Factors N_1 and N_2 account for these two types of disturbances, respectively. The change in stress state will depend on the sensitivity of the clay involved, whereas the amount of mechanical disturbance will depend on the method of obtaining the specimen.

(ii) Size of specimen—The typical specimen size used in a laboratory test is generally not large enough to include the influence of joints and fissures on the strength of clay. Hence, the correction factor N_3 will depend on the degree of fissures present in the in-situ clay. For example, for intact clay, a small size specimen is usually adequate.

(iii) Rate of shearing—Undrained tests are usually conducted at a rate of strain in which failure occurs in about 15 minutes. However, actual failures of slopes usually occur over a much longer period of time. The rapid tests generally tend to overestimate the in-situ strength. The magnitude of this discrepancy (as reflected by N_4) depends on the strain rate sensitivity of the soil as well as the loading duration of an actual failure at the site.

(iv) Sample orientation and anisotropy—In conventional soil sampling, soil samples are obtained from boreholes with vertical axes; also, the specimens are loaded vertically in laboratory tests. However, along a potential sliding surface, the soil may fail at various orientations from the vertical. The corresponding correction factor for this effect, N_5, can be evaluated as a function of the orientation of the failure plane for clays with various anisotropy.

(v) Plane strain failure—In many landslides, the stress and deformations during failure are close to plane strain conditions, whereas laboratory strengths are generally not measured under plane strain conditions; N_6 accounts for this effect.

(vi) Progressive failure effect—Because of stress concentrations at various parts of the soil along a failure surface, deformations will not be uniform along the failure surface. Hence, the peak strength will not be reached at all points simultaneously. At some points, the strength will be governed by the residual strength due to excessive deformations that have already occurred.

The overall correction $\bar{v}_s$ in Eq. (iii) can be estimated by combining the appropriate values of v_j for a specific clay layer. For practical design, soils should be classified into stiff-fissured and intact clays, also taking into account the strain-rate sensitivity of the soil and the type of anisotrophy.

A Specific Site Application

Consider the open cut at Congress Street, Chicago, with the soil profile as shown in Fig. E6.16b. Sampling was done by 2 in.-diameter Shelby tubes with samples taken at least 3 ft apart. The mean undrained strength and coefficient of variation as computed from the reported triaxial tests are summarized in Table E6.16b for each of the clay layers. The values of δ_s and Δ_o for each layer are also shown in Table E6.16b.

Among the various factors that contribute to the discrepancy between laboratory-measured and in-situ strengths, the effect of mechanical disturbance during sampling is most significant. A first estimate of the correction factor N_2 gives $\bar{v}_2'$ and Δ_2' as 1.30 and 0.13, respectively, based on the results of Table E6.16a. However, in a series of tests performed for the Chicago Subway project (Peck, 1940), the undrained strength of specimens obtained from large block samples are compared with those obtained from Shelby tube samples. Based on the results of 13 measurements, the correction factor N_2 is estimated to have $v_2'' = 1.39$ and $\Delta_2'' = 0.025$. The above two pieces of information can be combined (Tang et al, 1976), yielding $\bar{v}_2 = 1.38$ and $\Delta_2 = 0.024$.

Chicago clays have a medium sensitivity ($s_t = 4$). The value of $\bar{v}_1$ and Δ_1 are estimated to be 1.05 and 0.02, respectively.

Table E6.16a Summary of Uncertainty Factors

Factor	Effect	Soil Type	Range of N_j	v_j	Δ_j
N_1	Change in stress state	Low sensitivity ($s_t = 1 - 2$)	1.0 –1.1	1.03	0.02
		Medium sensitivity ($s_t = 2 - 4$)		1.05	0.02
		Sensitive ($s_t = 4 - 8$)		1.07	0.02
		Unknown sensitivity ($s_t = 1 - 8$)		1.05	0.03
N_2	Mechanical disturbance	Shelby tube specimen	1.0 –1.6	1.3	0.13
		Borehole specimen	1.15–2.25	1.7	0.19
N_3	Size of specimen	Stiff-fissured clay	0.55–0.85	0.70	0.12
		Intact clay	0.85–1.00	0.93	0.05
N_4	Rate of shearing	Slightly sensitive to sensitive (strength reduction 3–10% per log increment of time)	0.6 –1.0	0.80	0.14
		Very sensitive (10–14%)	0.45–0.7	0.58	0.12
		Unknown sensitivity (3–14%)	0.45–1.0	0.73	0.22
N_5	Sample orientation and anisotropy	Isotropic	0.97–1.08	1.0	0.03
		C-anisotropy	0.85–1.20	1.03	0.10
		M-anisotropy	0.8 –1.0	0.9	0.06
N_6	Plane strain failure	All soil types	1.0 –1.1	1.05	0.03
N_7	Progressive failure	Stiff clay	0.9 –1.0	0.93	0.03
		Medium clay	0.9 –1.0	0.97	0.03

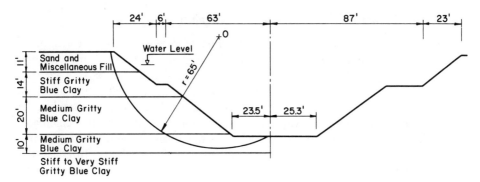

Figure E6.16b Approximate cross section of the cut and slip surface at Congress Street open cut.

The clays in the middle and lower layers are intact, whereas because of desiccations some cracks and joints exist in the upper layer. The cracks are judged to be not pronounced; an upper triangular distribution is chosen between $N_3 = 0.55$ to 0.85 (for stiff fissured clay) for the upper layer, which gives $\bar{v}_3 = 0.75$ and $\Delta_3 = 0.09$. For the two intact clay layers, a uniform distribution between $N_3 = 0.85$ to 1.0 yields conservative estimates of $\bar{v}_3 = 0.93$ and $\Delta_3 = 0.05$.

There is no information available on the effects of the rate of shearing on the undrained strength of Chicago clays. Nevertheless, if the sensitivity to rate of shearing in the three layers is assumed to lie between slightly sensitive to sensitive, values of $\bar{v}_4 = 0.80$ and $\Delta_4 = 0.14$ would be appropriate.

Chicago clay is believed to show little variation in strength resulting from anisotropy. Therefore, the values suggested for the isotropic case, with $\bar{v}_5 = 1.0$ and $\Delta_5 = 0.03$, may be used.

For the effect of plane strain, the mean correction and random error for the three layers are taken to be $\bar{v}_6 = 1.05$ and $\Delta_6 = 0.03$.

Some reduction in the undrained strength could be expected from the effect of progressive failure, especially in the stiff upper clay layer. Hence, $\bar{v}_7 = 0.93$ and $\Delta_7 = 0.03$ for the stiff upper clay layer are assumed as reasonable. Progressive failure effect is believed to be less dominant for the middle and lower clay layers, which consist of medium clay; thus, $\bar{v}_7 = 0.97$ and $\Delta_7 = 0.03$.

In summary, for the upper clay layer,

$$\bar{v}_s = 1.05 \times 1.38 \times 0.75 \times 0.80 \times 1.0 \times 1.05 \times 0.93 = 0.85$$

and

$$\Delta_s = \sqrt{0.03^2 + 0.02^2 + 0.024^2 + 0.09^2 + 0.14^2 + 0.03^2 + 0.03^2 + 0.03^2} = 0.20$$

Similarly, it is found that $\bar{v}_s = 1.10$ and $\Delta_s = 0.16$ for the middle clay layer, and $\bar{v}_s = 1.0$ and $\Delta_s = 0.17$ for the lower clay layer.

Table E6.16b Data for Undrained Strengths at Congress Street Project

Layers	$\bar{s}$	σ_s	δ_S	n_e	$\delta_s = \dfrac{\delta_S}{\sqrt{n_e}}$	n	$\Delta_0 = \dfrac{\delta_S}{\sqrt{n}}$
Upper clay layer	1.06	0.54	0.51	28	0.096	38	0.083
Middle clay layer	0.62	0.16	0.26	54	0.035	55	0.035
Lower clay layer	0.78	0.25	0.32	125	0.029	33	0.056

6.3.4 An Alternative Representation of Uncertainty

In Section 6.2, the calculated failure probability represents the significance of the total uncertainty; that is, the inherent variability and modeling errors are combined and their aggregate effects are reflected in the probability of failure.

Alternatively, the two types of uncertainty may be treated differently and their respective significance represented separately in dual terms; in particular, the significance of the inherent variability may be represented in the calculated probability (i.e. frequency) of failure, whereas the effect of the modeling error is represented with an "error bound" on the calculated probability. In effect, this means that the failure probability is calculated assuming (or conditional on) a given set of mean values; since the mean values are subject to modeling errors, a range of calculated probabilities is possible. The resulting range, or error bound, of the calculated probability, therefore, represents the significance of the underlying modeling errors.

Dual Representations Consider a performance function

$$Y = g(X_1, X_2, \ldots, X_n)$$

For the purpose of the alternative formulation, the uncertainties for the individual variables $X_1, X_2, \ldots, X_n$ as well as for Y must be evaluated separately in terms of the inherent variabilities and the modeling errors; that is, the methods described in

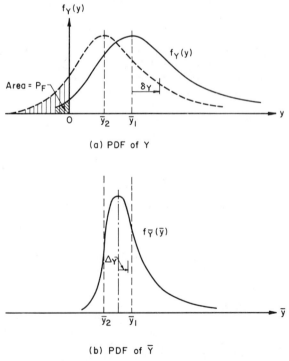

(a) PDF of Y

(b) PDF of $\bar{Y}$

Figure 6.12 PDF of Y and $\bar{Y}$.

Section 6.3 should be used to evaluate δ_{X_i} and Δ_{X_i}, $i = 1, 2, \ldots, n$, as well as δ_Y and Δ_Y.

Suppose that the estimated mean values and inherent variabilities of the variables X_i are $\bar{x}_i$ and δ_{X_i}, $i = 1, 2, \ldots, n$. Also, the uncertainties (c.o.v.) in the estimated mean values are Δ_{X_i}, $i = 1, 2, \ldots, n$, from which the variability and modeling uncertainty of Y, δ_Y and Δ_Y, may be evaluated as functions of δ_{X_i} and Δ_{X_i}, respectively, using Eq. 6.43 separately for δ_Y and Δ_Y.

For a given set of mean values, $\{\bar{x}_1, \bar{x}_2, \ldots, \bar{x}_n\}$, there is a corresponding mean value $\bar{y}$, Eq. 6.42, and the probability or *frequency* of failure, therefore, would be (see Fig. 6.12a)

$$P_F = P(Y < 0 | \bar{y}) \qquad (6.44)$$

This probability would be valid if there were no modeling errors in the mean value $\bar{y}$. However, in light of modeling errors, the mean value would also be a random variable, $\bar{Y}$, with PDF $f_{\bar{Y}}(\bar{y})$ as shown in Fig. 6.12b. Furthermore, since P_F is implicitly a function of $\bar{Y}$, it is also a random variable; its expected probability is then

$$E(P_F) = \int_{-\infty}^{\infty} P(Y < 0 | \bar{Y} = \bar{y}) f_{\bar{Y}}(\bar{y}) \, d\bar{y} \qquad (6.45)$$

Conceivably, there is also a CDF for P_F; that is, $F_{P_F}(p)$. Hence, the q-percentile value of the failure probability, namely $(p_F)_q$, may be obtained from

$$F_{P_F}[(p_F)_q] = q$$

or

$$(p_F)_q = F_{P_F}^{-1}(q) \qquad (6.46)$$

In practice, q may be the desired nonexceedance probability level (in %).

Depending on the complexity of the performance function and the form of the probability distributions involved, the expected and q-percentile values of P_F may be evaluated analytically (see Example 6.17 for normal variates and Problem 6.19 for lognormal variates) or through Monte Carlo simulation (see Example 6.18).

EXAMPLE 6.17

Consider the linear performance function

$$Y = C - D$$

where C and D are statistically independent normal random variables with mean values μ_C, μ_D and standard deviations σ_C, σ_D. Assume that μ_C and μ_D are also independent normal variates with mean values $\bar{C}$, $\bar{D}$ and standard deviations s_C, s_D. Since Y is also normal, the failure probability conditional on given values of μ_C and μ_D is

$$P_F = P(Y < 0 | \mu_C, \mu_D)$$

$$= \Phi\left(\frac{\mu_D - \mu_C}{\sqrt{\sigma_C^2 + \sigma_D^2}}\right)$$

For simplicity, let $\mu = \mu_D - \mu_C$ and $\sigma = \sqrt{\sigma_C^2 + \sigma_D^2}$. The above conditional probability then becomes

$$P_F = \Phi\left(\frac{\mu}{\sigma}\right)$$

The expected failure probability, therefore, is

$$E(P_F) = \int_{-\infty}^{\infty} \Phi\left(\frac{\mu}{\sigma}\right) f_\mu(\mu)\, d\mu$$

Since μ is a linear function of μ_C and μ_D, it is also a normal variate with mean $\bar{x} = \bar{D} - \bar{C}$ and standard deviation $s = \sqrt{s_C^2 + s_D^2}$. Hence,

$$E(P_F) = \int_{-\infty}^{\infty} \Phi\left(\frac{\mu}{\sigma}\right) \frac{1}{\sqrt{2\pi}s} e^{-(1/2)[(\mu - \bar{x})/s]^2}\, d\mu$$

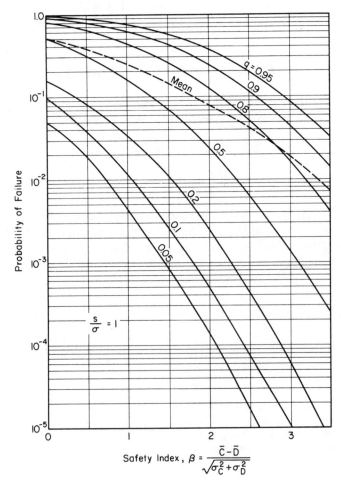

Figure E6.17a Mean and *q*-percentile failure probability as a function of safety index (for $s/\sigma = 1$).

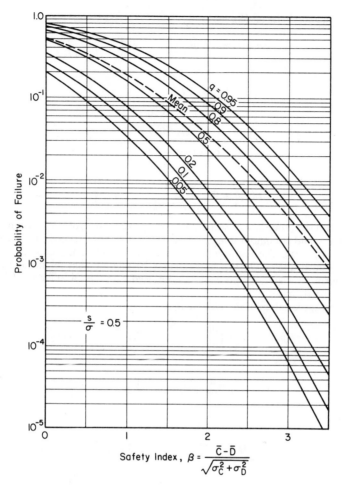

Figure E6.17b Mean and q-percentile failure probability as a function of safety index (for $s/\sigma = 0.5$).

This integral may be evaluated by first observing that

$$\Phi(a) = \frac{1}{2} + \frac{1}{2}\,\mathrm{erf}\!\left(\frac{a}{\sqrt{2}}\right)$$

where $\mathrm{erf}(-)$ is the *error function*. Hence,

$$E(P_F) = \frac{1}{2} + \frac{1}{2\sqrt{2\pi}\,s}\int_{-\infty}^{\infty}\mathrm{erf}\!\left(\frac{\mu}{\sqrt{2}\,\sigma}\right)e^{-(1/2)[(\mu-\bar{x})/s]^2}\,d\mu$$

Defining $y = \mu/\sqrt{2}\,\sigma$, we obtain

$$E(P_F) = \frac{1}{2} + \frac{\sigma}{2\sqrt{\pi}\,s}\int_{-\infty}^{\infty}\mathrm{erf}(y)e^{-[(\sigma y/s - \bar{x}/s\,\sqrt{2})^2]}\,dy$$

According to Ng and Geller (1968),

$$\int_{-\infty}^{\infty} \operatorname{erf}(y) e^{-(ay+b)^2} \, dy = -\frac{\sqrt{\pi}}{a} \operatorname{erf}\left(\frac{b}{\sqrt{a^2+1}}\right)$$

With this we obtain

$$E(P_F) = \frac{1}{2} + \frac{1}{2} \operatorname{erf}\left(\frac{\bar{x}/s}{\sqrt{(\sigma/s)^2+1}} \middle/ \sqrt{2}\right)$$

$$= \Phi\left[\frac{\bar{x}/s}{\sqrt{(\sigma/s)^2+1}}\right]$$

or

$$E(P_F) = \Phi\left[\frac{\bar{D}-\bar{C}}{\sqrt{\sigma_C^2 + \sigma_D^2 + s_C^2 + s_D^2}}\right]$$

Observe that with the procedure of Section 6.3.3, the prediction error and inherent variability of each variable would be combined first, obtaining mean values $\bar{C}$, $\bar{D}$ and combined variances $(\sigma_C^2 + s_C^2)$ and $(\sigma_D^2 + s_D^2)$. The resulting failure probability will be the same as the expected probability obtained above. On this basis, it may be inferred that for linear performance functions of normal variates, the probabilities obtained with the formulation of Sections 6.2 and 6.3 are expected values.

For the q-percentile values of P_F, observe that

$$P[P_F \leq (p_F)_q] = P\left[\Phi\left(\frac{\mu}{\sigma}\right) \leq (p_F)_q\right] = P[\mu \leq \sigma\Phi^{-1}\{(p_F)_q\}] = q$$

or

$$\Phi\left[\frac{\sigma\Phi^{-1}\{(p_F)_q\} - \bar{x}}{s}\right] = q$$

Thus,

$$(p_F)_q = \Phi\left\{\frac{1}{\sigma}[\bar{x} + s\Phi^{-1}(q)]\right\}$$

or

$$(p_F)_q = \Phi\left\{\frac{(\bar{D}-\bar{C}) + \sqrt{s_C^2 + s_D^2}\,\Phi^{-1}(q)}{\sqrt{\sigma_C^2 + \sigma_D^2}}\right\}$$

For example, for $q = 0.5$, the median failure probability is

$$(p_F)_{0.5} = \Phi\left\{\frac{\bar{D}-\bar{C}}{\sqrt{\sigma_C^2 + \sigma_D^2}}\right\}$$

Observe that this is different from the mean failure probability derived earlier.

The q-percentile failure probability will depend on the ratio of the modeling error to the inherent variability, that is, $s/\sigma = \sqrt{s_C^2 + s_D^2}/\sqrt{\sigma_C^2 + \sigma_D^2}$. This is demonstrated in Figs. E6.17a and E6.17b, respectively, for $s/\sigma = 1.0$ and $s/\sigma = 0.5$ as functions of the safety index defined as $\beta = (\bar{C} - \bar{D})/\sqrt{\sigma_C^2 + \sigma_D^2}$. The corresponding expected failure probability, $E(P_F)$, is also shown. The results for $\beta = 2.0$ may be observed as follows:

	$s/\sigma = 1.0$	$s/\sigma = 0.5$
Mean P_F	0.078	0.035
Median P_F	0.023	0.023
$(p_F)_{0.90}$	0.240	0.084
90% error bound	0.00014–0.370	0.0025–0.120

EXAMPLE 6.18

Consider Example 6.7 again with the performance function

$$g(\mathbf{X}) = YZ - M$$

However, in the present case, the inherent variabilities and modeling errors of the respective variables are represented separtely as follows:

| | | Mean Value | |
Variable	Inherent Variability (Standard Deviation)	Estimated Mean	Modeling Error (Standard Deviation)
Y(ksi)	4	40	3
Z(in.^{3})	2	50	1.5
M(in-kip)	160	1000	120

Assume that Y, Z, and M and their respective mean values are uncorrelated normal variates.

Since the performance function is nonlinear, the CDF of the failure probability would be difficult to determine analytically. Of course, for given values of $\boldsymbol{\mu} = \{\mu_Y, \mu_Z, \mu_M\}$, the corresponding conditional probability of failure may be evaluated with the second-moment procedure described in Section 6.2. Because of uncertainties (modeling errors) in the mean values, the probability of failure will be a random variable whose distribution may be determined through Monte Carlo simulation.

For this purpose, one thousand sets of mean values $\{\mu_Y, \mu_Z, \mu_M\}$ are first generated using the procedure described in Section 5.2.2 for normal variates. The failure probabilities corresponding to this one thousand sets of $\boldsymbol{\mu}$ are then evaluated with the second-moment approach. The resulting histogram of the calculated failure probabilities is plotted in Fig. E6.18a. The failure probability is observed to vary over a wide range, approximately from 10^{-9} to 10^{-1}. The corresponding sample cumulative distribution is shown in Fig. E6.18b, from which the q-percentile values of P_F may be obtained. Observe that (in the present case) the median failure probability is 0.0001, whereas the mean failure probability is 0.00118. The 90% error bounds range from 4×10^{-7} to 6×10^{-3}. It is interesting to compare the expected probability of failure obtained here, namely, 0.00118, with the failure probability of 0.00114 obtained previously in Example 6.7.

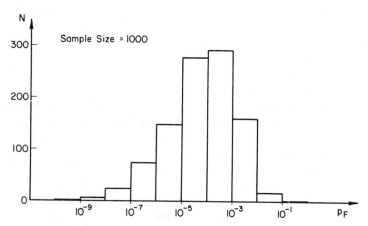

Figure E6.18a Histogram of failure probability.

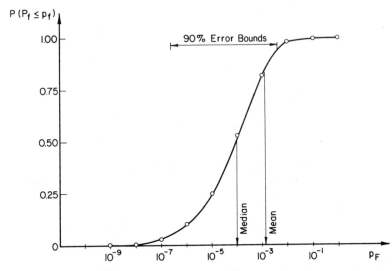

Figure E6.18b   CDF of failure probability.

6.4 PROBABILITY-BASED DESIGN CRITERIA

6.4.1 On Design and Design Criteria

The main objective of engineering planning and design is to insure the performance of an engineering system or product. As this must be accomplished under conditions of uncertainty, the assurance of performance is realistically possible only in terms of probability, namely p_S. In general, therefore, probabilistic analyses will be necessary in the development of such probability-based designs. However, designs satisfying such reliability requirements may also be developed without a complete probabilistic analysis—this may be accomplished through the adoption of appropriate deterministic design criteria (e.g., the use of traditional safety factors). In fact, probabilistic bases for design would be most effective if implemented in this form. This can be achieved if the required safety factors are predetermined on the basis of specified probability-based requirements. In particular, for the purpose of more routine designs, such as the proportioning of beams and columns of conventional buildings, criteria for design may be specified as provisions of a building code or standard. Properly, these criteria ought to be developed also on the basis of satisfying specified reliability.

For obvious reasons, design criteria should be as simple as possible; moreover, they should be developed in a form that is familiar to the users or designers. There is no unique format for a design criterion; irrespective of the format, a criterion can be developed on a probability basis. One traditional and common approach to design is through the use of *factors of safety*; for example, such factors of safety are widely used in geotechnical engineering, whereas safety factors are implicit in structural engineering when designs are developed through the use of allowable stresses. Another format is the use of load amplification factors and resistance reduction factors, known as the LRFD format (Ellingwood et al.,

1982); that is, the nominal design loads are amplified by appropriate *load factors* and the nominal resistances are reduced by corresponding *resistance factors*, and safety is assured if the "factored" resistance is at least equal to the "factored" load. In one format or another, the necessary *design factors* can be developed in order to obtain designs that achieve a prescribed level of reliability, p_S. The design factors are, of course, intended to cover uncertainties or compensate for the lack of complete information.

Basic Problem of Design   On the premise that an engineering design is intended to insure safety or performance with a given reliability p_S, the basic design problem, therefore, involves the determination of the position of the supply PDF, as shown in Fig. 6.2a, such that it is sufficiently separated from the demand PDF so that the failure probability p_F satisfies some target or acceptable value. Again, aside from the separation between $f_X(x)$ and $f_Y(y)$, the failure probability, p_F, is also a function of the degrees of dispersion (σ_X and σ_Y); a quantity that represents both of these influences is the safety index β. Therefore, specifying a value of β is equivalent to prescribing a *target reliability*, p_S, or *acceptable failure probability*, p_F.

One way to achieve the reliability of a given design is to use a sufficiently large mean margin of safety, $\mu_M = \mu_X - \mu_Y$. From Fig. 6.3, we see that to insure a target reliability, in terms of a given safety index β, the required mean safety margin is

$$\mu_M = \beta\sqrt{\sigma_X^2 + \sigma_Y^2}$$

Thus, the required mean supply (representing the required design) would be

$$\mu_X = \mu_Y + \mu_M$$

Alternatively, a given reliability may also be achieved by specifying a sufficiently large factor of safety. In this case, using the results of Example 6.2, the required median safety factor would be,

$$\theta_m = \exp[\beta\sqrt{\ln(1 + \delta_X^2)(1 + \delta_Y^2)}] \simeq e^{\beta\sqrt{\delta_X^2 + \delta_Y^2}}$$

from which the median supply is obtained as

$$x_m = \theta_m y_m$$

For codified design purposes, that is, to establish design code provisions, the most general format is the use of multiple (partial) load and resistance factors, as represented by the following requirement,

$$\phi R \geq \sum_{i=1}^{n} \gamma_i Q_i \tag{6.47}$$

where

ϕ = The resistance (supply) factor.

γ_i = The partial load (demand) factor to be applied to load Q_i.

The advantages of the multiple factor format of Eq. 6.47 are: (i) the individual factors, ϕ and γ_i, are relatively insensitive to changes in the design parameters (e.g., the load ratios between the various loads) and thus a given set of load and resistance factors should apply to a wide range of design conditions—a desirable attribute of a code provision; (ii) with the partial load and resistance factors, any other form of design factors, such as the safety factor, can be readily derived or evaluated (see Example 6.20).

6.4.2 Second-Moment Criteria

Even though a specific reliability-based design may be accomplished using the appropriate safety margin or safety factor, the design of a class of systems may require a more general form of design criteria. The most general and versatile form of criteria is to specify a design factor for each design variable. Consistent with practical situations, as described in Section 6.2.2, the necessary probability-based design criteria must often have to be developed also on the basis of the second-moment approach; that is, the required criteria have to be formulated on the basis of information for the first and second moments of the design variables.

In the space of the reduced variates, Section 6.3.2, designs at different levels of safety may be viewed as corresponding to satisfying different failure surfaces represented by varying distances to the origin, β, as shown in Fig. 6.13. Accordingly, the development of a design criterion is essentially tantamount to the determination of the design factors that will result in designs having failure surfaces that comply with a required safety index (i.e., the distance from the failure surface to the origin of the reduced variates satisfies some target value).

As indicated earlier, the most general design format is to apply a design factor on each of the basic design variables, known also as "partial factors." Without loss of generality, these factors may be applied to the respective mean values; thus,

$$g(\bar{\gamma}_1 \mu_{X_1}, \bar{\gamma}_2 \mu_{X_2}, \ldots, \bar{\gamma}_n \mu_{X_n}) = 0 \tag{6.48}$$

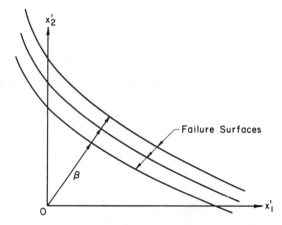

Figure 6.13   Designs corresponding to different failure surfaces.

Clearly, from the above Eq. 6.48, $\bar{y}_i \mu_{X_i}$ should be on the failure surface; in particular, it may be at the most probable failure point. Thus, the required partial design factors are

$$\bar{\gamma}_i = \frac{x_i^*}{\mu_{X_i}} \tag{6.49}$$

Therefore, the determination of the required design factors is also a problem of determining the most probable failure point, x_i^* (see Section 6.2).

In the space of the reduced variates, the most probable failure point, from Eq. 6.16a, is

$$x_i'^* = -\alpha_i^* \beta$$

where

$$\alpha_i^* = \frac{\left(\dfrac{\partial g}{\partial X_i'}\right)_*}{\sqrt{\displaystyle\sum_i \left(\dfrac{\partial g}{\partial X_i'}\right)_*^2}}$$

From this we obtain the original variates as

$$x_i^* = \mu_{X_i} - \alpha_i^* \beta \sigma_{X_i} = \mu_{X_i}(1 - \alpha_i^* \beta \Omega_{X_i})$$

Therefore, the required design factors are

$$\gamma_i = 1 - \alpha_i^* \beta \Omega_{X_i} \tag{6.50}$$

In Eq. 6.50, the direction cosines, α_i^*, must be evaluated at the most probable failure point x_i^*. In general, the determination of x_i^* requires an iterative solution. For this purpose, the following simple algorithm may be used:

(1) Assume x_i^* and obtain

$$x_i'^* = \frac{x_i^* - \mu_{X_i}}{\sigma_{X_i}}$$

(2) Evaluate $(\partial g/\partial X_i')_*$ and α_i^*.
(3) Obtain $x_i^* = \mu_{X_i} - \alpha_i^* \beta \sigma_{X_i}$.
(4) Repeat Steps (1) through (3) until convergence is achieved.

The required design factors are then obtained with Eq. 6.50. Again, for nonnormal variates, μ_{X_i} and σ_{X_i} should be replaced by the equivalent normal $\mu_{X_i}^N$ and $\sigma_{X_i}^N$ of Eqs. 6.25 and 6.26 in the above algorithm.

Linear Performance Function   For linear performance functions, the design factors, γ_i, are such that

$$a_o + \sum_i a_i \gamma_i \bar{x}_i = 0$$

In this case, the partial derivatives are independent of x_i, that is,

$$\frac{\partial g}{\partial X_i'} = a_i \sigma_{X_i}$$

and the corresponding direction cosines become

$$\alpha_i = \frac{a_i \sigma_{X_i}}{\sqrt{\sum_i (a_i \sigma_{X_i})^2}}$$

Then, the required design factors, Eq. 6.50, are

$$\gamma_i = 1 - \frac{a_i \sigma_{X_i}}{\sqrt{\sum_i (a_i \sigma_{X_i})^2}} \beta \Omega_{X_i} \qquad (6.50a)$$

EXAMPLE 6.19

Determine the mean resistance factor and the mean load factor (for total load) for the design of structural components to insure performance with a reliability of $\beta = 2.5$. Performance is defined with the function

$$g(\mathbf{X}) = \ln R/Q$$

Assume that the total load consists of statistically independent dead load, D, and live load, L, with mean load ratio $\bar{L}/\bar{D} = 2.0$, and respective c.o.v.'s $\Omega_D = 0.10$ and $\Omega_L = 0.25$. The corresponding c.o.v. of the resistance, R, is $\Omega_R = 0.11$.

The total load, therefore, is

$$Q = D + L$$

Thus,

$$\bar{Q} = \bar{D} + \bar{L} = \bar{D} + 2\bar{D} = 3\bar{D}$$

$$\sigma_Q = \sqrt{\sigma_D^2 + \sigma_L^2}$$

But, since $\bar{L}/\bar{D} = 2.0$

$$\frac{\sigma_L}{\Omega_L} \cdot \frac{\Omega_D}{\sigma_D} = \frac{0.10}{0.25} \frac{\sigma_L}{\sigma_D} = 2.0$$

From which,

$$\sigma_L = 5\sigma_D$$

Therefore,

$$\sigma_Q = \sqrt{26}\sigma_D$$

and

$$\Omega_Q = \frac{\sqrt{26}\sigma_D}{3\bar{D}} = \frac{\sqrt{26}}{3}(0.10) = 0.17$$

In this case,

$$\left(\frac{\partial g}{\partial R'}\right)_* = \left(\frac{\partial g}{\partial R}\right)_* \cdot \frac{dR}{dR'} = \frac{\sigma_R}{r^*}$$

$$\left(\frac{\partial g}{\partial Q'}\right)_* = \left(\frac{\partial g}{\partial Q}\right)_* \cdot \frac{dQ}{dQ'} = \frac{-\sigma_Q}{q^*}$$

But at the failure surface, $r^* = q^*$; therefore, the direction cosines are

$$\alpha_R^* = \frac{\sigma_R}{\sqrt{\sigma_R^2 + \sigma_Q^2}}$$

$$\alpha_Q^* = \frac{-\sigma_Q}{\sqrt{\sigma_R^2 + \sigma_Q^2}}$$

For the first trial, assume

$$r^* = \bar{R}; \qquad q^* = \bar{Q}$$

Since $r^* = q^*$, $\bar{R} = \bar{Q}$. Then,

$$\alpha_R^* = \frac{\Omega_R}{\sqrt{\Omega_R^2 + \Omega_Q^2}} = \frac{0.11}{\sqrt{(0.11)^2 + (0.17)^2}} = 0.543$$

$$\alpha_Q^* = \frac{-\Omega_Q}{\sqrt{\Omega_R^2 + \Omega_Q^2}} = \frac{-0.17}{\sqrt{(0.11)^2 + (0.17)^2}} = -0.840$$

Thus,

$$r^* = \bar{R}(1 - 0.543 \times 2.5 \times 0.11) = 0.851\bar{R}$$

$$q^* = \bar{Q}(1 + 0.840 \times 2.5 \times 0.17) = 1.357\bar{Q}$$

Again, since $r^* = q^*$,

$$\bar{R} = \frac{1.357\bar{Q}}{0.851} = 1.595\bar{Q}$$

Using this result for the second iteration, we obtain

$$\alpha_R^* = \frac{0.11 \times 1.595\bar{Q}}{\sqrt{(0.11 \times 1.595\bar{Q})^2 + (0.17\bar{Q})^2}} = \frac{0.175\bar{Q}}{0.244\bar{Q}} = 0.718$$

$$\alpha_Q^* = \frac{-0.17\bar{Q}}{0.244\bar{Q}} = -0.696$$

and

$$r^* = \bar{R}(1 - 0.718 \times 2.5 \times 0.11) = 0.803\bar{R}$$

$$q^* = \bar{Q}(1 + 0.696 \times 2.5 \times 0.17) = 1.296\bar{Q}$$

Thus,

$$\bar{R} = \frac{1.296}{0.803}\bar{Q} = 1.614\bar{Q}$$

Continuing, we obtain for the third iteration

$$\alpha_R^* = 0.722; \qquad \alpha_Q^* = -0.691; \qquad \bar{R} = 1.614\bar{Q}$$

Therefore, the required mean resistance and load factors, Eq. 6.50, are

$$\bar{\gamma}_R = 1 - 0.722 \times 2.5 \times 0.11 = 0.80$$

$$\bar{\gamma}_Q = 1 + 0.691 \times 2.5 \times 0.17 = 1.294$$

In other words, if we design with the requirement

$$\ln \frac{0.80\bar{R}}{1.294\bar{Q}} \geq 0$$

the reliability $\beta = 2.5$ will be satisfied. Observe that this requirement is also equivalent to

$$0.80\bar{R} \geq 1.294\bar{Q}$$
$$= 1.294(\bar{D} + \bar{L})$$

EXAMPLE 6.20   (*Linear Performance Function*)

A popular code format for the design of structural components is the linear inequality, for example, the ACI 318(81),

$$\phi R_n \geq \gamma_D D_n + \gamma_L L_n$$

in which the subscript n denotes the nominal design values of resistance and loads. The ratios of these nominal values to the respective mean values may be considered as the corresponding bias factors; that is,

$$v_R = \frac{R_n}{\bar{R}}; \qquad v_D = \frac{D_n}{\bar{D}}; \qquad v_L = \frac{L_n}{\bar{L}}$$

Determine the appropriate nominal resistance and load factors, that is, ϕ, γ_D, and γ_L, to achieve designs with a reliability of $\beta = 2.50$. Consider a mean live load to dead load ratio of $\bar{L}/\bar{D} = 2.0$. Assume, also,

$$\Omega_R = 0.11; \qquad \Omega_D = 0.10; \qquad \Omega_L = 0.25$$

and

$$v_R = 0.95; \qquad v_D = 0.95; \qquad v_L = 1.18$$

The above design equation implies a linear performance function; that is,

$$g(\mathbf{X}) = R - D - L$$

with partial derivatives

$$\frac{\partial g}{\partial R'} = \sigma_R$$

$$\frac{\partial g}{\partial D'} = -\sigma_D$$

$$\frac{\partial g}{\partial L'} = -\sigma_L$$

Then,

$$\frac{\bar{R} - \bar{D} - \bar{L}}{\sqrt{\sigma_R^2 + \sigma_D^2 + \sigma_L^2}} = \beta = 2.5$$

where

$$\bar{L} = 2\bar{D}$$

and

$$\sigma_D = 0.1\bar{D}$$

$$\sigma_L = \Omega_L \bar{L} = \Omega_L(2\bar{D}) = 0.5\bar{D}$$

$$\sigma_R = 0.11\bar{R}$$

Hence,

$$\frac{\bar{R} - \bar{D} - 2\bar{D}}{\sqrt{(0.11\bar{R})^2 + (0.1\bar{D})^2 + (0.5D)^2}} = 2.5$$

resulting in the following quadratic equation

$$\bar{R}^2 - 6.491\bar{R}\bar{D} + 7.978 = 0$$

The solution for $\bar{R}$ is

$$\bar{R} = 4.844\bar{D}$$

and

$$\sigma_R = 4.844\bar{D} \times 0.11 = 0.533\bar{D}$$

The direction cosines are then

$$\alpha_R = \frac{\sigma_R}{\sqrt{\sigma_R^2 + \sigma_D^2 + \sigma_L^2}} = \frac{0.533\bar{D}}{\sqrt{(0.533\bar{D})^2 + (0.1\bar{D})^2 + (0.5\bar{D})^2}} = 0.722$$

$$\alpha_D = \frac{-\sigma_D}{\sqrt{\sigma_R^2 + \sigma_D^2 + \sigma_L^2}} = \frac{-0.10\bar{D}}{0.738\bar{D}} = -0.136$$

$$\alpha_L = \frac{-\sigma_L}{\sqrt{\sigma_R^2 + \sigma_D^2 + \sigma_L^2}} = \frac{-0.5\bar{D}}{0.738\bar{D}} = -0.678$$

Hence, according to Eq. 6.50, the appropriate mean resistance and load factors are

$$\bar{\phi} = (1 - 0.722 \times 2.5 \times 0.11) = 0.80$$

$$\bar{\gamma}_D = (1 + 0.136 \times 2.5 \times 0.10) = 1.03$$

$$\bar{\gamma}_L = (1 + 0.678 \times 2.5 \times 0.25) = 1.42$$

These mean factors should be used with the corresponding mean resistance and mean loads; that is, the safety requirement would be

$$0.80\bar{R} \geq 1.03\bar{D} + 1.42\bar{L}$$

Observe that in this (linear) case, no iteration is necessary to obtain the design factors.

To determine the corresponding nominal resistance and load factors, we observe that

$$v_R = \frac{R_n}{\bar{R}} = 0.95$$

or

$$\bar{R} = \frac{R_n}{0.95}$$

Similarly,

$$\bar{D} = \frac{D_n}{0.95} \quad \text{and} \quad \bar{L} = \frac{L_n}{1.18}$$

Therefore, in terms of the nominal values, the above safety requirement becomes

$$0.80\left(\frac{R_n}{0.95}\right) \geq 1.03\left(\frac{D_n}{0.95}\right) + 1.42\left(\frac{L_n}{1.18}\right)$$

or

$$0.84R_n \geq 1.08D_n + 1.20L_n$$

The corresponding *total load factor* and *safety factor* may be evaluated as follows:

$$1.08D_n + 1.20L_n = \gamma_n(D_n + L_n)$$

or

$$\gamma_n = \frac{1.08D_n + 1.20L_n}{D_n + L_n}$$

For the load ratio, $\bar{L}/\bar{D} = 2$,

$$\frac{L_n}{D_n} = 2\left(\frac{1.18}{0.95}\right) = 2.48$$

Thus, the total load factor is

$$\gamma_n = \frac{1.08 + 1.20 \times 2.48}{1 + 2.48} = 1.17$$

whereas the corresponding safety factor is

$$\theta_n = \frac{1.17}{0.84} = 1.39$$

EXAMPLE 6.21   (Uncorrelated Nonnormals)

In Example 6.20, suppose that the probability distributions of the design variables R, D, and L are as follows:

> R is lognormal.
>
> D is normal.
>
> L is Type I asymptotic.

Determine the corresponding load and resistance factors for obtaining designs with a reliability of $\beta = 2.5$. Again, consider $\bar{L}/\bar{D} = 2.0$.

Since the variables are not normally distributed, the parameters of the corresponding equivalent normal distributions would first be determined. For the lognormally distributed R, using the results derived in Example 6.4, we get

$$\sigma_R^N = r^*\zeta_R; \qquad \mu_R^N = r^*(1 - \ln r^* + \lambda_R)$$

With $\Omega_R = 0.11$,

$$\sigma_R^N = 0.11r^*$$

$$\mu_R^N \simeq r^*\left(1 - \ln\frac{r^*}{\bar{R}}\right)$$

whereas for the Type I extremal distribution of L, according to Eqs. 6.25 and 6.26,

$$\sigma_L^N = \frac{\phi\{\Phi^{-1}[\exp(-e^{-\alpha(l^*-u)})]\}}{\alpha e^{-\alpha(l^*-u)}\exp[-e^{-\alpha(l^*-u)}]}$$

and

$$\mu_L^N = l^* - \sigma_L^N\Phi^{-1}\{\exp[-e^{-\alpha(l^*-u)}]\}$$

where

$$\alpha = \frac{\pi}{\sqrt{6}\sigma_L} = \frac{\pi}{\sqrt{6} \times 0.25 \times 2\bar{D}} = \frac{2.565}{\bar{D}}$$

$$u = \bar{L} - \frac{\gamma}{\alpha} = 2\bar{D} - \frac{0.5772}{2.565}\bar{D} = 1.775\bar{D}$$

For the first iteration, assume $r^* = 0.838\bar{R}$ and $l^* = 3.579\bar{D}$, obtaining

$$\sigma_R^N = 0.11 \times 0.838\bar{R} = 0.092\bar{R}$$

$$\mu_R^N = 0.838\bar{R}\left[1 - \ln\left(\frac{0.838\bar{R}}{\bar{R}}\right)\right] = 0.986\bar{R}$$

$$e^{-\alpha(l^*-u)} = e^{-(2.565/\bar{D})(3.579\bar{D}-1.775\bar{D})} = 0.0098$$

$$\sigma_L^N = \frac{\phi\{\Phi^{-1}[\exp(-0.0098)]\}}{\frac{2.565}{\bar{D}} \times 0.0098 \exp(-0.0098)} = 1.015\bar{D}$$

$$\mu_L^N = 3.579\bar{D} - 1.015\bar{D}\Phi^{-1}[\exp(-0.0098)] = 1.194\bar{D}$$

Since the performance function is linear, namely

$$g(\mathbf{X}) = R - D - L$$

the corresponding design equation is given by

$$\frac{\mu_R^N - \mu_D - \mu_L^N}{\sqrt{(\sigma_R^N)^2 + \sigma_D^2 + (\sigma_L^N)^2}} = \beta = 2.5$$

For the first iteration,

$$\frac{0.986\bar{R} - \bar{D} - 1.194\bar{D}}{\sqrt{(0.092\bar{R})^2 + (0.10\bar{D})^2 + (1.015\bar{D})^2}} = 2.5$$

yielding

$$\bar{R} = 6.533\bar{D}$$

The direction cosines become

$$\alpha_R^* = \frac{\sigma_R^N}{\sqrt{(\sigma_R^N)^2 + \sigma_D^2 + (\sigma_L^N)^2}} = \frac{0.092 \times 6.533\bar{D}}{\sqrt{(0.092 \times 6.533\bar{D})^2 + (0.10\bar{D})^2 + (1.015\bar{D})^2}} = 0.508$$

$$\alpha_D^* = -0.0845$$

$$\alpha_L^* = -0.857$$

The new failure point is

$$r^* = \mu_R^N - \alpha_R^* \sigma_R^N \beta = 0.986\bar{R} - 0.508 \times 0.092\bar{R} \times 2.5 = 0.869\bar{R}$$

$$d^* = \mu_D - \alpha_D^* \sigma_D \beta = \bar{D} + 0.0845 \times 0.1\bar{D} \times 2.5 = 1.0211\bar{D}$$

$$l^* = \mu_L^N - \alpha_L^* \sigma_L^N \beta = 1.194\bar{D} + 0.857 \times 1.015\bar{D} \times 2.5 = 3.369\bar{D}$$

Table E6.21 Summary of Iterations for Example 6.21

Iteration No.	Assumed Failure Point	σ_i^N	μ_i^N	$\bar{R}/\bar{D}$	α_i^*
	$l^* = 3.579\bar{D}$	$1.015\bar{D}$	$1.194\bar{D}$		-0.857
1	$d^* = \bar{D}$			6.533	-0.085
	$r^* = 0.838\bar{R}$	$0.092\bar{R}$	$0.986\bar{R}$		0.508
	$l^* = 3.369\bar{D}$	$0.966\bar{D}$	$1.307\bar{D}$		-0.889
2	$d^* = 1.021\bar{D}$			5.070	-0.095
	$r^* = 0.869\bar{R}$	$0.096\bar{R}$	$0.991\bar{R}$		0.448
	$l^* = 3.454\bar{D}$	$0.973\bar{D}$	$1.284\bar{D}$		-0.889
3	$d^* = 1.045\bar{D}$			5.056	-0.091
	$r^* = 0.883\bar{R}$	$0.097\bar{R}$	$0.993\bar{R}$		0.448

which can be used to start the second iteration. Table E6.21 summarizes the results of the first three iterations. With the results for α_i^* of the third iteration, the required mean resistance and load factors are, therefore,

$$\bar{\phi} = 1 - \alpha_R^* \beta \Omega_R = 1 - 0.448 \times 2.5 \times 0.11 = 0.877$$

$$\bar{\gamma}_D = 1 + \alpha_D^* \beta \Omega_D = 1 + 0.091 \times 2.5 \times 0.10 = 1.023$$

$$\bar{\gamma}_L = 1 + \alpha_L^* \beta \Omega_L = 1 + 0.889 \times 2.5 \times 0.25 = 1.556$$

Thus, the appropriate design equation is

$$0.877\bar{R} \geq 1.023\bar{D} + 1.556\bar{L}$$

In terms of nominal values,

$$\frac{0.877\bar{R}_n}{0.95} \geq \frac{1.023D_n}{0.95} + \frac{1.556L_n}{1.18}$$

or

$$0.92R_n \geq 1.08D_n + 1.32L_n$$

Correlated Variates   The method described above tacitly assumes that the random variables are uncorrelated. If some or all of the design variables are correlated, the determination of the failure point x* will require the orthogonal transformation of Eq. 6.31 to obtain the transformed variables **Y**. That is, the above algorithm may be applied to determine $\mathbf{y}^* = \{y_1^*, y_2^*, \ldots, y_n^*\}$, from which we obtain

$$\mathbf{x}'^* = \mathbf{T}\mathbf{y}^*$$

and

$$\mathbf{x}^* = [\sigma_X]\mathbf{x}'^* + \boldsymbol{\mu}_X$$

The required design factors are then obtained through Eq. 6.49, that is,

$$\gamma_i = \frac{x_i^*}{\mu_{X_i}}$$

EXAMPLE 6.22   (*Design of Travel Time*)

Consider a transportation network between cities A and B consisting of three branches as shown in Fig. E6.23.

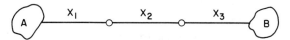

Figure E6.23   A transportation network.

Let X_1, X_2, and X_3 be the travel times over the three respective branches. Because of uncertainties in the traffic and weather conditions, these travel times are modeled as random variables as follows:

Travel Time	Mean	Standard Deviation
X_1	μ	σ
X_2	$c_2\mu$	$d_2\sigma$
X_3	$c_3\mu$	$d_3\sigma$

where c_2, c_3, d_2, and d_3 are constants dependent on the specific network. Furthermore, since the X_i's will be subject to common environmental factors, they are expected to be partially correlated. Suppose these correlations are

$$\rho_{12} = \rho_{13} = \rho_{23} = 0.5$$

The performance function may be defined as

$$g(\mathbf{X}) = T_o - (X_1 + X_2 + X_3)$$

where

$$T_o = \text{The scheduled total travel time between } A \text{ and } B.$$

The required travel time, T_o, may be determined using design factors as follows:

$$T_o = \bar{\gamma}_1 \bar{x}_1 + \bar{\gamma}_2 \bar{x}_2 + \bar{\gamma}_3 \bar{x}_3$$

where $\bar{\gamma}_1, \bar{\gamma}_2, \bar{\gamma}_3$ are the appropriate design factors necessary to insure that travel between A and B can be completed with a probability of $p_S = \Phi(\beta)$.

Since the travel times are correlated, transformation to a set of uncorrelated variates, or orthogonal coordinates, is required. The pertinent correlation matrix is

$$[C'] = \begin{bmatrix} 1.0 & 0.5 & 0.5 \\ 0.5 & 1.0 & 0.5 \\ 0.5 & 0.5 & 1.0 \end{bmatrix}$$

The eigenvalues of $[C']$ are the solutions of the following determinantal equations:

$$\det \begin{bmatrix} 1 - \lambda & 0.5 & 0.5 \\ 0.5 & 1 - \lambda & 0.5 \\ 0.5 & 0.5 & 1 - \lambda \end{bmatrix} = 0$$

or

$$(1 - \lambda)^3 - 0.5^2(1 - \lambda) \times 3 + 0.5^3 \times 2 = 0$$

The solutions are the eigenvalues

$$\lambda_1 = 0.5$$
$$\lambda_2 = 0.5$$
$$\lambda_3 = 2.0$$

with the corresponding eigenvectors

$$\{0.408, 0.408, -0.817\}$$
$$\{0.707, -0.707, 0\}$$
$$\{0.577, 0.577, 0.577\}$$

The orthogonal transformation matrix, **T**, therefore is

$$\mathbf{T} = \begin{bmatrix} 0.408 & 0.707 & 0.577 \\ 0.408 & -0.707 & 0.577 \\ -0.817 & 0 & 0.577 \end{bmatrix}$$

Thus, the transformed variates **Y** are

$$\mathbf{Y} = \mathbf{T}'\mathbf{X}'$$

and

$$\mathbf{X} = [\sigma_X]\mathbf{X}' + \boldsymbol{\mu}_X$$

where

$$[\sigma_X] = \begin{bmatrix} \sigma & 0 & 0 \\ 0 & d_2\sigma & 0 \\ 0 & 0 & d_3\sigma \end{bmatrix}$$

and

$$\boldsymbol{\mu}_X = \left\{ \begin{array}{c} \mu \\ c_2\mu \\ c_3\mu \end{array} \right\}$$

From which we obtain the components of **X**

$$X_1 = \sigma(0.408\,Y_1 + 0.707\,Y_2 + 0.577\,Y_3) + \mu$$
$$X_2 = d_2\sigma(0.408\,Y_1 - 0.707\,Y_2 + 0.577\,Y_3) + c_2\mu$$
$$X_3 = d_3\sigma(-0.817\,Y_1 + 0.577\,Y_3) + c_3\mu$$

The performance function then becomes

$$\begin{aligned} g(\mathbf{X}) &= T_0 - (X_1 + X_2 + X_3) \\ &= T_0 - \sigma(0.408 + 0.408d_2 - 0.817d_3)Y_1 \\ &\quad - \sigma(0.707 - 0.707d_2)Y_2 \\ &\quad - \sigma(0.577 + 0.577d_2 + 0.577d_3)Y_3 \\ &\quad - \mu(1 + c_2 + c_3) \end{aligned}$$

The partial derivatives are

$$\frac{\partial g}{\partial Y_1} = -\sigma(0.408)(1 + d_2 - 2d_3)$$

$$\frac{\partial g}{\partial Y_2} = -\sigma(0.707)(1 - d_2)$$

$$\frac{\partial g}{\partial Y_3} = -\sigma(0.577)(1 + d_2 + d_3)$$

giving the respective direction cosines

$$\alpha_{Y_1} = \frac{\dfrac{\partial g}{\partial Y_1} \sigma_{Y_1}}{\sqrt{\left(\dfrac{\partial g}{\partial Y_1} \sigma_{Y_1}\right)^2 + \left(\dfrac{\partial g}{\partial Y_2} \sigma_{Y_2}\right)^2 + \left(\dfrac{\partial g}{\partial Y_3} \sigma_{Y_3}\right)^2}}$$

$$= \frac{-0.408(1 + d_2 - 2d_3)\sqrt{0.5}}{K}$$

$$\alpha_{Y_2} = \frac{-0.707(1 - d_2)\sqrt{0.5}}{K}$$

$$\alpha_{Y_3} = \frac{-0.577(1 + d_2 + d_3)\sqrt{2}}{K}$$

where

$$K = \sqrt{[0.408(1 + d_2 - 2d_3)]^2(0.5) + [0.707(1 - d_2)]^2(0.5) + [0.577(1 + d_2 + d_3)]^2(2)}$$

Hence,

$$y_1^* = -\alpha_{Y_1}\beta\sigma_{Y_1} = \frac{0.408(1 + d_2 - 2d_3)(0.5)}{K}\beta$$

$$y_2^* = -\alpha_{Y_2}\beta\sigma_{Y_2} = \frac{0.707(1 - d_2)(0.5)}{K}\beta$$

$$y_3^* = -\alpha_{Y_3}\beta\sigma_{Y_3} = \frac{0.577(1 + d_2 + d_3)(2)}{K}\beta$$

and

$$x_1^* = \sigma(0.408y_1^* + 0.707y_2^* + 0.577y_3^*) + \mu$$

$$= \frac{\sigma\beta}{K}[0.408^2(1 + d_2 - 2d_3)(0.5) + 0.707^2(1 - d_2)(0.5) + 0.577^2(1 + d_2 + d_3)2] + \mu$$

$$x_2^* = \frac{d_2\sigma\beta}{K}[0.408^2(1 + d_2 - 2d_3)(0.5) - 0.707^2(1 - d_2)(0.5) + 0.577^2(1 + d_2 + d_3)2] + c_2\mu$$

$$x_3^* = \frac{d_3\sigma\beta}{K}[-(0.817)(0.408)(1 + d_2 - 2d_3)(0.5) + 0.577^2(1 + d_2 + d_3)2] + c_3\mu$$

According to Eq. 6.49, the required design factors are then

$$\bar{\gamma}_1 = \frac{x_1^*}{\mu_{X_1}} = \frac{x_1^*}{\mu}$$

$$= 1 + \frac{\sigma}{\mu}\beta\left[\frac{0.408^2(1+d_2-2d_3)(0.5)+0.707^2(1-d_2)(0.5)+0.577^2(1+d_2+d_3)2}{\sqrt{[0.408(1+d_2-2d_3)]^2(0.5)+[0.707(1-d_2)]^2(0.5)+2[0.577(1+d_2+d_3)]^2}}\right]$$

$$\bar{\gamma}_2 = 1 + \frac{\sigma\,d_2}{\mu\,c_2}\beta\left[\frac{0.408^2(1 + d_2 - 2d_3)(0.5) - 0.707^2(1 - d_2)(0.5) + 0.577^2(1 + d_2 + d_3)2}{\sqrt{[0.408(1+d_2-2d_3)]^2(0.5)+[0.707(1-d_2)]^2(0.5)+2[0.577(1+d_2+d_3)]^2}}\right]$$

$$\bar{\gamma}_3 = 1 + \frac{\sigma\,d_3}{\mu\,c_3}\beta\left[\frac{-0.817(0.408)(1+d_2-2d_3)(0.5)+0.577^2(1+d_2+d_3)2}{\sqrt{[0.408(1+d_2-2d_3)]^2(0.5)+[0.707(1-d_2)]^2(0.5)+2[0.577(1+d_2+d_3)]^2}}\right]$$

With these design factors, the scheduled travel time necessary to achieve a reliability p_S may then be obtained as

$$T_0 = \bar{\gamma}_1 \mu_{X_1} + \bar{\gamma}_2 \mu_{X_2} + \bar{\gamma}_3 \mu_{X_3}$$
$$= \bar{\gamma}_1 \mu + \bar{\gamma}_2 c_2 \mu + \bar{\gamma}_3 c_3 \mu$$

For illustration, consider the following numerical example. Assume

$$\mu_{X_1} = 40 \text{ min} \qquad \sigma_{X_1} = 10 \text{ min}$$
$$\mu_{X_2} = 15 \text{ min} \qquad \sigma_{X_2} = 5 \text{ min}$$
$$\mu_{X_3} = 50 \text{ min} \qquad \sigma_{X_3} = 10 \text{ min}$$

Hence,

$$c_2 = \frac{15}{40} = 0.375$$

$$c_3 = 1.25$$

$$d_2 = \frac{5}{10} = 0.5$$

$$d_3 = 1$$

Suppose further that the scheduled travel time should be such that 95% of all trips can be completed on time. The corresponding reliability index is $\beta = 1.96$. If we substitute these into the above expressions, the required design factors are:

$$\bar{\gamma}_1 = 1.39$$

$$\bar{\gamma}_2 = 1.45$$

$$\bar{\gamma}_3 = 1.31$$

Hence, the required travel time should be scheduled as

$$T_o = 1.39 \times 40 + 1.45 \times 15 + 1.31 \times 50 \simeq 143 \text{ min}$$

EXAMPLE 6.23   (Nonlinear Performance Function)

Determine the mean design factors for the nonlinear performance function of Example 6.7; namely,

$$g(\mathbf{X}) = YZ - M$$

in order to achieve a reliability of $\beta = 2.50$.

As this is a design problem, the purpose of the design factors is for determining the appropriate $\bar{Z}$ for any given $\bar{M}$ to satisfy the required reliability. With $\bar{Y} = 40$ ksi, $\Omega_Y = 0.125$; $\Omega_Z = 0.05$; and $\Omega_M = 0.20$, we determine the required design factors as follows.
At the limit-state,

$$y^* z^* - m^* = 0$$

and

$$\left(\frac{\partial g}{\partial Y'} \right)_* = z^* \sigma_Y; \qquad \left(\frac{\partial g}{\partial Z'} \right)_* = y^* \sigma_Z; \qquad \left(\frac{\partial g}{\partial M'} \right)_* = -\sigma_M$$

For the initial iteration, assume

$$y^* = \overline{Y} = 40 \text{ ksi}$$
$$z^* = \overline{Z}$$
$$m^* = \overline{M}$$

Then,

$$\left(\frac{\partial g}{\partial Y'}\right)_* = \overline{Z}(40 \times 0.125) = 5\overline{Z}$$

$$\left(\frac{\partial g}{\partial Z'}\right)_* = 40(0.05\overline{Z}) = 2\overline{Z}$$

$$\left(\frac{\partial g}{\partial M'}\right)_* = -0.20\overline{M} = -0.20(40\overline{Z}) = -8\overline{Z}$$

and

$$\overline{M} = \overline{Y}\,\overline{Z} = 40\overline{Z}$$

$$\alpha_Y^* = \frac{5}{9.644} = 0.5185$$

$$\alpha_Z^* = \frac{2}{9.644} = 0.2074$$

$$\alpha_M^* = \frac{-8}{9.644} = -0.8296$$

Therefore,

$$y^* = 40(1 - 0.5185 \times 2.5 \times 0.125) = 33.52$$
$$z^* = \overline{Z}(1 - 0.2074 \times 2.5 \times 0.05) = 0.9741\overline{Z}$$
$$m^* = \overline{M}(1 + 0.8296 \times 2.5 \times 0.20) = 1.4148\overline{M}$$

The results of the above and subsequent iterations are summarized in Table E6.23.
Comparing the last columns of iterations 3 and 4, we see that convergence has been achieved after the fourth iteration. The required design factors, therefore, are

$$\overline{\gamma}_Y = 1 - 0.7130 \times 2.5 \times 0.125 = 0.78$$
$$\overline{\gamma}_Z = 1 - 0.2216 \times 2.5 \times 0.05 = 0.97$$
$$\overline{\gamma}_M = 1 + 0.6652 \times 2.5 \times 0.20 = 1.33$$

The criterion for design (in terms of mean values) may then be given as

$$(0.78\,\overline{Y})(0.97\overline{Z}) - 1.33\overline{M} \geq 0$$

or

$$0.76\,\overline{Y}\,\overline{Z} - 1.33\overline{M} \geq 0$$

and for $\overline{Y} = 40$ ksi, the criterion gives the required section modulus for a given mean moment $\overline{M}$; that is,

$$\overline{Z} \geq 0.044\overline{M}$$

Table E6.23   Summary of Iterations for Example 6.23

Iteration No.	Variable	Assumed x_i^*	$\left(\dfrac{\partial g}{\partial X_i'}\right)_*$	$\alpha_{X_i}^*$	New x_i^*
1	Y	40	$5\bar{Z}$	0.5185	33.52
	Z	$\bar{Z}$	$2\bar{Z}$	0.2074	$0.9741\bar{Z}$
	M	$\bar{M}$	$-8\bar{Z}$	-0.8296	$1.4148\bar{M}$

Failure Equation: $33.52 \times 0.9741\bar{Z} = 1.4148\bar{M}$
$$\bar{M} = 23.08\bar{Z}$$

2	Y	33.52	$4.871\bar{Z}$	0.7052	31.19
	Z	$0.9741\bar{Z}$	$1.633\bar{Z}$	0.2364	$0.9705\bar{Z}$
	M	$1.4148\bar{M}$	$-4.616\bar{Z}$	-0.6683	$1.3342\bar{M}$

Failure Equation: $31.19 \times 0.9705\bar{Z} = 1.3342\bar{M}$
$$\bar{M} = 22.69\bar{Z}$$

3	Y	31.19	$4.853\bar{Z}$	0.7122	31.09
	Z	$0.9705\bar{Z}$	$1.513\bar{Z}$	0.2220	$0.9723\bar{Z}$
	M	$1.334\bar{M}$	$4.538\bar{Z}$	-0.6660	$1.333\bar{M}$

Failure Equation: $31.09 \times 0.9723\bar{Z} = 1.333\bar{M}$
$$\bar{M} = 22.68\bar{Z}$$

4	Y	31.09	$4.862\bar{Z}$	0.7130	31.09
	Z	$0.9723\bar{Z}$	$1.511\bar{Z}$	0.2216	$0.9723\bar{Z}$
	M	$1.333\bar{M}$	$-4.536\bar{Z}$	-0.6652	$1.333\bar{M}$

EXAMPLE 6.24   (*Design of Foundation for Settlement*)

Determine the mean partial factors for the design of a foundation such that the reliability against a settlement of 1 in. will be $\beta = 2.0$. Assume that the underlying clay stratum is normally consolidated and secondary consolidation settlement is neglected.

For mathematical convenience, the performance function in Example 6.11 can be rewritten as

$$g(\mathbf{X}) = 1 - NY \log(1 + M)$$

where

$$Y = \left(\frac{C_c}{1 + e_0}\right)H; \quad \text{a generalized flexibility factor}$$

and

$$M = \frac{\Delta p}{p_0}; \quad \text{the ratio of the design pressure to the initial overburden pressure.}$$

The statistics of the component variables are as follows:

Variable	Mean	c.o.v.
N	1.0	0.10
Y	$\bar{Y}$	0.26
M	$\bar{M}$	0.21

The partial derivatives of the performance function are:

$$\left(\frac{\partial g}{\partial N'}\right)_* = -y^* \log(1 + m^*)\sigma_N = -0.1y^* \log(1 + m^*)$$

$$\left(\frac{\partial g}{\partial Y'}\right)_* = -n^* \log(1 + m^*)\sigma_Y = -0.26\overline{Y}n^* \log(1 + m^*)$$

$$\left(\frac{\partial g}{\partial M'}\right)_* = -n^*y^* \frac{\log e}{1 + m^*}\sigma_M = -0.091 \frac{n^*y^*}{1 + m^*}\overline{M}$$

For the first iteration, assume $y^* = \overline{Y}$, $n^* = \overline{N} = 1.0$, and $m^* = \overline{M}$. Then,

$$\left(\frac{\partial g}{\partial N'}\right)_* = -0.1\overline{Y} \log(1 + \overline{M})$$

$$\left(\frac{\partial g}{\partial Y'}\right)_* = -0.26\overline{Y} \log(1 + \overline{M})$$

$$\left(\frac{\partial g}{\partial M'}\right)_* = -0.091 \frac{\overline{Y}\,\overline{M}}{1 + \overline{M}}$$

and, by first-order approximation (for the first iteration only)

$$\frac{1 - \overline{N}\,\overline{Y} \log(1 + \overline{M})}{\sqrt{[-0.1\overline{Y} \log(1 + \overline{M})]^2 + [-0.26\overline{Y} \log(1 + \overline{M})]^2 + \left(-0.091 \frac{\overline{Y}\,\overline{M}}{1 + \overline{M}}\right)^2}} = 2.0$$

yielding

$$\overline{Y} = \frac{1}{\log(1 + \overline{M}) + 2\sqrt{0.0776 \log^2(1 + \overline{M}) + 0.0083\left(\frac{\overline{M}}{1 + \overline{M}}\right)^2}}$$

Because of the complex relationship between $\overline{Y}$ and $\overline{M}$, further analytical solution is difficult to obtain; we evaluate the design partial factors assuming different values of $\overline{M}$. First consider $\overline{M} = 0.05$; the value of $\overline{Y}$ in the initial iteration is calculated from the above equation as $\overline{Y} = 27.9$. Therefore,

$$\left(\frac{\partial g}{\partial N'}\right)_* = -0.1 \times 27.9 \log(1 + 0.05) = -0.059$$

$$\left(\frac{\partial g}{\partial Y'}\right)_* = -0.26 \times 27.9 \log(1 + 0.05) = -0.154$$

$$\left(\frac{\partial g}{\partial M'}\right)_* = -0.091 \times \frac{27.9 \times 0.05}{1 + 0.05} = -0.121$$

and the direction cosines are

$$\alpha_N^* = \frac{-0.059}{\sqrt{(-0.059)^2 + (-0.154)^2 + (-0.121)^2}} = \frac{-0.059}{0.205} = -0.288$$

$$\alpha_Y^* = \frac{-0.154}{0.205} = -0.753$$

$$\alpha_M^* = \frac{-0.121}{0.205} = -0.592$$

Thus,

$$n^* = \bar{N} - \alpha_N^* \beta \sigma_N = 1.0 + 0.288 \times 2 \times 0.1 = 1.058$$
$$y^* = \bar{Y} - \alpha_Y^* \beta \sigma_Y = \bar{Y}(1 + 0.753 \times 2 \times 0.26) = 1.392\bar{Y}$$
$$m^* = \bar{M} - \alpha_M^* \beta \sigma_M = 0.05(1 + 0.592 \times 2 \times 0.21) = 0.0625$$

and on the basis of $g(\mathbf{X}^*) = 0$, we have

$$1 - 1.058 \times 1.392\bar{Y} \log(1 + 0.0625) = 0$$

which yields $\bar{Y} = 25.8$. Hence,

$$y^* = 1.392 \times 25.8 = 35.91$$

Using these updated values of n^*, y^*, and m^* for the second iteration, we obtain

$$\left(\frac{\partial g}{\partial N'}\right)_* = -0.1 \times 35.91 \log(1 + 0.0625) = -0.0945$$

$$\left(\frac{\partial g}{\partial Y'}\right)_* = -0.26 \times 25.8 \times 1.058 \log(1 + 0.0625) = -0.187$$

$$\left(\frac{\partial g}{\partial M'}\right)_* = -0.091 \times \frac{1.058 \times 35.91}{1 + 0.0625} \times 0.05 = -0.163$$

$$\alpha_N^* = \frac{-0.0945}{0.265} = -0.357$$

$$\alpha_Y^* = -0.706$$

$$\alpha_M^* = -0.615$$

and

$$n^* = 1 + 0.357 \times 2 \times 0.1 = 1.071$$
$$y^* = 1.367\bar{Y}$$
$$m^* = 0.0629$$

Thus,

$$1 - 1.071 \times 1.367\bar{Y} \log(1 + 0.0629) = 0$$

yielding $\bar{Y} = 25.8$; this is exactly the same value obtained in the first iteration. Hence, the required mean partial factors are

$$\bar{\gamma}_M = \frac{m^*}{\bar{M}} = \frac{0.0629}{0.05} = 1.258$$

$$\bar{\gamma}_Y = \frac{y^*}{\bar{Y}} = 1.367$$

$$\bar{\gamma}_N = \frac{n^*}{\bar{N}} = \frac{1.071}{1.0} = 1.071$$

The above factors are required to achieve a reliability of $\beta = 2.0$ for $\overline{M} = 0.05$ and $\overline{Y} = 25.8$. The above calculations were repeated for $\overline{M} = 0.01, 0.1$, and 0.5 and various $\overline{Y}$. The corresponding design factors for these cases are summarized in the following table:

$\overline{M}$	$\overline{Y}$	$\bar{\gamma}_M$	$\bar{\gamma}_Y$	$\bar{\gamma}_N$
0.01	126	1.262	1.362	1.071
0.05	25.8	1.258	1.367	1.071
0.10	13.3	1.253	1.370	1.072
0.50	3.22	1.228	1.391	1.075

Observe that the partial factors do not vary much over wide ranges of $\overline{M}$ and $\overline{Y}$, especially for $\overline{M} < 0.1$ or $\overline{Y} > 10.0$. Within these ranges, it may be satisfactory to use $\bar{\gamma}_M = 1.26$, $\bar{\gamma}_Y = 1.37$, and $\bar{\gamma}_N = 1.07$; with these the design requirement becomes

$$1 - 1.07 \times 1.37\overline{Y} \times \log(1 + 1.26\overline{M}) > 0$$

or

$$\overline{Y} < \frac{0.682}{\log(1 + 1.26\overline{M})}$$

In terms of the basic variables, this requirement would be

$$\overline{\Delta p} < \frac{0.682}{\log\left(1 + 1.26 \dfrac{\overline{C}_c}{1 + \bar{e}_o} \overline{H}\right)} \bar{p}_o$$

This is the allowable pressure for design in order to achieve a reliability of $\beta = 2$ against a settlement of 1 in. for an underlying normally consolidated clay stratum.

EXAMPLE 6.25   (Design of Drainage Pipe Section)

Determine the mean partial factors for the design of a pipe section to insure a reliability against overflow (inflow exceeds pipe capacity) of $\beta = 1.28$. Performance is defined with the function

$$g(\mathbf{X}) = 201.6Nf^{-1/2}Y - Q$$

where Q is the rate of inflow into the pipe; f is the friction factor; N is the correction for model error underlying the Darcy-Weisbach formula for pipe flow capacity; Y is the design variable equal to $S^{0.5}R^{2.5}$, in which S is the friction slope and R is the hydraulic radius. From the uncertainty analysis of Example 6.15, the statistics of the component variables are:

Variable	Mean	c.o.v.
f	0.0246	0.11
N	1.1	0.11
Y	$\overline{Y}$	0.04
Q	$\overline{Q}$	0.25

Assume that the inflow rate Q follows a Type I largest value distribution. Then, for a given $\overline{Q}$ and $\Omega_Q = 0.25$, the extremal parameters are

$$\alpha = \frac{1.282}{\sigma_Q} = \frac{1.282}{0.25\overline{Q}} = \frac{5.13}{\overline{Q}}$$

$$u = \overline{Q} - \frac{0.577}{\alpha} = \overline{Q} - \frac{0.577}{5.13}\overline{Q} = 0.888\overline{Q}$$

The corresponding CDF and PDF at the failure point q^* are, respectively,

$$F_Q(q^*) = \exp[-e^{-\alpha(q^*-u)}]$$

$$f_Q(q^*) = \alpha \exp[-\alpha(q^* - u) - e^{-\alpha(q^*-u)}]$$

For the first iteration, assume $q^* = \bar{Q}$, obtaining, with the above values of u and α,

$$F_Q(q^*) = \exp[-e^{-5.13(\bar{Q}-0.888\bar{Q})/\bar{Q}}] = 0.57$$

and

$$f_Q(q^*) = \frac{5.13}{\bar{Q}} \exp\left[-\frac{5.13}{\bar{Q}}(\bar{Q} - 0.888\bar{Q}) - e^{-5.13(\bar{Q}-0.888\bar{Q})/\bar{Q}}\right]$$

$$= \frac{1.645}{\bar{Q}}$$

The standard deviation and mean of the equivalent normal distribution at q^* are, according to Eqs. 6.25 and 6.26,

$$\sigma_Q^N = \frac{\phi[\Phi^{-1}(0.57)]}{1.645/\bar{Q}} = 0.24\bar{Q}$$

and

$$\mu_Q^N = \bar{Q} - \sigma_Q^N \Phi^{-1}(0.57) = 0.957\bar{Q}$$

Furthermore, substituting $y^* = \bar{Y}$, $n^* = \bar{N}$, and $f^* = \bar{f}$ into the partial derivatives, we have

$$\left(\frac{\partial g}{\partial N'}\right)_* = \frac{201.6 y^*}{f^{*1/2}}\sigma_N = \frac{201.6\bar{Y}}{0.0246^{1/2}}(0.11 \times 1.1) = 155.5\bar{Y}$$

$$\left(\frac{\partial g}{\partial Y'}\right)_* = \frac{201.6 n^*}{f^{*1/2}}\sigma_Y = \frac{201.6 \times 1.1}{0.0246^{1/2}}(0.04\bar{Y}) = 56.56\bar{Y}$$

$$\left(\frac{\partial g}{\partial f'}\right)_* = -\frac{100.8 n^* Y^*}{f^{*3/2}}\sigma_f = -\frac{100.8 \times 1.1\bar{Y}}{0.0246^{3/2}}(0.11 \times 0.0246) = -77.76\bar{Y}$$

$$\left(\frac{\partial g}{\partial Q'}\right)_* = -\sigma_Q^N = -0.24\bar{Q}$$

By first-order approximation, the reliability requirement is

$$\frac{201.6\bar{N}\bar{Y}\bar{f}^{-1/2} - \bar{Q}}{\sqrt{(155.5\bar{Y})^2 + (56.56\bar{Y})^2 + (-77.76\bar{Y})^2 + (-0.24\bar{Q})^2}} = \beta$$

or

$$\frac{1414\bar{Y} - 0.957\bar{Q}}{\sqrt{33,435\bar{Y}^2 + 0.058\bar{Q}^2}} = 1.28$$

which yields $\bar{Y} = 9.5 \times 10^{-4}\bar{Q}$. The partial derivatives become

$$\left(\frac{\partial g}{\partial N'}\right)_* = 155.5 \times 9.5 \times 10^{-4}\bar{Q} = 0.148\bar{Q}$$

$$\left(\frac{\partial g}{\partial Y'}\right)_* = 56.56 \times 9.5 \times 10^{-4}\bar{Q} = 0.053\bar{Q}$$

$$\left(\frac{\partial g}{\partial f'}\right)_* = -77.76 \times 9.5 \times 10^{-4}\bar{Q} = -0.074\bar{Q}$$

$$\left(\frac{\partial g}{\partial Q'}\right)_* = -0.24\bar{Q}$$

and the direction cosines are

$$\alpha_N^* = \frac{0.148\,\bar{Q}}{\sqrt{(0.148\,\bar{Q})^2 + (0.053\,\bar{Q})^2 + (-0.074\,\bar{Q})^2 + (-0.24\,\bar{Q})^2}}$$

$$= \frac{0.148\,\bar{Q}}{0.296\,\bar{Q}} = 0.500$$

$$\alpha_Y^* = \frac{0.053\,\bar{Q}}{0.296\,\bar{Q}} = 0.179$$

$$\alpha_f^* = \frac{-0.074\,\bar{Q}}{0.296\,\bar{Q}} = -0.250$$

$$\alpha_{Q'}^* = \frac{-0.24\,\bar{Q}}{0.296\,\bar{Q}} = -0.810$$

Thus,

$$n^* = \bar{N} - \alpha_N^*\beta\sigma_N = 1.1 - 0.500 \times 1.28 \times 0.11 \times 1.1 = 1.023$$

$$y^* = \bar{Y} - \alpha_Y^*\beta\sigma_Y = \bar{Y} - 0.179 \times 1.28 \times 0.04\bar{Y} = 0.991\,\bar{Y}$$

$$f^* = \bar{f} - \alpha_f^*\beta\sigma_f = 0.0246 + 0.250 \times 1.28 \times 0.0246 \times 0.11 = 0.0255$$

$$Q^* = \mu_Q^N - \alpha_Q^*\beta\sigma_Q^N = 0.957\,\bar{Q} + 0.810 \times 1.28 \times 0.24\bar{Q} = 1.206\,\bar{Q}$$

and

$$g(\mathbf{X}^*) = 201.6 n^* f^{*-1/2} y^* - q^*$$
$$= 201.6 \times 1.023 \times 0.0255^{-1/2} \times 0.991\,\bar{Y} - 1.206\,\bar{Q}$$

At the failure surface,

$$g(\mathbf{X}^*) = 0$$

Hence,

$$\bar{Y} = 9.4 \times 10^{-4}\,\bar{Q}$$

Using the above results, we perform the subsequent iterations; the results for the first three iterations are summarized in Table E6.25. Observe that convergence has been achieved after the third iteration. Hence, the required mean partial factors are

$$\bar{\gamma}_N = \frac{n^*}{\bar{N}} = \frac{1.04}{1.1} = 0.945$$

$$\bar{\gamma}_Y = \frac{y^*}{\bar{Y}} = 0.993$$

$$\bar{\gamma}_f = \frac{f^*}{\bar{f}} = \frac{0.0252}{0.0246} = 1.024$$

$$\bar{\gamma}_Q = \frac{q^*}{\bar{Q}} = 1.28$$

In other words, if we design using the requirement

$$201.6 \times 0.945\bar{N} \times \frac{0.993\,\bar{Y}}{(1.024\bar{f})^{1/2}} - 1.28\,\bar{Q} \geq 0$$

Table E6.25   Summary of Iterations for Example 6.25

Iteration No.	Variable	Assumed x_i^*	Parameters of Equivalent Normal	$\left(\dfrac{\partial g}{\partial X_i}\right)_*$	$\alpha_{X_i}^*$	New x_i^*
1	f	0.0246		$-0.074\bar{Q}$	-0.250	0.0255
	N	1.1		$0.148\bar{Q}$	0.500	1.023
	Y	$\bar{Y}$		$0.053\bar{Q}$	0.179	$0.991\bar{Y}$
			$\sigma_Q^N = 0.24\bar{Q}$			
	Q	$\bar{Q}$		$-0.24\bar{Q}$	-0.810	$1.206\bar{Q}$
			$\mu_Q^N = 0.957\bar{Q}$			

Failure Equation: $201.6 \times 1.023 \times 0.0255^{-1/2} \times 0.991\bar{Y} - 1.206\bar{Q} = 0$
$$\bar{Y} = 9.4 \times 10^{-4}\bar{Q}$$

2	f	0.0255		$-0.064\bar{Q}$	-0.180	0.0252
	N	1.023		$0.142\bar{Q}$	0.400	1.038
	Y	$0.991\bar{Y}$		$0.049\bar{Q}$	-0.138	$0.993\bar{Y}$
			$\sigma_Q^N = 0.315\bar{Q}$			
	Q	$1.206\bar{Q}$		$-0.315\bar{Q}$	-0.888	$1.273\bar{Q}$
			$\mu_Q^N = 0.915\bar{Q}$			

Failure Equation: $201.6 \times 1.038 \times 0.0252^{-1/2} \times 0.993\bar{Y} - 1.273\bar{Q} = 0$
$$\bar{Y} = 9.73 \times 10^{-4}\bar{Q}$$

3	f	0.0252		$-0.070\bar{Q}$	-0.178	0.0252
	N	1.038		$0.149\bar{Q}$	0.388	1.04
	Y	$0.993\bar{Y}$		$0.051\bar{Q}$	0.134	$0.993\bar{Y}$
			$\sigma_Q^N = 0.341\bar{Q}$			
	Q	$1.273\bar{Q}$		$-0.341\bar{Q}$	-0.894	$1.28\bar{Q}$
			$\mu_Q^N = 0.889\bar{Q}$			

Failure Equation: $201.6 \times 1.04 \times 0.0252^{-1/2} \times 0.993\bar{Y} - 1.28\bar{Q} = 0$
$$\bar{Y} = 9.76 \times 10^{-4}\bar{Q}$$

or in terms of the basic variables (substituting $\bar{N} = 1.1$, $\bar{f} = 0.0246$)

$$\bar{S}^{1/2}\bar{R}^{2.5} \geq 9.76 \times 10^{-4}\bar{Q}$$

the required reliability of $\beta = 1.28$ will be satisfied.

PROBLEMS

6.1   The steel beam shown in Fig. P6.1 is subjected to a uniformly distributed load with mean $\bar{w} = 2$ k/ft and c.o.v. $\Omega_w = 0.20$. The yield stress of the steel is $f_y = 47$ ksi and $\Omega_{f_y} = 0.10$; the section modulus of the beam I/c has a mean of 50 in.^{3} and c.o.v. of 0.10.

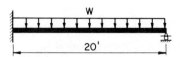

Figure P6.1   A beam structure.

If the probability distributions of all the variables were normal, what is the probability of collapse through plastic yielding of the beam?

6.2   Because of fabrication tolerance and errors, the members of the truss shown in Fig. P6.2

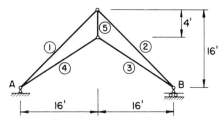

Figure P6.2   Truss structure.

may be shorter or longer than planned. Suppose the error in each member is a normal variate as follows:

$$\Delta_1 = \Delta_2 = N(0.25'', 0.10'')$$

$$\Delta_3 = \Delta_4 = N(0.30'', 0.10'')$$

$$\Delta_5 = N(0.20'', 0.08'')$$

(a)   Determine the probability that the location of the roller support B of the truss will be off by more than 3 in. Assume that the errors in the members are statistically independent.

(b)   Repeat part (a) if the distributions of the fabrication errors are lognormal; the means and standard deviations of the respective members remain the same as those of part (a).

(c)   Repeat part (a) if the fabrication errors between any two members are correlated with a uniform correlation of $\rho = 0.8$.

6.3   The platform in Example 6.13 can be subject to rotation caused by wave action in severe storms. Suppose the maximum rotation (in 10^{-4} radians) is given by

$$R = AM + BM^2 + \varepsilon$$

where M is the moment induced by the annual maximum wave, A and B are coefficients that are random variables, and ε is the error associated with the above quadratic model. Assume satisfactory performance requires that $\bar{R}$ should not exceed 4×10^{-4} radians. Suppose the following statistics for the respective variables:

Variable	Mean	c.o.v.	Distribution
M	1.5	0.25	lognormal
A	0.644	0.30	assume normal
B	0.571	0.30	assume normal
ε	0	$\sigma = 0.2$	assume normal

Determine the probability of unsatisfactory performance with respect to rotational displacement.

6.4   Figure P6.4 shows a vehicle B following another vehicle A in a traffic flow at a distance S apart. The driver in vehicle A suddenly applies the brake because of an unusual traffic condition ahead. Suppose the driver in vehicle B takes T sec to react to driver A's action and to his surprise he discovers that his brakes had failed, so he has to rely completely on the friction of the road pavement to slow down his vehicle. Collision between vehicles A and B will occur if the stopping distance of vehicle B exceeds the sum of the

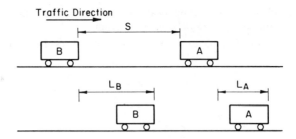

Figure P6.4   Two vehicles in a traffic flow.

gap S and the stopping distance of vehicle A. In other words, the performance function is given by

$$g(\mathbf{X}) = L_A + S - L_B$$

$$= \frac{V^2}{2\gamma} + S - \left(VT + \frac{V^2}{2f}\right)$$

where V is the speed of both vehicles, assumed to be 36.67 fps (25 mph) for an urban setting; γ is the deceleration of vehicle A as a result of braking; and f is the coefficient of friction of the pavement. Suppose the statistics of the random variables are as follows:

Variable	Mean	c.o.v.
γ	$0.2g$	0.10
S	200 ft	0.80
T	0.9 sec	$\sigma = 0.3$ sec
f	$0.05g$	0.05

Assume S follows a gamma distribution but no distribution information is available for the other variables (you may assume they are normal). Determine the probability of collision.

6.5   The maximum ground acceleration at a site, a_{max}, in cm/sec² , induced by an earthquake may be given by the following empirical formula (Donovan, 1973):

$$a_{max} = 1320 N e^{0.58M}(R + 25)^{-1.5}$$

where M is the earthquake magnitude in Richter scale; R is the distance (in km) between the earthquake source and the site; N denotes the error of the attenuation formula. Assume the following statistics of the variables:

Variable	Mean	c.o.v.	Distribution
R	25	0.2	lognormal
N	1.0	0.5	lognormal

Also the annual maximum earthquake magnitude has PDF as follows:

$$f_M(m) = 1.01 \exp(-ve^{-\beta m})v\beta e^{-\beta m}; \qquad 4 \le m \le 8.5$$

where $\beta = 1.57$ and $v = 6640$. Determine the CDF of the annual maximum ground acceleration and graphically show the results.

6.6   Consider a plate containing a crack of length $2a$ and subject to membrane stress σ as shown in Fig. P6.6. According to fracture mechanics, the crack will open if the *stress*

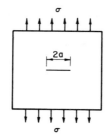

Figure P6.6   Plate under membrane stress σ.

intensity factor at the crack tip defined as

$$K_I = \sigma\sqrt{\pi a}$$

exceeds the plane strain fracture toughness of the material K_C.

Assume a, K_C, and σ are independent normal variates with mean values 0.6 in., 170 ksi$\sqrt{\text{in.}}$, 100 ksi, and c.o.v.'s 0.25, 0.07, and 0.52, respectively. Determine the probability of fracture failure of the plate.

6.7   Consider the cantilever beam shown in Fig. P6.7, loaded at its free end with a bending moment M and a torque T. The cross-sectional area of the beam is circular with radius

Figure P6.7   Cantilever beam.

$r = 0.20$m. M and T are random variables with mean values 26 kN-m, 17 kN-m and c.o.v.'s 0.18 and 0.14, respectively. Determine the reliability of the beam against yielding using the Tresca yield criterion, which in the present case is given by $\sigma^2/4 + \tau^2 = Y^2/4$, where Y is the yield stress in tension, σ and τ are the normal and shear stresses at a point. Assume that Y is normally distributed with $\mu_Y = 7000$ kN/m^2 and c.o.v. of 0.08.

6.8   Sliding of the earth mass along the circular surface about point O, as shown in Fig. P6.8, will occur if the resisting moment provided by the cohesive resistances of the soil

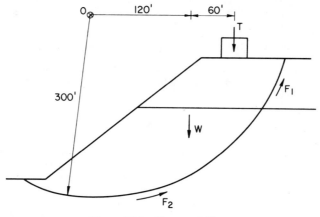

Figure P6.8   Slope stability.

$(F_1$ and $F_2)$ is exceeded by the driving moment caused by the weight of the soil W and the applied surcharge load T. In other words, the performance function is

$$g(\mathbf{X}) = 300F_1 + 300F_2 - 120W + 180T$$

With the statistics of the variables shown below:

Variable	Mean	c.o.v.
W	400	0.15
F_1	100	0.30
F_2	300	0.20
T	10	0.10

and assuming a correlation between F_1 and F_2 of $\rho_{F_1, F_2} = 0.5$, determine the probability of sliding of the slope along the specified circular surface.

6.9 In a pressure vessel subject to combined internal pressure and earthquake forces, suppose the state of stress at the most critical point is defined by the following principal stresses:

$$\sigma_1 = 1200A + 25P$$

$$\sigma_2 = 800A + 50P$$

$$\sigma_3 = -50P$$

where P is the internal pressure in psi and A is the earthquake acceleration in ft/sec^2. Assume P and A are independently normal random variables with mean values 200 psi, $0.25g$, and c.o.v.'s 0.20 and 0.50, respectively. Observe that the principal stresses, σ_1, σ_2, σ_3 are correlated. The yield strength of the pressure vessel material, Y, in pure tension has a mean of 47,000 psi and c.o.v. of 0.10.

Determine the probability of yielding failure of the pressure vessel at the critical point based on the following yield criteria:

(a) The von Mises yield criterion, which defines the limit state as

$$\sigma_1^2 + \sigma_2^2 + \sigma_3^2 - \sigma_1\sigma_2 - \sigma_2\sigma_3 - \sigma_3\sigma_1 = Y^2$$

(b) The Tresca yield criterion, which defines the limit state as

$$\sigma_{max} - \sigma_{min} = Y$$

where σ_{max} and σ_{min} denote the maximum and minimum principal stresses.

6.10 The performance of a pavement may be measured in terms of N, the number of equivalent 18-kip single-axle load application before failure, given by the following equation (Darter et al., 1973):

$$\log N = \log[\sqrt{5 - P_2} - \sqrt{5 - P_1}] + \log \alpha - 2 \log SCI + 4.27$$

where P_1 and P_2 are initial and terminal serviceability indices; SCI is the *stiffness coefficient index*; α is the temperature constant. Suppose the statistics of the variables are as follows:

Variable	Mean	Standard Deviation
P_1	4.0	0.36
P_2	3.0	0
α	31	4.24
SCI	0.15	0.02

Determine the probability that the pavement will fail to carry a 20-year design load of 6,000,000 equivalent 18-kip single-axle load applications.

6.11   The flow capacity of a culvert may be estimated by (Bodhaine, 1968)

$$Q = M\pi\sqrt{y_u - y_d}\left[\frac{46n^2L}{D^{16/3}} + (K_{ent} + K_{exit})\frac{8}{gD^4}\right]^{-1/2}$$

where y_u and y_d are the water surface elevations of the upstream and downstream locations; K_{ent} and K_{exit} are the entrance and exit loss coefficients; D and L are the diameter and length of the culvert; n is the friction coefficient in the Manning's formula; g is the gravitational constant; M is the model error of the flow capacity formula. With the following statistics of the component variables, determine the probability that the culvert can carry a design flood inflow of 300 cfs.

Variable	Mean	c.o.v.
$y_u - y_d$	4.5 ft	0.102
n	0.013	0.063
L	100 ft	0.001
D	6 ft	0.004
K_{ent}	0.35	0.197
K_{exit}	1.00	0.02
M	1.0	0.05

6.12   The area of a quadrilateral shown in Fig. P6.12 is to be estimated by the mean measurements of the four sides a, b, c, d and the two angles θ and ϕ. Suppose the standard

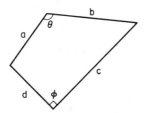

Figure P6.12   Quadrilateral area.

deviation of each measured length is 0.1 in., whereas that of each measured angle is 0.05°. The number of independent measurements made on a, b, c, d, θ, and ϕ are 10, 10, 6, 6, 5, and 9, respectively. Determine the probability that the error in estimating the area will exceed 1 sq in. Assume that the mean measured values of a, b, c, d, θ, and ϕ are, respectively, 50, 60, 100, 50, 120°, and 90°.

6.13   (a)   For the spillway capacity in Example 6.12, the following alternatives are being considered for increasing the spillway capacity of the existing dam.
      (I)   Extending the mean length of the existing spillway by an additional 20 ft.
      (II)   Lowering the crest of the existing spillway so that the mean total head on the crest H is increased by 1 ft.
    Assume all the other parameters remain the same as those given in Example 6.12. Determine the probability of unsatisfactory performance for each of the alternatives.
    (b)   Suppose the probability of unsatisfactory performance of the spillway capacity should not exceed 0.01, which could be achieved either by lengthening the existing spillway *or* by lowering the crest of the existing spillway. Assume the cost (in $ million) associated with lengthening the spillway by L ft is

$$C_1 = 0.5 + 0.08L$$

whereas that associated with lowering the spillway crest by D ft is

$$C_2 = 0.8 + 0.7D + 0.4D^2$$

Determine the optimal plan for meeting the performance criterion.

6.14 For a given level of earthquake acceleration and a number of equivalent cycles of loading, an element of a saturated sand deposit is predicted to liquefy if the earthquake-induced cyclic shear stress τ_A exceeds its shear resistance against cyclic loading τ_R. Suppose

$$\tau_A = S_L r_d \gamma h \frac{a_{max}}{g}$$

where S_L is the amplitude (in terms of fraction of peak stress) of equivalent uniform cycles, r_d is the stress reduction due to flexibility of the soil column, γ is the unit weight of the soil, h is the depth of the soil element, a_{max} is the maximum ground acceleration, and g is the gravitational acceleration.

The shear resistance is

$$\tau_R = N_f N_S C_r R \sigma'_v D_r$$

where C_r is the discrepancy between in-situ strength and laboratory strength; R is the normalized laboratory strength parameter for a given type of test and failure strain criterion; σ'_v is the effective vertical stress acting on the soil element in situ; D_r is the relative density of the soil in situ; N_S is the correction accounting for secondary factors such as frequency of cyclic loading, grain shape, etc., N_f is the correction for additional error associated with the simplified liquefaction model.

With the following statistics for a critical location at a site, determine the corresponding probability of liquefaction for case 1 when all the variables are assumed lognormal, and for case 2 when all the variables are assumed normal instead.

Variable	Mean	c.o.v.
a_{max}	0.1g	0
h	25 ft	0
γ	120 pcf	0.013
r_d	0.948	0.018
S_L	0.75	0
N_f	1	0.05
N_S	1	0.1
C_r	0.58	0.06
R	0.40	0.225
σ'_v	1625 psf	0.03
D_r	0.653	0.2

6.15 (a) The American Society for Testing and Materials (ASTM D2049-69) defines the relative density of sand as

$$D_r = \frac{\gamma_{max}}{\gamma} \frac{\gamma - \gamma_{min}}{\gamma_{max} - \gamma_{min}}$$

in which γ_{max}, γ_{min}, and γ are the maximum, minimum, and inplace dry density, respectively, of a cohesionless deposit.

Tavenas et al. (1972) estimated that the c.o.v. of γ_{min} and γ_{max} are 0.018 and 0.023, respectively, based on 62 tests. Moreover, it is suggested that the intra-laboratory (reproducibility) error is approximately one-third of the inter-laboratory error (in terms of c.o.v.) assumed above.

The uncertainties associated with estimating the in-situ density of sand, γ, depend on the methods used. For the nuclear method, the density may be determined within ± 5 to ± 9 pcf. However, based on current procedures for laboratory testing of undisturbed samples, the in-situ density generally cannot be estimated within ± 2 pcf, but more likely within ± 3 pcf.

Suppose the mean values of γ, γ_{max}, γ_{min} of the sand under consideration are 108, 115, and 95 pcf, respectively. Perform an uncertainty analysis to estimate the mean and overall uncertainty in D_r for both the nuclear method and the laboratory test method.

(b) An alternative procedure is to estimate the relative density of a sand deposit from Standard Penetration Test (SPT) results following the formula suggested by Gibbs and Holtz (1957), namely

$$N = 20D_r^{2.5} + 10D_r^2 P$$

where N is the SPT value, in blows per foot; D_r is the relative density; and P is the effective overburden pressure in kips per square foot. The prediction error associated with Gibbs and Holtz's formula may be evaluated based on field observations where D, the SPT value, and P are all measured. Based on three sets of observed data, the value of D_r actually measured in the field versus the D_r predicted from Gibbs and Holtz's equation are plotted in Fig. P6.15. An upper bound (line B), lower bound (line E), and a mean line are proposed as shown in the Fig. P6.15. For a given value of relative density calculated from Gibbs and Holtz's relationship, the actual relative density may be assumed to follow a triangular distribution between the upper and lower bounds, symmetrically about the mean line.

In addition to the prediction error in Gibbs and Holtz's relationship, error in D_r may also be induced by the variability of SPT values in the field. Tavenas

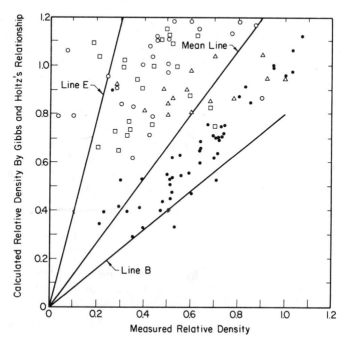

Figure P6.15 Comparison of measured versus calculated relative density using Gibbs and Holtz's relationship (from Haldar and Tang, 1979).

(1971) noted that for a homogeneous deposit, the c.o.v. of SPT values is about 25%.

Estimate the mean and total c.o.v. of the actual relative density for calculated relative densities of 0.4 and 0.7, respectively. Assume $P = 10$ psi.

6.16 A strip footing resting on the surface of a cohesionless soil is shown in Fig. P6.16. An inclined line load is applied off-center on the footing. The bearing capacity of the

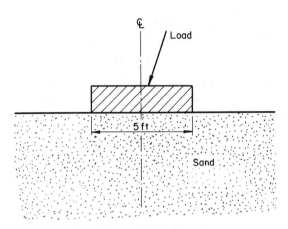

Figure P6.16   Strip footing

footing may be determined as (Bjerrum, 1973)

$$q = \tfrac{1}{2}\gamma \, B \, N_\gamma R \, I \, E_\gamma$$

where γ = the average density of sand; B = footing width; N_γ = bearing capacity factor; R, E_γ and I are correction factors for the effects of foundation size, load eccentricity, and load inclination, respectively. Regression analyses of model footing test data yield the following relationships (Baecher et al., 1980):

(i)
$$E(\ln N_\gamma|\bar{\phi}) = -1.667 + 0.173\bar{\phi}$$

and

$$\text{Var}(\ln N_\gamma|\bar{\phi}) = 0.0425$$

(ii)
$$E(E_\gamma|E/B) = 1.0 - 3.5(E/B) + 3.03(E/B)^2$$

and

$$\text{Var}(E_\gamma|E/B) = 0.0058$$

(iii)
$$E(I|H/V) = 1.0 - 2.41(H/V) + 1.36(H/V)^2$$

and

$$\text{Var}(I|H/V) = 0.0089$$

where $\bar{\phi}$ is the mean friction angle of the sand, E/B is the eccentricity ratio, H and V are the horizontal and vertical components of the inclined load.

(a)   Assume $\ln N_\gamma$ to be normally distributed at a given value of $\bar{\phi}$; show that

$$E(N_\gamma|\bar{\phi}) = \exp(-1.646 + 0.173\bar{\phi})$$

and

$$\text{Var}(N_\gamma|\bar{\phi}) = 0.0434 \exp(-3.292 + 0.346\bar{\phi})$$

(b) Suppose $\bar{\phi}$ has a mean $\mu_{\bar{\phi}}$ and variance $\sigma_{\bar{\phi}}^2$. By applying Eq. 3.81 of Vol. I, show that

$$\text{Var}(N_\gamma) = \exp(-3.292 + 0.345\mu_{\bar{\phi}})(0.0434 + 0.173^2\sigma_{\bar{\phi}}^2)$$

(c) Derive expressions for $\text{Var}(E_\gamma)$ and $\text{Var}(I)$ in terms of $\mu_{E/B}$, $\mu_{H/V}$, $\sigma_{E/B}^2$ and $\sigma_{H/V}^2$ which are the means and variances of E/B and H/V, respectively.

(d) Consider a strip footing with the following data:

$$B = 5\,\text{ft}$$

$$\mu_{\bar{\phi}} = 37° \qquad\qquad \sigma_{\bar{\phi}} = 1°$$

$$\gamma = 120\,\text{pcf} \qquad\quad \delta_\gamma = 0.05$$

$$\mu_{H/V} = 0.3 \qquad\quad \sigma_{H/V} = 0.033$$

$$\mu_{E/B} = 0.1 \qquad\quad \sigma_{E/B} = 0.01$$

$$R = 0.43$$

Determine the mean and total c.o.v. of the bearing capacity for this footing.

6.17 The ultimate bending capacity of the longitudinal girder of a single-deck ship hull (including the effects of buckling) may be expressed as

$$M_u = \phi Z \sigma_y$$

where

σ_y = The yield stress of the hull plate material.

Z = The fully plastic section modulus of the hull section.

ϕ = The "strength factor" that accounts for the effects of buckling.

Evaluate the total uncertainty in the estimation of the ultimate bending capacity. Assume the following (Ang, 1979):

The average c.o.v. for the yield strength of ship structural steel has been reported, for example, by Staugitis (1962); the average value of the c.o.v. is approximately 0.06.

For a single deck hull, the cross section of the hull girder may be idealized as a rectangular box girder with depth D and width B; on this basis, the section modulus, for a uniform material thickness t, is

$$Z = BDt + \tfrac{1}{6}D^2t$$

For uncertainty analysis, the second term may be neglected, whereas the c.o.v. of D and B are small. For the original thickness, t_o, assume a c.o.v. of 3%.

Corrosion, of course, will reduce the thickness; this reduction is additive, that is, it reduces the thickness by the same amount irrespective of the original thickness. Therefore, the actual thickness including corrosion is

$$t = t_o - t_c$$

where t_o is the original thickness and t_c is the reduction caused by corrosion. The average rate of corrosion of steel in sea water may be assumed (see Afanasief, 1974) to range between 10 and 14 mpy (mils per year).

The c.o.v. of the strength factor, ϕ, may be evaluated using the test results of stiffened plates reported in Smith (1975), giving the following values of ϕ:

0.76	0.82
0.73	0.83
0.91	0.72
0.83	0.49
0.69	0.65
0.61	

In addition to the above factors, the effects of residual stress and geometrical distortion are also significant. These effects may be included by applying a correction factor to the calculated M_u ranging from 0.75 to 1.00, with the higher values being more likely.

6.18 A circular tunnel structure, shown in Fig. P6.18, is proposed for a protective shelter against a blast pressure, q. The material of the circular plate and liner are A36 steel with a nominal yield strength of 36 ksi and a c.o.v. of 0.08. The nominal yield strength, y_n, corresponds to the 10-percentile value; that is, $F_Y(y_n) = 0.10$, in which $F_Y(y)$ is normal.

The reliability of the circular plate against yielding is of interest. The von Mises

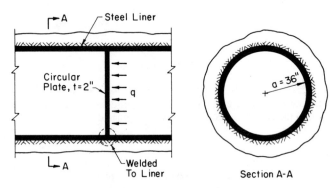

Figure P6.18   Circular tunnel structure.

yield condition may be used to determine the limit state. Under plane stress, this yield condition is

$$\sqrt{\sigma_x^2 - \sigma_x \sigma_y + \sigma_y^2} = Y$$

where

Y = The yield strength of the material under uni-axial stress.

σ_x, σ_y = The in-plane stresses.

A fixed support condition may be assumed for the circular plate; with such supports and subjected to a uniformly distributed load, q, the radial and tangential stresses in the plate are, respectively,

$$\sigma_r = \frac{3}{8}\left(\frac{q}{t^2}\right)[a^2(1 + v) - r^2(3 + v)]$$

$$\sigma_t = \frac{3}{8}\left(\frac{q}{t^2}\right)[a^2(1 + v) - r^2(1 + 3v)]$$

where

t = Plate thickness.

a = Radius of circular plate.

r = Radial distance from center.

v = The Poisson's ratio (0.3 for steel).

Under a rapid blast loading, the mean yield strength of A36 steel could be increased by as much as 30% over the static yield strength. Moreover, significant increases in the dynamic yield strength are more likely than not.

To evaluate any inaccuracy of the von Mises criterion, a number of fixed circular plates of different sizes and thicknesses were tested under static pressure until yielding occurs at the support. The results are as follows (hypothetical):

Plate Thickness, t (in.)	Plate Radius, a (in.)	Observed Static Yield Pressure, q (psi)
1	24	92
2	24	440
$1\frac{1}{2}$	18	430
2	30	250
1	18	200

(a) Evaluate the failure probability of the protective plate element with $t = 2''$, $a = 36''$ against a dynamic blast pressure of 150 psi.

(b) What plate thickness would be required to resist a blast pressure of 200 psi in order to maintain the same reliability as in (a)?

6.19 Consider the following performance function

$$Z = \frac{C}{D} - 1$$

where C and D are statistically independent lognormal random variables with mean values μ_C, μ_D and c.o.v. δ_C and δ_D. Assume that μ_C and μ_D are also independent lognormal variates with mean values $\bar{C}$, $\bar{D}$ and c.o.v. Δ_C and Δ_D. Using the dual representation approach of Section 6.3.4, show the following:

(a) The expected probability of failure is

$$E(P_F) = \Phi\left[\frac{\ln\left(\frac{\bar{D}}{\bar{C}}\sqrt{\frac{(1 + \Delta_C^2)(1 + \delta_C^2)}{(1 + \Delta_D^2)(1 + \delta_D^2)}}\right)}{\sqrt{\ln\left[(1 + \delta_C^2)(1 + \delta_D^2)(1 + \Delta_C^2)(1 + \Delta_D^2)\right]}}\right]$$

(b) The q-percentile value of the failure probability is

$$(p_F)_q = \Phi\left[\frac{1}{\sqrt{\ln[(1 + \delta_C^2(1 + \delta_D^2)]}}\left\{\ln\left(\frac{\bar{D}}{\bar{C}}\sqrt{\frac{1 + \delta_C^2}{1 + \delta_D^2}}\sqrt{\frac{1 + \Delta_C^2}{1 + \Delta_D^2}}\right)\right.\right.$$
$$\left.\left. + \sqrt{\ln[(1 + \Delta_C^2)(1 + \Delta_D^2)]}\,\Phi^{-1}(q)\right\}\right]$$

6.20 A flow of Q cfs, carrying a concentration C mg/l of radioactive waste, is continually discharged into a small lake whose total volume is V in ft³. Suppose the wind current conditions are such that the lake contents are completely mixed. The equilibrium concentration of radioactive waste in the lake is given by (Metcalf and Eddy, 1972)

$$E = \frac{QC}{\left(\frac{Q}{V} + K\right)V}$$

where K is the first order decay rate for the radioactive waste.

(a) Suppose the half-life of the radioactive waste is uncertain, such that K has a mean of 8.1×10^{-6} per sec (corresponds to the half-life of 1 day) and a c.o.v. of 10%.

Determine the probability that the equilibrium concentration will not exceed the acceptable level of 5 mg/l given the following data:

$$Q = 2 \text{ cfs}; \ V = 450{,}000 \text{ cu ft}; \ \bar{C} = 10 \text{ mg/l}; \ \delta_c = 0.20$$

(b) Derive the design factors for C and K, and the design equation such that a 99% probability of E meeting the acceptable requirement of 5 mg/l is achieved. Assume K and C have means $\bar{K}, \bar{C}$ and c.o.v. of 10% and 20%, respectively.

6.21 Suppose that steel tension members are designed in terms of nominal values according to the following equation:

$$0.90R_n \geq 1.10D_n + 0.40L_n + 1.70W_n$$

where

$$\frac{\bar{R}}{R_n} = 1.10; \qquad \Omega_R = 0.11$$

$$\frac{\bar{D}}{D_n} = 1.05; \qquad \Omega_D = 0.10$$

The live load is $L_n = \bar{L} = L_0(0.25 + 15/\sqrt{A_I})$. Assume the *influence area* $A_I = 800$ sq ft. Also, the arbitrary-point-in-time live load, L_{apt}, is a random variable with

$$\bar{L}_{apt} = 12 \text{ psf} \qquad \text{and} \qquad \Omega_{L_{apt}} = 0.60$$

whereas the ratio of the annual maximum wind load to nominal wind load, W_{an}/W_n, has a Type I distribution with parameters

$$u = 0.24; \qquad \alpha = 6.65$$

(a) Evaluate the level of safety of tension members designed according to the above requirement for a life of 25 years. Assume that the maximum combined load over the life of the structure may be approximated by $D + L_{apt} + W_{25}$, where W_{25} is the 25-year maximum wind load. Load ratios assumed for design are

$$\frac{L_n}{D_n} = 1.0 \qquad \frac{W_n}{D_n} = 2.0$$

(b) If instead of the previous design equation a total load factor is to be used for design, that is,

$$0.90R_n \geq \gamma(D_n + L_n + W_n)$$

determine the total load factor, γ, so as to maintain the same level of safety as in part (a).

(c) Suppose that the probability distribution of the variables are as follows:

R—Normal.
D—Normal.
L_{apt}—Gamma (approximate the resulting gamma function with the factorial of the closest integer).
W_{25}—Type I asymptotic.

Determine the appropriate nominal resistance and load factors to achieve an acceptable failure probability of 6.21×10^{-3} in 25 years. What will be the corresponding factor of safety θ; that is, determine the value of θ to be used in the design equation,

$$R_n \geq \theta S_n$$

where

$$S_n = D_n + L_n + W_n$$

Again, the lifetime maximum combined load may be approximated by $D + L_{apt} + W_{25}$.

6.22 In Problem 6.7, determine the load and resistance factors for design to achieve a 99% reliability against yielding in the beam.

6.23 In Problem 6.18, evaluate the required stress and strength factors for the codified design of blast-resistant circular plates to achieve a 90% reliability.

6.24 A system of parallel 2 in. × 4 in. timber stringers are used to provide the rigidity of a plywood formwork for an 8-in. thick reinforced concrete floor slab. The stringers are simply supported at an 8 ft span. The spacings between the stringers (beams) are to be designed so that the probability of the maximum deflection of the stringers exceeding 1 in. is no more than 10%. Determine the spacing B. Assume the following:

   (i)   The rafters are of structural grade Douglas fir, with an average modulus of elasticity $E = 2,500,000$ psi and a c.o.v. of 15%.

  (ii)   Moisture in the field could reduce the modulus of elasticity by as much as 10%.

 (iii)   The average density of wet concrete is 150 pcf with a c.o.v. of 5%.

 (iv)   Each of the cross-sectional dimensions of the rafters has a c.o.v. of 3%.

  (v)   A nominally 8-in. thick slab has been known to range from 7.5 to 9 in. thick.

Prescribe all probability distributions to be lognormal, and invoke additional assumptions as necessary.

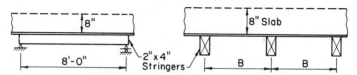

Figure P6.24   Construction formwork.

7. *Systems Reliability*

7.1 INTRODUCTION

The problems considered in Chapter 6 strictly involve a single failure mode, defined by a single limit state. Engineering problems, however, often involve multiple failure modes; that is, there may be several potential modes of failure, in which the occurrence of any one of the potential failure modes will constitute failure or nonperformance of the system or component. For example, a structural element may fail in flexure or shear or buckling, or combinations thereof, whereas for a multicomponent structural system, failures of different sets of components may constitute different failure modes. In the case of a building foundation, failure may be caused by inadequate bearing capacity or excessive settlement; whereas, in an environmental system, inadequate control of the different pollutants may be the respective modes of failure. Generally, the different failure modes may be correlated (partially at least).

Indeed, in a complex multicomponent engineering system, the possibilities of failure or the different ways in which failure of the system can occur may be so involved that a systematic scheme for identifying all the potential failure modes and their respective consequences may be necessary or useful. The *fault tree* and *event tree* models serve these purposes. A fault tree diagram essentially decomposes the main failure event (top event) into unions and intersections of subevents or combination of subevents; the process of decomposition continues until the probabilities of the subevents can be evaluated as single-mode failure probabilities. Calculations of the single-mode probabilities will require the methods described in Chapter 6. The consequence of a particular failure event (top event) may depend on the sequence of events following the top event. The systematic identification of the possible event sequences is accomplished through an event tree.

7.2 MULTIPLE FAILURE MODES

The reliability of a multicomponent system is essentially a problem involving multiple modes of failure; that is, the failures of different components, or different sets of components constitute distinct and different failure modes of the system. The consideration of multiple modes of failure, therefore, is fundamental to the problem of system reliability. The identification of the individual failure modes and the evaluation of the respective failure probabilities may be problems in themselves.

Consider a system with k potential failure modes. The different failure modes would have different performance functions; suppose that the respective performance functions may be represented as

$$g_j(\mathbf{X}) = g_j(X_1, X_2, \dots, X_n); \qquad j = 1, 2, \dots, k \tag{7.1}$$

such that the individual failure events are

$$E_j = [g_j(\mathbf{X}) < 0] \tag{7.2}$$

Then the compliments of E_j are the safe events; that is,

$$\bar{E}_j = [g_j(\mathbf{X}) > 0] \tag{7.3}$$

In the case of two variables, the above events may be portrayed as in Fig. 7.1, in which three failure modes represented by the limit-state equations $g_j(\mathbf{X}) = 0$, $j = 1, 2, 3$, are shown.

The safety of a system is the event in which none of the k potential failure modes occur; this means

$$\bar{E} = \bar{E}_1 \cap \bar{E}_2 \cap \dots \cap \bar{E}_k \tag{7.4}$$

Conversely, the failure event would be

$$E = E_1 \cup E_2 \cup \dots \cup E_k \tag{7.5}$$

Equation 7.5 literally means that one or more of the potential failure modes occurs. Theoretically, therefore, the probability of the safety of the system may be expressed as the volume integral

$$p_S = \int \cdots \int_{(\bar{E}_1 \cap \dots \cap \bar{E}_k)} f_{X_1, X_2, \dots, X_n}(x_1, \dots, x_n) \, dx_1 \cdots dx_n \tag{7.6}$$

whereas the probability of failure of the system would be

$$p_F = \int \cdots \int_{(E_1 \cup \dots \cup E_k)} f_{X_1, X_2, \dots, X_n}(x_1, \dots, x_n) \, dx_1 \cdots dx_n \tag{7.7}$$

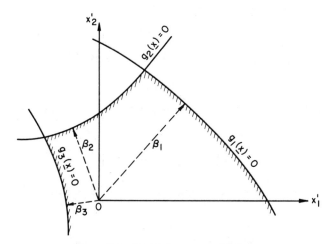

Figure 7.1   Multiple modes of failure.

7.2.1   Probability Bounds

The calculation of the probability of safety or failure of a system through Eq. 7.6 or 7.7 is generally difficult; approximations are almost always necessary. In this latter regard, lower and upper bounds of the corresponding probability are also useful.

Uni-modal Bounds   First, consider failure modes that are positively correlated; that is, $\rho_{ij} > 0$. This means, that for two events E_i and E_j,

$$P(E_j|E_i) \geq P(E_j)$$

which also means

$$P(\bar{E}_j|\bar{E}_i) \geq P(\bar{E}_j)$$

Therefore,

$$P(\bar{E}_i\bar{E}_j) \geq P(\bar{E}_i)P(\bar{E}_j)$$

For k events, this can be generalized to yield

$$P(\bar{E}) = P(\bar{E}_1\bar{E}_2 \ldots \bar{E}_k) \geq \prod_{i=1}^{k} P(\bar{E}_i) \tag{7.8}$$

Conversely, we observe that

$$\bar{E}_1\bar{E}_2 \ldots \bar{E}_k \subset \bar{E}_j; \qquad \text{for any } j$$

and, in particular,

$$\bar{E}_1\bar{E}_2 \ldots \bar{E}_k \subset \min_j \bar{E}_j; \qquad j = 1, 2, \ldots, k$$

Therefore,

$$P(\bar{E}) \leq \min_j P(\bar{E}_j) \tag{7.9}$$

If we denote the reliability against the ith failure mode as

$$p_{S_i} = P(\bar{E}_i)$$

and the reliability against all potential failure modes as

$$p_S = P(\bar{E})$$

p_S is, therefore, bounded according to Eqs. 7.8 and 7.9 as follows (Ang and Amin, 1968):

$$\prod_{i=1}^{k} p_{S_i} \leq p_S \leq \min_i p_{S_i} \tag{7.10}$$

Conversely, the corresponding bounds for the failure probability $p_F = P(E)$ would be

$$\max_i p_{F_i} \leq p_F \leq 1 - \prod_{i=1}^{k} (1 - p_{F_i}) \tag{7.11}$$

where $p_{F_i} = P(E_i)$, the failure probability of the ith mode. Observe also that for small p_{F_i}, the right-hand side of Eq. 7.11 is

$$1 - \prod_{i=1}^{k} (1 - p_{F_i}) \simeq \sum_{i=1}^{k} p_{F_i} \tag{7.12}$$

For negatively correlated failure modes, or events that are negatively dependent, that is, $\rho_{ij} < 0$, we have for two events E_i and E_j

$$P(E_j|E_i) \le P(E_j)$$

and

$$P(\bar{E}_j|\bar{E}_i) \le P(\bar{E}_j)$$

In particular, if E_i and E_j are perfectly negatively correlated; that is,

$$\rho_{ij} = -1.0; \qquad P(\bar{E}_j|\bar{E}_i) = P(E_j|E_i) = 0$$

Therefore,

$$P(\bar{E}_i\bar{E}_j) \le P(\bar{E}_i)P(\bar{E}_j)$$

and

$$P(\bar{E}) = P(\bar{E}_1\bar{E}_2 \dots \bar{E}_k) \le \prod_{i=1}^{k} P(\bar{E}_i)$$

Trivially, of course, $P(\bar{E}) \ge 0$. Thus,

$$p_S \le \prod_{i=1}^{k} p_{S_i} \tag{7.13}$$

Conversely,

$$p_F \ge 1 - \prod_{i=1}^{k} P(\bar{E}_i) \tag{7.13a}$$

Observe that for a set of events, $E_1, E_2, \dots, E_k$, the mutual correlations must be such that the correlation matrix $[C']$ of Eq. 6.30 is positive semi-definite; that is, $\det[C'] \ge 0$.

EXAMPLE 7.1

A reservoir may be designed for both flood control and water supply. Flooding may occur only in the spring caused by the melting of accumulated snow in the previous winter (event A) coupled with abundant rainfall in the spring (event B). On the other hand, water supply may be inadequate only in the summer and fall; water shortage may occur if the reservoir water is low in early summer (event C) because of a dry winter or spring coupled with a low rainfall over the summer (event D).

Then, inadequate flood control (event F) is

$$F = A \cap B$$

whereas insufficient water supply (event G) would be

$$G = C \cap D$$

Suppose that heavy spring rain may be expected to follow a wet winter, and dry summer generally follows a dry spring. Therefore, events A and B are positively dependent; similarly, events C and D are also positively dependent. However, event F and event G will be negatively dependent. Suppose the annual probabilities are as follows:

$$P(A) = 0.15$$

$$P(B) = 0.2$$

$$P(C) = 0.1$$

$$P(D) = 0.2$$

Then, the probability of unsatisfactory reservoir performance is

$$P(E) = P(F \cup G)$$
$$= P[(A \cap B) \cup (C \cap D)]$$

$P(E)$, of course, will depend on the correlations or degrees of dependency among the basic events. Without this information, we may be able only to evaluate the bounds of $P(E)$ in a given year as follows. According to Eq. 7.13a, where F and G are negatively dependent,

$$P(E) \geq 1 - P(\bar{F})P(\bar{G})$$

Since A and B are positively correlated, Eq. 7.10 yields

$$P(\bar{A})P(\bar{B}) \leq P(\bar{F}) \leq \min[P(\bar{A}), P(\bar{B})]$$

or

$$0.85 \times 0.8 \leq P(\bar{F}) \leq 0.8$$

Similarly,

$$P(\bar{C})P(\bar{D}) \leq P(\bar{G}) \leq \min[P(\bar{C}), P(\bar{D})]$$

or

$$0.9 \times 0.8 \leq P(\bar{G}) \leq 0.8$$

Therefore, the bounds of $P(E)$ may be given as

$$P(E) \geq 1 - 0.8 \times 0.8$$

or

$$P(E) \geq 0.36$$

The separation between the lower and upper bounds of Eq. 7.10 or 7.11 will obviously depend on the number of potential failure modes, and on the relative magnitudes of the individual mode probabilities. For example, if there is a dominant mode, the probability of safety or failure will be dominated by this mode, and indeed may be represented by the probability of this single dominant mode. In such cases, the bounds will also be narrow. In general, however, the bounds may be widely separated especially if the number of potential failure modes is large. The above bounds, Eqs. 7.10 through 7.13, may be called the "first-order" or "uni-modal" bounds on p_S and p_F, in the sense that the lower and upper bound probabilities involve single mode probabilities.

Bi-modal Bounds   The bounds described above can be improved by taking into account the correlation between pairs of potential failure modes; the resulting

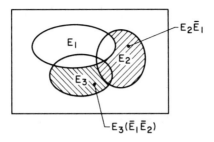

Figure 7.2 Decomposition of E.

improved bounds will necessarily require the probabilities of joint events, such as $E_i E_j$ or $\bar{E}_i \bar{E}_j$, and thus may be called "bi-modal" or "second-order" bounds.

The failure event $E = E_1 \cup E_2 \cup \ldots \cup E_k$ may be decomposed as follows (see Fig. 7.2 for $k = 3$), with E_1 assumed to be the largest set:

$$E = E_1 \cup E_2 \bar{E}_1 \cup E_3(\bar{E}_1 \bar{E}_2) \cup \cdots \cup E_k(\bar{E}_1 \bar{E}_2 \ldots \bar{E}_{k-1}) \tag{7.14}$$

Also, by deMorgan's rule,

$$\bar{E}_1 \bar{E}_2 \ldots \bar{E}_{i-1} = \overline{E_1 \cup E_2 \cup \cdots \cup E_{i-1}}$$

and for $i = 2, 3, \ldots, k$,

$$E_i(\bar{E}_1 \bar{E}_2 \ldots \bar{E}_{i-1}) = E_i(\overline{E_1 \cup E_2 \cup \cdots \cup E_{i-1}})$$

Observe that

$$E_i(\overline{E_1 \cup \cdots \cup E_{i-1}}) \cup E_i(E_1 \cup \cdots \cup E_{i-1}) = E_i$$

Thus,

$$P[E_i(\bar{E}_1 \bar{E}_2 \ldots \bar{E}_{i-1})] = P(E_i) - P(E_i E_1 \cup E_i E_2 \cup \cdots \cup E_i E_{i-1})$$

But

$$P(E_i E_1 \cup E_i E_2 \cup \cdots \cup E_i E_{i-1}) \leq P(E_i E_1) + P(E_i E_2) + \cdots + P(E_i E_{i-1})$$

Therefore,

$$P[E_i(\bar{E}_1 \bar{E}_2 \ldots \bar{E}_{i-1})] \geq P(E_i) - \sum_{j=1}^{i-1} P(E_i E_j)$$

Hence, with Eq. 7.14 we have

$$P(E) \geq P(E_1) + \max\left[\sum_{i=2}^{k} \left\{ P(E_i) - \sum_{j=1}^{i-1} P(E_i E_j) \right\}; 0 \right] \tag{7.15a}$$

On the other hand,

$$\bar{E}_1 \bar{E}_2 \ldots \bar{E}_{i-1} \subset \bar{E}_j; \qquad \text{for any } j$$

In particular,

$$\bar{E}_1 \bar{E}_2 \ldots \bar{E}_{i-1} \subset \min_{j<i} \bar{E}_j;$$

Therefore,

$$(\bar{E}_1\bar{E}_2\ldots\bar{E}_{i-1})E_i \subset \left(\min_{j<i}\bar{E}_j\right)E_i$$

Observe that

$$\left(\min_{j<i}\bar{E}_j\right)E_i \cup \left(\max_{j<i}E_j\right)E_i = E_i\left(\min_{j<i}\bar{E}_j \cup \max_{j<i}E_j\right) = E_i$$

Thus,

$$P(\bar{E}_1\bar{E}_2\ldots\bar{E}_{i-1}E_i) \leq P(E_i) - \max_{j<i}P(E_iE_j)$$

Hence, Eq. 7.14 yields

$$P(E) \leq P(E_1) + \sum_{i=2}^{k}\left[P(E_i) - \max_{j<i}P(E_iE_j)\right]$$

or

$$P(E) \leq \sum_{i=1}^{k}P(E_i) - \sum_{i=2}^{k}\max_{j<i}P(E_iE_j) \tag{7.15b}$$

Equations 7.15a and 7.15b, respectively, represent the "bi-modal" lower and upper bound failure probabilities when there are k potential failure modes (Kounias, 1968; Hunter, 1976); that is,

$$p_{F_1} + \max\left[\sum_{i=2}^{k}\left\{p_{F_i} - \sum_{j=1}^{i-1}P(E_iE_j)\right\}; 0\right] \leq p_F \leq \sum_{i=1}^{k}p_{F_i} - \sum_{i=2}^{k}\max_{j<i}P(E_iE_j) \tag{7.16}$$

In general, some of the failure modes may be correlated; hence, the calculations of the joint probabilities $P(E_iE_j)$ in Eqs. 7.15 and 7.16 remain difficult.

A weakened version of the above bi-modal bounds has been proposed by Ditlevsen (1979) for Gaussian variates. Suppose the basic design variables $\mathbf{X} = (X_1, X_2, \ldots, X_n)$ are normal (Gaussian) variates. Consider two potential failure modes E_i and E_j defined by the limit-state equations $g_i(\mathbf{X}) = 0$ and $g_j(\mathbf{X}) = 0$, respectively, with a positive mutual correlation $\rho_{ij} \geq 0$. In the space of the reduced variates $X_1', X_2', \ldots, X_n'$, these limit-state equations represent two intersecting hypersurfaces, which are approximated by the respective tangent planes, with corresponding distances β_i and β_j from the origin of the reduced variates. The geometrical representation for the two-variable case is shown in Fig. 7.3.

In Fig. 7.3, the cosine of the angle θ between $g_i(\mathbf{X}) = 0$ and $g_j(\mathbf{X}) = 0$ is the correlation coefficient ρ_{ij}; that is, $\cos\theta = \rho_{ij}$. This can be shown as follows.

Let

$$g_i(\mathbf{X}) = a_0 + a_1X_1 + a_2X_2$$

$$g_j(\mathbf{X}) = b_0 + b_1X_1 + b_2X_2$$

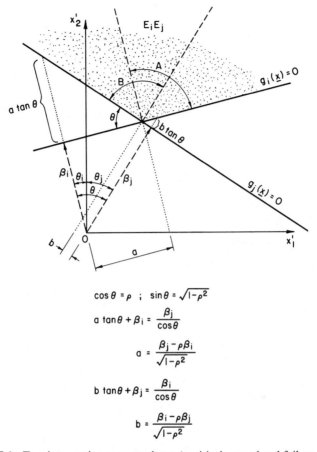

$$\cos\theta = \rho \quad ; \quad \sin\theta = \sqrt{1-\rho^2}$$

$$a\tan\theta + \beta_i = \frac{\beta_j}{\cos\theta}$$

$$a = \frac{\beta_j - \rho\beta_i}{\sqrt{1-\rho^2}}$$

$$b\tan\theta + \beta_j = \frac{\beta_i}{\cos\theta}$$

$$b = \frac{\beta_i - \rho\beta_j}{\sqrt{1-\rho^2}}$$

Figure 7.3   Two intersecting tangent planes (positively correlated failure events).

where X_1 and X_2 are uncorrelated; that is, $\mathrm{Cov}(X_1, X_2) = 0$. Then,

$$
\begin{aligned}
\mathrm{Cov}(g_i, g_j) &= E[(a_0 + a_1 X_1 + a_2 X_2)(b_0 + b_1 X_1 + b_2 X_2)] \\
&\quad - (a_0 + a_1\mu_{X_1} + a_2\mu_{X_2}) - (b_0 + b_1\mu_{X_1} + b_2\mu_{X_2}) \\
&= a_1 b_1 \sigma_{X_1}^2 + a_2 b_2 \sigma_{X_2}^2
\end{aligned}
$$

and

$$
\rho_{ij} = \frac{\mathrm{Cov}(g_i, g_j)}{\sigma_{g_i}\sigma_{g_j}} = \frac{a_1 b_1 \sigma_{X_1}^2 + a_2 b_2 \sigma_{X_2}^2}{\sqrt{(a_1^2\sigma_{X_1}^2 + a_2^2\sigma_{X_2}^2)(b_1^2\sigma_{X_1}^2 + b_2^2\sigma_{X_2}^2)}}
$$

From the direction cosines associated with $g_i(\mathbf{X}) = 0$ and $g_j(\mathbf{X}) = 0$, we have

$$
\cos\theta_i = \frac{a_2\sigma_{X_2}}{\sqrt{a_1^2\sigma_{X_1}^2 + a_2^2\sigma_{X_2}^2}}
$$

$$
\cos\theta_j = \frac{b_2\sigma_{X_2}}{\sqrt{b_1^2\sigma_{X_1}^2 + b_2^2\sigma_{X_2}^2}}
$$

Also, from Fig. 7.3,

$$\cos \theta = \cos (\theta_j - \theta_i) = \cos \theta_i \cos \theta_j + \sin \theta_i \sin \theta_j$$

$$= \frac{a_1 b_1 \sigma_{X_1}^2 + a_2 b_2 \sigma_{X_2}^2}{\sqrt{(a_1^2 \sigma_{X_1}^2 + a_2^2 \sigma_{X_2}^2)(b_1^2 \sigma_{X_1}^2 + b_2^2 \sigma_{X_2}^2)}}$$

Therefore, $\rho_{ij} = \cos \theta$.

The joint failure event, $E_i E_j$, is the shaded region shown in Fig. 7.3. Clearly,

$$E_i E_j \supset A$$

Also,

$$E_i E_j \supset B$$

where A and B are as defined in Fig. 7.3.

Therefore, observing Fig. 7.3 again, we obtain the following:

$$\max[P(A), P(B)] \le P(E_i E_j) \le P(A) + P(B) \qquad (7.17)$$

where, by reason of orthogonality (see Fig. 7.3),

$$P(A) = \Phi(-\beta_i)\Phi(-a) = \Phi(-\beta_i)\Phi\left(-\frac{\beta_j - \rho\beta_i}{\sqrt{1-\rho^2}}\right) \qquad (7.17a)$$

and

$$P(B) = \Phi(-\beta_j)\Phi(-b) = \Phi(-\beta_j)\Phi\left(-\frac{\beta_i - \rho\beta_j}{\sqrt{1-\rho^2}}\right) \qquad (7.17b)$$

The probabilities of the joint events, $P(E_i E_j)$, in Eq. 7.16 may then be approximated with the appropriate sides of Eq. 7.17; that is, for the lower bound (Eq. 7.15a), use

$$P(E_i E_j) = P(A) + P(B)$$

whereas for the upper bound (Eq. 7.15b), use

$$P(E_i E_j) = \max[P(A), P(B)]$$

Equation 7.17 actually applies to failure events involving n variables. In particular, the geometry of Fig. 7.3 is valid for any pair of positively correlated performance functions $Z_i = g_i(\mathbf{X})$ and $Z_j = g_j(\mathbf{X})$, where $\mathbf{X}$ is an n-dimensional vector of normal variates (Ditlevsen, 1979). For this generalized representation, the coordinates X_1' and X_2' in Fig. 7.3, of course, must be replaced by the corresponding uncorrelated standard normal variates, Z_i' and Z_j', which may be related to Z_i and Z_j through a rotation of coordinates, Eq. 6.31.

In general, the range of the bi-modal bounds of Eq. 7.16, with the approximations of Eq. 7.17, will improve (i.e., decrease) as the single-mode failure probabilities decrease; for example, for single-mode failure probabilities of order 10^{-4}, the bi-modal bounds can be very narrow (Ditlevsen, 1979). However, if the single-mode failure probabilities are all large (e.g., $> 10^{-2}$), the bounds of Eq. 7.16 could be wide (Ma and Ang, 1981).

The bi-modal bounds, Eq. 7.16, will depend on the ordering of the individual failure modes; that is, different orderings of the individual failure modes may yield different values for Eq. 7.15a and Eq. 7.15b, and thus the bounds corresponding to different orderings may have to be evaluated to determine the sharpest bounds. When the number of modes is large, the bounds corresponding to several random orderings may be examined.

From Eq. 7.17, we observe that the mutual correlations between failure modes are required in evaluating the bi-modal bounds of Eq. 7.16. Following the derivation of Eq. 6.18, we can show on the basis of the Taylor series expansion about the failure point $\mathbf{x}^* = (x_1^*, x_2^*, \ldots, x_n^*)$ that the covariance between two performance functions $g_i(\mathbf{X})$ and $g_j(\mathbf{X})$ is (for uncorrelated basic variates)

$$\text{Cov}(g_i, g_j) = \sum_{k=1}^{n} \left(\frac{\partial g_i}{\partial X_k'}\right)_* \left(\frac{\partial g_j}{\partial X_k'}\right)_*$$

Therefore, with Eq. 6.18, the required correlation coefficient is

$$
\begin{aligned}
\rho_{g_i, g_j} &= \frac{\text{Cov}(g_i, g_j)}{\sigma_{g_i} \sigma_{g_j}} \\
&= \frac{\sum_k \left(\frac{\partial g_i}{\partial X_k'}\right)_* \left(\frac{\partial g_j}{\partial X_k'}\right)_*}{\sqrt{\sum_k \left(\frac{\partial g_i}{\partial X_k'}\right)_*^2} \sqrt{\sum_k \left(\frac{\partial g_j}{\partial X_k'}\right)_*^2}} \\
&= \sum_{k=1}^{n} \alpha_{ik}^* \alpha_{jk}^*
\end{aligned}
\tag{7.18}
$$

Therefore, the required correlation may be evaluated using the direction cosines of the tangent planes at the most probable failure points. It should be observed that in the iterative process of solution for the reliability of the individual failure modes, these direction cosines have already been calculated.

If E_i and E_j are negatively dependent, that is, $\rho_{ij} < 0$, the limit-state equations would be as depicted in Fig. 7.4, where the joint failure event $E_i E_j$ is the shaded region. In such cases, it can be seen from Fig. 7.4 that

$$P(E_i E_j) \leq \min[P(A), P(B)]$$

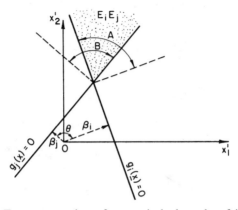

Figure 7.4 Two tangent planes for negatively dependent failure events.

where $P(A)$ and $P(B)$ are given by Eqs. 7.17a and 7.17b, respectively; the lower bound is (trivially) zero.

EXAMPLE 7.2

A simply supported beam subjected to a uniformly distributed load, w, as shown in Fig. E7.2, may fail in flexure or shear or in combined flexure and shear. Suppose the beam is a rolled 18 WF 70 section of A36 steel with mean tensile yield strength of $\bar{f}_y = 44$ ksi and a c.o.v. of $\Omega_{f_y} = 0.10$. Assume that the mean yield stress in shear is 50% of the tensile yield strength; that is, $\bar{\tau}_y = 0.50\bar{f}_y = 22$ ksi, and $\Omega_{\tau_y} = 0.15$; whereas the load w is $\bar{w} = 6$ kip/ft and $\Omega_w = 0.25$.

The section modulus of the beam is $S = 128.2$ in.^{3}. The bending capacity of the beam, therefore, is

$$\bar{M}_o = \frac{44(128.2)}{12} = 470 \text{ ft-kip}$$

with

$$\Omega_{M_o} = 0.10$$

The corresponding shear capacity may be approximated by (with $A_{\text{web}} = 0.438 \times 16.5$)

$$\bar{V}_o = \bar{\tau}_y A_{\text{web}} = 22(0.438 \times 16.5)$$
$$= 159 \text{ kips}$$

and

$$\Omega_{V_o} = 0.15$$

The sections along the beam may be assumed to be perfectly correlated. The performance functions for failures in bending and shear are, respectively,

$$g_1(\mathbf{X}) = M_o - \tfrac{1}{8}wL^2$$

$$g_2(\mathbf{X}) = V_o - \tfrac{1}{2}wL$$

Failure under combined bending and shear may be defined as $M/M_o + V/V_o > 1.0$; then, the performance function for this combined mode is

$$g_3(\mathbf{X}) = 1 - \left(\frac{M}{M_o} + \frac{V}{V_o}\right)$$

Clearly, failure in bending will occur at mid-span, whereas shear failure will occur adjacent to the supports. Under combined bending and shear, however, failure will occur at the location

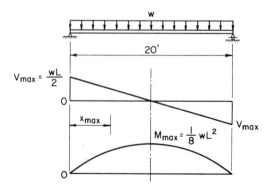

Figure E7.2   Simply-supported beam.

of minimum $g_3(\mathbf{X})$ or maximum $(M/M_o + V/V_o)$. It can be shown that this will occur at (see Fig. E7.2)

$$x_{\max} = \frac{L}{2} - \frac{M_o}{V_o}$$

at which

$$g_3(\mathbf{X}) = 1 - \left(\frac{wL^2}{8M_o} + \frac{wM_o}{2V_o^2}\right)$$

With the assumption that w, M_o, and V_o are uncorrelated, we obtain the failure probabilities in bending and shear, respectively, as follows:

$$p_{F_1} = \Phi\left[-\frac{\overline{M}_o - \frac{1}{8}\overline{w}L^2}{\sqrt{\sigma_{M_o}^2 + (\frac{1}{8}L^2\sigma_w)^2}}\right]$$

$$= \Phi\left[-\frac{470 - \frac{1}{8} \times 6(20)^2}{\sqrt{(47)^2 + (\frac{1}{8} \times 20^2 \times 1.5)^2}}\right] = \Phi(-1.92) = 2.74 \times 10^{-2}$$

$$p_{F_2} = \Phi\left[-\frac{\overline{V}_o - \frac{1}{2}\overline{w}L}{\sqrt{\sigma_{V_o}^2 + (\frac{1}{2}L\sigma_w)^2}}\right]$$

$$= \Phi\left[-\frac{159 - \frac{1}{2} \times 6 \times 20}{\sqrt{(23.85)^2 + (\frac{20}{2} \times 1.5)^2}}\right] = \Phi(-3.51) = 2.24 \times 10^{-4}$$

The iterative solution for the failure probability of the combined mode may be summarized as follows:

Table E7.2   Iterations for Failure Mode 3

Iteration No.	Variables x_i	Assumed x_i^*	$\left(\dfrac{\partial g}{\partial x_i'}\right)_*$	$\alpha_{x_i}^*$	New x_i^*
1	w	6	-0.174	-0.945	$6 + 1.418\beta$
	M_o	470	0.058	0.315	$470 - 14.81\beta$
	V_o	159	0.017	0.092	$159 - 2.194\beta$

Failure equation: $1 - \dfrac{(6 + 1.418\beta)(20)^2}{8(470 - 14.81\beta)} - \dfrac{(6 + 1.418\beta)(470 - 14.81\beta)}{2(159 - 2.194\beta)^2} = 0$

$\beta = 1.59$

2	w	8.255	-0.182	-0.893	$6 + 1.338\beta$
	M_o	446.5	0.089	0.436	$470 - 20.49\beta$
	V_o	155.5	0.023	0.113	$159 - 2.69\beta$

Failure equation: $1 - \dfrac{(6 + 1.340\beta)(20)^2}{8(470 - 20.49\beta)} - \dfrac{(6 + 1.340\beta)(470 - 20.49\beta)}{2(159 - 2.695\beta)^2} = 0$

$\beta = 1.57$

3	w	8.104	-0.185	-0.892	$6 + 1.338\beta$
	M_o	437.8	0.091	0.439	$470 - 20.63\beta$
	V_o	154.8	0.023	0.111	$159 - 2.647\beta$

Failure equation: $1 - \dfrac{(6 + 1.338\beta)(20)^2}{8(470 - 20.63\beta)} - \dfrac{(6 + 1.338\beta)(470 - 20.63\beta)}{2(159 - 2.647\beta)^2} = 0$

$\beta = 1.57$

Therefore,

$$p_{F_3} = \Phi(-1.57) = 5.82 \times 10^{-2}$$

The three failure modes are obviously correlated. The first-order (uni-modal) bounds on the failure probability of the beam are

$$5.82 \times 10^{-2} \leq p_F \leq 8.582 \times 10^{-2}$$

For the second-order (bi-modal) bounds, we evaluate the mutual correlations between the failure modes with Eq. 7.18 as follows:
The pertinent partial derivatives are

$$\frac{\partial g_1}{\partial M_o'} = \sigma_{M_o} = 0.10 \times 470 = 47$$

$$\frac{\partial g_1}{\partial w'} = -\frac{L^2}{8}\sigma_w = -\frac{(20)^2}{8} \times 1.5 = -75$$

$$\frac{\partial g_2}{\partial V_o'} = \sigma_{V_o} = 0.15 \times 159 = 23.85$$

$$\frac{\partial g_2}{\partial w'} = -\frac{L}{2}\sigma_w = -\frac{20}{2} \times 1.5 = -15$$

Thus, the direction cosines are

$$\alpha_{1M_o}^* = \frac{47}{\sqrt{(47)^2 + (75)^2}} = 0.531$$

$$\alpha_{1w}^* = \frac{-75}{\sqrt{(47)^2 + (75)^2}} = -0.847$$

$$\alpha_{2V_o}^* = \frac{23.85}{\sqrt{(23.85)^2 + (15)^2}} = 0.846$$

$$\alpha_{2w}^* = \frac{-15}{\sqrt{(23.85)^2 + (15)^2}} = -0.532$$

Therefore, the correlation between modes 1 and 2 is,

$$\rho_{g_1, g_2} = (-0.847)(-0.532) = 0.451$$

To obtain the correlation coefficients ρ_{g_1, g_3} and ρ_{g_2, g_3}, we make use of the direction cosines for mode 3 from Table E7.2. Thus, with Eq. 7.18

$$\rho_{g_1, g_3} = (0.531)(0.439) + (-0.847)(-0.892) = 0.989$$
$$\rho_{g_2, g_3} = (0.846)(0.111) + (-0.532)(-0.892) = 0.569$$

The bi-modal bounds corresponding to the ordering E_1, E_2, E_3 are then obtained as follows. For failure modes 1 and 2, we have

$$P(A) = \Phi(-1.92)\Phi\left[-\frac{3.51 - 0.451 \times 1.92}{\sqrt{1 - (0.451)^2}}\right]$$

$$= \Phi(-1.92)\Phi(-2.96) = (0.0274)(0.00154) = 4.22 \times 10^{-5}$$

$$P(B) = \Phi(-3.51)\Phi\left[-\frac{1.92 - 0.451 \times 3.51}{\sqrt{1 - (0.451)^2}}\right]$$

$$= \Phi(-3.51)\Phi(-0.38) = (2.24 \times 10^{-4})(0.351973) = 7.88 \times 10^{-5}$$

Therefore,

$$7.88 \times 10^{-5} \leq P(E_1 E_2) \leq 1.21 \times 10^{-4}$$

Similarly, for modes 1 and 3,

$$P(A) = \Phi(-1.92)\Phi\left[-\frac{1.57 - 0.989 \times 1.92}{\sqrt{1 - (0.989)^2}}\right]$$

$$= \Phi(-1.92)\Phi(2.22) = (0.0274)(0.9868) = 2.704 \times 10^{-2}$$

$$P(B) = \Phi(-1.57)\Phi\left[-\frac{1.92 - 0.989 \times 1.57}{\sqrt{1 - (0.989)^2}}\right]$$

$$= \Phi(-1.57)\Phi(-2.48) = (0.0582)(0.0066) = 3.84 \times 10^{-4}$$

Thus,

$$2.704 \times 10^{-2} \leq P(E_1 E_3) \leq 2.742 \times 10^2$$

and for modes 2 and 3,

$$P(A) = \Phi(-3.51)\Phi\left[-\frac{1.57 - 0.569 \times 3.51}{\sqrt{1 - (0.569)^2}}\right]$$

$$= \Phi(-3.51)\Phi(0.52) = (2.24 \times 10^{-4})(0.6984) = 1.564 \times 10^{-4}$$

$$P(B) = \Phi(-1.57)\Phi\left[-\frac{3.51 - 0.569 \times 1.57}{\sqrt{1 - (0.569)^2}}\right]$$

$$= \Phi(-1.57)\Phi(-2.18) = (0.0582)(7.40 \times 10^{-4}) = 4.30 \times 10^{-5}$$

Thus,

$$1.564 \times 10^{-4} \leq P(E_2 E_3) \leq 1.994 \times 10^{-4}$$

The bi-modal bounds of the failure probability obtained with the ordering E_1, E_2, E_3, therefore, are

$$8.582 \times 10^{-2} - [P(E_1 E_2) + P(E_1 E_3) + P(E_2 E_3)]$$
$$\leq p_F \leq 8.582 \times 10^{-2} - P(E_1 E_2) - \max[P(E_1 E_3), P(E_2 E_3)]$$

$$8.582 \times 10^{-2} - (1.21 \times 10^{-4} + 2.742 \times 10^{-2} + 1.99 \times 10^{-4})$$
$$\leq p_F \leq 8.582 \times 10^{-2} - 7.88 \times 10^{-5} - 2.704 \times 10^{-2}$$

Thus,

$$5.820 \times 10^{-2} \leq p_F \leq 5.870 \times 10^{-2}$$

EXAMPLE 7.3

Figure E7.3 shows a cantilever retaining wall and the corresponding soil profile of the site. The potential modes of geotechnical failure in this case are: (i) overturning of the wall; (ii) horizontal sliding of the wall, and (iii) bearing capacity failure of the wall foundation. For simplicity, assume that the bearing capacity is adequate; furthermore, the passive resistance of the 4-foot depth of soil at the toe of the structure may be neglected in evaluating the overturning and sliding failure modes.

In the overturning mode, the performance function is given by the difference between the resisting and driving moments. The driving moment is the result of the active earth pressure, which is

$$M_o = \tfrac{1}{2}\gamma \tan^2\left(45 - \frac{\phi}{2}\right) \times H^2 \times d$$

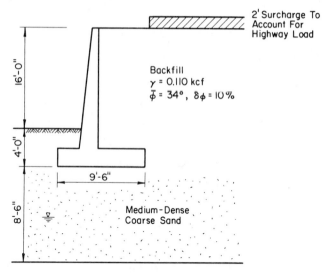

Figure E7.3 A retaining wall.

where γ is the backfill density, ϕ is the average friction angle of the backfill, H is the effective height of backfill (including surcharge) above the base of the wall, and d is the moment arm of the resultant horizontal force about the base. Suppose γ, H, and d are known deterministically, with values $\gamma = 0.11$ kcf, $H = 22$ ft, and $d = 7.33$ ft, respectively; ϕ is the only random variable with a mean value of 34° and a c.o.v. of 10%. The resisting moment is provided by the weight of the backfill behind the wall that has been calculated to be 112.5 kip. Hence, the performance function for the overturning mode (mode 1) is

$$g_1(\mathbf{X}) = 112.5 - 195.1 \tan^2\left(45 - \frac{\phi}{2}\right)$$

Sliding will occur if the horizontal force on the backfill exceeds the frictional resistance provided at the base of the retaining wall. The horizontal force is

$$F = \tfrac{1}{2}\gamma \tan^2\left(45 - \frac{\phi}{2}\right)H^2 = 26.6 \tan^2\left(45 - \frac{\phi}{2}\right)$$

whereas the frictional resistance is

$$R = W \tan \delta$$

in which W is the total pressure at the base, calculated to be 20.14 k, and δ is the average friction angle between the concrete base and the sand stratum. The mean and c.o.v. of δ are assumed to be 30° and 10%, respectively. Therefore, the performance function for the sliding mode (mode 2) is

$$g_2(\mathbf{X}) = 20.14 \tan \delta - 26.6 \tan^2\left(45 - \frac{\phi}{2}\right)$$

For mode 1, since ϕ is the only random variable, the direction cosine

$$\alpha_\phi^* = 1.0$$

and

$$\phi^* = \bar{\phi} - \alpha_\phi^*\beta_1\sigma_\phi = 34 - 3.4\beta_1$$

With $g_1(\mathbf{X}^*) = 112.5 - 195.1 \tan^2[45° - \frac{1}{2}(34 - 3.4\beta_1)°] = 0$, we obtain $\beta_1 = 5.42$. With the assumption of Gaussian distribution, the corresponding probability of failure is

$$p_{F_1} = \Phi(-5.42) = 0.30 \times 10^{-7}$$

For mode 2, assume $\phi^* = 34°$ and $\delta^* = 30°$ for the first iteration. Then,

$$\left(\frac{\partial g_2}{\partial \phi'}\right)_* = -26.6 \times 2 \tan\left(45° - \frac{34°}{2}\right) \sec^2\left(45° - \frac{34°}{2}\right) \times \left(-\frac{1}{2}\right)\left(\frac{3.4\pi}{180}\right)$$

$$= 1.078$$

$$\left(\frac{\partial g_2}{\partial \delta'}\right)_* = 20.14(\sec^2 30°)\left(\frac{3.0\pi}{180}\right) = 1.406$$

The direction cosines are

$$\alpha_\phi^* = 0.608; \qquad \alpha_\delta^* = 0.794$$

Thus,

$$\phi^* = \bar{\phi} - \alpha_\phi^* \beta_2 \sigma_\phi = 34° - 2.067\beta_2$$
$$\delta^* = \bar{\delta} - \alpha_\delta^* \beta_2 \sigma_\delta = 30° - 2.382\beta_2$$

and

$$g_2(\mathbf{X}^*) = 20.14 \tan(30° - 2.382\beta_2) - 26.6 \tan^2[45° - \frac{1}{2}(34° - 2.067\beta_2)] = 0$$

yielding $\beta_2 = 2.32$. For the second iteration, assume

$$\phi^* = 29.205; \qquad \delta^* = 24.474$$

Then,

$$\left(\frac{\partial g_2}{\partial \phi'}\right)_* = 1.246; \qquad \left(\frac{\partial g_2}{\partial \delta'}\right)_* = 1.273$$

and

$$\alpha_\phi^* = 0.699; \qquad \alpha_\delta^* = 0.715$$

Thus,

$$\phi^* = 34° - 2.377\beta_2; \qquad \delta^* = 30° - 2.145\beta_2$$

and

$$g_2(\mathbf{X}^*) = 20.14 \tan(30 - 2.145\beta_2) - 26.6 \tan^2[45 - \frac{1}{2}(34 - 2.377\beta_2)] = 0$$

yielding $\beta_2 = 2.31$, which is close enough to that obtained in the first iteration. Therefore,

$$p_{F_2} = \Phi(-2.31) = 0.01044$$

Since the random variable ϕ appears in both performance functions, the two failure modes can be expected to be correlated. The first-order bounds on the failure probability of the wall are:

$$0.01044 \leq p_f \leq 0.01044 + 0.3 \times 10^{-7}$$

In this case, because the sliding mode is dominant, the bounds for p_F are very narrow.

Therefore, for practical purposes, $p_F = 0.01044$. This example demonstrates that if there is a dominant failure mode (the sliding mode in this case), the failure probability of a system will be very close to that of the dominant mode.

EXAMPLE 7.4

The capacity of a concrete friction pile driven into a sand stratum, as shown in Fig. E7.4, may be limited by the supporting strength of the soil or by the compressive strength of the concrete. Suppose the concrete pile has a constant circular cross section; the performance function in failure mode 1, that is, inadequate soil capacity, is (see e.g., Teng, 1962)

$$g_1(\mathbf{X}) = 2\pi RL(\gamma L/2 + q)K \tan \phi - P$$

where R is the pile radius; L is the length of embedment of the pile; q is the permanent surcharge load; K is the coefficient of lateral earth pressure; ϕ is the average angle of internal friction of the soil; and P is the applied load on the pile. Suppose the following:

$R = 0.5$ ft	$\bar{K} = 1.5;$	$\delta_K = 0.10$
$L = 40$ ft	$\bar{\phi} = 30°;$	$\delta_\phi = 0.04$
$\gamma = 0.11$ kcf	$\bar{P} = 200$ kip;	$\delta_P = 0.10$
$q = 0.300$ ksf		

Then,

$$g_1(\mathbf{X}) = 314.16K \tan \phi - P$$

Similarly the performance function for the second failure mode, that is, inadequate concrete capacity, is

$$g_2(\mathbf{X}) = N\pi R^2 f'_c - P$$

where f'_c is the compressive strength of concrete, and N is to account for the modeling error of the in-situ effective compressive strength of concrete. For the given pile,

$$g_2(\mathbf{X}) = 113.1Nf'_c - P$$

Suppose

$$\overline{f'_c} = 3.0 \text{ ksi}; \qquad \delta_{f'_c} = 0.10$$
$$\bar{N} = 1.0; \qquad \delta_N = 0.10$$

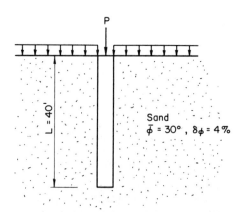

Figure E7.4 Concrete friction pile in sand stratum.

We first determine the reliability index for each failure mode. For mode 1, we observe that

$$\left(\frac{\partial g_1}{\partial K'}\right)_* = 314.16\,\sigma_K \tan\phi^*$$

$$\left(\frac{\partial g}{\partial \phi'}\right)_* = 314.16k^*\,\sigma_\phi \sec^2\phi^*$$

$$\left(\frac{\partial g}{\partial P'}\right)_* = -\sigma_P$$

With $k^* = \bar{K}$, $\phi^* = \bar{\phi}$, $p^* = \bar{P}$ for the first iteration, the calculations for two iterations are summarized in Table E7.4a. Therefore,

$$p_{F_1} = \Phi(-2.03) = 0.0212$$

For failure mode 2, observe that

$$\left(\frac{\partial g_2}{\partial f'_c}\right)_* = 113.1n^*\sigma_{f'_c}$$

$$\left(\frac{\partial g_2}{\partial N'}\right)_* = 113.1f'^*_c\sigma_N$$

$$\left(\frac{\partial g_2}{\partial P'}\right)_* = -\sigma_P$$

and again assume $f'^*_c = \bar{f}_c$, $n^* = \bar{N}$, and $p^* = \bar{P}$ in the first iteration. The calculations for two iterations are summarized in Table E7.4b. Therefore,

$$p_{F_2} = \Phi(-2.90) = 0.00186$$

Table E7.4a Summary of Iterative Calculations for Failure Mode 1 of Example 7.4

Iteration No.	Variable x_i	Assumed Failure Point x_i^*	$\left(\frac{\partial g}{\partial x_i}\right)_*$	$\alpha_{x_i}^*$	New x_i
1	K	1.5	27.207	0.751	$1.5 - 0.119\beta_1$
	ϕ	30	13.159	0.363	$30 - 0.436\beta_1$
	P	200	-20	-0.552	$200 + 11.04\beta_1$

Failure equation:

$314.16(1.5 - 0.119\beta_1)\tan(30 - 0.436\beta_1) - (200 + 11.04\beta_1) = 0$

By trial and error: $\beta = 2.0$

2	K	1.262	26.261	0.756	$1.5 - 0.113\beta_1$
	ϕ	29.13	10.882	0.313	$30 - 0.376\beta_1$
	P	222.08	-20	-0.575	$200 + 11.50\beta_1$

Failure equation:

$314.16(1.5 - 0.113\beta_1)\tan(30 - 0.376\beta_1) - (200 + 11.50\beta_1) = 0$

By trial and error: $\beta = 2.03$

Table E7.4b Summary of Iterative Calculations for Failure Mode 2 of Example 7.4

Iteration No.	Variable x_i	Assumed Failure Point x_i^*	$\left(\dfrac{\partial g}{\partial x_i'}\right)_*$	$\alpha_{x_i}^*$	New x_i^*
1	f_c'	3.0	33.929	0.653	$3.0 - 0.196\beta_2$
	N	1.0	33.929	0.653	$1 - 0.065\beta_2$
	P	200	-20	-0.385	$200 + 7.7\beta_2$

Failure equation:

$$113.1(1 - 0.065\beta_2)(3.0 - 0.196\beta_2) - (200 + 7.7\beta_2) = 0$$

yielding $\beta_2 = 2.92$

2	f_i'	2.428	27.482	0.629	$3.0 - 0.189\beta_2$
	N	0.81	27.460	0.628	$1 - 0.063\beta_2$
	P	222.484	-20	-0.458	$200 + 9.16\beta_2$

Failure equation:

$$113.1(1 - 0.063\beta_2)(3.0 - 0.189\beta_2) - (200 + 9.16\beta_2) = 0$$

yielding $\beta_2 = 2.90$

Since the two failure modes are correlated because of the common variable P, the first-order bounds on the failure probability of the pile are

$$0.0212 \le p_F \le 0.0212 + 0.00186$$

or

$$0.0212 \le p_F \le 0.0231$$

For the bi-modal bounds, we calculate the correlation between the two failure modes with Eq. 7.18, obtaining

$$\rho_{g_1, g_2} = (-0.575)(-0.458) = 0.263$$

Thus

$$P(A) = \Phi(-\beta_1)\Phi\left(-\frac{\beta_2 - \rho\beta_1}{\sqrt{1 - \rho^2}}\right)$$

$$= \Phi(-2.03)\Phi\left(-\frac{2.90 - 0.263 \times 2.03}{\sqrt{1 - 0.263^2}}\right)$$

$$= 0.0212 \times 0.0072 = 1.53 \times 10^{-4}$$

$$P(B) = \Phi(-\beta_2)\left(-\frac{\beta_1 - \rho\beta_2}{\sqrt{1 - \rho^2}}\right)$$

$$= \Phi(-2.90)\Phi\left(-\frac{2.03 - 0.263 \times 2.90}{\sqrt{1 - 0.263^2}}\right)$$

$$= 0.00186 \times 0.0951 = 1.77 \times 10^{-4}$$

from which

$$0.000177 \le P(E_1 E_2) \le 0.000330$$

Finally, the bi-modal bounds on the failure probability of the pile are

$$0.0231 - 0.000330 \le p_F \le 0.0231 - 0.000177$$

or

$$0.0228 \le p_F \le 0.0229$$

EXAMPLE 7.5

A sewer system consists of pipe 1 and pipe 2 flowing into pipe 3 as shown in Fig. E7.5. The inflows to pipes 1 and 2 are, respectively,

$$Q_{I_1} = NC_1 iA_1$$

and

$$Q_{I_2} = NC_2 iA_2$$

where A_1 and A_2 are the basin areas draining into the respective pipes; C_1 and C_2 are the respective runoff coefficients; i is the precipitation rate; and N is the model correction for the rational formula, assumed to be the same for both areas 1 and 2 (because of proximity). The inflow to pipe 3 consists of the flows from pipes 1 and 2 and the additional inflow from its own basin; thus,

$$Q_{I_3} = NC_3 iA_3 + K(Q_{I_1} + Q_{I_2})$$

where K is the factor accounting for the attenuation of peak flows Q_{I_1} and Q_{I_2} through the entire lengths of pipes 1 and 2.

The capacity of each pipe is given by the Darcy-Weisbach formula, namely,

$$Q_{ci} = \lambda \left(\frac{\pi}{4}\right)\left(\frac{2gS_i}{f_i}\right)^{1/2} (4R_i)^{2.5}; \quad i = 1, 2, 3$$

where S_i, f_i, and R_i are the friction slope, friction factor, and hydraulic radius of the ith pipe; λ is the correction for model error underlying the Darcy-Weisbach formula.

Suppose the following data are given:

$$A_1 = A_2 = A_3 = 435{,}600 \text{ ft}^2 = 10 \text{ acres}$$
$$S_1 = S_2 = S_3 = 0.001$$
$$R_1 = R_2 = 1.125 \text{ ft}$$
$$R_3 = 1.75 \text{ ft}$$

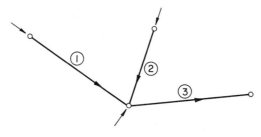

Figure E7.5   A sewer pipe system.

Assume the friction factors for the pipes are the same, namely, $f_1 = f_2 = f_3 = f$. For simplicity, let $W = Ni$ and $Y = \lambda f^{-1/2}$. The statistics of the random variables are given as follows:

Variable	Mean	c.o.v.
C_1	0.825	0.07
C_2	0.825	0.07
C_3	0.90	0.05
K	0.9	0.05
W	8.59×10^{-5} fps	0.233
Y	7.01	0.123

Suppose all the random variables are statistically independent. Satisfactory performance requires that each pipe will not have overflow (inflow exceeds pipe capacity). The performance function for each pipe is thus simplified as

$$g_1(\mathbf{X}) = 8.30Y - 435,420C_1 W$$

$$g_2(\mathbf{X}) = 8.30Y - 435,420C_2 W$$

$$g_3(\mathbf{X}) = 25.835Y - 435,420W(C_3 + KC_1 + KC_2)$$

For the failure of pipe 1, observe that

$$\left(\frac{\partial g_1}{\partial Y'}\right)_* = 8.30\sigma_Y$$

$$\left(\frac{\partial g_1}{\partial C_1'}\right)_* = -435,420w^*\sigma_{C_1}$$

$$\left(\frac{\partial g}{\partial W'}\right)_* = -435,420c_1^*\sigma_W$$

Then with $y^* = \bar{Y}$, $c_1^* = \bar{C}_1$, and $w^* = \bar{W}$ assumed for the first iteration, the calculations for the first two iterations are as summarized in Table E7.5a.

Table E7.5a   Summary of Iterative Calculations for Failure of Pipe 1 in Example 7.5

Iteration No.	Variable x_i	Assumed Failure Point x_i^*	$\left(\dfrac{\partial g}{\partial x_i'}\right)_*$	$\alpha_{x_i}^*$	New x_i^*
1	Y	7.01	7.155	0.690	$7.01 - 0.595\beta_1$
	C_1	0.825	-2.162	-0.209	$0.825 + 0.0121\beta_1$
	W	8.59×10^{-5}	-7.18	-0.693	$(8.59 + 1.386\beta_1) \times 10^{-5}$

Failure equation:

$$8.30(7.01 - 0.595\beta_1) - 4.3542(0.825 + 0.0121\beta_1)(8.59 + 1.386\beta_1) = 0$$

yielding $\beta_1 = 2.60$

2	Y	5.46	7.155	0.664	$7.01 - 0.572\beta_1$
	C_1	0.856	-3.069	-0.285	$0.825 + 0.0165\beta_1$
	W	12.194×10^{-5}	-7.454	-0.692	$(8.59 + 1.384\beta_1) \times 10^{-5}$

Failure equation:

$$8.30(7.01 - 0.572\beta_1) - 4.3542(0.825 + 0.0165\beta_1)(8.59 + 1.384\beta_1) = 0$$

yielding $\beta_1 = 2.58$

Thus,

$$p_{F_1} = \Phi(2.58) = 0.00494$$

Similarly, for pipe 2, we also have (since the statistics are the same as those for pipe 1)

$$\beta_2 = 2.58 \quad \text{and} \quad P_{F_2} = 0.00494$$

For pipe 3, we have

$$\left(\frac{\partial g_3}{\partial Y'}\right)_* = 25.835\sigma_Y$$

$$\left(\frac{\partial g_3}{\partial W'}\right)_* = -435,420(c_3^* + k^*c_1^* + k^*c_2^*)\sigma_W$$

$$\left(\frac{\partial g_3}{\partial C'_3}\right)_* = -435,420w^*\sigma_{C_3}$$

$$\left(\frac{\partial g_3}{\partial C'_1}\right)_* = -435,420w^*k^*\sigma_{C_1}$$

$$\left(\frac{\partial g_3}{\partial C'_2}\right) = -435,420w^*k^*\sigma_{C_2}$$

$$\left(\frac{\partial g_3}{\partial K'}\right)_* = -435,420w^*(c_1^* + c_2^*)\sigma_K$$

and with $y^* = \overline{Y}$, $w^* = \overline{W}$, $c_3^* = \overline{C}_3$, $c_1^* = \overline{C}_1$, $c_2^* = \overline{C}_2$, $k^* = \overline{K}$ for the initial iteration, the calculations for two iterations are summarized in Table E7.5b. Hence,

$$p_{F_3} = \Phi(-2.955) = 0.00156$$

It is obvious that the three failure modes are mutually correlated. The first-order bounds on the failure probability of the pipe system are

$$0.00494 \le p_F \le 0.01144$$

For the bi-modal bounds, we first calculate the correlation between the failure modes on the basis of Eq. 7.18. Using the appropriate direction cosines summarized in Tables E7.5a and E7.5b, we obtain

$$\rho_{g_1,g_2} = (0.664)^2 + (-0.692)^2 = 0.920$$

$$\rho_{g_1,g_3} = (0.664)(0.710) + (-0.285)(-0.092)$$
$$+ (-0.692)(-0.675) = 0.965$$

$$\rho_{g_2,g_3} = 0.965$$

Then with $\beta_1 = \beta_2 = 2.58$,

$$P(A) = \Phi(-2.58)\Phi\left(-\frac{2.58 - 0.920 \times 2.58}{\sqrt{1 - 0.920^2}}\right)$$

$$= \Phi(-2.58)\Phi(-0.53) = 0.00494 \times 0.2981 = 0.00147$$

Similarly,

$$P(B) = 0.00147$$

Hence,

$$0.00147 \le P(E_1 E_2) \le 0.00294$$

Table E7.5b   Summary of Iterative Calculations for Failure of Pipe 3 in Example 7.5

Iteration No.	Variable x_i	Assumed Failure Point x_i^*	$\left(\dfrac{\partial g}{\partial x_i'}\right)_*$	$\alpha_{x_i}^*$	New x_i^*
1	Y	7.01	22.27	0.724	$7.01 - 0.624\beta_3$
	W	8.59×10^{-5}	-20.77	-0.675	$8.59 + 1.35\beta_3$
	C_3	0.9	-1.683	-0.055	$0.9 + 0.002\beta_3$
	C_1	0.825	-1.946	-0.063	$0.825 + 0.004\beta_3$
	C_2	0.825	-1.946	-0.063	$0.825 + 0.004\beta_3$
	K	0.9	-2.777	-0.09	$0.9 + 0.004\beta_3$

Failure equation:

$$25.835(7.01 - 0.624\beta_3) - 4.3542(8.59 + 1.35\beta_3)$$
$$[(0.9 + 0.002\beta_3) + 2(0.825 + 0.004\beta_3)(0.825 + 0.004\beta_3)] = 0$$

yielding $\beta_3 = 2.96$

2	Y	5.16	22.27	0.710	$7.01 - 0.612\beta_3$
	W	12.59	-21.18	-0.675	$8.59 + 1.35\beta_3$
	C_3	0.906	-2.467	-0.079	$0.9 + 0.004\beta_3$
	C_1	0.837	-2.89	-0.092	$0.825 + 0.005\beta_3$
	C_2	0.837	-2.89	-0.092	$0.825 + 0.005\beta_3$
	K	0.912	-4.13	-0.132	$0.9 + 0.006\beta_3$

Failure equation:

$$25.835(7.01 - 0.612\beta_3) - 4.3542(8.59 + 1.35\beta_3)$$
$$[(0.9 + 0.004\beta_3) + 2(0.825 + 0.005\beta_3)(0.825 + 0.005\beta_3)] = 0$$

yielding $\beta_3 = 2.955$

With $\beta_1 = 2.58$ and $\beta_3 = 2.955$,

$$P(A) = \Phi(-2.58)\Phi\left[-\frac{2.955 - 0.965 \times 2.58}{\sqrt{1 - 0.965^2}}\right]$$

$$= \Phi(-2.58)\Phi(-1.77) = 0.00494 \times 0.0384 = 1.897 \times 10^{-4}$$

$$P(B) = \Phi(-2.955)\Phi\left[-\frac{2.58 - 0.965 \times 2.955}{\sqrt{1 - 0.965^2}}\right]$$

$$= \Phi(-2.955)\Phi(1.04) = 0.00156 \times 0.8508 = 0.00133$$

Hence,

$$0.00133 \leq P(E_1 E_3) \leq 0.00152$$

Also,

$$0.00133 \leq P(E_2 E_3) \leq 0.00152$$

Combining these results, we obtain the bi-modal bounds

$$0.01144 - (0.00294 + 0.00152 + 0.00152) \leq p_F \leq 0.01144 - 0.00147 - 0.00133$$

or

$$0.00546 \leq p_F \leq 0.00864$$

7.3 REDUNDANT AND NONREDUNDANT SYSTEMS

The reliability of a multicomponent system will be a function of the redundancy of the system; indeed, the analysis of reliability depends on whether the system is redundant or nonredundant. The failure of a component in a nonredundant system is tantamount to the failure of the entire system. Generally, the reliability of a system will be improved through the use of redundant components. Redundancy in a system may be of the *active* type or the *standby* type. In the case of active redundancies, all the components of a system are participating (e.g., carrying or sharing loads), whereas for systems with standby redundancies some of the (redundant) components are inactive and become activated only when some of the active components have failed.

In the evaluation of system reliability, there are also fundamental differences between an active redundant system and a standby redundant system. For systems with active redundancies, failures of the components will occur sequentially, unless the capacities of the components are perfectly correlated and identically distributed; therefore, for an active redundant system, all subsequent component failures (except the first one) will involve conditional probabilities. In contrast, for systems with standby redundancies, the component failure events may be statistically independent.

Standby redundancy will invariably increase the reliability of a system (unless the redundant components are perfectly correlated); in contrast, active redundancy may or may not be very effective in improving the reliability of a system. Consider the following for an example.

EXAMPLE 7.6

Three identical wires are used in an unbonded prestressed concrete beam. Suppose the prestressing force T is

$$\bar{T} = 120 \text{ kips}$$

$$\Omega_T = 0.25$$

The rupture strength of each prestressing wire has

$$\bar{R} = 75 \text{ kips} \quad \text{and} \quad \Omega_R = 0.15$$

The wires may be assumed to carry equal shares of the total force T, that is, each wire initially carries $T/3$. The probability of failure of each wire is then (assume T and R_i are statistically independent normal variates)

$$p_{F_i} = \Phi\left(\frac{75 - 40}{\sqrt{(11.25)^2 + (10)^2}}\right) = \Phi(-2.33) = 0.010$$

As there are three wires, the first-order bounds of the initial damage (i.e., failure of one of the three wires) is

$$0.010 \le P(D) \le 0.030$$

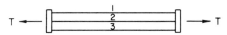

Figure E7.6   Three prestressing wires.

In this case, the forces in the wire are perfectly correlated. If the rupture strengths, R_i, can be assumed to be also perfectly correlated, the failures of the three wires will also be perfectly correlated and the probability of initial damage would be $P(D) = 0.010$. However, if the wire strengths, R_i, are uncorrelated, the failures of the wires will be partially correlated, in which case we obtain the second-order bounds as follows.

The performance function for each wire is

$$g_i(\mathbf{X}) = R_i - \frac{T}{3}$$

It can be shown that the correlation coefficient between the failures of two wires is

$$\rho_{ij} = \frac{\text{Var}(T/3)}{\sigma_{g_i} \cdot \sigma_{g_j}}$$

in which

$$\text{Var}(T/3) = 10^2$$

and

$$\sigma_{g_j} = \sigma_{g_i} = \sqrt{\sigma_{R_i}^2 + \sigma_{T/3}^2} = \sqrt{(11.25)^2 + (10)^2} = 15.05$$

Thus,

$$\rho_{ij} = \frac{(10)^2}{(15.05)^2} = 0.44$$

Equation 7.17 then becomes

$$P(A) = P(B) = \Phi(-2.33)\Phi\left(-2.33\sqrt{\frac{1-0.44}{1+0.44}}\right)$$

$$= (0.010)(0.074) = 0.74 \times 10^{-3}$$

and

$$0.74 \times 10^{-3} \le P(E_i E_j) \le 1.48 \times 10^{-3}$$

The bi-modal bounds of the initial damage probability, Eq. 7.16, are therefore

$$3 \times 0.010 - (1.48 + 1.48 + 1.48)10^{-3} \le P(D) \le 3 \times 0.010 - (0.74 + 0.74)10^{-3}$$

or

$$0.0256 \le P(D) \le 0.0285$$

If one of the wires should fail (rupture), the probability of a second wire failure would be as follows:

$$p_F(2|1) = \Phi\left(-\frac{75-60}{\sqrt{(11.25)^2 + (15)^2}}\right) = \Phi(-0.80) = 0.212$$

The first-order bounds of a second wire failure, therefore, are

$$0.212 \le P(D'|D) \le 0.379$$

Therefore, should one wire fail, the probability of a second wire failure is increased significantly such that for practical purposes the remaining two-wire system is no longer reliable. Moreover, if two of the wires should fail, the failure of the remaining last wire would be almost certain; that is,

$$p_F(3|1,2) = \Phi\left(-\frac{75-120}{\sqrt{(11.25)^2 + (30)^2}}\right) = \Phi(1.40) = 0.919$$

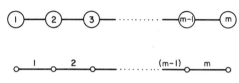

Figure 7.5 Representations of series system.

7.3.1 Series Systems

Many physical systems that are composed of multiple components can be classified as series-connected or parallel-connected systems, or combinations thereof. More generally, the failure events (e.g., in the case of multiple failure modes) may also be represented as events in series (union) or in parallel (intersection).

Systems that are composed of components connected in series (series systems) are such that the failure of any one or more of these components constitutes the failure of the system; such systems, therefore, have no redundancy and are also known as "weakest link" systems. In other words, the reliability or safety of the system requires that none of the components fail. A series system may be represented schematically as in Fig. 7.5.

If E_i denotes the failure of component i, then the failure of a series system is the event

$$E_s = E_1 \cup E_2 \cup \cdots \cup E_m \tag{7.19}$$

and the safety of the system would be the event

$$\bar{E}_s = \bar{E}_1 \cap \bar{E}_2 \cap \cdots \cap \bar{E}_m \tag{7.20}$$

Observe that E_i is equivalent to the failure of mode i (see Section 7.2); hence, E_s is tantamount to the failure of any one of the potential failure modes, that is, Eq. 7.7. Hence, the failure probability of the system, $p_F = P(E_s)$, is also bounded in accordance with Eq. 7.11 or 7.16.

EXAMPLE 7.7

A chain consisting of 20 identical links is a simple series system as shown in Fig. E7.7. Obviously, when subjected to an applied tension T, the failure of any one (or more) of the connected links will cause the failure of the entire chain.

If the failure probability of the individual links is

$$p_{F_i} = 10^{-4}$$

the failure probability of the system would depend on the correlation between the links, as follows:

Case A: Failures of the links are statistically independent.

$$p_F \simeq 20 \times 10^{-4} = 0.002$$

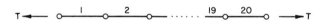

Figure E7.7 20-link chain system.

Case B: Failures of the links are perfectly correlated.

$$p_F = p_{F_i} = 10^{-4}$$

Case C: The mutual correlation between any two link failures is

$$\rho_{ij} = \rho \quad \text{for any } i \text{ and } j$$

In this last case (Case C), the first-order bounds for p_F are

$$0.0001 \le p_F \le 0.002$$

whereas the second-order bounds, Eq. 7.16, are as follows:

$$P(A) = P(B) = \Phi(-\beta)\Phi\left(-\beta\sqrt{\frac{1-\rho}{1+\rho}}\right)$$

in which corresponding to $p_{F_i} = 10^{-4}$, $\beta = -\Phi^{-1}(10^{-4}) = 3.72$. And (for example) if $\rho = 0.5$,

$$P(A) = P(B) = \Phi(-3.72)\Phi\left(-3.72\sqrt{\frac{1-0.5}{1+0.5}}\right)$$
$$= \Phi(-3.72)\Phi(-2.15) = 10^{-4}(1.58 \times 10^{-2})$$
$$= 1.58 \times 10^{-6}$$

and

$$1.58 \times 10^{-6} \le P(E_i E_j) \le 3.16 \times 10^{-6}$$

The second-order bounds, Eq. 7.16, are then

$$0.002 - 3.16 \times 10^{-6}(190) \le p_F \le 0.002 - 1.58 \times 10^{-6}(19)$$

or

$$0.00140 \le p_F \le 0.00197$$

7.3.2   Parallel Systems

Systems that are composed of components connected in parallel (parallel systems) are such that the total failure of the system requires failures of all the components; in other words, if any one of the components survives, the system remains safe. A parallel system is clearly a redundant system and may be represented schematically as shown in Fig. 7.6.

The failure of an m-component parallel system, therefore, is

$$E_s = E_1 \cap E_2 \cap \cdots \cap E_m \tag{7.21}$$

and the safety of the system is the event

$$\bar{E}_s = \bar{E}_1 \cup \bar{E}_2 \cup \cdots \cup \bar{E}_m \tag{7.22}$$

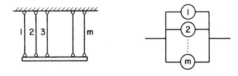

Figure 7.6   Representations of parallel systems.

In particular, observe that the failure of a series system is the union of the component failures, whereas the safety of a parallel system is the union of the survival events of the components; also, the safety of a series system is the intersection of the safety of the components, whereas the failure of a parallel system is the intersection of the component failures.

Since a parallel system is composed of redundant components, its reliability will depend on whether the redundancies are active or standby, as illustrated in the following examples.

EXAMPLE 7.8 *(Standby Redundancy)*

The emergency electric power for a nuclear power plant may be provided by diesel generators; emergency power may be necessary to operate the automatic safety systems of the plant during an earthquake. In order to increase the reliability of the required emergency power generating system, several generators may be installed, each of which has sufficient capacity to provide the emergency power requirement. Suppose five such generating units are installed, each of which is rated to withstand a (mean) peak earthquake acceleration of $0.20g$ (g = gravitational acceleration) with a c.o.v. of 0.25. Assume that the acceleration resistance, R_i, of each unit is normally distributed.

During a strong motion earthquake, the units will be subjected to random ground motions; suppose the peak ground acceleration, Q, at the site of the plant is also normally distributed with a mean of $0.10g$ and a c.o.v. of 0.30.

Determine the reliability of the emergency generating system during an earthquake.

Since any single unit can supply the emergency power requirement, the failure of the system requires failures of all five units during an earthquake; thus, the system is a parallel system, with standby redundancy, as shown in Fig. E7.8.

The performance function of the individual units may be defined as

$$g_i(\mathbf{X}) = R_i - Q$$

where

R_i = The capability of unit i to withstand earthquake acceleration, with probability distribution $N(0.20g, 0.05g)$.

Q = The peak ground acceleration of an earthquake, with distribution $N(0.10g, 0.03g)$.

It is reasonable to assume that R_i and Q are statistically independent. Then, the failure probability of the individual units would be

$$p_{F_i} = \Phi\left(-\frac{0.20g - 0.10g}{\sqrt{(0.05g)^2 + (0.03g)^2}}\right) = \Phi(-1.71) = 0.044$$

The probability of failure of the system is

$$p_F = P(E_1 \cap E_2 \cap \cdots \cap E_5)$$

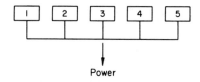

Power

Figure E7.8 Five generating units in parallel.

This probability is bounded (first-order) as follows:

$$\prod_{i=1}^{5} P(E_i) \leq p_F \leq \min_i P(E_i)$$

Thus,

$$1.65 \times 10^{-7} \leq p_F < 0.044$$

The above bounds correspond, respectively, to the assumption of statistical independence and perfect correlation among the failures of the individual units. These bounds are obviously much too wide to be useful. We seek a more accurate evaluation of the system failure probability as follows.

The performance function $g_i(\mathbf{X})$ is normally distributed with

$$\mu_{g_i} = 0.20g - 0.10g = 0.10g$$
$$\sigma_{g_i} = \sqrt{(0.05g)^2 + (0.03g)^2} = 0.058g$$

Since the five units will be subjected to the same ground motion during an earthquake, the failure events between the units will be correlated as follows.

Suppose R_i and R_j are statistically independent (independence of resistance among the units would be desirable in order to increase the reliability of the system). The correlation between $g_i(\mathbf{X})$ and $g_j(\mathbf{X})$ would then be

$$E(g_i g_j) = E[(R_i - Q)(R_j - Q)]$$
$$= E[R_i R_j - Q(R_i + R_j) + Q^2]$$

Since $\text{Cov}(R_i, Q) = 0$ and $\text{Cov}(R_i, R_j) = 0$,

$$\text{Cov}(g_i, g_j) = \sigma_Q^2$$

Moreover,

$$\sigma_{g_i} = \sqrt{\sigma_{R_i}^2 + \sigma_Q^2} \quad \text{and} \quad \sigma_{g_j} = \sqrt{\sigma_{R_j}^2 + \sigma_Q^2}$$

Then,

$$\rho_{ij} = \frac{(0.03g)^2}{(0.05g)^2 + (0.03g)^2}$$
$$= 0.265$$

The exact probability of failure requires the evaluation of the fifth order joint normal PDF; that is,

$$p_F = \int_{-\infty}^{0} \int_{-\infty}^{0} \cdots \int_{-\infty}^{0} f_{g_1, g_2, \ldots, g_5} \, dg_1 \ldots dg_5$$

Let

$$g_i' = \frac{g_i - \mu_{g_i}}{\sigma_{g_i}}$$

Then, $f_{g_1', g_2', \ldots, g_5'}$ is a joint standard normal PDF; the failure probability becomes

$$p_F = \int_{-\infty}^{-\beta} \int_{-\infty}^{-\beta} \cdots \int_{-\infty}^{-\beta} f_{g_1', g_2', \ldots, g_5'} \, dg_1' \ldots dg_5'$$

The above multiple integral may be evaluated using the method of Gupta (1963), yielding for the present case

$$p_F = \int_{0}^{\infty} \Phi^n \left(-\frac{\beta - u\sqrt{\rho}}{\sqrt{1 - \rho}} \right) \phi(u) \, du$$

where $n = 5$, $\beta = 1.71$, and $\rho = 0.265$. By numerical quadrature, we obtain $p_F = 0.00010$. Therefore, the probability of failure of the emergency generating system is $p_F = 0.00010$.

EXAMPLE 7.9   (Active Redundancy)

Redundancy in a structural system is invariably of the active type; that is, all the components in a structural system carry loads or are subject to the effects of the applied loads.

For illustration, consider again the three-wire prestressing system of Example 7.6. Assume that the failure of the system occurs only when all the three wires have ruptured. In this case, as one of the wires fails, the tensile loads on the remaining two wires are increased and, therefore, the probability of a second wire failure will also be increased accordingly.

If the strengths of the wires are perfectly correlated, the failure of one wire will automatically lead to the failures of all the other wires; thus, the failure of the three-wire system is represented by the failure of any one of the wires. Accordingly, in this case, the failure probability of the system is the same as the failure probability of one wire; that is,

$$p_F = p_{F_i} = 0.010$$

However, if the strengths of the wires are uncorrelated (or statistically independent), it should be observed that as one wire fails, the strengths of the remaining (surviving) wires will be higher than the rupture strength of the failed wire. Moreover, there is no assurance as to which of the wires will fail first; indeed, it is impossible to predetermine the sequence in which the three wires might fail. Consequently, all possible failure sequences must be considered; the different sequences, however, are mutually exclusive. In the present example, there are $3! = 6$ possible failure sequences of the three wires; as the wires are identical, the capacity of the system will be the same for all the six failure sequences.

With the above considerations, the CDF of the system capacity, R, would be (see also Shinozuka and Itagaki, 1966; Daniels, 1945.)

$$F_R(r) = 6 \int_0^{r/3} \int_{r_1}^{r/2} \int_{r_2}^{r} f_{R_1}(r_1) f_{R_2}(r_2) f_{R_3}(r_3) \, dr_3 \, dr_2 \, dr_1$$

$$= F^3\left(\frac{r}{3}\right) - 3F^2\left(\frac{r}{2}\right)F\left(\frac{r}{3}\right) - 3F(r)F^2\left(\frac{r}{3}\right) + 6F(r)F\left(\frac{r}{2}\right)F\left(\frac{r}{3}\right)$$

where $F(-)$ is the CDF of one wire and the failure sequence $(1 \to 2 \to 3)$ is presumed in the above integral; for this sequence, the strength $R_3 \geq r_2$ and $R_2 \geq r_1$ are reflected in the limits of the above integrals.

Under a load T, the probability of system failure may then be obtained through Eq. 6.2; that is,

$$p_F = \int_0^\infty F_R(t) f_T(t) \, dt$$

yielding, by numerical integration for the present example,

$$p_F = 0.0238$$

It may be emphasized from this example that many of the concepts relevant to a standby parallel system are not valid for an active parallel system. For example, the upper bound failure probability of a standby parallel system,

$$P(E_1 \cap E_2 \cap \cdots \cap E_m) \leq \min_i P(E_i),$$

is obviously not applicable for an active redundant system.

The calculation of the above system failure probability (for uncorrelated wire strengths) is obviously impractical in general. With the assumption that the failures of the surviving wires are statistically independent and neglecting the information that as one wire fails, the strengths

478   SYSTEMS RELIABILITY

of the surviving wires are higher than that of the failed wire, we obtain for the particular failure sequence $(1 \rightarrow 2 \rightarrow 3)$

$$p_{F_1} = \left[\int_0^\infty \int_0^{t/3} f_{R_1}(r_1) f_T(t)\, dr_1\, dt \right]\left[\int_0^\infty \int_0^{t/2} f_{R_2}(r_2) f_T(t)\, dr_2\, dt \right]\left[\int_0^\infty \int_0^{t} f_{R_3}(r_3) f_T(t)\, dr_3\, dt \right]$$

$$= P(1) \cdot P(2|1) \cdot P(3|1, 2)$$

where

$P(1) =$ Failure probability of wire #1.
$P(2|1) =$ Conditional failure probability of wire #2 given failure of wire #1.
$P(3|1, 2) =$ Conditional failure probability of wire #3 given failures of both wires #1 and #2.

Using the results of Example 7.6, we obtain

$$p_{F_1} = (0.010)(0.212)(0.919) = 1.95 \times 10^{-3}$$

As there are six possible failure sequences of the three wires, which are identical and mutually exclusive, the estimated failure probability of the system is

$$p_F = 6 p_{F_1} = 6 \times 1.95 \times 10^{-3} = 0.0117$$

This last result is, of course, approximate and in the present example the error is on the unsafe side, but of the order of magnitude as the correct result ($p_F = 0.0238$). For practical purposes, the approximate calculational method illustrated here for active redundant systems may be necessary, as any exact method is difficult to implement in practice.

7.3.3   Combined Series–Parallel Systems

Engineering systems may be composed of a combination of series-connected and parallel-connected components; however, not all systems can be decomposed into such series and parallel components. A general system may, nevertheless, be represented as a combination of failure or safe events in series (i.e., union) and/or in parallel (intersection); the pertinent events may not be simply the failures or survivals of the individual components.

EXAMPLE 7.10

The landing system of a Boeing 747 airplane consists of eighteen tires arranged as shown in Fig. E7.10a. The nose gear contains two tires in parallel, whereas each of the wing and rear gears contain four tires that are also in parallel.

Suppose the occurrence of any one of the following events during a landing could lead to a hazardous condition and thus constitutes failure of the system:

(i)   The nose gear tire system fails (blow out).
(ii) Either of the wing gear tire system fails.
(iii) Both of the rear gear tire systems fail.

The landing system is clearly composed of combined series and parallel arrangements of tires that may be modeled as shown in Fig. E7.10b.

Each of the subsystems contains active redundant components (tires); as one of the tires fail, the remaining tires in the subsystem must carry an additional load. Suppose that the total plane load, W, is distributed among the three gears as follows:

Gear	Percent of W
Nose	10
Wing	40
Rear	50

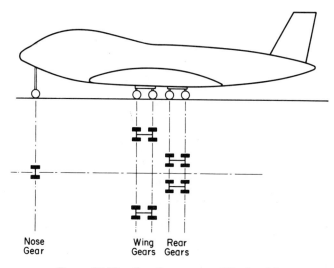

Figure E7.10a Landing system of Boeing 747.

The c.o.v. of the above distributed loads (including the c.o.v. in W and variations in load distribution) is 0.15.

Assume that the tire capacity in the different gears are as follows:

Gear	Mean Capacity of Each Tire	c.o.v. of Each Tire Capacity
Nose	$0.15\overline{W}$	0.20
Wing	$0.125\overline{W}$	0.20
Rear	$0.125\overline{W}$	0.20

Also, assume that the load, W, and the tire capacities are normally distributed.

The probability of failure of the tire system during a landing may be estimated as follows. If the failures of the individual tires in a gear can be assumed to be statistically independent, we obtain the first tire failure in the nose gear

$$p_{F_N}(1) = \Phi\left[-\frac{0.15 - 0.05}{\sqrt{(0.20 \times 0.15)^2 + (0.15 \times 0.05)^2}}\right] = \Phi(-3.23) = 6.19 \times 10^{-4}$$

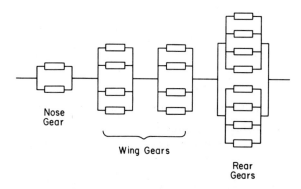

Figure E7.10b Series and parallel arrangement of tires.

whereas for the second tire failure

$$p_{F_N}(2|1) = \Phi\left[-\frac{0.15 - 0.10}{\sqrt{(0.20 \times 0.15)^2 + (0.15 \times 0.10)^2}}\right] = \Phi(-1.49) = 6.81 \times 10^{-2}$$

Similarly, for the tires in each of the wing gears, we have for a given failure sequence

$$p_{F_W}(1) = \Phi\left[-\frac{0.125 - 0.05}{\sqrt{(0.20 \times 0.125)^2 + (0.15 \times 0.05)^2}}\right] = \Phi(-2.87) = 2.05 \times 10^{-3}$$

$$p_{F_W}(2|1) = \Phi\left[-\frac{0.125 - 0.2/3}{\sqrt{(0.20 \times 0.125)^2 + (0.15 \times 0.2/3)^2}}\right] = \Phi(-2.17) = 1.5 \times 10^{-2}$$

$$p_{F_W}(3|1, 2) = \Phi\left[-\frac{0.125 - 0.10}{\sqrt{(0.20 \times 0.125)^2 + (0.15 \times 0.10)^2}}\right] = \Phi(-0.86) = 0.1949$$

$$p_{F_W}(4|1, 2, 3) = \Phi\left[-\frac{0.125 - 0.20}{\sqrt{(0.20 \times 0.125)^2 + (0.15 \times 0.20)^2}}\right] = \Phi(1.92) = 0.9726$$

And for the tires in each of the rear gears

$$p_{F_R}(1) = \Phi\left[-\frac{0.125 - 0.25/4}{\sqrt{(0.20 \times 0.125)^2 + (0.15 \times 0.25/4)^2}}\right] = \Phi(-2.34) = 9.64 \times 10^{-3}$$

$$p_{F_R}(2|1) = \Phi\left[-\frac{0.125 - 0.25/3}{\sqrt{(0.20 \times 0.125)^2 + (0.15 \times 0.25/3)^2}}\right] = \Phi(-1.49) = 6.81 \times 10^{-2}$$

$$p_{F_R}(3|1, 2) = \Phi\left[-\frac{0.125 - 0.125}{\sqrt{(0.20 \times 0.125)^2 + (0.15 \times 0.125)^2}}\right] = \Phi(0) = 0.500$$

$$p_{F_R}(4|1, 2, 3) = \Phi\left[-\frac{0.125 - 0.25}{\sqrt{(0.20 \times 0.125)^2 + (0.15 \times 0.25)^2}}\right] = \Phi(2.77) = 0.9972$$

The failure probabilities of the subsystems are then

$$P(E_N) = 2p_{F_N}(1)p_{F_N}(2|1)$$
$$= 2(6.19 \times 10^{-4})(6.81 \times 10^{-2}) = 8.431 \times 10^{-5}$$

$$P(E_{W_1}) = P(E_{W_2}) = 4!\, p_{F_W}(1)p_{F_W}(2|1)p_{F_W}(3|1, 2)p_{F_W}(4|1, 2, 3)$$
$$= 4!(2.05 \times 10^{-3} \times 1.50 \times 10^{-2} \times 0.1949 \times 0.9726)$$
$$= 1.399 \times 10^{-4}$$

and if E_{W_1} and E_{W_2} are statistically independent,

$$P(E_W) = P(E_{W_1} \cup E_{W_2}) \simeq 2.798 \times 10^{-4}$$

$$P(E_{R_1}) = P(E_{R_2}) = 4!\, p_{F_R}(1)p_{F_R}(2|1)p_{F_R}(3|1, 2)p_{F_R}(4|1, 2, 3)$$
$$= 4!(9.64 \times 10^{-3} \times 6.81 \times 10^{-2} \times 0.500 \times 0.9972)$$
$$= 7.856 \times 10^{-3}$$

The conditional failure probability of the rear system can be shown to be

$$P(E_{R_1}|E_{R_2}) = P(E_{R_2}|E_{R_1})$$
$$= 0.500 \times \Phi(1.18)\Phi(2.77)\Phi(4.74) = 0.439$$

and thus

$$P(E) = 2P(E_{R_1}|E_{R_2})P(E_{R_2}) = 2(0.439)(7.856 \times 10^{-3})$$
$$= 6.898 \times 10^{-3}$$

Therefore, the probability of failure of the landing system is

$$P_F = P(E_N \cup E_W \cup E_R)$$
$$\simeq 8.431 \times 10^{-5} + 2.798 \times 10^{-4} + 6.898 \times 10^{-3} = 7.262 \times 10^{-3}$$

EXAMPLE 7.11

A three-girder deck bridge is simply supported on two piers, each of which is supported by two pile groups of four piles each, as shown in Fig. E7.11a. During an earthquake, the bridge could fail (collapse) in one or more of the following modes:

(1) The girders dislodged from their bearing supports at either end; denote these failure events as R and L.
(2) The failure of both pile groups in a pier; denote the failures of the individual pile groups as events $P1, P2, P3, P4$.
(3) Any two of the girders, or all three girders, are overstressed beyond their yield capacities; denote the failure of any two girders as $G2$ and the failure of all three girders as event $G3$.

The collapse of the bridge during an earthquake may be represented as a combination of the above failure events in series and parallel as follows (see Fig. E7.11b). The system failure event, therefore, is

$$E_s = (R \cup L) \cup (P1 \cap P2) \cup (P3 \cap P4) \cup G2 \cup G3$$

Dislodging of the supports at one end of the girders may be highly correlated, whereas dislodging between the two ends of a girder may be uncorrelated. Then, if the probability of dislodging at one end of a girder is 10^{-4}, we have

$$P(R \cup L) = 2 \times 10^{-4}$$

It is reasonable to assume that each pile group carries one quarter of the total load; that is, $S/4$, and each pile carries $(S/4)/4 = S/16$. Assume $\Omega_S = 0.30$ and a mean pile capacity of $\bar{R} = \bar{s}/8$, with $\Omega_R = 0.20$. The failure probability of a pile group then is

$$P(P1) = p_F(1) \cdot p_F(2|1) \cdot p_F(3|1, 2) \cdot p_F(4|1, 2, 3)$$

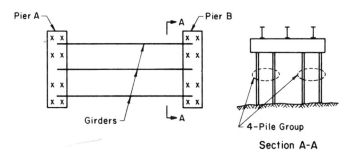

Figure E7.11a   Girder bridge system.

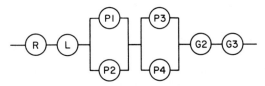

Figure E7.11b   Failure events in series and parallel.

where

$$p_F(1) = \Phi\left[-\frac{\bar{s}/8 - \bar{s}/16}{\sqrt{(0.2 \times \bar{s}/8)^2 + (0.3 \times \bar{s}/16)^2}}\right] = \Phi(-2.00) = 2.275 \times 10^{-2}$$

$$p_F(2|1) = \Phi\left[-\frac{\bar{s}/8 - \bar{s}/12}{\sqrt{(0.2 \times \bar{s}/8)^2 + (0.3 \times \bar{s}/12)^2}}\right] = \Phi(-1.18) = 0.1190$$

$$p_F(3|1, 2) = \Phi\left[-\frac{\bar{s}/8 - \bar{s}/8}{\sqrt{(0.2 \times \bar{s}/8)^2 + (0.3 \times \bar{s}/8)^2}}\right] = \Phi(0) = 0.5000$$

$$p_F(4|1, 2, 3) = \Phi\left[-\frac{\bar{s}/8 - \bar{s}/4}{\sqrt{(0.2 \times \bar{s}/8)^2 + (0.3 \times \bar{s}/4)^2}}\right] = \Phi(1.58) = 0.9429$$

As the four piles in a group may fail in $4! = 24$ different sequences, the failure probability of each pile group is

$$P(P1) = P(P2) = P(P3) = P(P4) = 24(2.275 \times 10^{-2})(0.1190)(0.5000)(0.9429)$$
$$= 3.063 \times 10^{-2}$$

Assume further that the conditional failure probabilities of the pile groups are

$$P(P2|P1) = P(P1|P2) = 0.90$$

Similarly,

$$P(P4|P3) = P(P3|P4) = 0.90$$

Then,

$$P(P1 \cap P2) = P(P3 \cap P4) = (3.063 \times 10^{-2})(0.90) = 2.757 \times 10^{-2}$$

For the girders, assume that the failure probability of the individual girders are as follows:

$$p_{F_i} = 10^{-3}$$
$$p_F(2|1) = 0.10$$
$$p_F(3|1, 2) = 0.70$$

Then,

$$P(G2) = \binom{3}{2}(10^{-3} \times 0.10)(1 - 0.70) = 0.9 \times 10^{-4}$$

and

$$P(G3) = 10^{-3} \times 0.10 \times 0.70 = 0.7 \times 10^{-4}$$

Hence, the failure probability of the bridge system during an earthquake is bounded as follows:

$$2.757 \times 10^{-2} \le p_F \le 2 \times 10^{-4} + 2(2.757 \times 10^{-2}) + 0.9 \times 10^{-4} + 0.7 \times 10^{-4}$$
$$2.757 \times 10^{-2} \le p_F \le 5.549 \times 10^{-2}$$

EXAMPLE 7.12   (Reliability-Based Decision Analysis)

Two transmission line support systems are being considered for power lines in a remote area. For purpose of illustration, the supporting systems are idealized as follows:

System A: A single prismatic pole securely anchored at the base, of an extra strong 10-in. diameter steel pipe with $\frac{1}{2}$-in. thickness. See Fig. E7.12a.

System B: System A with $\frac{3}{16}$-in. diameter guyed cables, as shown in Fig. E7.12a.

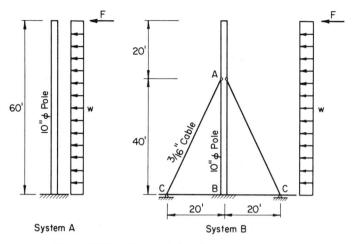

Figure E7.12a   Transmission line support systems.

The section modulus of the 10 in. $\phi \times \frac{1}{2}$ in. steel pipe is 39.4 in.^{3}; its yield strength is $N(40$ ksi; 4 ksi), whereas the ultimate strength of the cables is $N(200$ ksi; 24 ksi).

The critical loading on the structure will be from the maximum wind over a service life of 25 years. This will consist of the imbalanced wind-induced transmission line load, F, and a uniformly distributed wind pressure over the height of the pole. For a uniform pole spacing of 500 ft, the following maximum load statistics may be assumed:

Wind-Induced Loads	Mean	c.o.v.	Distribution
F	0.7 kips	0.35	Type I Extreme
w	0.0181 kip/ft	0.35	Type I Extreme

The additional cost of the cables (material plus installation) is $1000 per pole. In the event of a failure, the repair and/or replacement costs per pole are estimated to be $3000 and $4000, respectively, for Systems A and B; moreover, in such an event, the additional loss associated with a temporary power outage would be $10,000 for either system.

Reliability of System A

The maximum moment in the pole of System A will occur at the base, which is (in ft-kip),

$$M = 60F + \frac{(60)^2}{2} w$$

It is reasonable to assume that F and w are perfectly correlated; hence, M will also have a Type I asymptotic distribution, with a mean of 895.4 in.-kip and a c.o.v. of 0.35.

The performance function against yielding of the pole is

$$g_1(\mathbf{X}) = 39.4Y - M$$

in which Y is the yield strength of the pole.

Observe that

$$\left(\frac{\partial g}{\partial Y'}\right)_* = 39.4\sigma_Y$$

$$\left(\frac{\partial g}{\partial M'}\right)_* = -\sigma_M^N$$

For the first iteration let

$$y^* = \bar{Y} = 40 \text{ ksi}$$

$$m^* = \bar{M} = 895.4 \text{ in.-kip}$$

The calculational results for three iterations are as summarized in Table E7.12; from which we obtain the failure probability of the pole in yielding

$$p_{F_1} = \Phi(-1.75) = 0.040$$

Reliability of System B

The reliability of System B will obviously be higher than that of System A. Under the wind loads F and w, the tension (in kip) in the cables can be shown to be

$$T = 2.51F + 5.08w$$

whereas the maximum applied moment in the pole (at the base) is (in in.-kip)

$$M_B = 720F + 21{,}600w - \frac{480}{\sqrt{5}} T$$

Again, if we assume F and w are perfectly correlated, the distributions of T and M_B will also be Type I asymptotic, with mean values 2.86 kip and 281.5 in-kip, respectively, and c.o.v. of 0.35.

Conceivably, the failure of System B may occur through (i) yielding at the base of the pole followed by the rupture of the pertinent cable; or (ii) rupture of a cable followed by the yielding of the pole. However, since M_B is small ($\bar{M}_B = 281.5$ in.-kip), the yielding of the pole prior to the rupture of the cables is not likely (probability of order 10^{-9}); hence, the first mode of failure can be neglected. For the second mode, observe that the performance function of a cable is

$$g_2(\mathbf{X}) = \frac{\pi}{4}\left(\frac{3}{16}\right)^2 C - T$$

Table E7.12 Summary of Iterative Calculations for Failure of System A

Iteration No.	Variable x_i	Assumed Failure Point x_i^*	$\left(\dfrac{\partial g}{\partial x_i}\right)_*$	$\alpha_{x_i}^*$	New x_i^*
1	Y	40	157.6	0.465	$40 - 1.86\beta$
	M	895.4	-299.7	-0.885	$842.0 + 265.3\beta$
	Failure equation:				
	$39.4(40 - 1.86\beta) - (842.0 + 265.3\beta) = 0;$		$\beta = 2.17$		
2	Y	35.97	157.6	0.304	$40 - 1.216\beta$
	M	1417.2	-494.1	-0.953	$666.2 + 470.7\beta$
	Failure equation:				
	$39.4(40 - 1.216\beta) - (666.2 + 470.7\beta) = 0;$		$\beta = 1.75$		
3	Y	37.87	157	0.289	$40 - 1.156\beta$
	M	1490.0	-521.2	-0.957	$623.2 + 498.8\beta$
	Failure equation:				
	$39.4(40 - 1.156\beta) - (623.2 + 498.8\beta) = 0;$		$\beta = 1.75$		

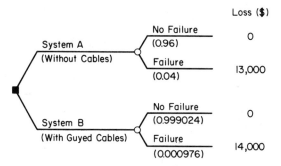

Figure E7.12b   Decision tree for transmission line support systems.

where C is the ultimate strength of a cable, assumed to be $N(200 \text{ ksi}; 24 \text{ ksi})$. On this basis, an iterative reliability analysis will yield $\beta = 1.97$. Hence,

$$p_{F_2} = \Phi(-1.97) = 0.0244$$

Upon the failure (rupture) of a cable, the conditional probability of failure of the pole is 0.040 (the same as that of System A). The failure probability of the second mode, therefore, is

$$p_F = 0.0244 \times 0.040 = 9.76 \times 10^{-4}$$

Decision Analysis

In order to determine which of the two systems is preferable, a decision analysis is performed; the appropriate decision tree is shown in Fig. E7.12b. On this basis, the expected total losses of the respective systems are as follows:

System	Expected Loss
A	$520
B	$1013

Therefore, based on the minimum expected monetary loss criterion, the unguyed System A is preferable; apparently, the additional cost for higher reliability provided by the cables is not cost-effective.

However, if the consequence (loss) of a failure is more serious (e.g., if the total loss exceeds $22,600), then the additional cost for the guyed cables will become cost-effective.

7.4 FAULT TREE, EVENT TREE ANALYSES

For a complex system, the adverse conditions or *faults* that could lead to the occurrence of one of the potential modes of failure may be quite involved, such that some systematic means of identifying the various faults and their interactive effects (if any) on the failure event would become necessary. The fault tree diagram serves exactly this purpose.

A fault tree analysis may include a quantitative evaluation of the probabilities of the various faults or failure events leading eventually to the calculation of the probability of the main failure event (top event). Used in a quantitative form, fault tree analysis (FTA) is a valuable diagnostic tool. Aside from identifying all potential paths that could lead to failure, it may serve also to single out the critical events that contribute significantly to the likelihood of failure of a system and reveal the weak links, if any, in the system.

The occurrence of a top event may or may not lead to a serious or adverse consequence. A number of potential consequences may be possible, the relative likelihoods of which will depend on the conditions or subsequent events that follow. The potential consequences may be systematically identified with the *event tree* diagram in which all the possible paths following a top event or *initiating event* may be traced through the subsequent events, each path leading to a different consequence.

7.4.1   The Fault Tree Diagram

The objective of a fault tree, therefore, is to identify and model the various system conditions (faults) that can result in the occurrence of a given undesired event, known as the "top event." A fault tree diagram is then a graphical decomposition of a top event (major fault) into the union and/or intersection of subevents (faults). The alternative faults that could lead to the top event are logically related to the top event by "OR" gates and "AND" gates. The symbol for the "OR" gate is $\boxed{+}$ as shown following the top event in Fig. 7.7, which indicates that the top event E is the union of subevents E_1 and E_2. In other words E will occur if either E_1 or E_2 (or both) occurs. Similarly, further down the fault tree, E_1 is the union of E_5 and E_6. The symbol for the "AND" gate is $\boxed{\cdot}$ as shown following E_2 in Fig. 7.7, which signifies that E_2 is the intersection of subevents E_3 and E_4, implying that the occurrences of both subevents E_3 and E_4 are necessary for the occurrence of E_2. Each of the subevents may be decomposed further until the sequence of events lead to basic causes of interest, called "basic events." The "basic events" are usually those whose occurrence probabilities may be readily assessed.

The symbols shown in Fig. 7.7 for events and subevents represent specific types of events in a fault tree analysis. The rectangle (e.g., E, E_1, E_2) defines a fault event that will be developed further into other subevents through logic gates. The circle (e.g., E_3 and E_6) defines a basic inherent failure of a system element. It is therefore a primary failure event. The diamond (e.g., E_4) represents a secondary failure event that is purposely not developed further; such secondary failure events are generally caused by excessive environmental or operational stress on the system element, and human error may be included in this category. The switch event (e.g., E_5) represents an event whose occurrence will change the operating condition of the system.

EXAMPLE 7.13

Consider again the collapse of the bridge discussed earlier in Example 7.11. The potential modes of failure of the bridge during an earthquake and the respective causes or faults leading to each of the failure modes may also be represented in terms of a fault tree as shown in Fig. E7.13.

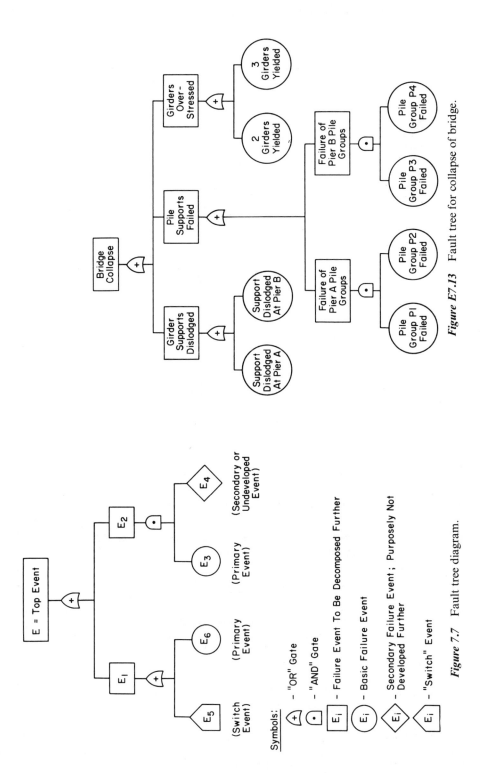

Figure E7.13 Fault tree for collapse of bridge.

Symbols:

⊕ – "OR" Gate

⊙ – "AND" Gate

□ E_i – Failure Event To Be Decomposed Further

○ E_i – Basic Failure Event

◇ E_i – Secondary Failure Event; Purposely Not Developed Further

⬠ E_i – "Switch" Event

Figure 7.7 Fault tree diagram.

487

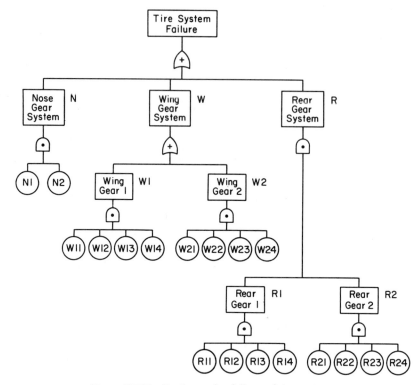

Figure E7.14   Fault tree for failure of tire system.

EXAMPLE 7.14

The failure of the landing tire system in a Boeing 747 airplane (Example 7.10) may also be represented in terms of a fault tree diagram shown in Fig. E7.14

EXAMPLE 7.15   (*Automotive Brake System Failure; Excerpted from Ang et al., 1979*)

The brake system of an automobile consists of independent front and rear brakes, and thus is a redundant system. An accident caused by the brake failure of an automobile requires the failure of both the front and rear brake systems as well as the inability of the driver to take proper evasive action. The failure of each brake system could be through the failure of the wheel cylinders, or brake lining, or master cylinder, or insufficient brake fluid. Furthermore, a wheel cylinder failure may occur in the left or right wheel. The fault tree representation of a brake failure accident is shown in Fig. E7.15.

EXAMPLE 7.16   (*Feed Water System Failure; Excerpted from Cornell and Newmark, 1981*)

In the event of an earthquake, the main feed water in a pressurized water nuclear reactor (PWR) may fail at a critical seismic level. The loss of the containment through over-pressure failure is virtually certain if the auxiliary feed water system fails to provide water to the steam generators. By neglecting other components such as piping, valves, control systems, etc., a coarse model of the auxiliary feed water supply system is shown in Fig. E7.16a. The model implicitly

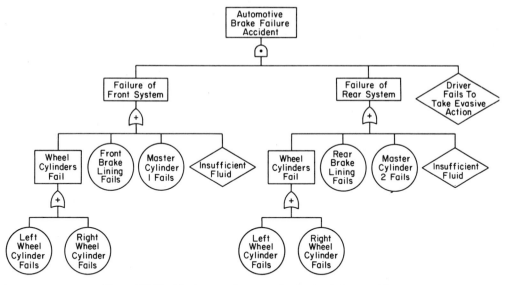

Figure E7.15 Fault tree of automotive brake failure accident.

assumes that failure will occur if (i) all three of the stored water facilities fail, or (ii) all three pumps fail to function. One of the pumps is turbine driven, whereas the other two are electric pumps that depend on the functioning of at least one of the two standby diesel generators.

The fault tree representation of the failure of this auxiliary feed water system is shown in Fig. E7.16*b*.

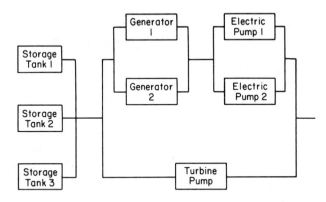

Figure E7.16a Auxiliary feed water system.

EXAMPLE 7.17 (Failure of Waste Storage Facility; Excerpted from Lambe, Marr, and Francisco, 1981)

In the safety assessment of a dam designed for waste storage, the most serious event will consist of the loss of contaminated fluid stored within the waste storage facility. The following five failure mechanisms may result in the loss of fluid:

(i) Excessive flow (i.e., exceeding the capacity of the perimeter collection ditch) through the dam as a result of highly permeable soil present in the dam and the poor performance of the cutoff system.

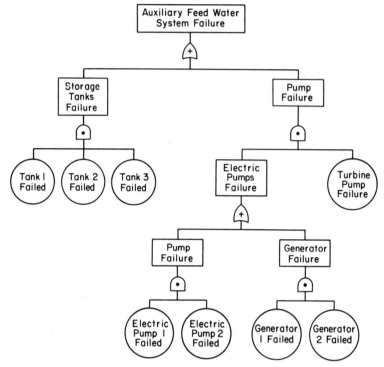

Figure E7.16b   Fault tree for auxiliary feed water system failure.

(ii) Overtopping caused by excessive rainfall. This event could occur if the rainfall exceeds 5.7 ft in a 24-hour period or 6.4 ft in a 48-hour period, provided the spillway is clean. On the other hand, if the spillway is plugged (e.g., by vegetation and debris), overtopping could occur if the 24-hour rainfall exceeds 5 ft.

(iii) A sinkhole occurs within the area of the dam that subsequently expands leading to a loss of containment.

(iv) Shear slide occurs through the crest of the dam.

(v) Piping (or internal erosion) developed through the silty sand foundation underlying the reservoir. Piping may have started with small cracks that gradually widen, eventually leading to a concentrated flow of containment in the absence of a well-designed cutoff system. The fault tree representation of the loss of contaminated fluid is shown in Fig. E7.17.

EXAMPLE 7.18 (*Nuclear   Reactor   Containment   Failure;   Excerpted   from Cummings, 1975*)

Fault tree analysis has been applied extensively in the analysis of safety of nuclear power plants (WASH 1400, 1975); an example is the fault tree analysis of the reactor containment system. The final reduced fault tree for a containment failure in a pressurized water reactor (PWR) containment system is presented in Fig. E7.18. Containment failure is defined as the leakage of radioactive material through an opening with a diameter greater than 4 in., the occurrence of such events could have severe consequences.

As shown in Fig. E7.18a, the top event could be caused by the passive structural failure of the containment shell or failure of the main appurtenances including nozzle weld fracture,

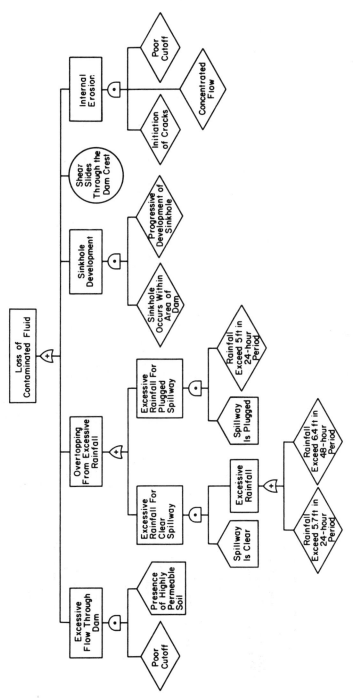

Figure E7.17 Fault tree of loss of contaminated fluid.

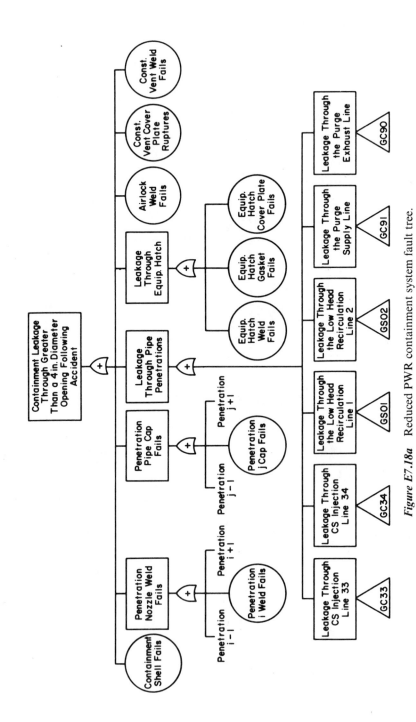

Figure E7.18a Reduced PWR containment system fault tree.

492

pipe cap rupture, equipment hatch failures, airlock weld fracture, construction vent cover plate ruptures, and construction vent weld fractures. The nozzle weld fracture could occur at any of the penetrations; similarly, the failure of the pipe cap could occur at any of the corresponding penetrations. Leakage through the equipment hatches could be due to failures of the welds or the gasket or the cover plate. The top event may also be caused by leakage through a pipe penetration, due to a failure in (i) the containment spray (CS) injection system, because of atmospheric back leakage through the two injection lines; (ii) the low-pressure recirculation system, because of leakage through the two pump suction lines; or (iii) the containment purge supply system because of leakage through the supply and exhaust lines. Each of these three faults are expanded in Figs. E7.18*b*, E7.18*c*, and E7.18*d*, respectively.

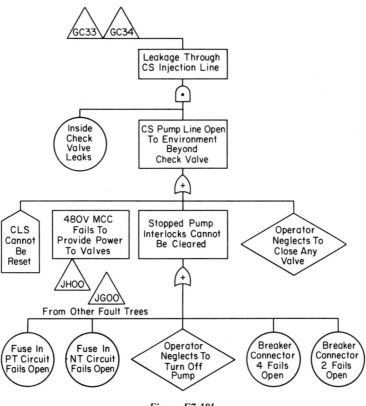

Figure E7.18b

Figure E7.18*b* shows that leakage through a containment spray (CS) injection line will occur if there is a leakage in the inside check valve and the pump line remains open to the environment beyond the check valve. This latter event could be due to the Consequence-Limiting Safeguards Initiation System (CLS) not operating, a power supply not available to close the valves, the operator neglecting to close any of the outer valves, or the inability to clear stopped pump interlocks. The last event may be caused by a failure in one or both of the two fuses, the failure of either breaker connectors, or simply the operator's negligence in turning off the pump.

Figure E7.18*c* shows that leakage through a low-pressure recirculation system may result from ruptures of a pipe section, a pump or valve body, or the pump sump vessel.

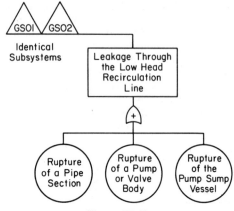

Figure E7.18c

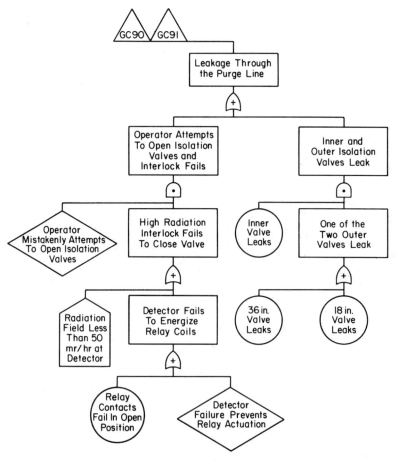

Figure E7.18d

Finally, in Fig. E7.18d, leakage through a purge line would occur if (i) the interlocks fail when the operator attempts to open the isolation valves, or (ii) there is leakage in the isolation valves. The first event will occur if the operator mistakenly opens the isolation valves and the interlock failed to close the valve caused by high radiation, whose presence was not detected either because of a low radiation field at the detector or the detector's failure to energize the relay coils. Furthermore, the failure to energize the relay coils could be caused by the relay contacts being in the open position, or relay actuation is prevented by a detector failure. The second event, namely leakage in the isolation valves, would occur if both the inner valves leak and at least one of the two outer valves leaks.

7.4.2 Probability Evaluation

Since the top event is related to the subevents and basic faults at subsequent levels of a fault tree through combinations of "AND" and "OR" logic gates, its probability can be expressed in terms of the probabilities of the subevents or basic events through the probabilities of unions and intersections of events. For example, the probability of the top event E in Fig. 7.7 can be determined as follows:

$$
\begin{aligned}
P(E) &= P(E_1 \cup E_2) \\
&= P[(E_5 \cup E_6) \cup E_3 E_4] \\
&= P(E_5 \cup E_6) + P(E_3 E_4) - P[E_3 E_4(E_5 \cup E_6)] \\
&= P(E_5) + P(E_6) - P(E_5 E_6) + P(E_3 E_4) - P(E_3 E_4 E_5) \\
&\quad - P(E_3 E_4 E_6) + P(E_3 E_4 E_5 E_6)
\end{aligned}
$$

If the basic events E_3 through E_6 are statistically independent with corresponding probabilities p_3 through p_6, then

$$
P(E) = p_5 + p_6 - p_5 p_6 + p_3 p_4 - p_3 p_4 p_5 - p_3 p_4 p_6 + p_3 p_4 p_5 p_6
$$

For a large fault tree, an efficient method of calculating the probability of occurrence of the top event is based on a Boolean representation for fault trees (see, e.g., Barlow and Lambert, 1975) and the use of indicator functions for the top event. The Boolean indicator functions can be determined from the minimum cut sets of basic events for a specified fault tree.

EXAMPLE 7.19 (*Basement Flooding*)

During heavy rainstorms, a building basement equipped with a sump pump system may be flooded if the rate of inflow of storm water exceeds the normal pump capacity. However, even smaller rainstorms can also cause flooding in the basement if the sump pump system is not operating properly. Suppose a system consists of a primary pump operated by utility power and a back-up battery-operated pump that will automatically be activated during times of power failure (see Fig. E7.19a). The failure of the sump pump system will occur if both the primary pump and the back-up pump fail to operate during periods of ground water inflow from a storm. The primary pump will fail if it breaks down or if there is power outage. The back-up pump will fail if it malfunctions or if the battery is drained. The battery will be drained if both the period of power outage and the period of inflow exceed the battery capacity. All the faults that can potentially cause flooding of the basement are summarized in the fault tree of Fig. E7.19b.

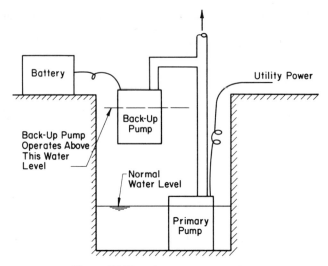

Figure E7.19a   A sump pump system.

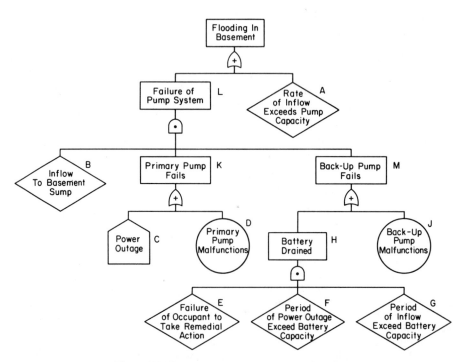

Figure E7.19b   Fault tree for basement flooding.

Suppose the basic events for this fault tree have the following annual probabilities of occurrence:

$$P(A) = 0.001$$
$$P(B) = 0.9$$
$$P(C) = 0.1$$
$$P(D) = 0.1$$
$$P(E) = 0.2$$
$$P(F) = 0.05$$
$$P(G) = 0.5$$
$$P(J) = 0.05$$

Assume that these basic failure events are statistically independent. Determine the annual probability of flooding of the basement.

The calculation of the probabilities starts from the last level of the fault tree and progresses upward in stages toward the top event. In the present example, we have

$$P(H) = P(E \cap F \cap G) = 0.2 \times 0.05 \times 0.5 = 0.005$$
$$P(M) = P(H \cup J) = 0.005 + 0.05 - 0.005 \times 0.05 = 0.055$$
$$P(K) = P(C \cup D) = 0.1 + 0.1 - 0.1 \times 0.1 = 0.19$$
$$P(L) = P(B \cap K \cap M) = 0.9 \times 0.19 \times 0.055 = 0.0094$$

Hence,

$$P(\text{basement flooding}) = P(A \cup L) = 0.001 + 0.0094 - 0.001 \times 0.0094$$
$$= 0.010$$

The basic events may be dependent. For example, two failure events could be caused by a common environment or have factors in common; these are called *common-mode* failures. Moreover, if two subsystems share a given load and if one of the subsystems fails, the remaining subsystem will experience an increase in load. In such cases, the evaluation of the probabilities will require conditional or joint probabilities. At times, the dependent events may be located far apart within the fault tree. Hence, the effect of dependence will not be apparent until further up the tree. Such joint probabilities or their bounds may be estimated using the methods developed in Section 7.2 for the reliability of correlated failure modes.

Qualitative Evaluation A fault tree provides an organized and systematic identification of the possible or potential faults of a system that can lead to the occurrence of a top event. Besides identifying the possible causes of an accident (top event), fault tree analysis may lead to the discovery of failure combinations that might otherwise be overlooked.

Instead of performing a quantitative evaluation of the probability of the top event, a purely qualitative evaluation of the fault tree (without accurate probability values) may also provide valuable information for design purposes. A fault tree may also be formulated as a network; in this context, a list of minimal cut sets can be identified using network analysis techniques (Barlow and Lambert, 1975). Each of the minimal cut sets represents an irreducible set of basic events whose occurrence may lead to the top event. In other words, the critical events that will significantly contribute to the probability of the top event can be identified. The likelihood of the occurrence of the top event may then be reduced by minimizing the probability of occurrence of the appropriate sets of critical events.

A large fault tree may be reduced or trimmed by first assigning order-of-magnitude probabilities for the basic events and then identifying the minimal cut sets with significant probabilities; the minimal cut sets with probabilities within two orders of magnitude of the maximum are retained for further quantitative probabilistic evaluation (Vesely, 1975). This reduction process can be effective for identifying the critical or dominant subevents in a large fault tree.

An objective of the fault tree evaluation is to determine if an acceptable level of risk is achieved for a proposed system or design. If the proposed design is found to be inadequate, the design may be improved by first identifying the critical events that significantly contribute to the occurrence probability of the top event. Decisions are then made to revise the design such that the effect of these critical events can be minimized. When all design changes are made, the fault tree is reevaluated to determine if the acceptable level of risk is achieved.

7.4.3   Event Tree Analysis

Usually, a number of possible consequences may result from the occurrence of an *initiating event* or failure of a system, some of which may be adversely serious. The top event of a fault tree may be only an initiating event. In general, whether a particular consequence of an initiating event is serious or not may depend on whether a sequence of other adverse events occur following the initiating event. In order to ascertain that all potentially dangerous or adverse consequences are considered following an initiating event, the different possible sequences of subsequent events need to be identified. This can be accomplished systematically and effectively through the use of the event tree diagram, as described below.

The Event Tree Model   A general event tree is shown in Fig. 7.8 with an initiating event, E, and a number of possible consequences, $C_{ij\ldots k}$. It can be observed that a particular consequence depends on the subsequent events following the initiating event; in other words, for a given consequence to occur, a sequence of subsequent events, or *path* in the event tree, must occur. Given an initiating event, there may be several "first subsequent events" that will follow. Obviously, these subsequent events are mutually exclusive. If we assume a particular first subsequent event, a mutually exclusive set of "second subsequent events" is possible. Each path in the event tree, therefore, represents a specific sequence of (subsequent) events, resulting in a particular consequence. The probability associated with the occurrence of a specific path is simply the product of the (conditional) probabilities of all the events on that path; for example, with reference to Fig. 7.8,

$$P(C_{ij}\ldots_k|E) = P(E_{1i}|E)P(E_{2j}|E_{1i}E)\cdots P(E_{nk}|E_{1i}E_{2j}\cdots E) \qquad (7.23)$$

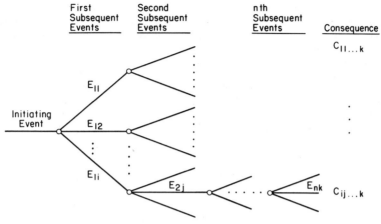

Figure 7.8 Event tree model.

EXAMPLE 7.20   (*Consequences of Automotive Steering Failure; Excerpted from Ang et al., 1979*)

Consider the event tree of Fig. E7.20 showing the potential consequences of a steering failure on a two-way highway. The initiating event I is vehicle #1 having a steering failure. Subsequent events pertain to the possible actions of the drivers in vehicle #1 and vehicle #2 (on-coming), and the roadway environment. For example, path no. 1 indicates that proper corrective action is taken by the driver in vehicle #1 and a collision is avoided. Path no. 2 indicates that no corrective action was taken by the driver in vehicle #1, but avoidance action is taken by the driver in the on-coming vehicle #2; again collision is avoided. However, roadside obstruction may cause some damage or even injury, even though the consequence

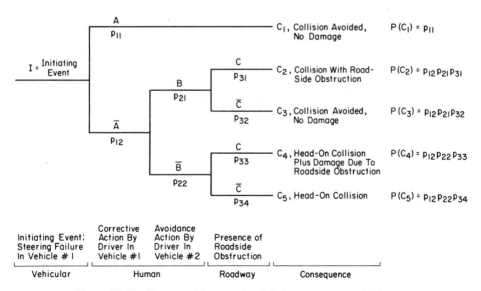

Figure E7.20   Event tree for steering failure on a two-way highway.

is not as serious as a head-on collision. Therefore, given the initiating event and depending on the action or inaction of the drivers, several potential consequences in terms of damage and injury may result, ranging from head-on collision in which injury (including death) is almost certain to no damage.

The probabilities of the subsequent events are generally conditioned on the prior events. For example, p_{11} is the conditional probability of corrective action taken by the driver in vehicle #1 given there is a steering failure, that is, $P(A|I)$. Similarly, $p_{21} = P(B|I\bar{A})$ is the conditional probability of the driver in vehicle #2 taking avoidance action given that both I and $\bar{A}$ occurred. These probabilities may be evaluated from data or through probabilistic modeling, or subjectively assessed. The probability of a given path, say path no. 2 in Fig. E7.20, is given by

$$P(\bar{A}BC|I) = P(C|I\bar{A}B)P(B|I\bar{A})P(\bar{A}|I) = p_{31}p_{21}p_{12}$$

Since the presence of roadside obstructions is independent of steering failure or any corrective actions by the two drivers,

$$p_{31} = P(C|I\bar{A}B) = P(C)$$

The probability associated with each path of the event tree is shown in the last column of Fig. E7.20. The conditional probability of no damage given the occurrence of a steering failure is, therefore, the combination of paths no. 1 and 3, namely

$$P(\text{no damage}|I) = p_{11} + p_{12}p_{21}p_{32}$$

If the damage consequences associated with paths no. 2, 4, and 5 are denoted by C_2, C_4, and C_5, respectively, the expected damage given the occurrence of a steering failure is

$$E(\text{Damage}|I) = p_{12}p_{21}p_{31}C_2 + p_{12}p_{22}p_{33}C_4 + p_{12}p_{22}p_{34}C_5$$

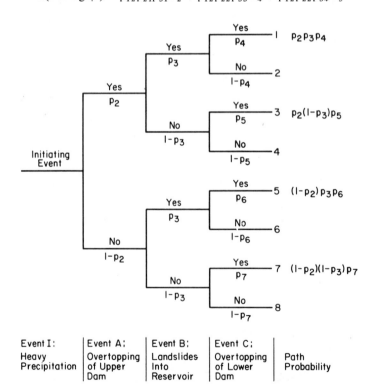

Event I:	Event A:	Event B:	Event C:	
Heavy Precipitation	Overtopping of Upper Dam	Landslides Into Reservoir	Overtopping of Lower Dam	Path Probability

Figure E7.21 Event tree for dam overtopping.

EXAMPLE 7.21 (Dam Overtopping)

The overtopping of a dam resulting from a heavy rainstorm over the watershed could lead to several failure paths. If another dam is located further downstream, the overtopping of the upper dam would intensify the flow into the lower reservoir, thus increasing the probability of overtopping of the second dam downstream. Moreover, heavy rain could induce landslides of the slopes adjacent to the reservoir, which in turn could generate waves that may be high enough to cause overtopping. Various sequences of events leading to the overtopping of a dam are shown in the event tree in Fig. E7.21. The initiating event in this case is a heavy precipitation. Given the initiating event, the conditional probability of the upper dam over-topping (event A) is assessed. The event of landslides into the reservoir (event B) will depend on the amount and rate of precipitation, but is generally not affected by the occurrence of event A. The overtopping of the lower dam (event C), however, will depend on the specific combination of events I, A, and B. For example, given the occurrence of heavy precipitation, the overtopping of the upper dam coupled with landslides into the reservoir will almost surely cause overtopping of the lower dam (large p_4); whereas if the upper dam was not overtopped and landslides do not occur, the probability (p_7) of overtopping the lower dam would be small. The probability of overtopping the lower dam given a heavy precipitation is the sum of the probabilities of paths no. 1, 3, 5, and 7; namely,

$$P(C|S) = p_2 p_3 p_4 + p_2(1 - p_3)p_5 + (1 - p_2)p_3 p_6 + (1 - p_2)(1 - p_3)p_7$$

EXAMPLE 7.22 (Train Derailment; Excerpted from Ang et al., 1979)

Defects in the track geometry of a rail could cause derailment of trains. Track geometry defects consist of two types: severe discrepancy in geometry (e.g., track buckling from thermal stresses) and harmonic defects in the track geometry. If a defect of the first kind exists, almost no train can travel over the spot at normal speed. For the second kind of defects, the chance of a de-railment accident will further depend on the train equipment used, whether it is defective or if harmonic vibration could be induced by the defects in the track geometry for the particular type of equipment.

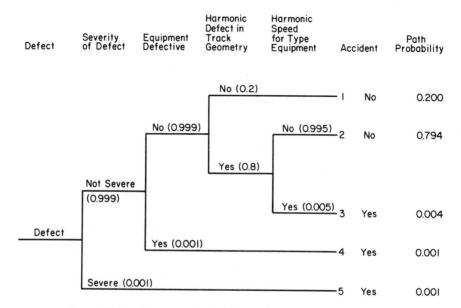

Figure E7.22 Event tree for derailment from track geometry defects.

Figure E7.22 shows the appropriate event tree. It may be observed that paths no. 3, 4, and 5 in the event tree may lead to an accident. Path no. 3 represents a speed-, equipment-, and track geometry-related accident; whereas in path no. 5, the defects are so severe that it is practically impossible for a car to travel over the spot at normal speed. Path no. 4 represents an accident caused by defective equipment; with defective equipment, even nonsevere geometry defects can derail the cars. An example of this sequence would be a car with a broken spring making it sensitive to irregularities in the track geometry.

Each event in a path will have a probability of success or failure. These numbers will be different for different sections of a track, levels of track maintenance, track densification, and equipment maintenance. The probabilities for each situation will have to be developed and applied for each case. For the probabilities assumed in Fig. E7.22 the probability that a track defect will cause an accident is 6×10^{-3}.

EXAMPLE 7.23   (Rail Accident Sequence; Excerpted from Ang et al., 1979)

Consider the stopping of a train on a track because of the temporary malfunction of the train. The probability that this action will cause a collision with another train is affected by a sequence of events, such as the condition of the automated signal system, flagmen on Train #1, the presence of Train #2, the alertness of Train Crew #2, and the sight distance condition. The possible sequences of events (paths) are identified in the event tree of Fig. E7.23; only two branches are assumed for each subsequent event. Partial event failures are considered complete failures. Thus, should a flagman go only half the necessary distance needed to protect his train, this would be considered a failure to provide flag protection. As used here, good sight distance denotes the situation in which an alert crew member in Train #2 can see Train #1 and stops the train.

The probability for each event in a sequence will be different from rail line to rail line, and from train crew to train crew. For example, the probability of a following train is much greater on a commuter line than on a light branch line. Not every path will lead to an accident; according to Fig. E7.23, only five of the paths may lead to accidents. If the conditional probabilities shown in Fig. E7.22 are assumed, the probability that a collision will result is 0.0502.

Aside from the initiating event, each of the subsequent failure events in the various paths of an event tree may be decomposed also into basic events through appropriate fault tree diagrams, from which the probabilities of the respective (subsequent) top events may be systematically evaluated. The various paths in the event tree then provide the bases for evaluating the probabilities of the different potential consequences of the initiating event.

EXAMPLE 7.24   (Residential Home Fire)

In the event of a fire in a residential home equipped with a smoke detector, the potential consequences of the fire to an occupant may be analyzed using event trees and fault trees as shown in Fig. E7.24. The consequence of the fire will depend on whether the smoke detector operates during the fire; moreover, it will depend also on whether the occupant is able to escape.

If the detector sounded and the occupant manages to escape safely, he may suffer no injury or only minor injury. On the other hand, if the detector fails to sound the alarm, the occupant may still be able to escape, but the fire would most likely be at an advanced stage; in this event, the occupant is likely to suffer severe injury. If the occupant fails to escape, irrespective of the working condition of the smoke detector, the consequence would be death.

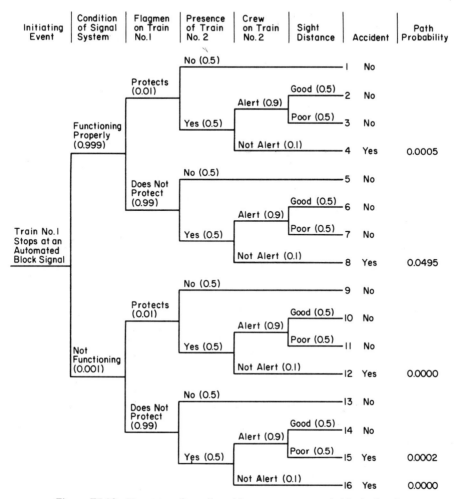

Initiating Event	Condition of Signal System	Flagmen on Train No. I	Presence of Train No. 2	Crew on Train No. 2	Sight Distance		Accident	Path Probability

Figure E7.23   Event tree for rail accident at an automatic block signal.

The faults or events E_1, E_2, and E_3 in the event tree of Fig. E7.24a may be analyzed further using fault trees as shown in Fig. E7.24b, where E_1, E_2, and E_3 are the respective top events. Suppose the basic events in the fault trees have the following probabilities:

$$P(A) = 0.1 \qquad\qquad P(F) = 0.2$$
$$P(B) = 0.02 \qquad\qquad P(G) = 0.5$$
$$P(C) = 0.5 \qquad\qquad P(H) = 0.4$$
$$P(D) = 0.1 \qquad\qquad P(J) = 0.5$$
$$P(E) = 0.5 \qquad\qquad P(K) = 0.3$$

Assume that the above events are mutually statistically independent. Then,

$$P(E_1) = 0.1 + 0.02 - 0.1 \times 0.02 = 0.118$$

$$P(E_2) = 0.5 \times (0.1 + 0.5 \times 0.2 - 0.1 \times 0.5 \times 0.2) = 0.095$$

$$P(E_3) = 0.5 \times (0.4 + 0.5 \times 0.3 - 0.4 \times 0.5 \times 0.3) = 0.245$$

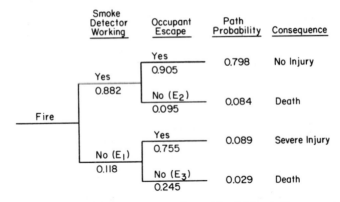

Figure E7.24a Event tree for residential home fire.

These probabilities are then used in the event tree of Fig. E7.24a. The probabilities of the respective consequences may then be evaluated through the corresponding paths of the event tree. For example, the probability that the occupant will be a fatality in a residential home fire would be

$$P(\text{death}|\text{fire}) = 0.882 \times 0.095 + 0.118 \times 0.245 = 0.113$$

whereas

$$P(\text{severe injury}|\text{fire}) = 0.118 \times 0.755 = 0.089$$

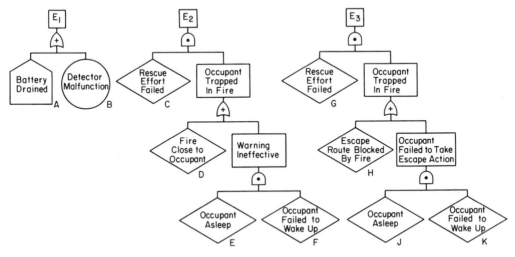

Figure E7.24b Fault trees for events E_1, E_2, and E_3.

7.5 APPROXIMATE METHODS

For systems with large numbers of potential failure modes, many of which may be partially correlated, the evaluation of the system reliability could become formidable; the bi-modal bounds of the failure probability, Eq. 7.16, may become

widely separated, particularly when the individual single-mode failure probabilities are not small, whereas Monte Carlo calculations may be prohibitively
expensive. Approximate methods may then be necessary, or may be the most
practically feasible approach. One crude approximation is the first-order lower
bound of the failure probability, that is, the left-hand side of Eq. 7.11; this, in fact,
is the basis of the method of PERT (MacCrimmon and Ryavec, 1964) that is
widely used for evaluating activity network problems. The same approximation
has been suggested also for structural system reliability (e.g., Cornell, 1971), in
which the reliability of a system is represented by the reliability of the weakest
mode. This approximation would be reasonable when there is a dominant failure
mode; in general, however, results obtained on this basis are invariably on the
unsafe or optimistic side.

To obtain a better point estimate of the failure probability, the other modes of
failure must obviously be included; moreover, the mutual correlations between
the failure modes would generally be significant and must also be considered.
The method of PNET, which stands for *probabilistic network evaluation technique*
(Ang, Abdelnour, and Chaker, 1975), is an approximate method developed on
these bases. Other approximate methods, primarily for structural systems, have
been developed; for example, Stevenson and Moses (1970), Vanmarcke (1971),
and Grimmelt and Schuëller (1982).

7.5.1 The PNET Method

From the first-order uni-modal bounds of the system failure probability, Eq. 7.11,
we recall that the lower bound (for positively correlated failure events)

$$P(E) = \max_i P(E_i)$$

is based on the assumption that the failure events $E_1, E_2, \ldots, E_n$ are perfectly
correlated, whereas the upper bound

$$P(E) = 1 - \prod_i P(\bar{E}_i)$$

is based on the assumption that the failure events are statistically independent.

An approximation, therefore, can be developed based on the premise that those
failure events that are highly correlated, with $\rho_{ij} \geq \rho_o$, may be assumed to be
perfectly correlated, whereas failure events with low mutual correlations, $\rho_{ij} < \rho_o$,
may be assumed to be statistically independent, in which ρ_o is a prescribed *demarcating correlation* (see below). In particular, any subgroup of the failure events,
say $E_1, E_2, \ldots, E_k$, that are highly correlated with E_r (i.e., $\rho_{rj} \geq \rho_o; j = 1, 2, \ldots, k$)
may be "represented" by the event E_r. That is, the failure probability of the entire
subgroup is represented by the probability of the single "representative" event,
E_r, or

$$P(E_1 \cup E_2 \cup \cdots \cup E_k) \simeq P(E_r)$$

in which, $P(E_r) = \max_j P(E_j; j = 1, 2, \ldots, k)$. Moreover, among the different
"representative" events E_r, the mutual correlations will be low, that is, for two
representative events E_q and E_r, $\rho_{q,r} < \rho_o$, and thus may be assumed to

be statistically independent. Hence, the failure probability of the system may be approximated by

$$P(E) \simeq 1 - \prod_{\text{all } r} P(\bar{E}_r) \tag{7.24}$$

The PNET method may be implemented as follows:

(1) Prescribe a value of ρ_o.
(2) Arrange the failure events in decreasing order of the individual single-mode failure probabilities; for example, $E_1, E_2, \ldots, E_n$. The event E_1 is automatically a "representative" failure event.
(3) Evaluate the correlation coefficients $\rho_{12}, \rho_{13}, \ldots, \rho_{1n}$; those events with $\rho_{1j} \geq \rho_o$ are "represented" by E_1.
(4) The events with $\rho_{1j} < \rho_o$ are again rearranged in decreasing order of the single-mode failure probabilities; suppose these are $E_2', E_3', \ldots, E_k'$ and the mutual correlations are $\rho_{23}', \rho_{24}', \ldots, \rho_{2k}'$. E_2' is the next representative mode. Those events with $\rho_{2j}' \geq \rho_o$ are represented by E_2'; whereas for those with $\rho_{2j}' < \rho_o$, the procedure is continued to search for other representative modes.
(5) The "representative" failure events are therefore E_1, E_2', etc., and the system failure probability is obtained through Eq. 7.24.

The results of the PNET method will obviously depend on the value of the *demarcating correlation* ρ_o. The appropriate value for ρ_o appears to depend on the level of reliability (Ma and Ang, 1981). Experience has shown that for systems in which the failure probabilities of the individual modes are high, $P(E_i)$ of order 10^{-1}, a value of $\rho_o = 0.5$ gives satisfactory results (Ang et al., 1975). On the other hand, for systems with small $P(E_i)$, for example, $P(E_i)$ of order 10^{-3}, a value of $\rho_o = 0.7$ is appropriate, and $\rho_o = 0.8$ is appropriate for systems with $P(E_i)$ of order 10^{-4} (Ma and Ang, 1981).

To illustrate the PNET method, consider the following examples.

EXAMPLE 7.25   (Beam-Cable System)

Consider the simple beam-cable system shown in Fig. E7.25a, which is subjected to a uniformly distributed load w with $\bar{w} = 2$ kip/ft and $\Omega_w = 0.20$. The areas of the cables are 1.0 in.^{2} and 0.5 in.^{2} as shown in Fig. E7.25a and the mean yield strength of the cables is $\bar{f}_y = 60$ ksi with

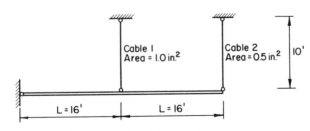

Figure E7.25a   A beam cable system.

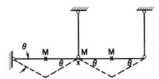

Figure E7.25b Mechanism 1.

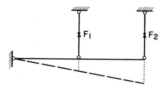

Figure E7.25c Mechanism 2.

$\Omega_{f_y} = 0.10$; whereas the mean fully plastic moment capacity of the prismatic beam is $\overline{M} = 100$ ft-kip with $\Omega_M = 0.15$.
Assume the following:

(i) The cables and beam have elastic-prefectly plastic behavior, with unlimited ductility.

(ii) The yield capacities of cables 1 and 2 are statistically independent and are also statistically independent of the beam capacity, whereas the capacities at all sections along the beam are perfectly correlated. The load w is statistically independent of the cable and beam capacities.

In light of the above assumptions, the failure of the beam-cable system will occur through the formation of plastic yield mechanisms; that is, as a cable yield or as a plastic hinge is formed in the beam, it continues to sustain its yield capacity. Moreover, plastic hinges in the beam will occur only at points of high applied bending moments, which may be assumed to be at the location of cable 1 and at mid-spans of the beam.

The plastic yield mechanisms of the system are shown in Figs. E7.25b through E7.25e. The respective performance functions (which are linear) can be formulated through the principle

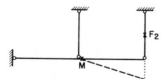

Figure E7.25d Mechanism 3.

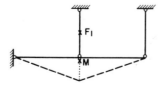

Figure E7.25e Mechanism 4.

of virtual work (e.g., Gaylord and Gaylord, 1972), from which the corresponding failure probabilities are obtained.

With a virtual rotation, θ, the total virtual work yields for mechanism 1 shown in Fig. E7.25b

$$g_1(\mathbf{X}) = 6M\theta - 1/2\left(2Lw \times \frac{L}{2}\theta\right)$$

or

$$g_1(\mathbf{X}) = 6M - \frac{L^2}{2}w$$

Thus,

$$\beta_1 = \frac{6 \times 100 - 1/2 \times 16^2 \times 2}{\sqrt{(6 \times 15)^2 + (1/2 \times 16^2 \times 0.4)^2}} = 3.32$$

$$p_{F_1} = \Phi(-3.32) = 4.5 \times 10^{-4}$$

Similarly, for mechanisms 2 through 4 (see Figs. E7.23c through E7.23e),

$$g_2(\mathbf{X}) = F_1L + 2F_2L - 2wL^2$$

$$\beta_2 = \frac{60 \times 16 + 2 \times 30 \times 16 - 2 \times 2 \times 16^2}{\sqrt{(6 \times 16)^2 + (2 \times 3 \times 16)^2 + (2 \times 16^2 \times 0.4)^2}} = 3.65$$

$$p_{F_2} = \Phi(-3.65) = 1.33 \times 10^{-4}$$

$$g_3(\mathbf{X}) = M + F_2L - 1/2wL^2$$

$$\beta_3 = \frac{100 + 30 \times 16 - 1/2 \times 2 \times 16^2}{\sqrt{(15)^2 + (3 \times 16)^2 + (1/2 \times 16^2 \times 0.4)^2}} = 4.51$$

$$p_{F_3} = \Phi(-4.51) = 3.25 \times 10^{-6}$$

$$g_4(\mathbf{X}) = 2M + F_1L - wL^2$$

$$\beta_4 = \frac{2 \times 100 + 60 \times 16 - 2 \times 16^2}{\sqrt{(2 \times 15)^2 + (6 \times 16)^2 + (16^2 \times 0.4)^2}} = 4.51$$

$$p_{F_4} = \Phi(-4.51) = 3.25 \times 10^{-6}$$

The above mechanisms have already been rearranged in decreasing order of p_{F_i}. Prescribing $\rho_0 = 0.8$, we determine the "representative" mechanisms by first evaluating the correlations between the mechanisms starting with the first one. As the performance functions are linear, we obtain the correlation coefficients, taking into account the statistical independence between w, M, F_1, and F_2, as follows:

$$\rho_{12} = \frac{L^4\sigma_w^2}{\sigma_{g_1} \cdot \sigma_{g_2}} = \frac{(16)^4 \times (0.4)^2}{\sqrt{(6\times15)^2 + (\frac{1}{2}\times16^2\times0.4)^2} \cdot \sqrt{(6\times16)^2 + (2\times3\times16)^2 + (2\times16^2\times0.4)^2}}$$

$$= 0.412$$

$$\rho_{13} = \frac{6\sigma_M^2 + \frac{1}{4}L^4\sigma_w^2}{\sigma_{g_1} \cdot \sigma_{g_3}}$$

$$= \frac{6 \times 15^2 + \frac{1}{4} \times 16^4 \times (0.4)^2}{\sqrt{(6\times15)^2 + (\frac{1}{2}\times16^2\times0.4)^2}\sqrt{(15)^2 + (3\times16)^2 + (\frac{1}{2}\times16^2\times0.4)^2}}$$

$$= 0.534$$

$$\rho_{14} = \frac{12\sigma_M^2 + \frac{1}{2}L^4\sigma_w^2}{\sigma_{g_1} \cdot \sigma_{g_4}}$$

$$= \frac{12 \times 15^2 + \frac{1}{2} \times 16^4 \times (0.4)^2}{\sqrt{(6 \times 15)^2 + (\frac{1}{2} \times 16^2 \times 0.4)^2}\sqrt{(2 \times 15)^2 + (6 \times 16)^2 + (16^2 \times 0.4)^2}}$$

$$= 0.534$$

Therefore, with $\rho_o = 0.8$, none of the mechanisms 2, 3, and 4 are represented by mechanism 1. Continuing, we obtain

$$\rho_{23} = 0.856$$

$$\rho_{24} = 0.856$$

Therefore, mechanisms 3 and 4 are represented by mechanism 2, and the "representative" mechanisms are 1 and 2. Accordingly, by the PNET method, the failure probability of the system is approximately

$$p_F \simeq p_{F_1} + p_{F_2}$$
$$= 4.5 \times 10^{-4} + 1.33 \times 10^{-4}$$
$$= 5.83 \times 10^{-4}$$

We may observe that the uni-modal bounds for this problem are

$$4.5 \times 10^{-4} \le p_F \le 5.895 \times 10^{-4}$$

whereas the corresponding bi-modal bounds can be shown to be

$$5.822 \times 10^{-4} \le p_F \le 5.855 \times 10^{-4}$$

EXAMPLE 7.26 *(Activity Network)*

Activity networks are often used in the planning and scheduling of projects, including construction, and research and development projects. In the case of activity networks, the branches between node points represent specific activities. Clearly, there may be certain activities that must be completed before subsequent ones can be started; hence, in an activity network, certain precedence logic must be observed.

The *critical path method* (CPM) is often used to plan and schedule construction projects. This is based on the assumption that the durations of the individual activities in a network are deterministic (i.e., no uncertainty), on the basis of which the completion time of the project is determined by the "critical path" which is the activity path with the longest duration.

However, as there are invariably uncertainties in the estimation of the durations of the individual activities, the project completion time may be evaluated only with an associated probability. In this regard, the probability determined solely on the basis of the critical path is equivalent to the lower bound failure probability of Eq. 7.11, which is also the basis of the method of PERT (McCrimmon and Ryavec, 1964).

For purposes of illustration, consider a construction project involving the paving of 2.2 miles of roadway pavement and the construction of appurtenant drainage structures, excavation to grade, placement of the macadam shoulders, erection of guardrails, and landscaping. The plan and profile of the project are shown in Fig. E7.26a; further details of the problem can be found in Brooks et al. (1967).

The various activities of the project are identified and described in Table E7.26a, where the respective mean durations and standard deviations are also indicated. The project activity network is shown in Fig. E7.26b. The various possible paths of the network (a total of nine) from start to completion of the project are listed in Table E7.26b, in decreasing order of mean durations.

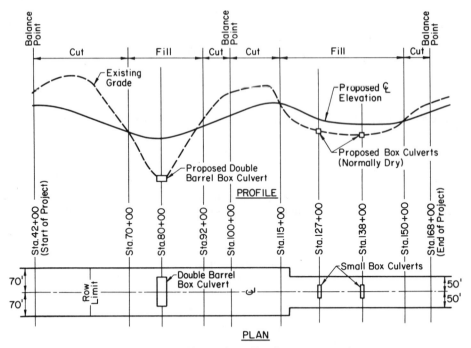

Figure E7.26a   Plan and profile of pavement project.

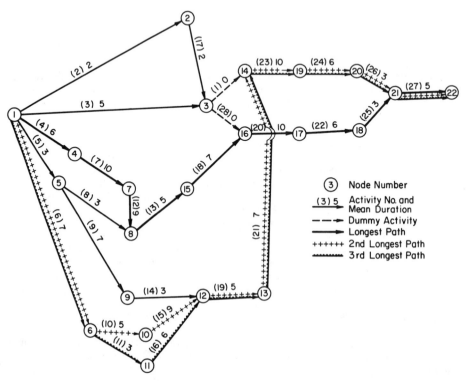

Figure E7.26b   Network of pavement project.

Table E7.26a   Activities and Estimated Durations (Pavement Project)

Activity no.	Description of Activities	Mean Duration (days)	Std. Dev. (days)
1	Dummy	0	0
2	Set-up batch plant	2	0.5
3	Order and deliver paving mesh	5	1.0
4	Deliver rebars for double barrel culvert	6	1.5
5	Move in equipment	3	0.5
6	Deliver rebars for small box culvert	7	4.0
7	Build double barrel culvert	10	2.0
8	Clear and grub from sta. 42 to sta. 100	3	1.0
9	Clear and grub from sta. 100 to sta. 158	7	1.5
10	Build box culvert at sta. 127	5	2.0
11	Build box culvert at sta. 138	3	1.5
12	Cure double barrel culvert	9	2.0
13	Move dirt between sta. 42 and sta. 100	5	1.5
14	Start moving dirt between sta. 100 and sta. 158	3	0.5
15	Cure box culvert at sta. 127	9	4.5
16	Cure box culvert at sta. 138	6	2.0
17	Order and stockpile paving material	2	0.5
18	Place subbase from sta. 42 to sta. 100	7	1.73
19	Finish moving dirt between sta. 100 and sta. 158	5	2.0
20	Pave from sta. 42 to sta. 100	10	2.0
21	Place subbase from sta. 100 to sta. 158	7	3.31
22	Cure pavement from sta. 42 to sta. 100	6	1.5
23	Pave from sta. 100 to sta. 158	10	4.5
24	Cure pavement from sta. 100 to sta. 158	6	1.5
25	Place shoulders from sta. 42 to sta. 100	3	1.0
26	Place shoulders from sta. 100 to sta. 158	3	1.0
27	Place guardrail and landscape	5	1.5
28	Dummy	0	0

Table E7.26b   Ordered Paths and Duration Statistics (Pavement Project)

Path no.	Activities in Path	Mean Duration (days)	Std. Dev. (days)
1	4, 7, 12, 13, 18, 20, 22, 25, 27...	61	5.0
2	6, 10, 15, 19, 21, 23, 24, 26, 27...	57	9.0
3	6, 11, 16, 19, 21, 23, 24, 26, 27...	52	7.94
4	5, 9, 14, 19, 21, 23, 24, 26, 27...	49	6.54
5	5, 8, 13, 18, 20, 22, 25, 27...	42	4.0
6	3, 28, 20, 22, 25, 27...	29	3.24
7	3, 1, 23, 24, 26, 27...	29	5.19
8	2, 17, 28, 20, 22, 25, 27...	28	3.16
9	2, 17, 1, 23, 24, 26, 27...	28	5.12

In the present case, the total project duration along a given path is simply the sum of the activity durations in that path or

$$T_i = \sum_{j=1}^{n_i} t_{ij}$$

where n_i = the number of activities in path i; and t_{ij} = the duration of activity j. It follows then that the mean project duration along path i is

$$\overline{T}_i = \sum_{j=1}^{n_i} \overline{t}_{ij}$$

and if we assume that the individual activities are uncorrelated, the variance of T_i would be

$$\sigma_{T_i}^2 = \sum_{j=1}^{n_i} \sigma_{t_{ij}}^2$$

The correlation coefficient between T_i and T_j is

$$\rho_{ij} = \frac{\sum_k \sigma_{t_{ik}}^2}{\sigma_{T_i}\sigma_{T_j}}$$

where k includes all the activities common to T_i and T_j.

In Table E7.26b, the paths are arranged in order of decreasing mean path durations. It may be obvious on the basis of the mean project durations that the first five paths are significant and only these need to be considered in the evaluation of the relevant probability. For these first five paths, we obtain the correlation coefficients as follows:

$$\rho_{12} = 0.05$$

$$\rho_{13} = 0.06$$

$$\rho_{14} = 0.07$$

$$\rho_{15} = 0.74$$

According to Ang et al (1975), a demarcating correlation of $\rho_o = 0.5$ is appropriate for construction networks. On this basis, path #5 is therefore represented by path #1. Continuing, we obtain

$$\rho_{23} = 0.79$$

$$\rho_{24} = 0.69$$

which mean that the paths #3 and #4 are represented by path #2. Therefore, the "representative" paths for this problem are paths #1 and #2. Then, according to Eq. 7.24, the probability of the project being completed within a target time t is

$$P(T \le t) = P(T_1 \le t)P(T_2 \le t)$$

and if the distributions of t_{ij} are individually Gaussian, the CDF of the completion time becomes

$$P(T \le t) = \Phi\left(\frac{t - \overline{T}_1}{\sigma_{T_1}}\right)\Phi\left(\frac{t - \overline{T}_2}{\sigma_{T_2}}\right)$$

Conversely, the probability of not completing the project within t is

$$P(T > t) = 1 - P(T \le t)$$

The completion time function, $P(T \le t)$, obtained with PNET as well as with PERT are shown in Fig. E7.26c for the entire range of probabilities of interest. Results obtained with Monte

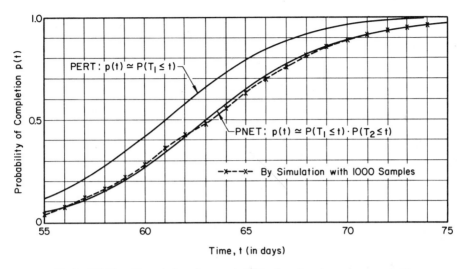

Figure E7.26c Completion time probability function (pavement project).

Carlo calculations (using a sample size of 1000) are also shown for comparison. From Fig. E7.26c, it can be seen that the results obtained with the PERT method consistently err on the "optimistic" side.

Suppose the scheduled completion time of the construction project is 68 days, and the contract stipulates that the penalty (in dollars) because of any delay is

$$D = 20,000 + 2000t^2; \quad t > 0$$

where t is the length of delay (in days). Observing (from Fig. E7.26c) that the probability of delay is about 18%, the contractor may contemplate crashing (i.e., speeding up) some of the critical activities. Two options are available, namely:

Crash Program 1

Shorten the duration of the paving operation in activity 20 by renting more equipment and hiring additional personnel. Suppose this will shorten the mean duration of activity 20 to eight days and revise its standard deviation to 1.5 days; the additional cost involved will be $1500.

Crash Program 2

Apply crashing to the paving operation of the entire roadway, that is, both activities 20 and 23 will be shortened to have mean durations of eight days and standard deviations of 1.5 and 3 days, respectively; the additional cost incurred will be $2500.

PNET analyses for the modified networks corresponding to the two proposed crash programs yield the corresponding CDFs of the project completion time as shown in Fig. E7.26d. A decision tree is drawn in Fig. E7.26e for the three alternatives. The expected loss (i.e., additional cost, including penalty for delay) for the alternative of "no crashing" is

$$E(\text{loss}|\text{no crash}) = 0.18 \times 20,000 + \int_{68}^{\infty} 2000(t - 68)^2 f_{T_1}(t)\, dt$$

$$= \$4695$$

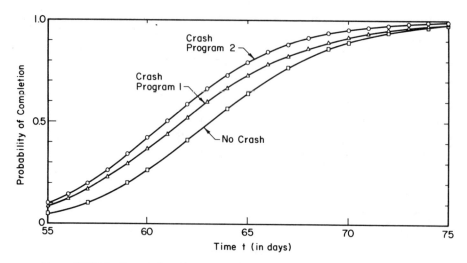

Figure E7.26d Completion time probability function for the three alternatives.

where $f_{T_1}(t)$ is the PDF of the project duration without crashing. Similarly, the expected loss of the other alternatives are shown in the following table:

Alternatives	Crash Cost	Expected Penalty	Expected Total Loss
No crashing	$0	$4695	$4695
Crash program 1	$1500	$3693	$5193
Crash program 2	$2500	$2009	$4509

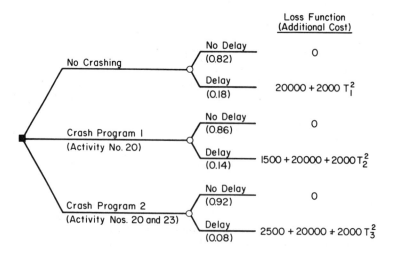

T_1, T_2 and T_3 Are the Delays (in days) Associated With Each of the Three Alternatives

Figure E7.26e Decision tree for crash options.

Based on the maximum expected monetary value criterion, the contractor should opt for a full-scale crash program to speed up the paving operations of the entire 2.2 miles of roadway. Observe that although activity 23 is not on the mean critical path (see Table E7.26b), the crashing of activity 23 significantly reduces the probability of delay in the project completion time (see Fig. E7.26d) and thus reduces the total loss.

EXAMPLE 7.27   (Flow Network)

Network problems may also involve the flow through a network. In such a case, the branches between the node points are characterized by their respective flow capacities; the flow capacity of the network is of interest. The flow through the network will depend on the specific "cut set" of the network (see examples in Fig. E7.27). If the flow capacities of the individual branches are deterministic, the flow capacity of the network will be represented by the capacity of the minimum cut set. However, if there is uncertainty in the capacities of the individual branches, the network capacity can only be evaluated in terms of probability; in this case, all the possible cut sets should be considered.

The flow capacity of a cut set is simply the sum of the branch capacities in that cut set; that is,

$$C_i = \sum_{j=1}^{n_i} c_{ij}$$

in which n_i = the number of branches in cut set i, and c_{ij} = the capacity of branch j. Since the capacity of a cut is a linear function (i.e. sum) of the branch capacities, the corresponding statistics, such as the means, variances, and mutual correlations, of the cut sets may be readily obtained. For example, if the individual branch capacities, c_{ij}, are uncorrelated, we obtain for cut set C_i

$$\bar{C}_i = \sum_j \bar{c}_{ij}$$

$$\sigma_{C_i}^2 = \sum_j \sigma_{c_{ij}}^2$$

and, for two cut-sets C_i and C_j

$$\rho_{ij} = \frac{\sum_k \sigma_{c_k}^2}{\sigma_{C_i} \sigma_{C_j}}$$

where k includes all the branches common to C_i and C_j.

For numerical illustration, consider the water distribution network between node points A and B as shown in Fig. E7.27a, consisting of seven pipes; the arrows show the direction of the flow. The mean capacities and corresponding c.o.v. of the branch pipes are summarized in Table E7.27a.

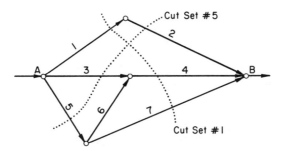

Figure E7.27a   Water distribution network.

Table E7.27a   Statistics of Pipe Capacities

Pipe No.	Mean Branch Capacity	c.o.v. of Branch Capacity
1	70 cfs	0.15
2	80	0.15
3	90	0.15
4	100	0.15
5	100	0.15
6	40	0.15
7	70	0.15

In this case, there are a total of eight possible cut sets as listed in Table E7.27b; for example, cut set #1 involves the capacities of pipes 1, 4, and 7, whereas cut set #5 contains pipes 2, 3, and 5.

Since the capacity of a cut set is a linear function of the branch capacities, the mean and standard deviation of a cut set as well as the corresponding failure probability may be readily obtained. The results are summarized in Table E7.27b in which the cut sets are also listed in order of decreasing p_{F_i}.

Suppose a minimum flow of 200 cfs is required at node point B; the probability of less than 200 cfs flow is therefore of interest. Prescribing $\rho_o = 0.70$, we determine the "representative" cut sets as follows:

$$\rho_{12} = 0.73$$

$$\rho_{13} = 0.23$$

$$\rho_{14} = 0.67$$

$$\rho_{15} = 0$$

$$\rho_{16} = 0.44$$

$$\rho_{17} = 0.50$$

$$\rho_{18} = 0.24$$

Table E7.27b   Statistics of Cut Sets

Ordered Cut Set No.	Branches in Cut Set	μ (cfs)	σ (cfs)	β_i	$p_{F_i} = P(C_i < 200)$
1	1, 4, 7	240	21.11	1.89	2.94×10^{-2}
2	2, 4, 7	250	21.89	2.28	1.13×10^{-2}
3	1, 3, 5	260	22.75	2.64	4.14×10^{-3}
4	1, 4, -6^a, 5	270	23.67	2.96	1.54×10^{-3}
5	2, 3, 5	270	23.48	2.98	1.44×10^{-3}
6	2, 4, 5, -6^a	280	24.37	3.28	5.19×10^{-4}
7	1, 3, 6, 7	270	20.95	3.34	4.19×10^{-4}
8	2, 3, 6, 7	280	21.74	3.68	1.17×10^{-4}

[a] In this cut set, -6 denotes the fact that the water in pipe 6 flows in an opposite direction, namely from B to A; hence its capacity is not included when calculating the total capacity of the cut set.

Therefore, only cut set #2 is represented by cut set #1. Continuing, we obtain

$$\rho_{34} = 0.62$$
$$\rho_{35} = 0.76$$
$$\rho_{36} = 0.41$$
$$\rho_{37} = 0.61$$
$$\rho_{38} = 0.37$$

which mean that cut set #5 is represented by cut set #3. Furthermore,

$$\rho_{46} = 0.78$$
$$\rho_{47} = 0$$
$$\rho_{48} = 0$$

and

$$\rho_{78} = 0.72$$

Therefore, cut set #6 is represented by #4, and #8 is represented by #7. Hence, the "representative" cut sets are #'s 1, 3, 4, and 7. With the PNET method, we obtain the probability that the network flow will be less than 200 cfs as

$$P(C < 200) = 2.94 \times 10^{-2} + 4.14 \times 10^{-3} + 1.54 \times 10^{-3} + 4.19 \times 10^{-4}$$
$$= 3.55 \times 10^{-2}$$

In this case, we observe that the uni-modal bounds are

$$2.94 \times 10^{-2} \le P(C < 200) \le 4.82 \times 10^{-2}$$

PROBLEMS

7.1 The splicing of large reinforcing bars are often done by using sleeves, as shown in Fig. P7.1.

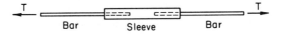

T   Bar   Sleeve   Bar   T

Figure P7.1 Spliced reinforcing bar.

(a) Suppose each bar has a mean yield strength of 60 ksi with a 15% coefficient of variation, and the mean capacity of the sleeve is 140 kips with a coefficient of variation of 20%. The applied tension has a mean value of 60 kips with a standard deviation of 15 kips. If the net bar area is 2.0 in.², determine the first and second order bounds of the reliability of the entire bar system. Assume normal distributions and statistically independent resistances.

(b) For design purposes, some low value is usually designated as the nominal design value. Statistically, this may be specified as the p-percentile value. In the above bar system, assume that the strength of each bar has a Rayleigh distribution with a mode of 100 kips; the capacity of the sleeve is also Rayleigh with a modal value of 120 kips. Determine the 10-percentile value of the bar system. Assume statistical independence.

7.2   The cable in the pulley system shown in Fig. P7.2 consists of a 2-wire strand. Each wire has a mean rupture strength of 15 kips and a coefficient of variation of 10%; assume normal distribution. The system is used to lift weights that are $N(30$ kips, 6 kips$)$. What is the probability of failure of the pulley system (through rupture of the cable)?

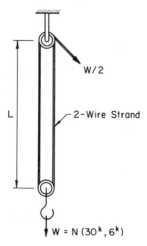

Figure 7.2   Pulley system.

Assume the two wires to be statistically independent; also, between nonoverlapping lengths L of the same wire the strengths are also statistically independent, but within any length L the strength is perfectly correlated.

7.3   For a pair of adjacent column footings in a building, suppose the settlement of each footing has a mean of 5 cm and a c.o.v. of 30%. Because of similar loading and soil conditions associated with a pair of footings, the settlements are correlated with $\rho = 0.7$. Unsatisfactory performance will result if the differential settlement between the pair of footings exceeds 2 cm.

(a)   Determine the probability of excessive differential settlement if the settlement of the individual footing follows a normal distribution.

(b)   Repeat part (a) if the footing settlement follows a lognormal distribution.

7.4   Suppose that levees built for flood protection along a river could fail in the following two modes (see Fig. P7.4):

(i)   Overtopping failure; that is, when the level of flood water exceeds the levees elevation.

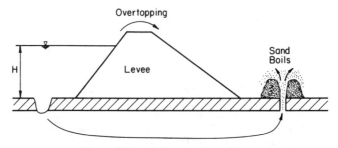

Figure P7.4   Failure modes of levee.

    (ii)   Sand boiling failure; that is, when the level of flood water creates a sufficient head to cause an uplift of the sand on the dry side of the levee.

    In a given year, let H be the maximum flood level. Suppose H_1 is the elevation of the levee and H_2 is the allowable flood elevation before a sand boil occurs. For the following statistical information on the variables, determine the probability of failure of the levee in a given year.

Variable	Mean	c.o.v.	Distribution
H	18	0.20	lognormal
H_1	25	0.02	normal
H_2	26	0.10	normal

7.5   A braced excavation system is shown in Fig. P7.5. The horizontal distance between struts is 18 feet. The bracing system consists of struts at various levels in addition to wales, sheetings, connections, etc. Suppose failure of the struts is of concern in the failure of the excavation system. The strut could fail either by yielding or by buckling.

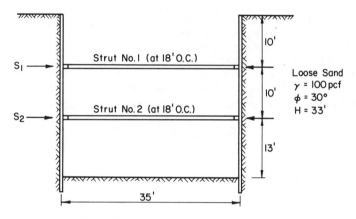

Figure P7.5   Braced excavation supports.

    The load applied to each strut from the earth pressure may be calculated from

$$S_i = m_i \lambda K_A \gamma H; \qquad i = 1, 2$$

where H is the depth of excavation, γ is the average density of the sand, K_A is the active pressure coefficient, λ is the coefficient dependent on the earth pressure distribution, and m_i is the load coefficient obtained from stress analysis. The yield strength of each strut may be calculated from

$$R_i^Y = N_y \sigma_y A_i; \qquad i = 1, 2$$

where σ_y is the yield stress of steel, A_i is the cross-sectional area of strut i, and N_y is the correction for error associated with estimating the yield strength.

    The buckling capacity of each strut may be calculated from

$$R_i^b = N_b \frac{\pi^2 E I_i}{(KL_i)^2}; \qquad i = 1, 2$$

where E is the modulus of elasticity, I is the moment of inertia, KL_i is the effective length of strut i, and N_2 the correction for errors associated with the buckling formula.

Suppose 12 WF 120 and 12 WF 96 sections have been selected for struts 1 and 2, respectively. Assume the following for the pertinent variables:

Variable	Mean	c.o.v.
m_1	270 ft^2	0
m_2	207 ft^2	0
λ	0.65	0.21
K_A	0.33	0.039
γ	100 pcf	0
H	33 ft	0
N_y	1.0	0.1
σ_y	36 ksi	0.1
E	30,000 ksi	0
K	1.0	0.1
N_b	1.0	0.1

Determine:
(a) The probability of failure of strut 1 (with two potential failure modes).
(b) First- and second-order bounds on the probability of failure of one section of the braced excavation system as shown in Fig. P7.5. Assume σ_y and KL_i to be statistically independent between struts.
(c) If there is a sequence of ten similar sections in an excavation project, estimate the probability of system failure. Assume that failures between adjacent sections are correlated with $\rho = 0.5$, whereas the correlation between nonadjacent sections may be assumed to be negligible (i.e., $\rho \simeq 0$).

7.6   The cables in the system shown in Fig. P7.6 are identically distributed with the following properties:

$$\text{Fracture strength:} \quad \mu_{F_i} = 50 \text{ ksi}$$
$$\Omega_{F_i} = 0.15$$
$$\text{Area:} \quad A = 0.80 \text{ in.}^2$$
$$\text{Modulus of elasticity:} \quad E = 30,000 \text{ ksi}$$

The load W is as follows:

$$\mu_W = 45 \text{ kips}$$
$$\Omega_W = 0.25$$

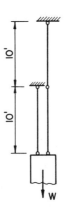

Figure P7.6   A parallel cable system.

Note that the load W is not equally distributed between the cables of the original system.

Evaluate the collapse probability of the system.

7.7 The structure consisting of the continuous beam ABC and column BD, shown in Fig. P7.7, is subjected to a uniformly distributed load w, as follows:

$$\mu_w = 5 \text{ kip/ft}$$

$$\Omega_w = 0.25$$

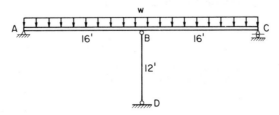

Figure P7.7   Beam and column system.

The fully plastic capacities of the beam and column are, respectively, as follows:

$$\mu_{M_Y} = 1000 \text{ ft-kip}; \qquad \Omega_{M_y} = 0.15$$

$$\mu_{R_Y} = 35 \text{ kip}; \qquad \Omega_{R_y} = 0.20$$

Assume that the capacities along the beam are perfectly correlated, whereas those of the beam and the column are statistically independent. Evaluate the probability of failure of the structural system, through plastic collapse mechanisms.

7.8 The beam and cable system shown in Fig. P7.8 is subjected to a concentrated load P with

$$\mu_p = 80 \text{ kips}$$

$$\Omega_p = 30\%$$

The fully plastic capacities of the components are as follows:

$$\text{Beam yield moment: } \mu_{M_y} = 6000 \text{ in.-kips}$$

$$\Omega_{M_y} = 20\%$$

$$\text{Cable yield strength: } \mu_{f_y} = 44 \text{ ksi}$$

$$\Omega_{f_y} = 10\%$$

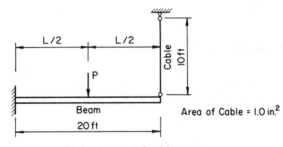

Figure P7.8   Beam and cable system.

Assume the following:
(i)   The capacities or strengths along the entire length of a member are identically distributed and perfectly correlated.
(ii) The capacities of the beam and cable are statistically independent.
(a)   Determine the first-order bounds for the probability of collapse of the structure through plastic yielding.
(b)   Determine the corresponding second-order bounds.

7.9   The structural system shown in Fig. P7.9, consisting of a beam *AB* and a truss *BCD*, is subjected to a uniformly distributed vertical load *w*. Assume that the beam *AB* will fail through plastic yield mechanism, whereas the truss members will fail through

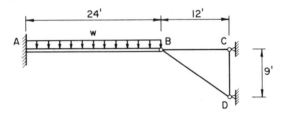

Figure P7.9   Structural system.

fracture or buckling. The following statistics are given for the load and resistance variables:

Variable	Mean	c.o.v.
w	2 kips/ft	0.20
M	2000 in.-kips	0.12
F	48 kips	0.15
B	60 kips	0.15

where *w* is the uniformly distributed load, *M* is the bending moment capacity of the beam *AB*, *F* is the fracture strength of member *BC*, and *B* is the buckling capacity of member *BD*.

Assume (i) the moment capacities along the length of the beam are perfectly correlated, and (ii) failures of the truss members are statistically independent. The beam and truss members are also statistically independent.

Evaluate the first- and second-order bounds of the probability of collapse of the system,

7.10   The frame and pulley system shown in Fig. P7.10*a* is used for lifting heavy weights *W*. Member *BC* is a single W12 × 26, which has area $A = 7.65$ in.^{2}, and radius of gyration about the weak axis $r = 1.51$ in.

Member *ACD* is composed of four channels as shown in Fig. P7.10*b*.

Each of the channels is A36 steel MC 12 × 50 section with section modulus $S = 44.9$ in.^{3}.

Assume the following material properties for the above rolled section members:

	Mean	c.o.v.
Yield strength	44 ksi	0.10
Young's modulus	29,000 ksi	0.05
Radius of gyration	—	0.05
Area	—	0.05

The pulley system is as shown in Fig. P7.10*c*.

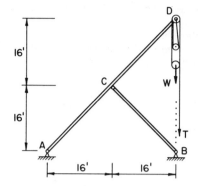

Figure P7.10a Frame and pulley system.

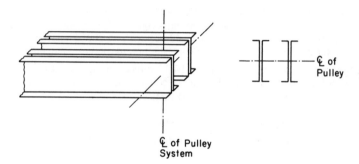

Figure P7.10b Channel system for member *ACD*.

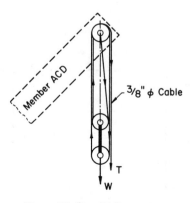

Figure P7.10c Pulley system.

The high strength cables has a 3/8-in. diameter with a mean fracture strength of 80 ksi and a c.o.v. of 0.10.

Assume also the following:

(i) The critical failure mode for member *BC* is Euler buckling; that is, the critical buckling stress may be determined by

$$\sigma_{cr} = \frac{\pi^2 E}{\left(k\dfrac{L}{r}\right)^2}$$

where

$$\bar{k} = 1.0$$

$$\Omega_k = 0.12$$

(ii) The only failure mode of concern for member *ACD* is flexural yielding. Failure of one side of the member requires yielding of both channels on the same side. However, yielding of the channels on one side constitutes failure of the member.

(iii) The resistances of the member elements including the cable are statistically independent, but the resistance along a member are perfectly correlated. The resistance among the four channels of *ACD* are also statistically independent.

(a) Evaluate the failure probabilities of the individual failure modes of the frame pulley system if the mean maximum weight is $\bar{W} = 20$ kips with a c.o.v. of 0.20. Assume that all the connections and pulleys have 100% reliability.

(b) Determine the corresponding first-order bounds of the system failure probability.

7.11 For the frame structure shown in Fig. P7.11 subject to the given loads, determine the second-order bounds of the probability of collapse (through plastic yield mechanisms) of the frame system with the following statistics:

Variable	Mean	c.o.v.
W	100 kips	0
K	0.3	0.1
M_1	300 ft-kips	0.15
M_2	450 ft-kips	0.10

where M_1 and M_2 are the bending moment capacities of the columns and beam, respectively. Assume that M_1 is the same for both columns whereas M_1 and M_2 are statistically independent. The properties along a member are perfectly correlated, and the connections have 100% reliability.

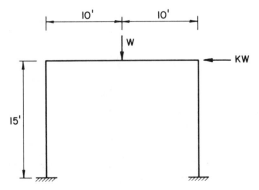

Figure P7.11   Frame system.

7.12   A building is supported on four identical first-story columns, as shown in Fig. P7.12. The shear and bending capacities of each column are

Shear: $\bar{V} = 60$ kips; $\Omega_V = 20\%$.

Bending: $\bar{M} = 400$ ft-kips; $\Omega_M = 15\%$.

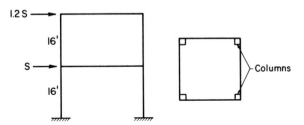

Figure P7.12   Building system.

The safety of the building to earthquakes is to be examined. Suppose earthquake effects can be represented by equivalent lateral forces S applied at the floor levels, with $\bar{S} = 60$ kips and c.o.v. = 25%. Assume that all the variates are normal.
  (a)   Determine the probability of horizontal shear failure of the building. Assume the following: the failure of any two of the four columns is tantamount to the failure of the building; failures of the columns are statistically independent; and once failed in shear, a column can no longer carry any shear force.
  (b)   Determine the probability of failure of the building by plastic yield mechanism. For simplicity, consider that plastic hinges can form in the first-story columns only; also, the fully plastic moments between columns are statistically independent, but are perfectly correlated within a column.
  (c)   What is the probability of failure of the building, assuming the above are the only failure modes?

7.13   A passenger car is used for a 1000-mile trip. Suppose the remaining life in each of the tires is lognormally distributed with a mean of 2000 miles and a c.o.v. of 30%. There is a spare tire in the trunk that is limited to 200 miles of travel. Assume the lives between tires to be statistically independent, and no other tire services are available.
  (a)   Determine the probability that the car will complete the trip without problems.
  (b)   Repeat part (a) if the car is not equipped with a spare tire.
  (c)   Repeat part (a) if the car is equipped with two spare tires.

7.14   A system of bridges connecting three cities A, B, and C is shown in Fig. P7.14. Each of the five bridges can handle 1000 vehicles per hour each way. Normally, on a given day, the peak volume of traffic using each of the bridges follows a normal distribution as follows:

Bridge	Direction	Mean Peak Traffic (vph)
1	A–B	800
2	A–B	800
3	A–B	800
4	B–C	700
5	B–C	700

Assume the c.o.v. of traffic on each bridge is 25% and the traffic on the bridges are statistically independent. A "jamming" condition will develop in a bridge if its capacity is exceeded by the traffic, or if an accident occurs on the bridge. The probability of an

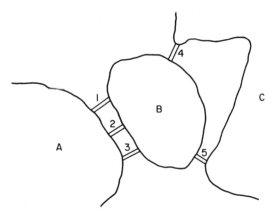

Figure P7.14   Bridge network connecting cities A, B, and C.

accident on a given bridge is 0.02 during peak traffic. If one of the bridges is jammed, the traffic in the other bridges between two cities will be increased 10%; however, if two bridges connecting two cities are jammed, the traffic on the remaining bridge will be increased by 30%. Suppose the flow of traffic from A to C will be seriously impaired if at least two of the bridges 1, 2, and 3 or at least one of the bridges 4 and 5 are "jammed."

    (a)   Determine the probability of a serious traffic condition from City A to City C.

    (b)   Determine the probability that all bridges will be jammed.

7.15 Suppose 1% of the wooden ties along a railroad are defective. Assume defective ties are randomly located along the railroad.

    (a)   If a group of five ties in sequence is selected at random, what is the probability that they will all be defective?

    (b)   A hazardous condition will exist if there is a sequence of at least five consecutive defective ties. For a 10-mile stretch of railroad, estimate the probability that such a hazardous condition exists. Suppose the ties are 2 ft apart.

    (c)   Repeat parts (a) and (b) if the defective ties are not randomly located; that is, there is a 10% probability that a defective tie is followed by another defective tie.

7.16 In Example 7.15, for the following probabilities,

Component	P(Failure)
Each tank	0.05
Each electric pump	0.02
Each generator	0.01
Turbine pump	0.005

determine the probability of an auxiliary feed water system failure in the event of an earthquake. Assume that failures between any two components are statistically independent.

7.17 In a risk analysis of nuclear power plant structures (Collins and Hudson, 1981) subject to earthquake hazards, the fault and event trees in Fig. P7.17 were suggested to determine the probabilities of various levels of radioactive releases.

    Suppose the probability of failure of each component i, $i = 1$ to 8, is equal to 0.02. Determine the probabilities associated with the three levels of release for each of the following assumptions:

    (a)   All component failures are statistically independent.

    (b)   The performance functions of each component pair (i.e., components 3 and 4, 5 and 6, 7 and 8) are correlated with $\rho = 0.6$. There is no other correlation.

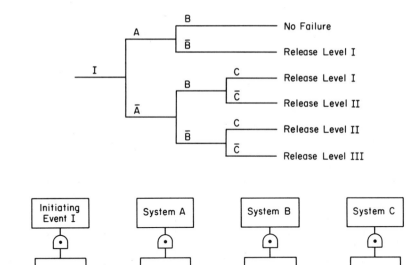

Figure P7.17 Event and fault trees.

7.18 The event of a bad crop yield in a year (event E) could be caused by drought in the growing season (event A) or by very wet weather during the harvest season (event D). A wet condition may result from the saturation of the ground (event B) or from a long period of rainfall during the harvest season (event C). Since the basic events B and C are both due to an unusually wet year, they are expected to be positively dependent. On the other hand, events A and D would be negatively dependent because wet and dry spells usually extend through a period of a year or longer. Suppose the following annual probabilities are given

$$P(A) = 0.1$$
$$P(B) = 0.2$$
$$P(C) = 0.2$$

(a) Draw the fault tree for the top event of a bad crop yield.

(b) Determine the bounds on the probability of bad crop yield in a given year.

7.19 Develop the fault tree for the event of unsatisfactory reservoir performance (event E) in Example 7.1

7.20 The major modes of an earthdam failure are (i) overtopping, (ii) the instability of the dam structure, (iii) an excessive deformation of the dam proper, (iv) an excessive loss of water from the reservoir, and (v) the physical destruction of the dam because of either sabotage or act of war.

Overtopping could be caused by a large flood whose rate of inflow exceeds the spill-way capacity. Sometimes, human and/or mechanical errors in operating the control gates could reduce the spillway capacity to handle a large inflow. Earthquakes could also cause the structural damage of the spillway and outlet works such that the design spillway capacity is significantly reduced. When the reservoir is full, a strong wind could generate waves that may topple the crest of the dam; overtopping can also occur during earthquakes in a full reservoir through seiches caused by landslides of the surrounding slopes or the subsidence of the dam structure.

Under normal operating conditions, the dam structure may become unstable from the hydrostatic pressure of the stored water; as a result, horizontal sliding or rotational sliding at the down stream slope may occur. Furthermore, during a rapid drawdown of the reservoir level, rotational failure at the upstream slope may also occur. During an earthquake, the strong horizontal acceleration could cause sliding failure of the dam; moreover, if a stratum of saturated sand is present in the dam foundation, the liquefaction of sand could be induced causing instability of the dam structure. Sometimes, erosion at the toe of the dam, for example through stream action, may cause progressive slope failure. The major cause of dam instability, however, is induced by seepage. Seepage-induced instability can occur in three different modes, namely; (i) heaving at the toe of the dam, (ii) piping through the dam, and (iii) the phreatic (or top) line of flow rises above the slope surface. Piping is generally started with small cracks that gradually widen, eventually leading to a concentrated flow of water (in the absence of a properly designed filter in the dam for arresting the flow). The cracks may have existed initially in the dam or its foundation because of improper construction; more often, the cracks may have been initiated through the differential settlement of different parts of the dam, differential movement between the conduits and earth material, hydraulic fracturing, or earthquake motions.

Draw the fault tree representing the failure of an earth dam, consisting of the different potential failure modes described above.

7.21 The probability of loss of containment by a large shear slide in the downstream section of the dam discussed in Example E7.17 depends on the following factors:

(a)   The effectiveness of the core trench to cutoff seepage—which could be divided into two conditions, perfect and imperfect cutoff (thus allowing drainage), with relative likelihoods of 1 to 9.

(b)   The amount of pore pressure developed along critical sliding surfaces. Four discrete states of pore pressure condition, namely I, II, III, and IV, are considered, each corresponds to specific assumptions about the permeabilities of the soil materials. The probabilities of each pore pressure state is estimated subjectively to be 0.19, 0.06, 0.56, and 0.19 for conditions I, II, III, and IV, respectively, regardless of the cutoff effectiveness in (a).

(c)   The probability of shear slide failure depends on the cutoff effectiveness and the pore pressure conditions as follows:

	Pore Pressure States			
	I	II	III	IV
Perfect Cutoff	0.75	0.95	0.98	1.0
Imperfect Cutoff	0.008	0.03	0.16	0.23

Draw the appropriate event tree and calculate the probability of shear slide for the dam.

7.22 The network shown in Fig. P7.22 represents the sequence of activities required in a hypothetical project. The mean durations and corresponding standard deviations of

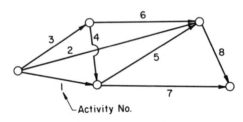

Figure P7.22   Activity network.

the various activities, except activity no. 4, are listed below:

Activity No.	Mean Duration (in days)	Standard Deviation (in days)
1	10	2
2	16	2
3	8	2
4	(to be determined)	
5	8	2
6	7	1
7	15	4
8	12	3

The duration of activity no. 4 is a function of the weather and the efficiency of the workers, as follows:

$$\tau_4 = \frac{5}{RE^2}$$

where

R = Ratio of good to bad weather; this ranges from 0.6 to 0.9.

E = Efficiency of workers; ranges from 0.70 to 0.90.

It is believed, however, that the above equation for τ_4 tends to consistently underestimate the duration of activity no. 4 by about 10%; moreover, in the judgment of the engineer, the uncertainty associated with the error of the above equation is $\pm 15\%$.
(a) Evaluate the mean and standard deviation of the duration of activity no. 4.
(b) If the target time for completing the project is 40 days, determine the probability of achieving this objective by the PNET method.

7.23 Two alternative cable systems, shown in Fig. P7.23, are being considered for carrying a tension, T, that is normally distributed with a mean value of 4000 kg and a c.o.v. of 20%. One system is a single wire cable as shown in Fig. P7.23a, whereas the other

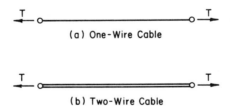

(a) One-Wire Cable

(b) Two-Wire Cable

Figure P7.23 Competing cable systems.

system is composed of two parallel wires (Fig. 7.23b). The rupture strength of a wire is also a normal variate with a mean of 6000 kg and a c.o.v. of 15%.

Assume that the strengths between wires are statistically independent and identically distributed; also, the wires in the 2-wire system share the applied tension equally.

If the cost of each wire is $200 and the consequence of the failure of the cable system is $10,000, which system would be preferable on the basis of its expected monetary value?

Appendix A. Tables

Table A.1   Values of the Gamma Function

$$\Gamma(x) = \int_0^\infty t^{x-1} e^{-t}\, dt$$

x	$\Gamma(x)$	x	$\Gamma(x)$	x	$\Gamma(x)$	x	$\Gamma(x)$
1.000	1.00000	1.250	0.90640	1.500	0.88623	1.750	0.91906
1.005	0.99714	1.255	0.90539	1.505	0.88640	1.755	0.92021
1.010	0.99433	1.260	0.90440	1.510	0.88659	1.760	0.92137
1.015	0.99156	1.265	0.90344	1.515	0.88680	1.765	0.92256
1.020	0.98884	1.270	0.90250	1.520	0.88704	1.770	0.92376
1.025	0.98617	1.275	0.90160	1.525	0.88729	1.775	0.92499
1.030	0.98355	1.280	0.90072	1.530	0.88757	1.780	0.92623
1.035	0.98097	1.285	0.89987	1.535	0.88786	1.785	0.92749
1.040	0.97844	1.290	0.89904	1.540	0.88818	1.790	0.92877
1.045	0.97595	1.295	0.89824	1.545	0.88851	1.795	0.93007
1.050	0.97350	1.300	0.89747	1.550	0.88887	1.800	0.93138
1.055	0.97110	1.305	0.89672	1.555	0.88924	1.805	0.93272
1.060	0.96874	1.310	0.89600	1.560	0.88964	1.810	0.93408
1.065	0.96643	1.315	0.89531	1.565	0.89005	1.815	0.93545
1.070	0.96415	1.320	0.89464	1.570	0.89049	1.820	0.93685
1.075	0.96192	1.325	0.89400	1.575	0.89094	1.825	0.93826
1.080	0.95973	1.330	0.89338	1.580	0.89142	1.830	0.93969
1.085	0.95757	1.335	0.89278	1.585	0.89191	1.835	0.94114
1.090	0.95546	1.340	0.89222	1.590	0.89243	1.840	0.94261
1.095	0.95339	1.345	0.89167	1.595	0.89296	1.845	0.94410
1.100	0.95135	1.350	0.89115	1.600	0.89352	1.850	0.94561
1.105	0.94935	1.355	0.89066	1.605	0.89409	1.855	0.94714
1.110	0.94740	1.360	0.89018	1.610	0.89468	1.860	0.94869
1.115	0.94547	1.365	0.88974	1.615	0.89529	1.865	0.95025
1.120	0.94359	1.370	0.88931	1.620	0.89592	1.870	0.95184
1.125	0.94174	1.375	0.88891	1.625	0.89657	1.875	0.95345
1.130	0.93993	1.380	0.88854	1.630	0.89724	1.880	0.95507
1.135	0.93816	1.385	0.88818	1.635	0.89793	1.885	0.95672
1.140	0.93642	1.390	0.88785	1.640	0.89864	1.890	0.95838
1.145	0.93471	1.395	0.88755	1.645	0.89937	1.895	0.96006
1.150	0.93304	1.400	0.88726	1.650	0.90012	1.900	0.96177
1.155	0.93141	1.405	0.88700	1.655	0.90088	1.905	0.96349
1.160	0.92980	1.410	0.88676	1.660	0.90167	1.910	0.96523
1.165	0.92823	1.415	0.88655	1.665	0.90247	1.915	0.96699
1.170	0.92670	1.420	0.88636	1.670	0.90330	1.920	0.96877
1.175	0.92520	1.425	0.88618	1.675	0.90414	1.925	0.97058
1.180	0.92373	1.430	0.88604	1.680	0.90500	1.930	0.97240
1.185	0.92229	1.435	0.88591	1.685	0.90588	1.935	0.97424
1.190	0.92089	1.440	0.88581	1.690	0.90678	1.940	0.97610
1.195	0.91951	1.445	0.88572	1.695	0.90770	1.945	0.97798
1.200	0.91817	1.450	0.88566	1.700	0.90864	1.950	0.97988
1.205	0.91686	1.455	0.88562	1.705	0.90960	1.955	0.98180
1.210	0.91558	1.460	0.88560	1.710	0.91057	1.960	0.98374
1.215	0.91433	1.465	0.88561	1.715	0.91157	1.965	0.98570
1.220	0.91311	1.470	0.88563	1.720	0.91258	1.970	0.98768
1.225	0.91192	1.475	0.88568	1.725	0.91361	1.975	0.98969
1.230	0.91075	1.480	0.88575	1.730	0.91467	1.980	0.99171
1.235	0.90962	1.485	0.88584	1.735	0.91574	1.985	0.99375
1.240	0.90852	1.490	0.88595	1.740	0.91683	1.990	0.99581
1.245	0.90745	1.495	0.88608	1.745	0.91793	1.995	0.99790
1.250	0.90640	1.500	0.88623	1.750	0.91906	2.000	1.00000

Table A.2   CDF of the Standard Extremal Variate *S*

s	CUM. PROB. F(s)	s	CUM. PROB. F(s)	s	CUM. PROB. F(s)
-3.00	.1892179D-08	-2.00	.6179790D-03	-1.00	.6598804D-01
-2.98	.2816371D-08	-1.98	.7153469D-03	-0.98	.6963721D-01
-2.96	.4159081D-08	-1.96	.8256604D-03	-0.96	.7340990D-01
-2.94	.6094700D-08	-1.94	.9502828D-03	-0.94	.7730618D-01
-2.92	.8863825D-08	-1.92	.1090675D-02	-0.92	.8132593D-01
-2.90	.1279584D-07	-1.90	.1248398D-02	-0.90	.8546887D-01
-2.88	.1833832D-07	-1.88	.1425112D-02	-0.88	.8973452D-01
-2.86	.2609487D-07	-1.86	.1622582D-02	-0.86	.9412226D-01
-2.84	.3687376D-07	-1.84	.1842673D-02	-0.84	.9863127D-01
-2.82	.5174952D-07	-1.82	.2087353D-02	-0.82	.1032606D+00
-2.80	.7214075D-07	-1.80	.2358693D-02	-0.80	.1080090D+00
-2.78	.9990751D-07	-1.78	.2658864D-02	-0.78	.1128752D+00
-2.76	.1374724D-06	-1.76	.2990133D-02	-0.76	.1178578D+00
-2.74	.1879697D-06	-1.74	.3354866D-02	-0.74	.1229552D+00
-2.72	.2554288D-06	-1.72	.3755520D-02	-0.72	.1281655D+00
-2.70	.3449964D-06	-1.70	.4194642D-02	-0.70	.1334868D+00
-2.68	.4632063D-06	-1.68	.4674860D-02	-0.68	.1389171D+00
-2.66	.6183018D-06	-1.66	.5198866D-02	-0.66	.1444543D+00
-2.64	.8206217D-06	-1.64	.5769502D-02	-0.64	.1500959D+00
-2.62	.1083056D-05	-1.62	.6389557D-02	-0.62	.1558396D+00
-2.60	.1421585D-05	-1.60	.7061962D-02	-0.60	.1616828D+00
-2.58	.1855905D-05	-1.58	.7789678D-02	-0.58	.1676229D+00
-2.56	.2410160D-05	-1.56	.8575712D-02	-0.56	.1736571D+00
-2.54	.3113787D-05	-1.54	.9423108D-02	-0.54	.1797826D+00
-2.52	.4002479D-05	-1.52	.1033494D-01	-0.52	.1859965D+00
-2.50	.5119294D-05	-1.50	.1131429D-01	-0.50	.1922956D+00
-2.48	.6515905D-05	-1.48	.1236425D-01	-0.48	.1986771D+00
-2.46	.8254009D-05	-1.46	.1348794D-01	-0.46	.2051377D+00
-2.44	.1040691D-04	-1.44	.1468842D-01	-0.44	.2116741D+00
-2.42	.1306127D-04	-1.42	.1596877D-01	-0.42	.2182833D+00
-2.40	.1631907D-04	-1.40	.1733201D-01	-0.40	.2249618D+00
-2.38	.2029973D-04	-1.38	.1878115D-01	-0.38	.2317063D+00
-2.36	.2514249D-04	-1.36	.2031911D-01	-0.36	.2385135D+00
-2.34	.3100889D-04	-1.34	.2194878D-01	-0.34	.2453799D+00
-2.32	.3808560D-04	-1.32	.2367297D-01	-0.32	.2523022D+00
-2.30	.4658731D-04	-1.30	.2549439D-01	-0.30	.2592769D+00
-2.28	.5675990D-04	-1.28	.2741569D-01	-0.28	.2663005D+00
-2.26	.6888383D-04	-1.26	.2943940D-01	-0.26	.2733697D+00
-2.24	.8327757D-04	-1.24	.3156793D-01	-0.24	.2804810D+00
-2.22	.1003014D-03	-1.22	.3380361D-01	-0.22	.2876310D+00
-2.20	.1203612D-03	-1.20	.3614860D-01	-0.20	.2948163D+00
-2.18	.1439123D-03	-1.18	.3860497D-01	-0.18	.3020335D+00
-2.16	.1714639D-03	-1.16	.4117462D-01	-0.16	.3092792D+00
-2.14	.2035828D-03	-1.14	.4385931D-01	-0.14	.3165501D+00
-2.12	.2408979D-03	-1.12	.4666064D-01	-0.12	.3238429D+00
-2.10	.2841041D-03	-1.10	.4958009D-01	-0.10	.3311543D+00
-2.08	.3339668D-03	-1.08	.5261892D-01	-0.08	.3384811D+00
-2.06	.3913259D-03	-1.06	.5577827D-01	-0.06	.3458201D+00
-2.04	.4570996D-03	-1.04	.5905908D-01	-0.04	.3531682D+00
-2.02	.5322885D-03	-1.02	.6246213D-01	-0.02	.3605223D+00

Table A.2 (Continued)

s	CUM. PROB. F(s)	s	CUM. PROB. F(s)	s	CUM. PROB. F(s)
0.00	.3678794D+00	1.00	.6922006D+00	2.00	.8734230D+00
0.02	.3752365D+00	1.02	.6972614D+00	2.02	.8757668D+00
0.04	.3825907D+00	1.04	.7022578D+00	2.04	.8780702D+00
0.06	.3899392D+00	1.06	.7071901D+00	2.06	.8803339D+00
0.08	.3972791D+00	1.08	.7120583D+00	2.08	.8825585D+00
0.10	.4046077D+00	1.10	.7168626D+00	2.10	.8847445D+00
0.12	.4119223D+00	1.12	.7216033D+00	2.12	.8868924D+00
0.14	.4192205D+00	1.14	.7262805D+00	2.14	.8890028D+00
0.16	.4264996D+00	1.16	.7308945D+00	2.16	.8910764D+00
0.18	.4337573D+00	1.18	.7354456D+00	2.18	.8931136D+00
0.20	.4409910D+00	1.20	.7399341D+00	2.20	.8951149D+00
0.22	.4481986D+00	1.22	.7443602D+00	2.22	.8970810D+00
0.24	.4553778D+00	1.24	.7487245D+00	2.24	.8990124D+00
0.26	.4625264D+00	1.26	.7530271D+00	2.26	.9009095D+00
0.28	.4696424D+00	1.28	.7572686D+00	2.28	.9027729D+00
0.30	.4767237D+00	1.30	.7614492D+00	2.30	.9046032D+00
0.32	.4837684D+00	1.32	.7655695D+00	2.32	.9064009D+00
0.34	.4907746D+00	1.34	.7696298D+00	2.34	.9081664D+00
0.36	.4977405D+00	1.36	.7736306D+00	2.36	.9099003D+00
0.38	.5046645D+00	1.38	.7775724D+00	2.38	.9116031D+00
0.40	.5115448D+00	1.40	.7814556D+00	2.40	.9132753D+00
0.42	.5183799D+00	1.42	.7852807D+00	2.42	.9149173D+00
0.44	.5251683D+00	1.44	.7890483D+00	2.44	.9165297D+00
0.46	.5319086D+00	1.46	.7927588D+00	2.46	.9181129D+00
0.48	.5385993D+00	1.48	.7964128D+00	2.48	.9196674D+00
0.50	.5452392D+00	1.50	.8000107D+00	2.50	.9211937D+00
0.52	.5518271D+00	1.52	.8035532D+00	2.52	.9226922D+00
0.54	.5583617D+00	1.54	.8070408D+00	2.54	.9241634D+00
0.56	.5648421D+00	1.56	.8104740D+00	2.56	.9256077D+00
0.58	.5712671D+00	1.58	.8138533D+00	2.58	.9270257D+00
0.60	.5776358D+00	1.60	.8171795D+00	2.60	.9284177D+00
0.62	.5839474D+00	1.62	.8204530D+00	2.62	.9297841D+00
0.64	.5902008D+00	1.64	.8236743D+00	2.64	.9311254D+00
0.66	.5963954D+00	1.66	.8268442D+00	2.66	.9324421D+00
0.68	.6025305D+00	1.68	.8299632D+00	2.68	.9337345D+00
0.70	.6086053D+00	1.70	.8330317D+00	2.70	.9350030D+00
0.72	.6146193D+00	1.72	.8360506D+00	2.72	.9362481D+00
0.74	.6205718D+00	1.74	.8390203D+00	2.74	.9374702D+00
0.76	.6264625D+00	1.76	.8419414D+00	2.76	.9386696D+00
0.78	.6322907D+00	1.78	.8448145D+00	2.78	.9398467D+00
0.80	.6380562D+00	1.80	.8476403D+00	2.80	.9410020D+00
0.82	.6437585D+00	1.82	.8504193D+00	2.82	.9421357D+00
0.84	.6493973D+00	1.84	.8531521D+00	2.84	.9432484D+00
0.86	.6549725D+00	1.86	.8558393D+00	2.86	.9443402D+00
0.88	.6604836D+00	1.88	.8584815D+00	2.88	.9454117D+00
0.90	.6659307D+00	1.90	.8610793D+00	2.90	.9464632D+00
0.92	.6713135D+00	1.92	.8636334D+00	2.92	.9474949D+00
0.94	.6766319D+00	1.94	.8661441D+00	2.94	.9485074D+00
0.96	.6818859D+00	1.96	.8686123D+00	2.96	.9495008D+00
0.98	.6870755D+00	1.98	.8710384D+00	2.98	.9504756D+00

Table A.2 (Continued)

s	CUM. PROB. F(s)	s	CUM. PROB. F(s)	s	CUM. PROB. F(s)
3.00	.9514320D+00	4.00	.9818511D+00	5.00	.9932847D+00
3.02	.9523704D+00	4.02	.9822072D+00	5.02	.9934172D+00
3.04	.9532912D+00	4.04	.9825565D+00	5.04	.9935472D+00
3.06	.9541946D+00	4.06	.9828989D+00	5.06	.9936745D+00
3.08	.9550809D+00	4.08	.9832347D+00	5.08	.9937994D+00
3.10	.9559504D+00	4.10	.9835639D+00	5.10	.9939218D+00
3.12	.9568036D+00	4.12	.9838867D+00	5.12	.9940418D+00
3.14	.9576405D+00	4.14	.9842032D+00	5.14	.9941594D+00
3.16	.9584616D+00	4.16	.9845136D+00	5.16	.9942748D+00
3.18	.9592672D+00	4.18	.9848179D+00	5.18	.9943878D+00
3.20	.9600574D+00	4.20	.9851163D+00	5.20	.9944986D+00
3.22	.9608326D+00	4.22	.9854089D+00	5.22	.9946073D+00
3.24	.9615931D+00	4.24	.9856957D+00	5.24	.9947138D+00
3.26	.9623391D+00	4.26	.9859769D+00	5.26	.9948182D+00
3.28	.9630709D+00	4.28	.9862527D+00	5.28	.9949205D+00
3.30	.9637887D+00	4.30	.9865231D+00	5.30	.9950208D+00
3.32	.9644929D+00	4.32	.9867882D+00	5.32	.9951192D+00
3.34	.9651836D+00	4.34	.9870481D+00	5.34	.9952156D+00
3.36	.9658611D+00	4.36	.9873029D+00	5.36	.9953101D+00
3.38	.9665256D+00	4.38	.9875528D+00	5.38	.9954028D+00
3.40	.9671775D+00	4.40	.9877977D+00	5.40	.9954936D+00
3.42	.9678168D+00	4.42	.9880379D+00	5.42	.9955826D+00
3.44	.9684439D+00	4.44	.9882734D+00	5.44	.9956699D+00
3.46	.9690590D+00	4.46	.9885042D+00	5.46	.9957555D+00
3.48	.9696623D+00	4.48	.9887306D+00	5.48	.9958393D+00
3.50	.9702540D+00	4.50	.9889525D+00	5.50	.9959216D+00
3.52	.9708343D+00	4.52	.9891700D+00	5.52	.9960022D+00
3.54	.9714035D+00	4.54	.9893834D+00	5.54	.9960812D+00
3.56	.9719618D+00	4.56	.9895925D+00	5.56	.9961586D+00
3.58	.9725092D+00	4.58	.9897975D+00	5.58	.9962345D+00
3.60	.9730462D+00	4.60	.9899985D+00	5.60	.9963090D+00
3.62	.9735728D+00	4.62	.9901956D+00	5.62	.9963819D+00
3.64	.9740893D+00	4.64	.9903888D+00	5.64	.9964534D+00
3.66	.9745957D+00	4.66	.9905782D+00	5.66	.9965235D+00
3.68	.9750925D+00	4.68	.9907639D+00	5.68	.9965923D+00
3.70	.9755796D+00	4.70	.9909460D+00	5.70	.9966596D+00
3.72	.9760573D+00	4.72	.9911244D+00	5.72	.9967257D+00
3.74	.9765258D+00	4.74	.9912994D+00	5.74	.9967904D+00
3.76	.9769852D+00	4.76	.9914710D+00	5.76	.9968538D+00
3.78	.9774358D+00	4.78	.9916391D+00	5.78	.9969160D+00
3.80	.9778776D+00	4.80	.9918040D+00	5.80	.9969770D+00
3.82	.9783109D+00	4.82	.9919657D+00	5.82	.9970368D+00
3.84	.9787357D+00	4.84	.9921241D+00	5.84	.9970954D+00
3.86	.9791524D+00	4.86	.9922795D+00	5.86	.9971528D+00
3.88	.9795609D+00	4.88	.9924318D+00	5.88	.9972091D+00
3.90	.9799616D+00	4.90	.9925811D+00	5.90	.9972643D+00
3.92	.9803544D+00	4.92	.9927274D+00	5.92	.9973184D+00
3.94	.9807397D+00	4.94	.9928709D+00	5.94	.9973714D+00
3.96	.9811174D+00	4.96	.9930116D+00	5.96	.9974234D+00
3.98	.9814879D+00	4.98	.9931495D+00	5.98	.9974744D+00

Table A.2 (Continued)

s	CUM. PROB. F(s)	s	CUM. PROB. F(s)	s	CUM. PROB. F(s)
6.00	.9975243D+00	7.00	.9990885D+00	8.00	.9996646D+00
6.02	.9975733D+00	7.02	.9991066D+00	8.10	.9996965D+00
6.04	.9976213D+00	7.04	.9991243D+00	8.20	.9997254D+00
6.06	.9976683D+00	7.06	.9991416D+00	8.30	.9997515D+00
6.08	.9977144D+00	7.08	.9991586D+00	8.40	.9997752D+00
6.10	.9977596D+00	7.10	.9991752D+00	8.50	.9997966D+00
6.12	.9978040D+00	7.12	.9991916D+00	8.60	.9998159D+00
6.14	.9978474D+00	7.14	.9992076D+00	8.70	.9998334D+00
6.16	.9978900D+00	7.16	.9992232D+00	8.80	.9998493D+00
6.18	.9979317D+00	7.18	.9992386D+00	8.90	.9998636D+00
6.20	.9979726D+00	7.20	.9992537D+00	9.00	.9998766D+00
6.22	.9980127D+00	7.22	.9992685D+00	9.10	.9998883D+00
6.24	.9980520D+00	7.24	.9992829D+00	9.20	.9998990D+00
6.26	.9980906D+00	7.26	.9992971D+00	9.30	.9999086D+00
6.28	.9981284D+00	7.28	.9993111D+00	9.40	.9999173D+00
6.30	.9981654D+00	7.30	.9993247D+00	9.50	.9999252D+00
6.32	.9982017D+00	7.32	.9993381D+00	9.60	.9999323D+00
6.34	.9982373D+00	7.34	.9993512D+00	9.70	.9999387D+00
6.36	.9982721D+00	7.36	.9993640D+00	9.80	.9999445D+00
6.38	.9983063D+00	7.38	.9993766D+00	9.90	.9999498D+00
6.40	.9983398D+00	7.40	.9993889D+00	10.00	.9999546D+00
6.42	.9983727D+00	7.42	.9994010D+00	10.10	.9999589D+00
6.44	.9984049D+00	7.44	.9994129D+00	10.20	.9999628D+00
6.46	.9984364D+00	7.46	.9994245D+00	10.30	.9999664D+00
6.48	.9984674D+00	7.48	.9994359D+00	10.40	.9999696D+00
6.50	.9984977D+00	7.50	.9994471D+00	10.50	.9999725D+00
6.52	.9985274D+00	7.52	.9994580D+00	10.60	.9999751D+00
6.54	.9985566D+00	7.54	.9994687D+00	10.70	.9999775D+00
6.56	.9985851D+00	7.56	.9994793D+00	10.80	.9999796D+00
6.58	.9986131D+00	7.58	.9994896D+00	10.90	.9999815D+00
6.60	.9986406D+00	7.60	.9994997D+00	11.00	.9999833D+00
6.62	.9986675D+00	7.62	.9995096D+00	11.10	.9999849D+00
6.64	.9986938D+00	7.64	.9995193D+00	11.20	.9999863D+00
6.66	.9987197D+00	7.66	.9995288D+00	11.30	.9999876D+00
6.68	.9987450D+00	7.68	.9995381D+00	11.40	.9999888D+00
6.70	.9987698D+00	7.70	.9995473D+00	11.50	.9999899D+00
6.72	.9987942D+00	7.72	.9995562D+00	11.60	.9999908D+00
6.74	.9988181D+00	7.74	.9995650D+00	11.70	.9999917D+00
6.76	.9988414D+00	7.76	.9995736D+00	11.80	.9999925D+00
6.78	.9988644D+00	7.78	.9995821D+00	11.90	.9999932D+00
6.80	.9988868D+00	7.80	.9995903D+00	12.00	.9999939D+00
6.82	.9989089D+00	7.82	.9995985D+00	12.20	.9999950D+00
6.84	.9989305D+00	7.84	.9996064D+00	12.40	.9999959D+00
6.86	.9989516D+00	7.86	.9996142D+00	12.60	.9999966D+00
6.88	.9989724D+00	7.88	.9996218D+00	12.80	.9999972D+00
6.90	.9989927D+00	7.90	.9996293D+00	13.00	.9999977D+00
6.92	.9990127D+00	7.92	.9996367D+00	13.25	.9999982D+00
6.94	.9990322D+00	7.94	.9996439D+00	13.50	.9999986D+00
6.96	.9990514D+00	7.96	.9996509D+00	13.75	.9999989D+00
6.98	.9990701D+00	7.98	.9996578D+00	14.00	.9999992D+00

Table A.3 The Three Types of Asymptotic Extremal Distributions

Asymptotic Type	Tail Characteristics of Initial Variate	Extreme	Cumulative Distribution Function	Mean Value	Standard Deviation	Standard Extremal Variate, S
I	Exponential	Largest	$\exp\left[-e^{-\alpha_n(x_n-u_n)}\right]$	$u_n + \dfrac{0.577}{\alpha_n}$	$\dfrac{\pi}{\sqrt{6}\,\alpha_n}$	$\alpha_n(X_n - u_n)$
		Smallest	$\exp\left[-e^{\alpha_1(x_1-u_1)}\right]$	$u_1 - \dfrac{0.577}{\alpha_1}$	$\dfrac{\pi}{\sqrt{6}\,\alpha_1}$	$-\alpha_1(X_1 - u_1)$
II	Polynomial	Largest	$\exp\left[-\left(\dfrac{v_n}{y_n}\right)^k\right]$	$v_n\Gamma\left(1-\dfrac{1}{k}\right)$	$v_n\left[\Gamma\left(1-\dfrac{2}{k}\right)-\Gamma^2\left(1-\dfrac{1}{k}\right)\right]^{1/2}$	$k\ln\dfrac{Y_n}{v_n}$
		Smallest	$\exp\left[-\left(\dfrac{v_1}{y_1}\right)^k\right]$	$v_1\Gamma\left(1-\dfrac{1}{k}\right)$	$v_1\left[\Gamma\left(1-\dfrac{2}{k}\right)-\Gamma^2\left(1-\dfrac{1}{k}\right)\right]^{1/2}$	$k\ln\dfrac{v_1}{Y_1}$
III	Bounded (in direction of extreme)	Largest	$\exp\left[-\left(\dfrac{\omega-z_n}{\omega-w_n}\right)^k\right]$	$\omega - (\omega-w_n)\Gamma\left(1+\dfrac{1}{k}\right)$	$(\omega-w_n)\left[\Gamma\left(1+\dfrac{2}{k}\right)-\Gamma^2\left(1+\dfrac{1}{k}\right)\right]^{1/2}$	$-k\ln\left(\dfrac{\omega-z_n}{\omega-w_n}\right)$
		Smallest	$\exp\left[-\left(\dfrac{z_1-\varepsilon}{w_1-\varepsilon}\right)^k\right]$	$\varepsilon + (w_1-\varepsilon)\Gamma\left(1+\dfrac{1}{k}\right)$	$(w_1-\varepsilon)\left[\Gamma\left(1+\dfrac{2}{k}\right)-\Gamma^2\left(1+\dfrac{1}{k}\right)\right]^{1/2}$	$-k\ln\left(\dfrac{z_1-\varepsilon}{w_1-\varepsilon}\right)$

Table A.4   Weights a_i and b_i for Leiblein's Order Statistics Estimator[a] ($p \geq 0.90$)

		\multicolumn{6}{c}{Weights a_i and b_i for $x_1 \leq x_2 \leq \cdots \leq x_n$}
n	$i =$	1	2	3	4	5	6
2	a_i	0.91637	0.08363				
	b_i	−0.72135	0.72135				
3	a_i	0.65632	0.25571	0.08797			
	b_i	−0.63054	0.25582	0.37473			
4	a_i	0.51100	0.26394	0.15368	0.07138		
	b_i	−0.55862	0.08590	0.22392	0.24880		
5	a_i	0.41893	0.24628	0.16761	0.10882	0.05835	
	b_i	−0.50313	0.00653	0.13045	0.18166	0.18448	
6	a_i	0.35545	0.22549	0.16562	0.12105	0.08352	0.04887
	b_i	−0.45928	−0.03599	0.07319	0.12673	0.14953	0.14581

[a] Reproduced from J. Leiblein, "A New Method of Analyzing Extreme-Value Data," *NACA TN 3053*, 1954.

Table A.5 Relations Among Parameters of the Type III Asymptotic Distribution

$1/k$	θ_Z	$A(k)$	$B(k)$	$1/k$	θ_Z	$A(k)$	$B(k)$
0.010	−1.08107	0.44815	78.98172	0.510	0.65781	0.24038	2.11963
0.020	−1.02485	0.44611	39.98904	0.520	0.68445	0.23516	2.08181
0.030	−0.97070	0.44392	26.98621	0.530	0.71103	0.22994	2.04511
0.040	−0.91845	0.44160	20.48081	0.540	0.73755	0.22471	2.00949
0.050	−0.86797	0.43915	16.57435	0.550	0.76404	0.21947	1.97489
0.060	−0.81910	0.43657	13.96734	0.560	0.79049	0.21424	1.94124
0.070	−0.77174	0.43386	12.10286	0.570	0.81690	0.20900	1.90851
0.080	−0.72577	0.43104	10.70245	0.580	0.84330	0.20377	1.87666
0.090	−0.68110	0.42810	9.61140	0.590	0.86968	0.19854	1.84563
0.100	−0.63764	0.42504	8.73689	0.600	0.89605	0.19331	1.81538
0.110	−0.59530	0.42188	8.01986	0.610	0.92241	0.18809	1.78590
0.120	−0.55400	0.41861	7.42093	0.620	0.94877	0.18288	1.75713
0.130	−0.51369	0.41524	6.91285	0.630	0.97514	0.17767	1.72905
0.140	−0.47429	0.41178	6.47613	0.640	1.00153	0.17247	1.70162
0.150	−0.43574	0.40822	6.09651	0.650	1.02793	0.16729	1.67482
0.160	−0.39800	0.40456	5.76326	0.660	1.05435	0.16211	1.64863
0.170	−0.36101	0.40082	5.46821	0.670	1.08081	0.15695	1.62302
0.180	−0.32473	0.39700	5.20498	0.680	1.10730	0.15180	1.59796
0.190	−0.28911	0.39309	4.96856	0.690	1.13382	0.14667	1.57343
0.200	−0.25411	0.38910	4.75490	0.700	1.16039	0.14156	1.54942
0.210	−0.21970	0.38504	4.56077	0.710	1.18701	0.13646	1.52590
0.220	−0.18583	0.38090	4.38350	0.720	1.21368	0.13138	1.50286
0.230	−0.15249	0.37669	4.22088	0.730	1.24042	0.12632	1.48027
0.240	−0.11963	0.37242	4.07108	0.740	1.26721	0.12128	1.45813
0.250	−0.08724	0.36808	3.93258	0.750	1.29407	0.11626	1.43641
0.260	−0.05527	0.36368	3.80405	0.760	1.32100	0.11126	1.41511
0.270	−0.02372	0.35922	3.68440	0.770	1.34801	0.10629	1.39420
0.280	0.00746	0.35470	3.57267	0.780	1.37510	0.10134	1.37368
0.290	0.03827	0.35013	3.46805	0.790	1.40228	0.09642	1.35354
0.300	0.06874	0.34551	3.36982	0.800	1.42955	0.09152	1.33375
0.310	0.09889	0.34083	3.27736	0.810	1.45690	0.08664	1.31431
0.320	0.12874	0.33611	3.19015	0.820	1.48436	0.08180	1.29522
0.330	0.15831	0.33135	3.10769	0.830	1.51192	0.07698	1.27645
0.340	0.18761	0.32654	3.02957	0.840	1.53959	0.07219	1.25800
0.350	0.21665	0.32169	2.95543	0.850	1.56736	0.06743	1.23987
0.360	0.24546	0.31681	2.88492	0.860	1.59525	0.06271	1.22203
0.370	0.27405	0.31189	2.81777	0.870	1.62326	0.05801	1.20449
0.380	0.30244	0.30693	2.75370	0.880	1.65140	0.05334	1.18723
0.390	0.33063	0.30195	2.69247	0.890	1.67966	0.04871	1.17026
0.400	0.35863	0.29693	2.63389	0.900	1.70804	0.04411	1.15355
0.410	0.38647	0.29189	2.57775	0.910	1.73657	0.03954	1.13711
0.420	0.41415	0.28683	2.52389	0.920	1.76523	0.03500	1.12092
0.430	0.44168	0.28173	2.47214	0.930	1.79404	0.03050	1.10499
0.440	0.46907	0.27662	2.42236	0.940	1.82299	0.02604	1.08930
0.450	0.49634	0.27149	2.37443	0.950	1.85209	0.02161	1.07385
0.460	0.52349	0.26634	2.32823	0.960	1.88135	0.01721	1.05863
0.470	0.55054	0.26117	2.28365	0.970	1.91077	0.01285	1.04364
0.480	0.57748	0.25599	2.24058	0.980	1.94034	0.00853	1.02888
0.490	0.60434	0.25080	2.19895	0.990	1.97009	0.00425	1.01433
0.500	0.63111	0.24560	2.15866	1.000	2.00000	0.00000	1.00000

Appendix B. Transformation of Non-normal Variates to Independent Normal Variates

B.1 THE ROSENBLATT TRANSFORMATION

In Section 6.2.4, probabilities involving nonnormal distributions were calculated using equivalent normal distributions. In effect, this involves the transformation of a general set of correlated random variables into an equivalent set of independent normal or Gaussian variates. A general transformation for this purpose is the *Rosenblatt transformation* (Rosenblatt, 1969).

Suppose a set of n random variables $\mathbf{X} = (X_1, X_2, \ldots, X_n)$ with a joint CDF $F_{\mathbf{X}}(\mathbf{x})$. A set of statistically independent standard normal variates $\mathbf{U} = (U_1, U_2, \ldots, U_n)$ can be obtained from the following equations:

$$\Phi(u_1) = F_1(x_1)$$
$$\Phi(u_2) = F_2(x_2|x_1)$$
$$\vdots \qquad \vdots \qquad\qquad\qquad\qquad (\text{B.1})$$
$$\Phi(u_n) = F_n(x_n|x_1, \ldots, x_{n-1})$$

Inverting the above equations successively, we obtain the desired normal variates $\mathbf{U}$

$$u_1 = \Phi^{-1}[F_1(x_1)]$$
$$u_2 = \Phi^{-1}[F_2(x_2|x_1)]$$
$$\vdots \qquad \vdots \qquad\qquad\qquad\qquad (\text{B.2})$$
$$u_n = \Phi^{-1}[F_n(x_n|x_1, \ldots, x_{n-1})]$$

Equation B.2 constitutes the *Rosenblatt transformation*.

The conditional CDFs in Eq. B.2 may be obtained from the joint PDFs as follows. Since

$$f(x_i|x_1, \ldots, x_{i-1}) = \frac{f(x_1, \ldots, x_i)}{f(x_1, \ldots, x_{i-1})}$$

the required CDF may be obtained as

$$F(x_i|x_1, \ldots, x_{i-1}) = \frac{\int_{-\infty}^{x_i} f(x_1, \ldots, x_{i-1}, s_i)\, ds_i}{f(x_1, \ldots, x_{i-1})}$$

The Inverse Transformation

The inverse transformation of Eq. B.1 can be obtained by sequentially inverting the one-dimensional relations; that is,

$$
\begin{aligned}
x_1 &= F_1^{-1}[\Phi(u_1)] \\
x_2 &= F_2^{-1}[\Phi(u_2|x_1)] \\
&\quad\vdots \qquad\qquad\vdots \\
x_n &= F_n^{-1}[\Phi(u_n|x_1, \ldots, x_{n-1})]
\end{aligned}
\tag{B.3}
$$

In general, these inverse relations may only be obtained numerically.

EXAMPLE B.1

The annual maximum mile wind speed may be modeled by the Type II asymptotic distribution of largest values as follows:

$$
F_V(v) = \exp\left[-\left(\frac{v}{\beta}\right)^{-\gamma}\right]; \qquad v \geq 0
$$

where β and γ are the parameters. A typical value of γ is 9.0, whereas β will depend on the site. In this case, the Rosenblatt transformation, Eq. B.2, yields the standard normal variate

$$
u = \Phi^{-1}[F_V(v)] = \Phi^{-1}[e^{-(v/\beta)^{-\gamma}}]
$$

For $\gamma = 9.0$ and $\beta = 55$ mph, the above relationship is depicted in Fig. B.1, which shows in graphic terms the transformation from the variable V to the equivalent standard normal variate U; for example, for $v = 60$ mph, $u = 0.34$.

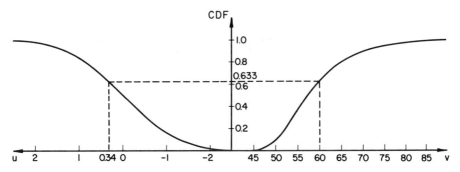

Figure B.1 Rosenblatt transformation of an extreme Type II distribution.

EXAMPLE B.2

Consider a random vector $\mathbf{X} = (X_1, X_2)$, in which the distribution of X_1 is normal with $\mu = 10$ and $\sigma = 2$, whereas the distribution of X_2 is lognormal with parameters $\lambda = 3.0$, $\zeta = 0.20$. Also assume a correlation coefficient $\rho_{X_1, X_2} = 0.60$. In this case, all the transformations can be obtained analytically. The first of Eq. B.1 is

$$
F_{X_1}(x_1) = \Phi\left(\frac{x_1 - \mu}{\sigma}\right)
$$

Also, it can be shown (see Chapter 4, Vol. 1) that

$$f_{X_2|X_1}(x_2|x_1) = \frac{1}{\sqrt{2\pi}\zeta\sqrt{1-r^2}} \exp\left[-\frac{1}{2}\left\{\frac{\ln x_2 - \lambda - r(\zeta/\sigma)(x_1 - \mu)}{\zeta\sqrt{1-r^2}}\right\}\right]$$

where

$$r = \frac{\rho\delta_{x_2}}{\zeta} \simeq \rho, \qquad \text{for small } \delta_{x_2}$$

Therefore,

$$F_{X_2|X_1}(x_2|x_1) = \Phi\left[\frac{\ln x_2 - \lambda - r(\zeta/\sigma)(x_1 - \mu)}{\zeta\sqrt{1-r^2}}\right]$$

The Rosenblatt transformation, Eq. B.2, then yields the independent standard normal variates

$$u_1 = \Phi^{-1}[F_{X_1}(x_1)] = \frac{x_1 - \mu}{\sigma}$$

and

$$u_2 = \Phi^{-1}[F_{X_2|X_1}(x_2|x_1)] = \frac{\ln x_2 - \lambda - r(\zeta/\sigma)(x_1 - \mu)}{\zeta\sqrt{1-r^2}}$$

With the distribution parameters given above,

$$u_1 = \frac{x_1 - 10}{2}$$

and

$$u_2 = \frac{\ln x_2 - 3 - 0.06(x_1 - 10)}{0.16}$$

For example, for $x_1 = 12$ (mean plus one standard deviation) and $x_2 = e^3 = 20.086$ (the median), the values of the corresponding standard normal variates are $u_1 = 1.0$ and $u_2 = -0.75$.

In the present case, that is, where X_1 is normal and X_2 is lognormal, the required transformations may also be obtained without the conditional CDF. First, define two standard normal variates (Y_1, Y_2)

$$Y_1 = \frac{X_1 - \mu}{\sigma}$$

$$Y_2 = \frac{\ln X_2 - \lambda}{\zeta}$$

with correlation $\rho_{Y_1, Y_2} = r \simeq \rho$.

The variables Y_1 and Y_2 can be written in terms of two independent standard normal variates (U_1, U_2) as follows:

$$Y_1 = U_1$$

$$Y_2 = \alpha_{21}U_1 + \alpha_{22}U_2$$

where α_{21} and α_{22} are obtained as solutions to the following equations:

$$\rho_{Y_1, Y_2} = \alpha_{21}$$

$$\sigma_{Y_2}^2 = 1 = \alpha_{21}^2 + \alpha_{22}^2$$

or

$$\alpha_{22} = \sqrt{1 - \alpha_{21}^2}$$

Therefore,

$$Y_1 = U_1$$
$$Y_2 = 0.6U_1 + 0.8U_2$$

Again, at the point ($x_1 = 12$, $x_2 = e^3$), $y_1 = 1.0$ and $y_2 = 0$; from which the corresponding independent standard normal variates are $u_1 = 1.0$ and $u_2 = -0.75$, as before.

The latter transformation illustrated in Example B.2 can be generalized for any vector composed entirely of normal and lognormal component variables. The generalized transformations would be as follows:

$$Y_1 = \alpha_{11}U_1$$
$$Y_2 = \alpha_{21}U_1 + \alpha_{22}U_2$$
$$\vdots \qquad \vdots$$
$$Y_n = \alpha_{n1}U_1 + \alpha_{n2}U_2 + \cdots + \alpha_{nn}U_n$$

(B.4)

where

$$\alpha_{11} = 1.0$$

$$\alpha_{i1} = \rho_{Y_iY_1}$$

$$\alpha_{ik} = \frac{1}{\alpha_{kk}}\left(\rho_{Y_iY_k} - \sum_{j=1}^{k-1}\alpha_{ij}\alpha_{kj}\right); \qquad 1 < k < i$$

$$\alpha_{ii} = \sqrt{1 - \sum_{j=1}^{i-1}\alpha_{ij}^2}$$

Successive forward substitutions would yield the independent standard normal variates

$$U_1 = Y_1$$

$$U_2 = \frac{1}{\alpha_{22}}(Y_2 - \alpha_{21}U_1)$$

(B.5)

$$\vdots \qquad \vdots$$

$$U_n = \frac{1}{\alpha_{nn}}(Y_n - \alpha_{n1}U_1 - \cdots - \alpha_{n,n-1}U_{n-1})$$

If X_i is normal and X_j is lognormal,

$$\rho_{Y_iY_j} = \frac{\rho_{X_iX_j}\delta_{X_j}}{\zeta_{X_j}} \simeq \rho_{X_iX_j}; \qquad \text{for small } \delta_{X_j}$$

whereas if X_i and X_j are both lognormals,

$$\rho_{Y_iY_j} = \frac{\ln(\rho_{X_iX_j}\delta_{X_i}\delta_{X_j} + 1)}{\zeta_{X_i}\zeta_{X_j}} \simeq \rho_{X_iX_j}; \qquad \text{for small } \delta_{X_i} \text{ and } \delta_{X_j}$$

EXAMPLE B.3

Consider the following joint exponential CDF for two variates X_1 and X_2

$$F_{X_1, X_2}(x_1, x_2) = 1 - e^{-x_1} - e^{-x_2} + e^{-(x_1 + x_2 + \theta x_1 x_2)}; \qquad x_1 \geq 0, x_2 \geq 0$$
$$= 0; \qquad \text{otherwise}$$

The corresponding joint PDF can be shown to be

$$f_{X_1, X_2}(x_1, x_2) = [1 - \theta(1 - x_1 - x_2) + \theta^2 x_1 x_2] e^{-(x_1 + x_2 + \theta x_1 x_2)}$$

whereas the marginal PDFs are

$$f_{X_1}(x_1) = e^{-x_1}$$
$$f_{X_2}(x_2) = e^{-x_2}$$

In this case,

$$F_{X_1}(x_1) = 1 - e^{-x_1}$$

whereas

$$F_{X_2|X_1}(x_2|x_1) = \frac{\int_0^{x_2} f_{X_1, X_2}(x_1, s) \, ds}{f_{X_1}(x_1)}$$

in which

$$\int_0^{x_2} f_{X_1, X_2}(x_1, s) \, ds = e^{-x_1} - (1 + \theta x_2) e^{-(x_1 + x_2 + \theta x_1 x_2)}$$

Hence,

$$F_{X_2|X_1}(x_2|x_1) = 1 - (1 + \theta x_2) e^{-(x_2 + \theta x_1 x_2)}$$

Therefore, according to Eq. B.2, the appropriate independent normal variates are

$$u_1 = \Phi^{-1}(1 - e^{-x_1})$$
$$u_2 = \Phi^{-1}[1 - (1 + \theta x_2) e^{-(x_2 + \theta x_1 x_2)}]$$

It should be observed that in the present case the variates X_1 and X_2 are correlated for $\theta \neq 0$; for example, for $\theta = 1$, it can be shown that

$$\rho_{X_1, X_2} = -0.40$$

whereas (trivially), for $\theta = 0$, $\rho_{X_1, X_2} = 0$.

B.2   DETERMINATION OF SAFETY INDEX

In Section 6.2.4 (specifically Example 6.10), we indicated that the rotation-of-coordinates method will yield approximate values of β for correlated nonnormal variates. Properly, for such nonnormal variates, the safety index may be evaluated in the space of the independent normal variates obtained through the Rosenblatt transformation. For this purpose, the following algorithm (Hohenbichler and Rackwitz, 1981) may be used:

(1)   Assume failure point $\mathbf{x}_o^* = \mathbf{x}_o$.
(2)   Using the Rosenblatt transformation, obtain the corresponding failure point in the u-space; that is, $\mathbf{u}_o$.

(3) Determine the Jacobian matrix

$$J = \frac{\partial(u_1, u_2, \ldots, u_n)}{\partial(x_1, x_2, \ldots, x_n)}$$

$$= \begin{bmatrix} \dfrac{\partial u_1}{\partial x_1} & \dfrac{\partial u_1}{\partial x_2} & \cdots & \dfrac{\partial u_1}{\partial x_n} \\[2ex] \dfrac{\partial u_2}{\partial x_1} & \dfrac{\partial u_2}{\partial x_2} & \cdots & \dfrac{\partial u_2}{\partial x_n} \\[2ex] \vdots & \vdots & & \vdots \\[2ex] \dfrac{\partial u_n}{\partial x_1} & \dfrac{\partial u_n}{\partial x_2} & \cdots & \dfrac{\partial u_n}{\partial x_n} \end{bmatrix}$$

evaluated at x_o.

(4) Evaluate the performance function and gradient vector at u_o

$$g(\mathbf{u}_o) = g(\mathbf{x}_o)$$

$$\mathbf{G}_{u_o} = (\mathbf{J}^{-1})^t \mathbf{G}_{x_o}$$

(5) Obtain a new failure point

$$\mathbf{u}^* = \frac{1}{\mathbf{G}_{u_o}^t \mathbf{G}_{u_o}} [\mathbf{G}_{u_o}^t \mathbf{u}_o - g(\mathbf{u}_o)] \mathbf{G}_{u_o}$$

and in the space of the original variables, the failure point is (in first-order approximation)

$$\mathbf{x}^* \simeq \mathbf{x}_o + \mathbf{J}^{-1}(\mathbf{u}^* - \mathbf{u}_o)$$

(6) Calculate $\beta = (\mathbf{u}^{*t}\mathbf{u}^*)^{1/2}$.

(7) Repeat Steps 2 through 6 using the above $\mathbf{x}^*$ as the new failure point until convergence is achieved.

Through implicit differentiation, it can be shown that the partial derivatives in the Jacobian of Step 3 are

$$\frac{\partial u_i}{\partial x_j} = \frac{\partial \Phi^{-1}[F(x_i|\ldots)]}{\partial x_j} = \frac{1}{\phi(u_i)} \frac{\partial [F(x_i|\ldots)]}{\partial x_j}$$

Also, since $\partial u_i / \partial x_j = 0$ for $i < j$, the Jacobian will be a lower triangular matrix, and thus its inverse $\mathbf{J}^{-1}$ is easily obtained through back substitution.

EXAMPLE B.4

In Example 6.10, the rotation of coordinates method was used to obtain an approximate solution for a problem involving correlated nonnormal variates. Properly, the solution for β requires the Rosenblatt transformation. Specifically, the performance function is

$$g(\mathbf{X}) = YZ - M$$

where the respective distributions are:

$$Y = \text{Lognormal}$$

$$Z = \text{Lognormal}$$

$$M = \text{Type I asymptotic}$$

with

$$\bar{Y} = 40 \text{ ksi}; \qquad \Omega_Y = 0.125$$

$$\bar{Z} = 50 \text{ in.}^3; \qquad \Omega_Z = 0.05$$

$$\bar{M} = 1000 \text{ in.-kip}; \qquad \Omega_M = 0.20$$

and the correlation matrix

$$[C] = \begin{array}{c} \\ Y \\ Z \\ M \end{array} \begin{array}{ccc} Y & Z & M \\ \left[\begin{array}{ccc} 1.0 & 0.4 & 0 \\ 0.4 & 1.0 & 0 \\ 0 & 0 & 1.0 \end{array} \right] \end{array}$$

Since M is independent of Y and Z, the Rosenblatt transformation may be obtained for Y and Z independently of M. Furthermore, since Y and Z are both lognormal variates, the joint PDF is

$$f_{Y,Z}(y,z) = \frac{1}{2\pi yz\zeta_y\zeta_z\sqrt{1-r^2}} \exp\left[-\frac{1}{2}\left(\frac{\ln y - \lambda_y}{\zeta_y}\right)^2 - \frac{1}{2}\left\{\frac{\ln z - \lambda_z - r(\zeta_z/\zeta_y)(\ln y - \lambda_y)}{\zeta_z\sqrt{1-r^2}}\right\} \right]$$

where

$$r = \frac{1}{\zeta_y\zeta_z}\ln(\rho\delta_y\delta_z + 1)$$

$$= \frac{1}{0.125 \times 0.05}\ln(0.4 \times 0.125 \times 0.05 + 1) = 0.40$$

The Rosenblatt transformations, therefore, are

$$\Phi(u_1) = \Phi\left(\frac{\ln y - \lambda_y}{\zeta_y}\right)$$

$$\Phi(u_2) = \int_{-\infty}^{z} f_{Z|Y}(z|y)\, dz$$

$$= \int_{-\infty}^{z} \frac{1}{\sqrt{2\pi}\,\zeta_z z\sqrt{1-r^2}} \exp\left[-\frac{1}{2}\left\{\frac{\ln z - \lambda_z - r(\zeta_z/\zeta_y)(\ln y - \lambda_y)}{\zeta_z\sqrt{1-r^2}}\right\}^2 \right] dz$$

$$= \Phi\left[\frac{\ln z - \lambda_z - r(\zeta_z/\zeta_y)(\ln y - \lambda_y)}{\zeta_z\sqrt{1-r^2}}\right]$$

$$\Phi(u_3) = \exp[-e^{-\alpha(m-u)}]$$

Inversion of the above, we obtain the independent standard normal variates as

$$u_1 = \frac{\ln y - \lambda_y}{\zeta_y} \qquad \text{assume } \lambda_y = \ln \bar{Y}$$

$$u_2 = \frac{\ln z - \lambda_z - r(\zeta_z/\zeta_y)(\ln y - \lambda_y)}{\zeta_z\sqrt{1-r^2}} \qquad \lambda_z = \ln \bar{Z}$$

$$u_3 = \Phi^{-1}[\exp\{-e^{-\alpha(m-u)}\}]$$

and the Jacobian matrix is

$$
\mathbf{J} =
\begin{bmatrix}
\dfrac{1}{y\zeta_y} & 0 & 0 \\[2ex]
\dfrac{-r}{y\zeta_y\sqrt{1-r^2}} & \dfrac{1}{z\zeta_z\sqrt{1-r^2}} & 0 \\[2ex]
0 & 0 & \dfrac{1}{\phi(u_3)}\dfrac{\alpha}{}[\exp\{-\alpha(m-u)-e^{-\alpha(m-u)}\}]
\end{bmatrix}
$$

The above algorithm for evaluating β then yields the following:

(1) Assume $\mathbf{x}_0^* = \{40, 50, 1000\}$.
(2) $\mathbf{u}_o = \{0.063, -0.005, 0.176\}$.
(3) The Jacobian evaluated at $\mathbf{x}_0^*$ is

$$
\mathbf{J}_o =
\begin{bmatrix}
0.200 & 0 & 0 \\
-0.087 & 0.436 & 0 \\
0 & 0 & 0.00523
\end{bmatrix}
$$

and

$$
\mathbf{J}_o^{-1} =
\begin{bmatrix}
5.00 & 0 & 0 \\
1.00 & 2.291 & 0 \\
0 & 0 & 191.3
\end{bmatrix}
$$

$$-\frac{5(.087)}{.436} = 1.00$$

(4) $g(\mathbf{x}_0^*) = 1000$

and with $\mathbf{G}_{x_o} = \{50, 40, -1\}$,

$$\mathbf{G}_{u_o} = (\mathbf{J}^{-1})^t\mathbf{G}_{x_o} = \{290, 91.6, -191.3\}$$

(5) The new failure point in the u-space is

$$
u^* = \frac{1}{[(290)^2 + (91.6)^2 + (191.3)^2]}\left\{[290\ \ 91.6\ \ -191.3]
\begin{bmatrix} 0.063 \\ -0.005 \\ 0.176 \end{bmatrix} - 1000\right\}
\begin{bmatrix} 290 \\ 91.6 \\ -191.3 \end{bmatrix}
$$

$$= \{-2.282, -0.721, 1.505\}$$

and in the original space

$$\mathbf{x}^* = \{28.28, 46.01, 1254.2\}$$

(6) For this (second) iteration, the safety index is therefore

$$\beta = \sqrt{(2.282)^2 + (0.721)^2 + (1.505)^2} = 2.827$$

Five iterations were required to obtain convergence, as summarized in Table B.4. From which we obtain the safety index $\beta = 2.663$.

Table B.4 Summary of Iterations for Example B.4

Iteration No.	$\mathbf{x}^{*t}$	$\mathbf{u}^t$	$g(\mathbf{x}^*)$	β
1	(40, 50, 1000)	(0.063, -0.005, 0.176)	1000	0.188
2	(28.3, 46.0, 1254)	(-2.705, -0.617, 1.257)	47.0	2.827
3	(32.6, 47.4, 1540)	(-1.574, -0.444, 2.110)	6.13	2.742
4	(33.7, 47.8, 1612)	(-1.319, -0.414, 2.275)	0.32	2.665
5	(33.7, 47.8, 1612)	(-1.298, -0.410, 2.289)	0.00	2.663

$$P_F = 1 - \Phi(2.66) = 4.0\times10^{-3} = .004$$

With normals $\beta = 3.05$

$$P_F = 1.14\times10^{-3} = .00114$$

References

1. Abramowitz, M., and Stegun, I.A. (eds.), *Handbook of Mathematical Functions*, Dover Publications, New York, 1965.
2. Afanasieff, L., "Corrosion Mechanisms, Corrosion Defense and Wastage Allowance," in *Ship Structural Design Concepts*, J. Harvey Evans (ed.), Cornell Maritime Press, Cambridge, Massachusetts, 1974.
3. Ali, E.M., Wu, T.H., and Chang, N., "Stochastic Model of Flow Through Stratified Soils," *J. of Geotechnical Engineering Division*, Proc. of ASCE, Vol. 106, No GT6, June 1980, pp. 593–610.
4. Anderson, K.H., and Aas, P.M., "Foundation Performance of the Brent B Condeep Platform," *Proc. Sym. on Brent Instrumentation Project*, Soc. for Underwater Technology, London, 1979.
5. Ang, A.H-S., "Structural Risk Analysis and Reliability-Based Design," *J. of Structural Division*, ASCE, Vol. 99, No. ST9, September 1973, pp. 1891–1910.
6. Ang, A.H-S., "Probability Bases for Ship Structural Analysis and Design," Class Notes, U.S. Coast Guard, Washington, D.C., June 1979.
7. Ang, A.H-S., Abdelnour, J., and Chaker, A.A., "Analysis of Activity Networks Under Uncertainty," *J. of Engineering Mechanics Division*, ASCE, Vol. 101, No. EM4, August 1975, pp. 373–387.
8. Ang, A.H-S., and Amin, M., "Reliability of Structures and Structural Systems," *J. of Engineering Mechanics Division*, ASCE, Vol. 94, No. EM2, April 1968, pp. 671–691.
9. Ang, A.H-S., and Cornell, C.A., "Reliability Bases of Structural Safety and Design," *J. of Structural Division*, ASCE, Vol. 100, No. ST9, September 1974, pp. 1755–1769.
10. Ang, A.H-S., et al., "Development of a Systems Risk Methodology for Single and Multi-Modal Transportation Systems," Final Report, Contract DOT-OS-50238, Washington, D.C., July 1979.
11. Ang, A.H-S., and Ma, H-F., "On the Reliability of Structural Systems," *Proc. 3rd Int. Conf. on Structural Safety and Reliability*, Trondheim, Norway, June 1981.
12. Ang, A.H-S., and Tang, W.H., "Statistical Decision Methods for Survivability and Vulnerability Assessments of Strategic Structures," Report No. DNA 4718F, Defense Nuclear Agency, Washington, D.C., 1980.
13. ASTM, "A Guide for Fatigue Testing and Statistical Analysis of Fatigue Data," Special Publ. No. 91-A, 1963.
14. Baecher, G.B., Chan, M., Ingra, T.S., Lee, T., and Nucci, L.R., "Geotechnical Reliability of Offshore Gravity Platform," MIT Sea Grant College Program, Report No. MIT SG80-20, December 1980.

15. Barlow, R.E., Fussell, J.B., and Singpurwalla, N.D. (eds.), *Reliability and Fault Tree Analysis*, SIAM, 1975.
16. Barlow, R.E., and Lambert, H.E., "Introduction to Fault Tree Analysis," in *Reliability and Fault Tree Analysis*, R.E. Barlow et al. (eds.), SIAM, 1975, pp. 7–35.
17. Bjerrum, L., "Geotechnical Problems Involved in Foundations of Structures in the North Sea," *Geotechnique*, Vol. 23, No. 3, 1973.
18. Bodhaine, G.L., "Measurement of Peak Discharge at Culvert by Indirect Methods," *Tech. Water-Resource Investigation*, Book 3, Chapter 43, U.S. Geological Survey, 1968.
19. Box, G.E.P., and Muller, M.E., "A Note on the Generation of Random Normal Deviates," *Annals of Math. Stat.*, 29, 1958, pp. 610–611.
20. Brooks, A.C., Leahy, J.P., and Shaffer, L.R., "A Man-Machine CPM System for Decision Making in the Construction Industry," Univ. of Ill., C.E. Studies, Const. Eng. Ser. No. 9, June 1967.
21. Bureau of Reclamation, *Design of Small Dams*, U.S. Dept. of Int., Water Resources Tech. Publ., Washington, D.C., 1977.
22. Carlin, A., and Park, R.E., "Model of Long Delays at Busy Airports," AD691860, The Rand Corporation, Santa Monica, California, 1969.
23. Chamberlain, S.G., "Applications of Statistical Decision Theory to Great Lakes Management," *Proc. of 13th Conference on Great Lakes Research*, International Association of Great Lakes Research, Ann Arbor, April 1970.
24. Chow, V.T., "Runoff," in *Handbook of Applied Hydrology*, V.T. Chow (ed.), McGraw-Hill, New York, 1964.
25. Chow, V.T., "Statistical and Probability Analysis of Hydrologic Data," in Section 8 of *Handbook of Applied Hydrology*, V.T. Chow (ed.), McGraw-Hill, New York, 1964.
26. Chow, V.T., and Takase, N., "Design Criteria for Hydrologic Extremes," *J. of Hydraulic Division*, ASCE, Vol. 103, No. HY4, April 1977, pp. 425–436.
27. Churchman, C.W., and Ackoff, R.L., "An Approximate Measure of Value," *Operations Research*, ORSA, Vol. 2, 1954, pp. 172–181.
28. Churchman, C.W., Ackoff, R., Arnoff, E., *Introduction to Operation Research*, J. Wiley and Sons, New York, 1957.
29. Collins, J.D., and Hudson, J.M., "Application of Risk Analysis to Nuclear Structures," *Proc. Symposium on Probabilistic Methods in Structural Engineering*, ASCE National Convention, St. Louis, October 1981, pp. 157–178.
30. Cornell, C.A., "Bounds on the Reliability of Structural Systems," *J. of Structural Division*, ASCE, Vol. 93, No. ST1, February 1967.
31. Cornell, C.A., "Structural Safety Specification Based on Second-Moment Reliability," Sym. Int. Assoc. of Bridge and Struct, Engr., London, 1969.
32. Cornell, C.A., and Newmark, N.M., "On the Seismic Reliability of Nuclear Power Plants," ANS Topical Mtg. on Probabilistic Reactor Safety, Newport Beach, California, May 1978.
33. Cramer, H., *Mathematical Methods of Statistics*, Princeton University Press, 1946.

34. Cummings, G.E., "Application of Fault Tree Techniques to a Nuclear Reactor Containment Systems," *Reliability and Fault Tree Analysis*, R.E. Barlow et al. (eds.), SIAM, 1975, pp. 805–825.

35. Dalkey, N. C., "The Delphi Method: An Experimental Study of Group Opinion," RM-5888-PR, The Rand Corporation, Santa Monica, California, 1969.

36. Dalkey, N.C., "Studies in the Quality of Life: Delphi and Decision Making," Lexington Books (Heath), Lexington, Massachusetts, 1972.

37. Daniels, H.E., "The Statistical Theory of the Strength of Bundles of Threads. I," *Proc. Royal Stat. Soc. of London*, Ser. A, vol. 183, No. 995, June 1945, pp. 405–435.

38. Darter, M.I., Hudson, W.R., and Brown, J.L., "Statistical Variation of Flexible Pavement Properties and Their Consideration in Design," Annual Meeting of the Association of the Asphalt Paving Technologists, 1973.

39. Davidson, R., "An Application of Utility and Decision Theory In A Planning Environment," 39th National Meeting, Operations Research Society of America, Dallas, Texas, May 1971.

40. Davis, D.R., and Dvoranchik, W.M., "Evaluation of the Worth of Additional Data," *Water Resources Bulletin*, American Water Resources Association, Paper No. 71062, Vol. 7, No. 4, August 1971, pp. 700–707.

41. de Neufville, R., and Keeney, R.L., "Use of Decision Analysis in Airport Development for Mexico City," in *Analysis of Public Systems*, Drake et al. (eds.), MIT Press, Massachusetts, 1972.

42. Ditlevsen, O., "Generalized Second Moment Reliability Index," *J. of Structural Mechanics*, Vol. 7, No. 4, 1979, pp. 435–451.

43. Ditlevsen, O., "Narrow Reliability Bounds for Structural Systems," *J. of Structural Mechanics*, Vol. 7, No. 4, 1979, pp. 453–472.

44. Donovan, N.C., "A Statistical Evaluation of Strong Motion Data Including the February 9, 1971 San Fernando Earthquake," *Proc. Fifth World Conference on Earthquake Engineering*, Rome, Italy, 1973.

45. Einstein, H.H., Labreche, D.A., Markow, M.J., and Baecher, G.B., "Decision Analysis Applied to Rock Tunnel Exploration," in *Near Surface Underground Opening Design*, W.R. Judd (ed.), *Engineering Geology*, Vol. 12, No. 1, 1978, pp. 143–161.

46. Ellingwood, B.R., MacGregor, J.G., Galambos, T.V., and Cornell, C.A., "Probability Based Load Criteria: Load Factors and Load Combinations," *J. of Structural Division*, ASCE, Vol. 108, No. ST5, May 1982, pp. 978–997.

47. Engi, D., and Solbey, J., "The Analysis of Intra-Building Traffic," ORSA/TIMS National Meeting, Boston, 1974.

48. Epstein, B., "Elements of the Theory of Extreme Values," *Technometrics*, Vol. 2, No. 1, February 1960, pp. 27–41.

49. Epstein, B., and Lomnitz, C., "A Model for the Occurrence of Large Earthquakes," *Nature*, Vol. 211, August 1966, pp. 954–956.

50. Erskine, T.C., and Shih, C.S., "Subjective Decision-Making for Water Resources Development," 41st National ORSA Meeting, New Orleans, Louisiana, April 1972.

51. Fiering, M.B., "Queueing Theory and Simulation in Reservoir Design," *J. of the Hydraulics Division*, ASCE, Vol. 87, No. HY6, November 1961, pp. 39–69.

52. Fiessler, B., Neumann, H-J., and Rackwitz, R. "Quadratic Limit States in Structural Reliability" *J. of the Engineering Mechanics Division*, ASCE, Vol. 105, No. EM4, August, 1979. pp. 661–676.

53. Fishburn, P.C., *Utility Theory for Decision Making*, Vol. 18, Publications in Operations Research Series, David B. Hertz (ed.), J. Wiley and Sons, New York, 1970.

54. Fisher, R.A., and Tippett, L.H.C., "Limiting Forms of the Frequency Distribution of the Largest or Smallest Number of a Sample," *Proc. Cambridge Philosophical Society*, XXIV, Part II, 1928, pp. 180–190.

55. Freese, Nichols, and Endress Consulting Engineers, "Report on Sewage and Waste Collection, Sewage Treatment and Related Water Quality from Tom Green County," Report to the City of San Angelo, Texas, 1971; "Report on Supplemental Gound Water Supply Study," Report to the City of San Angelo, Texas, August 1971; "Study Report—Wastewater Treatment Plants for Wastewater Reuse or Disposal," Report to Concho Valley Council of Governments, April 8, 1971; "Water Supply Plan: 1970–2000," Report to the Concho Valley Council of Governments, San Angelo, Texas, 1970.

56. Freudenthal, A.M., "The Safety of Structures," ASCE *Transactions*, Vol. 112, 1947, pp. 125–159.

57. Freudenthal, A.M., "Safety and the Probability of Structural Failure," ASCE *Transactions*, Vol. 121, 1956, pp. 1337–1397.

58. Freudenthal, A.M., Garrelt, J.M., and Shinozuka, M., "The Analysis of Structural Safety," *J. of Structural Division*, ASCE, Vol. 92, No. ST1, February 1966.

59. Galambos, T.V., Ellingwood, B.R., MacGregor, J.G., and Cornell, C.A., "Probability-Based Load Criteria: Assessment of Current Design Practice," *J. of Structural Division*, ASCE, Vol. 108, No. ST5, May 1982, pp. 959–977.

60. Gaver, D.P., and Thompson, G.L., *Programming and Probability Models in Operation Research*, Brooks/Cole, Monterey, California, 1973.

61. Gaylord, E.H., and Gaylord, C.N., *Design of Steel Structures*, 2nd Ed., McGraw-Hill, New York, 1972.

62. Gibbs, H.J., and Holtz, W.G., "Research of Determining the Density of Sand by Spoon Penetration Test," *Proc. Fourth International Conference on Soil Mechanics and Foundations Engineering*, Vol. 1, 1957, pp. 35–39.

63. Greenberger, M., "An A Priori Determination of Serial Correlation in Computer Generated Random Numbers," *Mathematics of Computation*, 15, 1961, pp. 383–389.

64. Griffis, F.H., "Optimizing Haul Fleet Size Using Queueing Theory," *J. of the Construction Division*, ASCE, Vol. 94, No. CO1, January 1968, pp. 75–88.

65. Grimmelt, M.J., and Schuëller, G.I., "Benchmark Study on Methods to Determine Collapse Failure Probabilities of Redundant Structures," in *Structural Safety*, Vol. 1, No. 2, Scientific, Elsevier Publ. Co., Amsterdam, Dec. 1982, pp. 93–106.

66. Gumbel, E.J., "The Return Period of Flood Flows," *Annals of Math Stat.*, Vol. 12, No. 2, June 1941, pp. 163–190.

67. Gumbel, E.J., "The Statistical Theory of Extreme Values and Some Practical Applications," *Applied Mathematics Series 33*, National Bureau of Standards, Washington, D.C., February 1954.

68. Gumbel, E., *Statistics of Extremes*, Columbia Univ. Press, New York, 1958.

69. Gupta, S.S., "Probability Integrals of Multivariate Normal and Multivariate *t*," *Annals of Math. Stat.*, Vol. 34, No. 3, September 1963, pp. 792–828.

70. Haefner, L.E., and Morlok, E.K., "Optimal Geometric Design Decisions for Highway Safety," *Highway Research Record No. 371*, 1971, pp. 12–23.

71. Haight, F., *Mathematical Theories of Traffic Flow*, Mathematics in Science and Engineering, Volume 7, Academic Press, New York, 1963.

72. Haldar, A. and Tang, W.H., "Uncertainty Analysis of Relative Density," *Jour. of Geotechnical Engineering Division*, Proc. of ASCE, Vol. 105, No. GT7, July 1979, pp. 899–904.

73. Hall, S., "Air Traffic Simulation, One and Two-Runway Airports," *Transportation Engineering J.*, ASCE, Vol. 100, No. TE1, February 1974, pp. 221–235.

74. Hammersley, J.M., and Morton, K.W., "A New Monte Carlo Technique Antithetic Variates," *Proc. Cambridge Phil. Soc.*, 52, 1956, pp. 449–474.

75. Hasofer, A.M., "Reliability Index and Failure Probability," *J. of Structural Mechanics*, Vol. 3, No. 1, 1974, pp. 25–27.

76. Hasofer, A.M., and Lind, N., "An Exact and Invariant First-Order Reliability Format," *J. of Engineering Mechanics*, ASCE, Vol. 100, No. EM1, February 1974, pp. 111–121.

77. Hershfield, D.M., "Rainfall Frequency Atlas of the United States," *U.S. Weather Serv. Tech. Paper*, No. 40, 1963.

78. Hillier, F.S., and Lieberman, G.J., *Introduction to Operations Research*, Holden-Day, Inc., 1967.

79. Hoel, P.G., Sidney, C.P., and Stone, C.J., *Introduction to Probability Theory*, Houghton Mifflin Company, Boston, 1971.

80. Hohenbichler, M., and Rackwitz, R., "Non-Normal Dependent Vectors in Structural Safety," *J. of the Engineering Mechanics Division*, ASCE, Vol. 107, No. EM6, December 1981, pp. 1227–1238.

81. Horowitz, G.F., and Lee N.N., "Effect of Mechanical Cleaning Equations on Flow in Large Conduits," *J. Hydraulic Division*, ASCE, Vol. 97, No. HY5, May 1971, pp. 677–689.

82. Howard, R.A., Matheson, J.E., and North, D.W., "The Decision to Seed Hurricanes," *Science*, American Association for the Advancement of Science, Vol. 176, No. 4040, June 16, 1972, pp. 1191–1202.

83. Hunter, D., "An Upper Bound for the Probability of a Union," *J. of Applied Probability*, Vol. 3, No. 3, 1976, pp. 597–603.

84. Ibbs, Jr., C.W., and Crandall, K.C., "Construction Risk: Multiattribute Approach," *J. of the Construction Division*, Proc. of ASCE, Vol. 108, No. CO2, June 1982, pp. 187–200.

85. Johnk, M.D., "Erzengung von Betraverteilten and Gammaverteilten Zuffalszahten," *Metrika*, 8, 1964, pp. 5–15.

86. Keeney, R.L., "Multiplicative Utility Functions," Technical Report No. 70, Operation Research Center, MIT, Cambridge, Massachusetts, March 1972.

87. Keeney, R.L., "Utility Functions for Multi-Attributed Consequences," *Management Science*, Vol. 18, 1972, pp. 276–87.

88. Keeney, R.L., and Raiffa, H., *Decision With Multiple Objectives: Preferences and Value Tradeoffs*, J. Wiley and Sons, New York, 1976.

89. Keeney, R.L., and Wood, E.F., "An Illustrative Example of the Use of Multiattribute Utility Theory for Water Resource Planning," *Water Resources Research*, Vol. 13, No. 4, August 1977, pp. 705–712.

90. Kimball, B.F., "Sufficient Statistical Estimation Functions for the Parameters of the Distribution of Maximum Values," *Annals of Math. Stat.*, Vol. 17, No. 3, September 1946, pp. 299–309.

91. Knuth, D.E., *The Art of Computer Programming: Seminumerical Algorithms*, Vol. 2, Addison-Wesley, Massachusetts, 1969.

92. Kounias, E.G., "Bounds for the Probability of a Union with Applications," *Annals of Math. Stat.*, Vol. 39, No. 6, 1968, pp. 2154–2158.

93. Lambe, T.W., Marr, W.A., and Francisco, S., "Safety of a Constructed Facility: Geotechnical Aspects," *J. of Geotechnical Engineering Division*, ASCE, Vol. 107, No. GT3, March 1981, pp. 339–352.

94. Leiblein, J., "A New Method of Analysing Extreme-Value Data," NACA Tech. Note 3053, 1954.

95. Linsley, R.K., and Franzini, J.B., *Water Resources Engineering*, McGraw-Hill, New York, 1964.

96. Little, J.D.C., "A Proof for the Queueing Formula: $L = \lambda W$," *Operations Research*, ORSA, Vol. 9, No. 3, 1961, pp. 383–387.

97. Ma, H-F., and Ang, A.H-S., "Reliability Analysis of Redundant Ductile Structural Systems," Univ. of Ill., C.E. Studies, Str. Res. Ser. No. 494, August 1981.

98. MacCrimmon, K.R., and Ryavec, C.A., "Analytic Study of PERT Assumptions," *Operations Research*, ORSA, Vol. 12, 1964, pp. 16–37.

99. Marsaglia, G., "Random Variables and Computers," *Proc. Third Prague Conference in Probability Theory*, 1962.

100. Metcalf and Eddy, Inc., *Wastewater Engineering: Collection, Treatment, Disposal*, McGraw-Hill, New York, 1972.

101. Mitchell, W.D., "Floods in Illinois: Magnitude and Frequency," Division of Waterways, Department of Public Works and Buildings, Illinois, 1954.

102. Morris, D., "Inclusion of Social Values in Facility Location Planning," *Jour. of the Urban Planning and Development Division*, Proc. of ASCE, Vol. 98, No. UP1, July 1972, pp. 17–31.

103. Ochi, M.K., "On Prediction of Extreme Values," SNAME *Jour. of Ship Research*, Vol. 17, No. 1, March 1973, pp. 29–37.

104. Paloheimo, E., and Hannus, H., "Structural Design Based on Weighted Fractiles," *J. of Structural Division*, ASCE, Vol. 100, No. ST7, 1974, pp. 1367–1378.

105. Papoulis, A., *Probability, Random Variables, and Stochastic Processes*, McGraw-Hill, New York, 1965.

106. Parzen, E., *Stochastic Processes*, Holden-Day, Inc., San Francisco, 1962.

107. Peck, R.B., "Sampling Methods and Laboratory Tests for Chicago Subway Soils," *Proc. Purdue Conference on Soil Mech. and Its Appl.*, 1940.

108. Peck, R.B., Hanson, W.E., and Thornburn, T.H., *Foundation Engineering*, J. Wiley and Sons, New York, 1974.

109. Press, H. "An Application of the Statistical Theory of Extreme Values to Gust Load Problems," NACA Report 991, 1950.

110. Pugsley, A.G., "Concepts of Safety in Structural Engineering," *J. Institute of Civil Engineers*, Vol. 36, No. 5, London, March 1951.

111. Pugsley, A.G., "Structural Safety," *J. Royal Aeronautical Society*, Vol. 58, England, 1955.

112. Pugsley, A.G., *The Safety of Structures*, E. Arnold Publ. Ltd., London, 1966.

113. Rackwitz, R., "Practical Probabilistic Approach to Design," *Bulletin 112*, Comite European du Beton, Paris, France, 1976.

114. Rackwitz, R., and Fiessler, B., "Structural Reliability Under Combined Random Load Sequences," *Computers and Structures*, Pergamon Press, Vol. 9, November 1978, pp. 489–494.

115. Raiffa, H., "Preferences for Multi-Attributed Alternatives," RM-5868-DOT/RC, The Rand Corporation, Santa Monica, California, April 1969.

116. Raiffa, H., *Decision Analysis*, Addison-Wesley, Reading, Massachusetts, 1970.

117. Rand Corporation, *A Million Random Digits with 1,000,000 Normal Deviates*, Free Press, Glencoe, Illinois, 1955.

118. Resendis, D., and Herrera, D., "A Probabilistic Formulation of Settlement Controlled Design," *Proc. 7th ICSMFE*, Mexico, 1969.

119. Revell, R.W., (Chmn.), "Reevaluating Spillway Adequacy of Existing Dams, by the Task Committee on the Reevaluation of the Adequacy of Spillways of Existing Dams," *J. of Hydraulic Division*, Vol. 99, No. HY2, February 1973, pp. 337–372.

120. Rosenblatt, M., "Remarks on a Multivariate Transformation," *Annals of Math. Stat.*, Vol. 23, No. 3, September 1952, pp. 470–472.

121. Rubinstein, R.Y., *Simulation and the Monte Carlo Method*, J. Wiley and Sons, New York, 1981.

122. Russell, L., and Schueller, G., "Probabilistic Models for Texas Gulf Coast Hurricane Occurrences," *Proc. 3rd Annual Offshore Technology Conference*, Houston, Texas, OTC Paper 1344, April 1971.

123. Saaty, T.L., *Elements of Queueing Theory With Applications*, McGraw-Hill, New York, 1961.

124. Schlaifer, R., *Analysis of Decisions Under Uncertainty*, McGraw-Hill, New York, 1969.

125. Shaffer, L.R., "Production Forecasts Via Mathematical Models," Department of Civil Engineering, Univ. of Ill. at Urbana-Champaign, Construction Research Series No. 29, 1965.

126. Shinozuka, M., "Basic Analysis of Structural Safety," *J. of Structural Division*, ASCE, Vol. No. 3, 109, Mar. 1983.

127. Shinozuka, M., and Itagaki, H., "On the Reliability of Redundant Structures," *Annals of Reliability and Maintainability*, Vol. 5, 1966, pp. 605–610.

128. Shooman, M.L., *Probabilistic Reliability: An Engineering Approach*, McGraw-Hill Book Co., New York, NY, 1968.

129. Shuler, J.B., "Business Failures in Construction," *J. of Construction Division*, Vol. 93, No. CO2, September 1967, p. 73.

130. Simiu, E., Changery, M.J., and Filliben, J.J., "Extreme Wind Speeds at 129 Stations in the Contiguous United States," *Building Science Series 118*, National Bureau of Standards, Washington, D.C., March 1979.

131. Smith, C.S., "Compressive Strength of Welded Steel Ship Grillages," RINA, Spring Meeting, 1975.

132. Spetzler, C.S., and Stahl von Holstein, C.S., "Probability Encoding in Decision Analysis," *Management Science*, Vol. 22, 1975, pp. 340–358.

133. Staugitis, C.L., "Mill Sampling Techniques for the Quality Determination of Ship Steel Plates," Ship Structures Committee, Report SSC-141, February 1962.

134. Stevenson, J., and Moses, F., "Reliability Analysis of Frame Structures," *J. of Structural Division*, ASCE, Vol. 96, No. ST11, November 1970, pp. 2409–2427.

135. Stimson, D.H., "Utility Measurement in Public Health Decision Making," *Management Science*, Vol. 16, 1959.

136. Tang, W.H., Yucemen, M.S., and Ang, A.H-S., "Probability-Based Short Term Design of Soil Slopes," *Canadian Geotechnical J.*, Vol. 13, 1976, pp. 201–215.

137. Tavenas, F.A., "Validity of the Correlations Between the Standard Penetration Index and the Relative Density of Sands," Fourth Pan-American Conference on Soil Mechanics and Foundation Engineering, Puerto Rico, Vol. III, 1971, pp. 64–70.

138. Tavenas, F.A., Ladd, R.S., and LaRochelle, P., "The Accuracy of Relative Density Measurements: Results of a Comparative Test Program," *Relative Density Involving Cohesionless Soils*, STP523, American Society for Testing and Materials, 1972, pp. 18–60.

139. Teng, W.C., *Foundation Engineering*, Prentice-Hall, Englewood Cliffs, New Jersey, 1962.

140. Thatcher, R.M., "Optimal Channel Service Policies for Stochastic Arrivals," Report Number ORC 68-16, Operations Research Center, University of California, Berkeley, June 1968.

141. Ting, Harold M., "Aggregation of Attributes for Multiattributed Utility Assessment," Technical Report No. 66, Operation Research Center, MIT, August 1971.

142. Vanmarcke, E.H., "Matrix Formulation of Reliability Analysis and Reliability Based Design," *Computers and Structures*, Vol. 13, 1971, pp. 757–770.

143. Vanmarcke, E.H., "Probabilistic Modeling of Soil Profiles," *J. of the Geotechnical Engineering Division*, Proceeding, ASCE, Vol. 103, No. GT11, November 1977, pp. 1237–1246.

144. Vesely, W. E., "Reliability Quantification Techniques Used in the Rasmussen Study," in *Reliability and Fault Tree Analysis*, R.E. Barlow et al. (eds.), SIAM, 1975, pp. 775–803.

145. WASH 1400, "Reactor Safety Study," U.S. Nuclear Regulatory Commission, NUREG–75/014, Oct. 1975.

146. Weibull, W., "A Statistical Distribution of Wide Applicability," *J. Applied Mechanics*, ASME, Vol. 18, 1951.

147. Whitman, R.V., Biggs, J.M., Brennan, III, J.E., Cornell, C.A., de Neufville, R.L., and Vanmarcke, E.H., "Seismic Design Decision Analysis," *J. of Structural Division*, Proc. of ASCE, Vol. 101, No. ST5, May 1975, pp. 1067–1084.

148. Wilks, S.S., "Order Statistics," *Bulletin American Math. Soc.*, Vol. 54, 1948.

149. Yen, B.C., and Sevuk, A.S., "Design of Storm Sewer Networks," *J. Environmental Engineering Division*, ASCE, Vol. 101, No. EE4, August 1975, pp. 535–553.

150. Yen, B.C., and Tang, W.H., "Risk-Safety Factor Relation for Storm Sewer Design," *J. of Environmental Engineering Division*, ASCE, Vol. 102, No. EE2, April 1976, pp. 509–516.

index

SI METRIC UNITS

CONVERSION FACTORS

		Customary to SI
inches (in.)	meters (m)	0.0254
inches (in.)	centimeters (cm)	2.54
inches (in.)	millimeters (mm)	25.4
feet (ft)	meters (m)	0.305
yards (yd)	meters (m)	0.914
miles (miles)	kilometers (km)	1.609
degrees (°)	radians (rad)	0.0174
acres (acre)	hectares (ha)	0.405
acre-feet (acre-ft)	cubic meters (m^3)	1233
gallons (gal)	cubic meters (m^3)	3.79×10^{-3}
gallons (gal)	liters (l)	3.79
pounds (lb)	kilograms (kg)	0.4536
tons (ton, 2000 lb)	kilograms (kg)	907.2
pound force (lbf)	newtons (N)	4.448
pounds per sq in. (psi)	newtons per sq m (N/m^2)	6895
pounds per sq ft (psf)	newtons per sq m (N/m^2)	47.88
foot-pounds (ft-lb)	joules (J)	1.356
horsepowers (hp)	watts (W)	746
British thermal units (BTU)	joules (J)	1055
British thermal units (BTU)	kilowatt-hours (kwh)	2.93×10^{-4}